Dynamics of Pelagic Fish Distribution and Behaviour: Effects on Fisheries and Stock Assessment

Pierre Fréon
ORSTOM, BP 5045, 3402 Montpellier, France

Ole Arve Misund
IMR, PO Box 1870, N-5024 Bergen, Norway

Fishing News Books

Fishing News Books
A division of Blackwell Science Ltd
Editorial Offices:
Osney Mead, Oxford OX2 0EL
25 John Street, London WC1N 2BL
23 Ainslie Place, Edinburgh EH3 6AJ
350 Main Street, Malden
MA 02148 5018, USA
54 University Street, Carlton
Victoria 3053, Australia
10, rue Casimir Delavigne
75006 Paris, France

Other Editorial Offices:

Blackwell Wissenschafts-Verlag GmbH
Kurfürstendamm 57
10707 Berlin, Germany

Blackwell Science KK
MG Kodenmacho Building
7–10 Kodenmacho Nihombashi
Chuo-ku, Tokyo 104, Japan

First published 1999

Set in 10/12.5pt Times
by DP Photosetting, Aylesbury, Bucks
Printed and bound in Great Britain at
the University Press, Cambridge

DISTRIBUTORS
Marston Book Services Ltd
PO Box 269
Abingdon
Oxon OX14 4YN
(*Orders:* Tel: 01865 206206
Fax: 01865 721205
Telex: 83355 MEDBOK G)

USA
Blackwell Science, Inc.
Commerce Place
350 Main Street
Malden, MA 02148 5018
(*Orders:* Tel: 800 759 6102
781 388 8250
Fax: 781 388 8255)

Canada
Login Brothers Book Company
324 Saulteaux Crescent
Winnipeg, Manitoba R3J 3T2
(*Orders:* Tel: 204 837 2987
Fax: 204 837 3116)

Australia
Blackwell Science Pty Ltd
54 University Street
Carlton, Victoria 3053
(*Orders:* Tel: 03 9347 0300
Fax: 03 9347 5001)

A catalogue record for this title is available from the British Library

ISBN 0-85238-241-3

For further information on
Fishing News Books, visit our website:
http://www.blacksci.co.uk/fnb/

Contents

Foreword ix
Preface xi

1 Introduction **1**
1.1 Recent history of animal behaviour studies and focus on fish behaviour 1
1.2 Brief history of stock assessment methods 5
1.3 Scope of the book 8

2 Pelagic Fisheries **10**
2.1 World catch of small pelagic species 10
2.2 World catch of large mackerel and tuna 14
2.3 Fishing methods for pelagic species 15
2.3.1 Purse seining 15
2.3.2 Midwater trawling 18
2.3.3 Line fishing 19
2.4 Summary 20

3 Habitat Selection and Migration **21**
3.1 Introduction 21
3.2 Habitat selection according to different time and space scales 22
3.3 Influence of abiotic (physical) factors 27
3.3.1 Temperature 28
3.3.2 Salinity 32
3.3.3 Dissolved oxygen 33
3.3.4 Water transparency 36
3.3.5 Light intensity 36
3.3.6 Current, turbulence and upwelling 38
3.3.7 Depth of the fish in the water column 40
3.3.8 Bottom depth and the nature of the sea bed 41
3.3.9 Floating objects 44
3.4 Influence of biotic factors 44
3.4.1 Conspecifics and the ideal free distribution 44
3.4.2 Other species 48
3.4.3 Prey 48
3.4.4 Predators 52
3.5 Conclusion 54

4 Schooling Behaviour **56**
4.1 Introduction 56
4.2 School definitions 56
4.3 Genetic basis of schooling 57
4.4 Ontogeny of schooling 58
4.5 Schooling and shoaling 59
4.6 Functions of schooling 60
4.6.1 Surviving predatory attack 60
4.6.2 Effective feeding 63
4.6.3 Hydrodynamic advantages 63
4.6.4 Migration 64
4.6.5 Reproduction 64
4.6.6 Learning 65
4.7 School size 65
4.8 School organisation 68
4.8.1 Study methods 68
4.8.2 Minimum approach distance 70
4.8.3 Nearest neighbour distance 71
4.8.4 Spatial distribution 72
4.8.5 Internal synchrony 75
4.8.6 Individual preferences and differences 77
4.8.7 Packing density 78
4.8.8 Packing density structure 79
4.8.9 School shape 80
4.8.10 Factors affecting school structure 82
4.8.11 Factors selecting for homogeneity, structure and synchrony 86
4.9 Mixed-species schools 88
4.10 Spatial distribution (clustering) 93
4.11 Communication 94
4.11.1 Vision 95
4.11.2 Lateral line 97
4.11.3 Hearing 98
4.12 Schooling, modern fishing and natural selection 98
4.13 Conclusion 100

5 Avoidance **102**
5.1 Avoidance of sounds from vessel and gear 102
5.1.1 Ambient noise in the sea 102
5.1.2 Noise from vessels 103
5.1.3 Fish hearing 108
5.1.4 Fish reactions to noise 116
5.2 Visually elicited avoidance 122
5.2.1 Light in the sea 122
5.2.2 Fish vision 123

5.2.3 Avoidance reactions elicited by visual stimuli of vessel and gear 124
5.3 Conclusion 126

6 Attraction and Association 128
6.1 Attraction to light 128
6.2 Attraction to bait 131
6.3 Associative behaviour 133
6.3.1 Association with floating objects 133
6.3.2 Association with other species 154
6.3.3 Summary of associative behaviour 157

7 Learning 159
7.1 Introduction 159
7.2 Learning in fish predation 159
7.2.1 Individual level 160
7.2.2 Interactions at the group level 163
7.2.3 Interactions among the prey 164
7.2.4 Key stimuli in learning 164
7.2.5 Hypothesis on interactive learning 167
7.2.6 From natural predators to fishing gear 167
7.3 Other learning processes 171
7.4 Conclusion 173

8 Effects of Behaviour on Fisheries and Stock Assessment using Population Dynamic Models 174
8.1 Introduction 174
8.2 Stock assessment by population dynamic models 174
8.2.1 Stock assessment by surplus-production models 174
8.2.2 Stock assessment by age-structured models and yield per recruit 179
8.2.3 Stock-recruitment relationship 182
8.3 Habitat selection and its influence on catchability and population parameters 183
8.3.1 Yearly changes in abundance, density, habitat and catchability in relation to exploitation and the environment 183
8.3.2 Some attempts at modelling the yearly variability of catchability in relation to habitat selection 193
8.3.3 Seasonal variability of catchability in relation to habitat selection 200
8.3.4 Circadian variation of catchability related to habitat selection 204
8.3.5 Spatial variability of catchability in relation to habitat selection 208

8.3.6 Influence of habitat selection on growth estimate 214
8.3.7 Influence of habitat selection on mortality estimate by ASA models 214
8.4 Influence of aggregation on fisheries and population dynamic models 215
8.4.1 Influence of aggregation on catchability 215
8.4.2 Influence of mixed-species schools on abundance indices 230
8.4.3 Influence of aggregation on growth estimates and age–length key 232
8.5 Influence of avoidance on abundance indices 235
8.6 Influence of attraction and associative behaviour on population modelling 236
8.6.1 Influence of attraction 236
8.6.2 Influence of association 236
8.7 Influence of learning on stock assessment 239
8.7.1 Influence of learning on surplus-production models 240
8.7.2 Influence of learning on intraspecific diversity and stock identification 244
8.7.3 Influence of learning on structural models 247
8.7.4 The strength of paradigms 248
8.8 Conclusion 249

9 Effects of Behaviour on Stock Assessment using Acoustic Surveys 251
9.1 Introduction 251
9.2 The hydroacoustic assessment method 251
9.3 Effects of habitat selection 252
9.4 Effects of social behaviour 254
9.4.1 Distribution function of densities 254
9.4.2 Target strength 254
9.4.3 Acoustic shadowing 256
9.5 Effects of avoidance 258
9.5.1 Methods to study vessel avoidance 258
9.5.2 Observations of vessel avoidance during surveys 260
9.5.3 Avoidance of vessel light 266
9.5.4 Vessel avoidance during sampling by trawls 267
9.5.5 Instrument avoidance 269
9.5.6 Sampling gear avoidance 270
9.6 Replicability of acoustic survey estimates 271
9.7 Conclusion 272

10 Other Methods of Stock Assessment and Fish Behaviour 274
10.1 Fishing gear surveys 274
10.1.1 Methodology and assumptions 274
10.1.2 Influence of fish behaviour on fishing gear surveys 275
10.2 Aerial surveys 276
10.2.1 Methodology and assumptions 276

10.2.2 Influence of fish behaviour on aerial survey 277
10.3 Ichthyoplankton surveys 279
10.3.1 Objectives, methodology and assumptions 279
10.3.2 Influence of fish behaviour on ichthyoplankton surveys 281
10.4 Capture-recapture 282
10.4.1 Methodology and assuptions 282
10.4.2 Influence of fish behaviour on capture-recapture methods 283
10.5 Conclusion 284

11 Conclusion **285**

References 290
Index 339

Foreword

By Alain Laurec and Anne E. Magurran

The arrival of this book is timely. A split has occurred in fisheries research, between those population dynamicians who use only indirect methods derived from cohort analysis and those who prefer direct assessment. Fréon and Misund show how this division is unnecessary, and how behavioural techniques can put a different light on the study of fisheries resources.

The authors are well qualified to establish such a link between the two camps. Their individual experience is complementary, particularly at the geographical level where their knowledge covers a wide area from tropical to Arctic seas and includes fresh water, pelagic coastal fish, and large migratory species.

The book will form a useful reference guide for scientific research, and it is hoped that it will ease the dialogue between scientists and fisheries professionals. The book explains clearly the problems and, in particular, how fishermen's observations can be reconciled with scientific findings.

Fisheries biologists are warmly advised to read this innovative book, which is based on the authors' experience and a comprehensive reference list. They will find here much information of interest to give a new perspective on their subject. A non-specialist reader will also find the book contains key information about fisheries research and management.

Alain Laurec
Director, DG XIV
European Commission
Brussels

Fish are often seen as unsophisticated creatures, immortalised in the myth about the goldfish which has a memory span about as long as the time it takes to circumnavigate its bowl. However, recent investigations by behavioural ecologists and evolutionary biologists have revealed that these creatures have complex abilities. For example, fish can recognise and remember the identities of particular individuals, and choose with whom to associate on the basis of past encounters. Fish schools, one of the most impressive and indeed widespread of animal social groupings (over 80% of species school at some phase of their lives), represent the sum of these individual behaviours. Fish schools are also of immense economic importance. Pelagic schooling species, such as anchovy, herring, mackerel and tuna, underpin the global fishing industry, and in 1994 alone the world catch of these fish was around 45 million tonnes.

To date, there have been few attempts to make links between fish biologists interested in behavioural mechanisms and function and fisheries biologists concerned with stock assessment. Yet effective stock assessment and fisheries management demand a good understanding of fish behaviour. Pelagic fish will, for instance, show avoidance behaviour to survey vessels thus biasing density and distribution estimates. Pierre Fréon and Ole Arve Misund recognise this need and in this book bring together these previously isolated domains of science for the first time, making a compelling case for strengthened interactions between the disciplines. Fisheries biologists will be persuaded to pay increased attention to behaviour while behavioural biologists will be reminded of applied problems outside their normal remit.

Fréon and Misund also emphasise a major limitation in applying current knowledge of fish behaviour to the questions that interest fisheries biologists. Most investigations of behaviour use freshwater species, small schools, and take place in laboratories. Such a bias is inevitable given the need for replicated and well-controlled experiments to test specific hypotheses about schooling tendency or mate choice or predator avoidance. Pelagic species, by contrast, often live in schools much larger than those studied in the laboratory. We know little about the consequences of this difference or the extent to which the dynamics of small schools can be extrapolated to large ones.

It is sobering to note that schooling, the behaviour that protects fish against their natural predators, has facilitated their over-exploitation by human ones. Large numbers no longer confer safety, but are instead a liability because modern technology can readily detect and capture entire schools. As Fréon and Misund observe, the test of this new dialogue between those who study fish and those who exploit them, will ultimately be the preservation of the resource for future generations.

Anne E. Magurran
School of Environmental & Evolutionary Biology
University of St Andrews
Scotland

Preface

Knowledge of fish behaviour has always been a basic prerequisite for successful fishing. Fishermen often find that changes in fish location, vertical distribution, aggregation behaviour, reaction to fishing gear, etc., influence catch rates drastically. The effects of fish behaviour may influence stock assessment by indirect methods that rely on catch statistics, especially if fishermen's experience is not properly taken into account. There is also an increasing awareness that effects of fish behaviour may greatly influence stock assessment by direct methods, such as hydroacoustics.

Pioneer fisheries scientists identified effects of fish behaviour on fisheries at the beginning of the twentieth century, and during the last two decades there have been substantial scientific efforts to study fish behaviour in relation to fisheries and stock assessment. However, there has been no comprehensive text that summarises the effects of fish behaviour on catch rates and stock assessment by indirect and direct methods. Our intention with this book is therefore to provide a review of dynamics of fish distribution and behaviour and its effects on fisheries and stock assessment. Throughout the book, examples are given for pelagic species from all over the world. We limit the book to small pelagic fish because such species contribute about one third of the total world catch, show substantial plasticity in their behaviour, and are assessed by both indirect and direct methods.

The book is written mainly for scientists working on or interested in stock assessment and fish behaviour, and may serve as an introductory text for graduate students. One of our aims in this book has been to bridge the gap between academic work on fish behaviour and fisheries research. We have tried to show how advances in fish behaviour can improve stock estimation and our wish is to see workers on fish behaviour directing their efforts towards remaining problems.

Over 1200 references are provided. There are several reasons. First, the book intends to cover two fields of research: fish behaviour and fish assessment. Second, field studies of fish behaviour are performed under uncontrolled conditions and therefore provide contrasting results, often difficult to interpret by themselves. There is a need to compare these different results among themselves and with controlled laboratory studies to validate some conclusions. As far as field studies of fish behaviour are concerned, there is a large body of 'grey' literature that covers this point and that we have cited. Some of these papers are excellent, others (including some of ours) are just working group papers, suffering from poor presentation or a lack of appropriate statistical analysis, weak discussion, etc. Nevertheless, these papers contain valid field observations that deserve to be taken into account within the framework of the comparative approach.

We are grateful to several people for helping us in developing this book. The first scientific content was discussed by the authors with François Gerlotto, who is warmly acknowledged for his input and for the revision of some chapters. The preparation of the book was made possible through a 6 month employee contract from ORSTOM, France, to Ole Arve Misund in 1992, and a guest scientist scholarship for 2 months from the Norwegian Research Council to Pierre Fréon in 1993. We are also grateful to many colleagues who took time to revise several chapters and to provide useful comments. Among them we extend special thanks to P. Cayré, A. Fonteneau and F. Marsac. Additional comments were provided by D. Binet, F. Conand, V. Csányi, P. Cury, E. Cillaurren, F. Laloë, D. Gaertner, P. Petitgas, M. Soria, J.-M. Stretta and D. Reid. Anne Brit Tysseland and Jofrid Øvredal prepared the figures, and Virgine Delcourt, Elen Hals and Laurence Vicens helped us with the reference list. Special thanks go to Chuck Hollingworth for his skilful editorial comments and suggestions for improving our English. Finally we would like to thank Blackwell Science for publishing our book, and for patience during its preparation.

P. Fréon
O.A. Misund

Chapter 1
Introduction

1.1 Recent history of animal behaviour studies and focus on fish behaviour

The foundation of animal behaviour studies occurred only in the nineteenth century and presently studies of animal behaviour or ethology can be subdivided into four major approaches (Drickamer and Vessey, 1992):

(1) Comparative psychology investigates the mechanisms controlling behaviour by systematic and objective observation or controlled experiments (Dewsbury, 1984).
(2) Ethology (*sensu stricto*) is mainly based on the principle that behavioural traits can be studied from the evolutionary viewpoint; it is at the origin of the definition of basic concepts such as the distinction between appetitive and consummatory behaviours. This period is dominated by two famous ethologists: Konrad Lorenz (1903-1989) and Niko Tinbergen (1907-1988), who developed their concepts mainly from *in situ* observations.
(3) Behavioural ecology, which emerged in the 1950s, reinforces the studies of interactions between animals and their environment (including other animals, conspecifics or predators). It focuses on the implication of behaviour for survival. For instance, the foraging advantage of flocking in birds was demonstrated by Krebs *et al.* (1972), and the antipredator advantage of schooling in fishes by Neill and Cullen (1974). One of the most famous findings of behavioural ecology is the optimal foraging theory which predicts how an animal should proceed to achieve a maximal rate of energy intake in the most economic way (Stephens & Krebs, 1987). Investigations are generally conducted both in the field and in the laboratory. In the introduction of a multi-author book on the behavioural ecology of fishes (Huntingford & Torricelli, 1993), Huntingford (1993) reviewed the evolution of behavioural ecology through the changes in the three editions of the major book of Krebs and Davies (1978, 1984, 1991). The first edition focused on space use and territoriality, foraging, predator avoidance, mating, sexual selection, evolution of sociality and of cooperative breeding, and finally life history strategies. The second edition covered more or less the same topics but marked the end of the 'romantic era' of behavioural ecology. Some accepted theory, like optimal foraging, was found to be too simple and the usual analytical techniques were criticised (e.g. the comparative approach).

Huntingford (1993) spelled out six points characterising the third edition that we classify as follows according to the scope of this book:

(a) Bridging of the gap between the analysis of behavioural adaptations at the individual level and processes in population and community ecology. (We will distinguish population (the whole set of individuals which exchange genes regularly) from the stock which is the exploited fraction of (ideally) one population.)
(b) Extension of optimality models to trade-off conflicting demands (e.g. reduction of predation risk and maximum rate of food intake) in order to maximise fitness
(c) Introduction of stochastic dynamic modelling to bridge the gap between behaviour now and fitness later
(d) Links between behaviour and structural/physiological traits, and the need to understand the mechanisms (e.g. proximate cues that elicit feeding, role of learning in foraging)
(e) Increase in the precision of genetic relatedness due to the development of DNA fingerprinting
(f) Interest in parasitism behaviour (especially sexual displays as indicators of disease resistance).

(4) Sociobiology emerged from the 1970s onwards and is often associated with the work of Wilson (1975). It focuses on the study of social systems of animals living in groups from the perspective of evolutionary biology (Drickamer & Vessey, 1992).

From a genetic point of view, Mayr (1976) distinguishes two kinds of ethological programmes according to the degree of plasticity of the corresponding phenotype. 'Closed programmes' do not allow modification by experience, as do 'open programmes'. Most of the behavioural traits related to intra- or inter-species behaviour are related to closed programmes because they are linked to the emission and/or reception of signal from other individuals. On the other hand non-communicative behaviour (feeding, habitat selection) is related to open programmes.

The history of animal behaviour is largely dominated by the observation of mammals and birds. Nevertheless, fishes have retained the attention of many workers (see Barlow (1993) for a review), but until recently this was mainly the case for freshwater and/or demersal fish, as for instance the famous three-spined stickleback (*Gasterosteus aculeatus*), or different species of coral fish. Reproductive behaviour (Dulzetto, 1928), especially mating behaviour (Clark *et al.*, 1954; Constantz, 1974; Farr, 1977) and parental care (Fryer & Iles, 1972; review in Baylis, 1981) were studied early in fish. The advantages of group life were also investigated early in fish, especially the antipredator function (von Frisch, 1938; Neill & Cullen, 1974) and feeding (Roberston *et al.*, 1976). Observations on fish also contributed to testing optimality theory (Werner & Hall, 1974; Kislalioglou & Gibson, 1976), exemplified the use of rules of thumb during foraging (O'Brien *et al.*, 1976) and revealed physiological constraints on diet selection (Werner, 1977). Milinski (1979) used sticklebacks to confirm the 'ideal free distribution' theory initially developed in birds by Fretwell and Lucas (1970). The trade-off between habitat selection, feeding and predator avoidance was studied in fishes (Milinski & Heller, 1978; Werner *et al.*, 1983), and Milinski (1985) demonstrated how parasites modify fish behaviour to promote transmission to the main host. Fishes were used also in experimental tests of the predictions of game theory (Turner & Huntingford, 1986; Enquist *et al.*, 1990). The cost and benefits approach, pioneered in terrestrial animals by MacArthur and Pianka (1966), was also applied to the territoriality in fish (Ebersole, 1980; McNicol & Noakes, 1984). Finally, fishes also supported numerous studies on alternative reproductive strategies (Jones, 1959; Gross & Charnov, 1980).

Pelagic fish are difficult to study because they are not easy to manipulate without being damaged and because *in situ* observation is made difficult by their low contrast with the environment, their great numbers in schools and their often strong avoidance reaction to human observers or to photographic light. New technologies have allowed easier observation of natural behaviour, especially acoustic devices and low-light-level cameras. Pioneer tank or aquarium observations and experiments on pelagic fish behaviour were conducted by Parr (1927), Breder (1954) and later Shaw (1969) on their most typical behavioural trait: schooling (Chapter 4). Breder also made pioneer observations on the flight of flying fish by using advanced flash photography (Breder, 1929). During the same period, fishery biologists accumulated, and still continue to accumulate, a large number of field observations on pelagic fish behaviour (especially horizontal and vertical migrations), but this work remained mainly descriptive. References of these pioneer works are available in the selected bibliography compiled by Russel and Bull (1932).

Researchers from the USSR were probably the earliest to develop a real school on fish behaviour studies from the 1950s as did, to a certain extent, Keenleyside (1955) in Canada. Radakov (1973) and co-workers developed advanced techniques to observe school behaviour and most of their interpretations are still in use. For instance, the trade-off between hunger and gear avoidance was already noticed by Radakov (1973) for sardines, which were observed to detect a trawl later when in the presence of prey. Lebedev (1969) gave a synthesis of his own work and of his colleagues' and proposed a theory of elementary populations which has not been completely supported by recent findings.

In North America, Strasburg and Yuen (1958) made pioneer visual underwater observations on skipjack tuna (*Katsuwonus pelamis*) behaviour in the wild. Winn and Olla (1972) edited a book gathering different papers on behaviour of marine vertebrates, but it is limited to US works. During the last decades of the twentieth century, Japanese, Canadian and US scientists participated increasingly in International Council for the Exploration of the Sea (ICES) working groups and committees and in other international events on fish behaviour (review in Bardach *et al.*, 1980).

Even though other aspects of fish behaviour (distribution, growth, natural mortality) were addressed by a symposium held in 1958 (Kesteven, 1960), early on a different approach to fish behaviour studies was related to the recognition of fish behaviour as an important factor in improvements to fishing technology (1957 first FAO international fishing gear congress, Kristjonsson, 1959). In the early 1960s, scientists from France, Germany, Netherlands, Norway, Sweden and the UK founded IF (International Fishing Technology Working Group), which was orientated toward trawl construction but also took up applied behavioural studies. The second FAO international fishing gear congress took place in 1963 (Finn, 1964) and the third one in 1967 (Ben-Tuvia & Dickson, 1968). This third FAO international congress was co-organised with ICES and named Conference on Fish Behaviour in Relation to Fishing Techniques and Tactics; it represented a turning point in research in this field. Finally, in 1968 an All-Union Conference was held in the USSR (Alekseev, 1968) which indicates the importance given by the international community to these aspects during this 11 year period (1957–1968).

The IF working group was recognised and adopted by ICES and renamed the Fishing Gear Technology Working Group in 1969. In 1972, a new working group on sound and vibration in relation to fish capture was established with the aim of improving the understanding of acoustic stimuli in fish behaviour during fishing operations. Both working groups were connected to the ICES Fish Capture Committee and were later merged into a single group renamed the Fish Technology and Fish Behaviour (FTFB) group. In 1979 a specific working group on acoustics was created, the Fish Acoustic Sciences Technology (FAST) group. These two working groups held joint sessions, recognising the importance of improving the understanding of fish behaviour in relation to their work. Even though ICES working groups devoted much time to technical and methodological improvements, there was a clear trend towards the use of acoustics as a scientific tool for studying fish behaviour and stock abundance. This trend, and a growing interest in applied fish behavioural studies, were also observed in a symposium held in 1992 in Bergen on Fish Capture and Fish Behaviour (Wardle & Hollingworth, 1993) and in the recent book of Fernö and Olsen (1994). Japanese scientists attended the Bergen symposium and gave a larger audience to Japanese works previously published in national journals.

Mainly from *in situ* studies, great progress took place in the fields of the ICES FTFB and FAST working groups. At the beginning, most of the scientific production consisted of the design of gear or technical acoustic devices and appeared mostly in the grey literature. As underlined by Fernö (1993), often fisheries behavioural studies lack rigorously defined units for measuring behaviour and remain mainly qualitative. But from the 1980s, the scientific production shifted towards the primary literature, and the above-mentioned symposia were edited by ICES after conventional refereeing. Unfortunately, behaviourists (*sensu stricto*) never attend these meetings, and fisheries biologists working on behaviour do not participate in behavioural ecology events; these two related branches of investigation seem to ignore each other, despite the clear needs expressed by the fisheries biologists (Harden Jones, 1978). Many behaviourists are investigating conceptual aspects of fish behaviour, especially from the perspective of modern behavioural ecology, mainly from studies in space-limited and controlled environments (Huntingford & Torricelli, 1993; Pitcher, 1993); as far as pelagic fish are concerned, schooling behaviour received a greater attention (e.g. Pitcher & Parrish, 1993) and will be largely developed in this book.

Independently, some fishery scientists intend to evaluate the effect of fish behaviour on direct or indirect stock assessment methods by direct measurements or by modelling (review in Fréon *et al.*, 1993a; Chapters 8 and 9 of this book). Behavioural ecology studies are sometimes considered more academic than fisheries studies. This is less and less true and, as in many branches of science, academic and applied research are complementary and can take advantage of each other. On this line, the recent approach of behavioural studies by artificial life (Langton, 1989) is promising for modelling individual fish behaviour by oriented object languages. Huth and Wissel (1990, 1994) applied it successfully in the case of schooling behaviour, confirming the fact that a leader is not useful for the cohesion of a moving school.

1.2 Brief history of stock assessment methods

After a long debate on the causes of fluctuations in stock abundance (natural fluctuation *vs.* influence of fisheries), which culminated in 1883 during the London Exposition, it was recognised that better knowledge was needed to manage the fisheries (review in Smith, 1988, 1994). This task was assigned to the International Council for the Exploration of the Sea (ICES), officially created in 1902 after the analysis of the data from the first experimental analysis of stock depletion by trawling, named the Garland experiment (Fulton, 1896), along with commercial data analysis of the English fisheries (Garstang, 1900). While the ICES overfishing committee, chaired by William Garstang, focused initially on the causes of stock abundance, the migration committee, chaired by Johann Hjort, paid attention to fish movements and the availability of fish to fishermen. In 1907 Hjort (1908) first suggested that fish age distribution be studied in order to understand the variability in catches. His suggestion of an international programme for collecting herring scales for ageing was finally adopted in 1909, but unfortunately he did not convince ICES of the value of this approach. One of the main criticisms of Hjort's approach, which delayed its wide acceptance for many years, was dealing with fish behaviour. D.W. Thompson, one of his primary antagonists, argued that schools of herring are likely to be composed of fish of the same age, so the age composition of samples would not represent the ages in the population.

Fish tagging started with the pioneer work of Fulton and Petersen on plaice at the end of the nineteenth century, but for the more abundant and fragile pelagic fish, this technique was not really applicable before the 1960s with the use of small internal magnetic tags. These tags were used for tagging small Atlantic herring (Dragesund & Hognestad, 1960) and menhaden (*Brevoortia tyrannus*) over 100 mm (Carlson & Reintjes, 1972; Pristas & Willis, 1973). They are automatically detected during fish processing (Parker, 1972). Egg surveys were initiated on demersal species at the beginning of the 1920s to back-calculate the size of the spawning stock, knowing the fecundity of females (review in Gunderson, 1993). This method was more difficult to apply to pelagic stocks because the spawning areas of these are wider and eggs suffer advection processes. In addition, many pelagic fishes are indeterminate spawners which are continuously maturing broods of eggs during the spawning season. This last problem was only overcome in the 1990s by the daily production method (Hunter *et al.*, 1993).

After Hjort's pioneer work, the idea of structured models was again proposed independently in the USSR, mainly by Baranov (1918) and Derzhavin (1922); (review in Ricker, 1971, 1975). Then it was developed by Fry (1949), but Beverton (1954) and later Beverton and Holt (1957) and Paloheimo (1958) emphasised estimation of mortality rates given catch and effort data. From that period, natural mortality was included in the models and fishing mortality was defined as the product of fishing effort and catchability. Throughout this book we will see how catchability is related to fish behaviour. The other improvements to the structured models will be developed in Chapter 8, along with the surplus-production models which appeared during the same period (Hjort *et al.*, 1933; Graham, 1935; Sette, 1943; Schaefer, 1954, 1957). These models were applied mainly to pelagic stocks. The surplus-production models consider the change in abundance of the whole population and do not require the age composition of the catches to be known, but are more sensitive to knowledge of the catchability, which is assumed to be constant in most of these models.

During the period following the Second World War, two analytical approaches were studied: the yield per recruit and theoretical mechanisms relating stock and recruitment (Ricker, 1954; Beverton & Holt, 1957). Even though the second approach remains largely theoretical owing to the difficulty of estimating the recruitment and to the large variability of the relationship between stock and recruitment, these two approaches allow us to distinguish between two sorts of overfishing: recruitment overfishing, when catches affect the reproductive capacity of the stock, and growth overfishing, when fish are caught before reaching an optimal weight. In contrast to demersal species, pelagic species have a fast growth and usually suffer recruitment overfishing, which is responsible for many stock collapses (often in conjunction with environmental changes) as reviewed by Csirke (1988). Nevertheless in the case of the Atlantic menhaden (*Brevoortia tyrannus*) fishery, Ahrenholz *et al.* (1987) suggested a growth overfishing. This exception is probably due to the relatively slow growth of this species in relation to its long life span (> 8 years).

Direct methods of stock assessment making use of fishing gears started at the beginning of the twentieth century and were at first limited to demersal fish caught by bottom trawl (swept-area method). They were much later extended to midwater trawl, but in only a few instances (Parmanne & Sjöblom, 1988) owing to the social behaviour of pelagic fish resulting in an extremely patchy distribution and therefore a large variability in the results. Recently, specific midwater trawls have been designed for the estimation of pelagic postlarval abundance of demersal species (Potter *et al.*, 1990) or juveniles (Godø & Valdemarsen 1993). Planes and helicopters have been used for many years by commercial fleets to locate schools of coastal pelagic fish or tunas (for instance in the menhaden fishery of the south-eastern USA or in the Atlantic tuna fishery). But aerial surveys with the aim of fish stock abundance estimation are not common (but often used for mammals). The reason for this poor success is that aerial detection is limited to surface schools and there is usually a large variability in the vertical distribution of schools. Visually based surveys of coastal pelagic fish are conducted during the day, but use of low-light-level video enables surveying during the night (Squire, 1972; Cram, 1974; Williams, 1981). New techniques which are less dependent on fish depth (light detecting and ranging – LIDAR;

Kronman, 1992) or on cloud coverage (compact airborne spectrographic imager – CASI; Borstad *et al.*, 1989; Nakashima & Borstad, 1993) are still on the methodological stage.

Following the pioneer work on stock assessment, where the main basis of population dynamics was established, a long period of improvement and refinement of methods took place and is still continuing. This was facilitated by the increasing use of electronic computation, which permitted complex and time consuming mathematical and statistical methods such as fitting by iteration. Computers also allowed for an increasing use of simulations in population dynamics. Simulations are not considered to be assessment methods, but are of great help for understanding and improving these methods, especially to estimate the risk of collapse in terms of probability. Nevertheless, as remarked by Francis (1980), during this period:

> 'More attention has been paid to developing mathematically sophisticated methods of fitting various analytic models than to the basic structure and assumptions of the models themselves ... The inference is made that, because the analytic models (gross abstraction of the reality) have long-term equilibrium properties, so too must the populations to which they are applied. This seems rather a backwards approach.'

The great novelty in the second half of the twentieth century has been the use of hydro-acoustics first by fishermen and then by fishery biologists. After the first world war, hydro-acoustic detection improved considerably for military purposes (Le Danois, 1928; Marti, 1928). The technology was soon applied to experimental fish detection (Kimura, 1929; Sund, 1935). During the second world war, the military technology improved again and after the war it started to be used by fishermen. Fisheries biologists started to use it for describing the horizontal and vertical distribution of pelagic fish (e.g. Cushing, 1952; Trout *et al.*, 1952; Harden Jones & McCartney, 1962; Mais, 1977). The next step was the quantification of biomass by echointegration, which started around 1970, following the pioneer work of Dragesund and Olsen (1965). The technique was rather imprecise at the beginning due to problems of electronic calibration and imprecise knowledge of the intensity of fish echo energy (target strength), and as a result it was not well accepted by the scientific community familiar with indirect methods of stock assessment. In the 1970s and 1980s, considerable technical and methodological improvements took place (reviews in Foote *et al.*, 1987; MacLennan & Simmonds, 1992; Simmonds *et al.*, 1992). Five symposia (four of them under the auspices of ICES) present milestones in the history of fisheries acoustics and behaviour: the 1979 Cambridge (Massachusetts) symposium as part of a joint programme between the USA and USSR (Suomala, 1981), the 1973 Bergen symposium (Margetts, 1977), the 1982 Bergen symposium (Venema & Nakken, 1983), the 1987 Seattle symposium (Karp, 1990) and the 1995 Aberdeen symposium (Simmonds & MacLennan, 1996).

The imprecision of the former equipment and methodology is now replaced by the possibility of performing absolute measurements of fish abundance (Foote & Knudsen, 1994) and the method is now used for providing fishery-independent estimates of many economically important stocks around the world. Nowadays the technique has reached a high degree of sophistication with the use of dual-beam (Ehrenberg, 1974) or split-beam (Carlson & Jackson, 1980; Foote *et al.*, 1984) echosounders which allow the length distribution of dispersed fish to be estimated. Moreover, automatic echo-classification permits the identification and characterisation of schools and layers (Rose & Leggett, 1988; Souid, 1988; and additional references in section 4.8.3). The last important improvement is the use of omnidirectional multibeam sonar for counting the schools in a large range around the boat, with a concomitant estimation of their size and preliminary attempts at biomass estimation (Misund, 1993a; Gerlotto *et al.*, 1994). Nevertheless, the acoustic method, like any other, still suffers from some limitations, mainly due to fish behaviour, and the quantification of the bias still represents an important research field as observed during the 1995 Aberdeen symposium.

1.3 Scope of the book

The brief histories of fish behaviour studies and stock assessment methods tell us that these two fields of science were developed relatively independently of each other. Former researchers in population dynamics had clearly identified the main problems related to fish behaviour, but this was done mainly from a theoretical point of view. Most of the fish behavioural studies have been performed on the one hand by animal behaviourists, working on a small scale with the aim of fundamental understanding of behaviour, and on the other hand by fishery technologists working in the field to improve gear efficiency. Recently, fisheries biologists involved in direct methods of stock assessment (mainly acoustic surveys and secondarily tagging experiments and aerial survey) have been experimenting *in situ* with the effect of fish behaviour on the accuracy of their estimates. But research teams including both animal behaviourists and fishery biologists remain scarce. The authors of this book are fisheries biologists and partly self-taught in fish behaviour. Throughout this book we will try to bridge the gap between these two fields of investigation and show how fish behaviour interacts with the different stock assessment methods.

In most cases, fish behaviour introduces bias into assessment methods because it is not fully taken into account. In indirect estimation of stock abundance through population dynamics, the central problem is the variation of catchability due to different fish behaviours. As far as direct estimations are concerned, and particularly during acoustic surveys or aerial survey, knowledge of fish behaviour is necessary at different levels: vertical and horizontal avoidance, packing density, geographic distribution, etc. We will try to identify the biases, quantify them and present some available methods to limit them when possible.

Chapter 2 offers a brief review of the major pelagic fisheries of the world, subdivided into an analysis of the catches and a short description of the main gears used to catch pelagic fish. Chapter 3 describes habitat selection and migration. We devote Chapter 4 to fish schooling, a major trait of pelagic fishes, including among other topics the function of schooling, school organisation, and mixed species schools. Chapter 5 focuses on avoidance reactions, elicited either by sounds (vessel or gear) or by visual cues. Chapter 6, in contrast, discusses attraction, by light, bait or by floating objects in the case of the tunas (associative behaviour). Chapter 7 deals with learning processes and their effects on fish behaviour and fisheries. Chapter 8 is devoted to the effects of fish behaviour on fisheries and stock assessment by population dynamics models. Similarly, Chapter 9 briefly presents the acoustic assessment methods and details the influence of habitat selection, social behaviour, avoidance and learning on these direct methods of stock assessment. Chapter 10 reviews the importance of fish behaviour on other stock assessment methods: midwater trawl surveys, aerial surveys, egg surveys and capture-recapture from tagging experiments.

We limit the scope of this book to marine pelagic species, including both small coastal species (herring, sardine, anchovies, mackerels, horse mackerels, etc.) and various oceanic species (mainly tunas). Demersal fishes (especially gadoids) will be considered only during their pelagic stage when they can be investigated with the usual tools applied to pelagic species. Examples from temperate and tropical areas are presented to illustrate the different problems and make use of the comparative approach.

Chapter 2
Pelagic Fisheries

The following remarks on world fisheries are based on the catch statistics provided in the FAO yearbook, *Fishery Statistics, Catches and Landings* (FAO, 1997), and are limited to the period 1985–1994.

The total world catch of aquatic organisms, including marine and freshwater fishes, crustaceans and molluscs, exceeded 100 million tonnes in 1993, and climbed to about 110 million tonnes in 1994. This is a rise in catches of aquatic organisms of about 10 million tonnes since 1988. The increase in catches of marine organisms in the period 1988–1994 was about 4.7 million tonnes, and that of freshwater organisms in the same period was about 5.7 million tonnes.

2.1 World catch of small pelagic species

The total world catch of small pelagic, marine species such as anchovies, herrings, jacks, mackerels, mullets, sardines and sauries amounted to about 40 million tonnes or about 36 % of the total world catch in 1994. This is an increase in the catch of these species of about 2 million tonnes since 1988. Here we present catch statistics of 14 small pelagic species that have produced an annual average catch of more than 0.5 million tonnes each during the decade between the mid 1980s and mid 1990s (Table 2.1). For most of these species the annual catches varied by a factor of about two between the mid 1980s and mid 1990s, in the extreme case of anchoveta by a factor of 12.

In terms of catch quantity, the most important of the small pelagic species between the mid 1980s and mid 1990s is the anchoveta or the Peruvian anchovy (*Engraulis ringens*). The catch of this species climbed enormously from about 1 million tonnes in 1985 to nearly 12 million tonnes in 1994 (Table 2.1). About 80% of the catch is landed in Peru, making the country the second largest fishing nation in the world. The rest of the catches of Peruvian anchovy are landed in Chile. The south-eastern Pacific is also the habitat of other important small pelagic species such as the Chilean jack mackerel (*Trachurus murphyi*) and the South American pilchard (*Sardinops sagax*). The catch of the Chilean jack mackerel doubled from 2.1 million tonnes in 1985 to 4.2 million tonnes in 1994, while the catch of South American pilchard declined from 6.5 million tonnes to 1.8 million tonnes in the same period (Table 2.1). Since the mid 1990s about 95% of the catch of Chilean jack mackerel has been landed in Chile, the rest in Peru. In the mid 1980s, about 25% of the catch was taken by distant water trawlers from the

Table 2.1 Total annual world catch (1985–1994) of the 14 most important small pelagic species (FAO, 1997).

Species	Latin name	Total annual catch (thousand tonnes)				
		1985	1988	1990	1992	1994
Anchoveta	*Engraulis ringens*	987	3613	3772	5488	11896
Chilean jack mackerel	*Trachurus murphyi*	2148	3245	3828	3371	4254
Atlantic herring	*Clupea harengus*	1450	1685	1535	1536	1886
South American pilchard	*Sardinops sagax*	6509	5383	4254	3043	1793
Chub mackerel	*Scomber japonicus*	1742	1825	1328	958	1507
Japanese pilchard	*Sardinops melanostictus*	4773	5429	4732	2488	1294
European pilchard	*Sardina pilchardus*	926	1366	1549	1188	1208
Capelin	*Mallotus villosus*	2216	1142	980	2114	884
Scads	*Decapterus* spp.	553	570	827	915	910
Atlantic mackerel	*Scomber scombrus*	597	709	660	783	857
Gulf menhaden	*Brevortia patronus*	884	639	520	433	768
European anchovy	*Engraulis encrasicolus*	599	859	539	412	534
Sardinellas spp.	*Sardinella* spp.	671	602	608	733	814
Japanese anchovy	*Engraulis japonicus*	349	303	536	662	820

former USSR, but this activity stopped around 1990. The South American pilchard was fished about equally by Chile and Peru in the middle of the 1980s, but in the 1990s this species has been most available to Peruvian purse seiners which in 1994 took about 90% of the catch.

The chub mackerel (*Scomber japonicus*) is caught in sub-tropical and tropical regions worldwide, but most of the catch (70–80%) is taken by Japan, China and Korea in the north-west Pacific. The total catch of chub mackerel varied between 1.0 and 1.8 million tonnes during the period from the mid 1980s to the mid 1990s. These countries also have major fisheries of Japanese pilchard (*Sardinops melanostictus*) and Japanese anchovy (*Engraulis japonicus*) in the same region. The total catch of Japanese pilchard, which is fished mainly by Japan, declined from 4.7 million tonnes to 1.3 million tonnes between the mid 1980s and mid 1990s (Table 2.1). On the other hand, the total catch of Japanese anchovy increased from 0.35 million tonnes to 0.82 million tonnes in the same period.

The Atlantic herring (*Clupea harengus*) has discrete stock units in the north-western Atlantic off Canada and USA, off Iceland (Icelandic summer-spawning herring), off Norway (Norwegian spring-spawning herring), and several stock units in the North Sea and around the British Isles (North Sea herring), in the Skagerrak (north of Denmark), and in the Baltic. The catch of the different stock units fluctuated substantially, but the total landings of Atlantic herring were remarkably stable at around 1.5 million tonnes between the mid 1980s and mid 1990s. The catch of this species is expected to increase in the years to come since the stock of Norwegian spring-spawning herring has recovered from its collapse of the late 1960s, and the total quota for this stock alone was 1.5 million tonnes in 1997. Between the mid 1980s and mid 1990s the major catches of Atlantic herring were taken by Norway (0.2–0.5 million tonnes), Canada (0.2–0.3 million tonnes), Denmark (0.1–0.2 million tonnes) and

Sweden (0.1–0.2 million tonnes). Finland, Germany, Netherlands, Poland, Russia, and the UK each took annual catches up to around 0.1 million tonnes of Atlantic herring.

The Atlantic mackerel (*Scomber scombrus*) and the capelin (*Mallotus villosus*) are also caught on both sides of the northern Atlantic, but the major fisheries for these species are conducted in the north-eastern Atlantic. The total catch of Atlantic mackerel increased to 0.9 million tonnes between the mid 1980s and mid 1990s, and the major fisheries are conducted by the UK, Norway and Ireland in the North Sea and west of the British Isles. The total catch of capelin varied between 0.8 and 2.1 million tonnes, and the major catches were taken by Norway and Russia on the Barents Sea stock, which spawns on the coast of northern Norway, by Iceland on the Icelandic stock, which spawns on the coast of southern Iceland, and by Canada on the Newfoundland stocks. Due to low stock levels, there were no quotas for capelin in the Barents Sea in 1988–1990, and from 1994 onwards.

The scads (*Decapterus* spp.) are caught mainly in eastern and south eastern Asia. The total annual catch increased from 0.6 million tonnes to 0.9 million tonnes between the mid 1980s and mid 1990s. The major catches of scads are landed by China, the Philippines, and Indonesia.

The sardinellas (*Sardinella* spp.) are fished in tropical waters off Africa and Asia. The total annual catch remained remarkably stable between the mid 1980s and mid 1990s, varying only from 0.6 to 0.8 million tonnes. The main fisheries are off Senegal, off Thailand and along the Philippines.

The Gulf menhaden (*Brevortia patronus*) is caught along the coast of the southern USA in the Caribbean Gulf. Between the mid 1980s and mid 1990s the annual catches have been rather stable, varying from 0.4–0.9 million tonnes.

The European pilchard (*Sardina pilchardus*) and European anchovy (*Engraulis encrasicolus*) are fished in coastal areas of the eastern Atlantic and in the Mediterranean. Total catches of these species were rather stable between the mid 1980s and mid 1990s. The total catch of European pilchard varied from 0.9 million tonnes (1985, Table 2.1) up to 1.6 million tonnes (1990), while the total catch of European anchovy varied from 0.4 million tonnes (1992) to 0.9 million tonnes. The major fishery for European pilchard is along the coast of north-west Africa, where about 55% of the total catch is taken, mainly by Morocco and Spain. About 25% of the total catch is taken in the Mediterranean, and about 20% is taken in the southern North Sea, in the Bay of Biscay and along the Iberian Peninsula. The major fishery for European anchovy is in the Mediterranean, where up to 90% of the total catch is taken. There is also an important fishery for this species off north-western Africa.

The fisheries for the 14 species listed in Table 2.1 contribute nearly 75% of the total world catch of small pelagic species. Other small pelagic species that sustain important fisheries with annual catches between 0.1 and 0.9 million tonnes are given in Table 2.2. Also for these species there are substantial annual variations in the catches, which in most cases varied by a factor of about two between the mid 1980s and mid 1990s. In the extreme case, for the Arauchanian herring in Chile, the catches varied by a factor of 15 for that decade. However, there are also examples of species that have

Table 2.2 Minimum – maximum total annual catch of small pelagic species that have given annual catches of about 0.1–0.9 million metric tonnes during the last decade (1985–1994) (FAO, 1997).

Species	Latin name	Fishing area	Min – max total catch (tonnes)
Pacific herring	*Clupea pallasii*	Northern Pacific, coast	184 000–349 000
Goldstripe sardinella	*Sardinella gibbosa*	South-east Asia, coast	108 000–158 000
Indian oil sardine	*Sardinella longiceps*	Southern Asia, coast	204 000–338 000
Round sardinella	*Sardinella aurita*	Central Atlantic, coast	193 000–438 000
California pilchard	*Sardinops caeruleus*	Central eastern Pacific coast	194 000–509 000
Southern African pilchard	*Sardinops ocellatus*	Southern Africa coast	88 000–210 000
Atlantic menhaden	*Brevoortia tyrannus*	South-eastern USA, coast	256 000–428 000
Bonga shad	*Ethmalosa fimbriata*	Central Africa, coast	100 000–134 000
European sprat	*Spattus sprattus*	North Sea – Mediterranean	220 000–581 000
Arauchanian herring	*Strangomera bentincki*	Chile, coast	38 000–583 000
Southern African anchovy	*Engraulis capensis*	Southern Africa, coast	167 000–969 000
Pacific achoveta	*Cetengraulis mysteticus*	Central eastern Pacific, coast	72 000–254 000
Stolephorus anchovies	*Stolephorus* spp.	South-east Asia, coast	240 000–281 000
Pacific saury	*Cololabis saira*	North-western Pacific	227 000–436 000
Mullets	Mugilidae	Worldwide, coast	169 000–240 000
Atlantic horse mackerel	*Trachurus trachurus*	Eastern Atlantic – Mediterranean	216 000–563 000
Japanese jack mackerel	*Trachurus japonicus*	North-western Pacific	122 000–371 000
Cape horse mackerel	*Trachurus capensis*	Southern Africa	284 000–584 000
Jacks	*Caranx* spp.	Tropical waters	124 000–193 000
Carangids	Carangidae	Tropical waters	210 000–294 000
Japanese amberjack	*Seriola quinqeradiata*	Japan	142 000–166 000

given a remarkably stable yield, such as the Stolephorus anchovies (*Stolephorus* spp.) and the Japanese amberjack (*Seriola quinqueradiata*) for which the annual catches varied by no more than 15%.

There are also fisheries on other small pelagic species which give annual catches in the range from 10 000 to 100 000 tonnes, and that is of vital importance to specific regions. An example is the fishery of Bali sardinella (*Sardinella lemuru*) in Indonesia which yielded catches of 54 000–145 000 tonnes between the mid 1980s and mid 1990s.

In many regions there are pelagic fisheries for species that are distributed in deep waters, that aggregate in a limited season during spawning, or that are semi-pelagic. An example of an important pelagic fishery on a species in deep waters is that of Alaska pollock (*Theragra calcogr噢ama*) which is distributed in deep waters over large areas in the northern Pacific. West of the British Isles there is a large pelagic fishery of blue whiting (*Micromesistius poutassou*) that aggregates for spawning. In Nowegian fjords there is a purse seine fishery with an annual yield of about 50 000 tonnes for saithe (*Pollachius virens*) that is schooling pelagically. The sand lance (*Ammodytidae*) in the North Sea, which often takes refuge by burying in the sand, is fished by high opening bottom trawls when schooling close to bottom. The annual yield of this species can be up to about 1 million tonnes.

2.2 World catch of large mackerel and tuna

The total world catch of large mackerel, tunas, bonitos and billfishes amounted to about 4.5 million tonnes or about 4% of the total world catch in 1994 (FAO, 1997) (Note that statistics provided by the FAO are slightly lower than those provided by international bodies devoted to tuna management.). This figure is an increase in the catch of these species of about 1.3 million tonnes since 1985. The total catch of the fish species in this category that have produced average annual catches exceeding 100 000 tonnes in the period 1985–1994 is given in Table 2.3. The ten species of large mackerel and tunas listed in Table 2.3 contributed about 90% of the total world catch of tunas, bonitos and billfishes in 1994. Compared with the substantial variations in the annual catches of small pelagic fishes considered in the preceding section, the annual catches of tunas are much more stable. For example, the annual catches of narrow barred Spanish mackerel (*Scomberomorus commerson*) varied by at most 25 000 tonnes or by a factor of 1.23 during this decade. The catches of skipjack tuna (*Katsuwonus pelamis*) increased by about 550 000 tonnes or about 60% during the same decade.

Table 2.3 Total annual world catch (1985–1994) of the most important tuna species (FAO, 1997).

Species		Total annual catch (tonnes)				
		1985	1988	1990	1992	1994
Narrow barred Spanish mackerel	*Scomberomorus commerson*	109 000	134 000	108 000	113 000	120 000
Japanese Spanish mackerel	*Scomberomorus niphonius*	121 000	170 000	247 000	172 000	228 000
Frigate and bullet tunas	*Auxis thazard, Auxis rochei*	153 000	183 000	194 000	224 000	212 000
Kawakawa	*Euthynnus affinis*	141 000	151 000	144 000	168 000	179 000
Skipjack tuna	*Katsuwonus pelamis*	914 000	1 285 000	1 306 000	1 428 000	1 463 000
Longtail tuna	*Thunnus tonggol*	912 000	141 000	166 000	112 000	101 000
Albacore	*Thunnus alalunga*	190 000	226 000	231 000	216 000	194 000
Yellowfin tuna	*Thunnus albacares*	724 000	909 000	1 065 000	1 124 000	1 075 000
Bigeye tuna	*Thunnus obesus*	242 000	231 000	274 000	271 000	293 000
Tuna-like fishes	Scombridae	178 000	205 000	241 000	240 000	251 000

The narrow-barred Spanish mackerel is caught in the Indian Ocean and in Polynesia, and the largest catches are taken by Indonesia, the Philippines and India. The Japanese Spanish mackerel (*Scomberomorus niphonius*) is caught in the northern Pacific by China, Korea and Japan. The Kawakawa (*Euthynnus affinis*) and longtail tuna (*Thunnus tonggol*) are caught both in the Pacific and in the Indian Ocean, and the Philippines, Thailand and Malaysia land the largest catches of these species. Frigate tuna (*Auxis thazard*), bullet tuna (*Auxis rochei*), skipjack tuna (*Katsuwonus pelamis*), albacore (*Thunnus alalunga*), yellowfin tuna (*Thunnus albacares*), and bigeye tuna (*Thunnus obesus*) are caught in subtropical and tropical areas worldwide.

The annual catches of skipjack and yellowfin tuna exceeded 1 million tonnes during most of the decade between the mid 1980s and mid 1990s, and are about one order of magnitude larger than the annual catches of the other tuna species. About 70% of the total catch of skipjack tuna is taken in the Pacific, and the largest catches are landed about equally by Indonesia, Japan, Korea, the Philippines and the USA (Plate 1) (opposite page 20). Similarly, about 60% of the annual total world catch of yellowfin tuna is taken in the Pacific, and the biggest catches are taken by Mexico, Indonesia, Japan, Korea, the Philippines and the USA. The catches of skipjack and yellowfin tuna in the Indian Ocean amount to about 20% of the total world catch of these species, and the Maldives, Spain and France take most of the catches of these species in the region. In the Atlantic fishery for skipjack and yellowfin tuna, France and Spain also take the biggest catches.

2.3 Fishing methods for pelagic species

The main fishing methods for catching small pelagic species are purse seining and midwater trawling (von Brandt, 1984). Tunas are caught by purse seining, longlining and pole-and-line.

2.3.1 *Purse seining*

In principle, a purse seine is a large net that is set from the aft of a purse seiner in an approximate circle to surround fish shoals. The top is kept floating by a line of floats at the surface, and the lower part of the net sinks by the force of a heavy leadline along the ground. The net will thus be stretched out as a circular wall surrounding the fish shoal (Fig. 2.1). The mesh size is so small that the net wall acts as an impenetrable fence preventing escape. When the purse seine has been set out and allowed to sink for some minutes, so that it reaches deeper than the target fish shoals, it can be closed by hauling the purse line, so that the ground of the purse seine will be confined, and pulled to the surface. When this operation is finished it is impossible for fish shoals to escape if the net is not torn. However, flying fishes (*Exocoetidae*) can still jump over the floatline.

The size of the purse seine depends on the behaviour of the fish to be captured and the size of the vessel from which it is operated. For catching fast-swimming, deep fish shoals, purse seines must be long, deep, have a high hanging ratio and be heavily leaded. Hanging ratio is the length of mounted net divided by length of stretched net. For catching slower-swimming, near-surface distributed fish shoals, purse seines can be shorter, shallow, and have a low hanging ratio. The relationships between fish species, vessel size and purse seine characteristics are given in Table 2.4.

Modern purse seining is mostly dependent on detection and location of fish shoals by hydroacoustic instruments (Misund, 1997). Hydroacoustic fisheries instruments have a transducer mounted to the hull under the vessel. The transducer transmits sequential sound pulses, at a frequency between 18 kHz and 180 kHz and at a duration of 3–50 milliseconds (ms), that propagate through the sea (MacLennan & Simmonds, 1992). The pulses are reflected by objects like fish that have acoustic

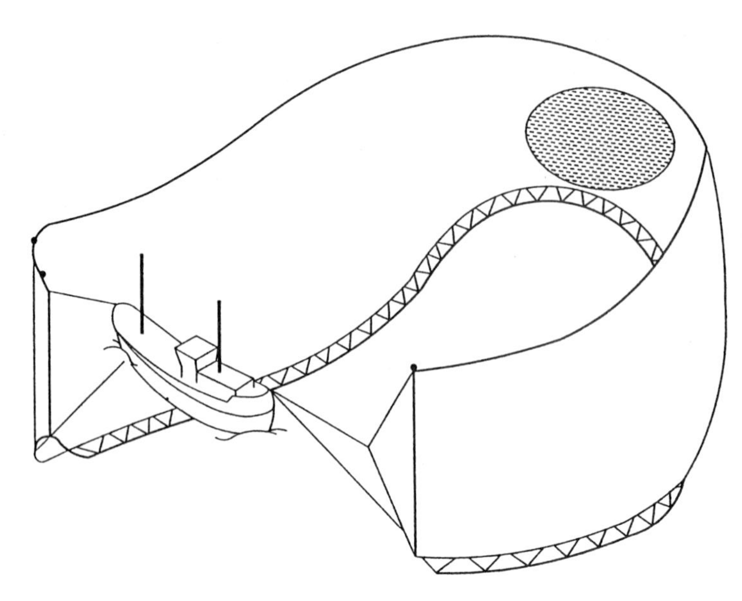

Fig. 2.1 Schematic drawing of a purse seine operation when the net is set around a fish school and pursing is about to start.

properties different from the surrounding sea water. Air-filled cavities like swimbladders give strong reflections or echoes. A horse mackerel which has a swimbladder will thus give a much stronger echo than an Atlantic mackerel which lacks a swimbladder. The sea bottom gives especially strong reflections. The transducer switches sequentially between transmission and reception mode, and echoes from fish or sea bottom can thus be recorded in reception mode. The echoes are converted to electric signals that are digitised, amplified and presented on a display in the wheelhouse.

Purse seiners are equipped with sonars that can train and tilt the transducer under the hull so that the transmitted sound pulse searches through large volumes in the sea. To enhance the volume coverage, some sonars can transmit sound pulses omnidirectionally, others transmit multiple beams, and the most advanced can be set to

Table 2.4 Relationship between target fish species, approximate purse seine characteristics (length, depth, mesh size, hanging ratio, leadline weight, sinking speed) and vessel size.

Species	Length	Depth	Mesh size	Hanging ratio	Leadline weight	Sinking speed	Vessel size
	(m)	(m)	(mm)		(kg/m)	(m/min)	(m)
Anchoveta	600	80	15	0.20	5.0		45
Capelin	600	160	22	0.45	7.0	15	60
Herring	650	160	35	0.45	6.0	18	60
Chilean jack mackerel	1000	200	35	0.40	7.0	18	60
Atlantic mackerel	650	160	35	0.45	6.0	18	60
Saithe	650	150	60	0.45	1.5	14	27
Tuna	2000	200	150	0.25	1.5		70

transmit in a single beam, a multi-beam or an omnidirectional mode. The detection capability is another important characteristic of fisheries sonars. In sea water, sound absorption increases exponentially with sound frequency (MacLennan & Simmonds, 1992). The detection range of fish shoals is therefore much longer for a low-frequency (18–34 kHz) than for a high-frequency (95–180 kHz) sonar. Many purse seiners have a low frequency sonar for detection of fish shoals at long range. However, the resolution of low-frequency sonars is usually rather limited because the width of the beam of such sonars is usually about 10°. High-frequency sonars have beam widths of about 5°, and thus better resolution. Many purse seiners therefore have a high-frequency sonar for more detailed mapping of shoal size and fish behaviour in relation to the vessel and the net. On the larger purse seiners (> 40 m) it is common to have both a low- and a high-frequency sonar to optimise fish detection and to make possible the detailed imaging of fish schools.

Purse seining is conducted on fish aggregated in dense shoals (Pitcher, 1983) or on fish occurring in distinct schools, in which the density is much higher than in shoals. Normally, purse seining on shoals takes place in darkness during night-time, while school fishing is limited to the daylight hours. In some fisheries, profitable purse seining is possible both when the fish are schooling during daytime and when they are shoaling at night. For example, this is usually the case during the winter fishery for capelin on the coast of northern Norway, and on the spawning grounds of Norwegian spring-spawning herring on the coast of western Norway in winter. Other purse seine fisheries are profitable only when the fish are shoaling at night or when they are schooling during daytime. An example of the former is the large Chilean fishery for Chilean jack mackerel, which normally is conducted when the fish occur in dense shoals near the surface at night (Hancock *et al.*, 1995). Most of the purse seine fisheries for herring and mackerel in the North Sea in summertime are conducted when the fish are schooling during the daylight hours.

The fishing capacity of purse seiners is normally proportional to vessel size. In the Chilean jack mackerel fishery where there is no limitation set by fishing quotas, the total annual catch of purse seiners was related to the hold capacity of the vessel through the equation:

$$\text{total annual catch (tonnes)} = 33.3 \times \text{hold capacity } (m^3) + 18.2$$

(Hancock *et al*, 1995)

In the 1992 season, a purse seiner with a hold capacity of 1350 m^3 was able to land about 65 000 tonnes of Chilean jack mackerel!

In some regions, artificial light is used to attract fish at night-time (see section 6.1). When sufficient fish have aggregated near the light source, they are caught by purse seining (Ben-Yami, 1976). This technique is probably of greatest importance for purse seine fisheries in Asian countries where it is used offshore. The technique is also common in the Mediterranean, the Black Sea, and in the Russian and African lakes. In other regions the technique is mostly used inshore, as during the sprat, herring and saithe fisheries in the fjords of southern Norway.

Tuna purse seining is conducted by large vessels (mostly > 60 m) with large nets (Table 2.4) in tropical/subtropical regions in the Atlantic and Indian Ocean

(Mozambique Channel, Seychelles), off Australia, and in the western and eastern Pacific. In the Atlantic and Indian Ocean, the tuna is caught by set made on free-swimming schools or by set made on fish associated with logs (floating objects of natural or artificial origin, mostly trees, branches, etc.). In the eastern Pacific, tuna are also caught by set made on dolphin herds with which the tuna is associated (Anon, 1992). The dolphin herds are visible on the surface and rather easy to encircle by the fastgoing tuna seiners. Usually large tuna are present underneath the dolphins. When the purse seine is closed around the encircled dolphins and tuna, an attempt is made to release the dolphins by the backdown procedure. This causes the floatline of the distant part of the purse seine to submerge so that the dolphins can swim and jump over. However, dolphins frequently become entangled in the purse seine and drown. The tuna pursers operating in this region were therefore under pressure to change the fishing strategy or fishing operation to decrease the accidental killing of dolphins, which is now achieved. The total dolphin mortality decreased from more than 130 000 individuals in 1986 to 3274 in 1995 (Anon, 1992, 1997). The association behaviour of tuna is considered especially in Chapter 6.

2.3.2 *Midwater trawling*

Pelagic species are also caught in large quantities by midwater trawling (also called pelagic trawling), both single boat and pair trawling. In the Atlantic fisheries, capelin, herring, horse mackerel, mackerel, sardines and sprat are caught by midwater trawling. Off Ireland there is a large pelagic trawl fishery for blue whiting, which aggregate for spawning during winter and spring. In the northern Pacific, there is a large pelagic trawl fishery for Alaska pollock.

In principle, a pelagic trawl is a net bag towed after a single vessel (Fig. 2.2) or between two vessels operating together. The two boats pull the trawl and open it horizontally by travelling parallel but some distance apart (500–1000 m). A single boat trawl is kept open by the lateral forces of two large trawl doors (5–15 m^2) in front of the trawl. The doors are attached to the warps from the vessel, and the trawl is connected to the doors via a pair of two or more sweeps. The length of the sweeps depends on the vertical opening of the trawl, and is about 180 m for a 30 m high trawl. A pair of weights (50–2000 kg), attached to the lower wings, and the weight of the doors pull the trawl downwards. The fishing depth of the trawl is adjusted by the warp

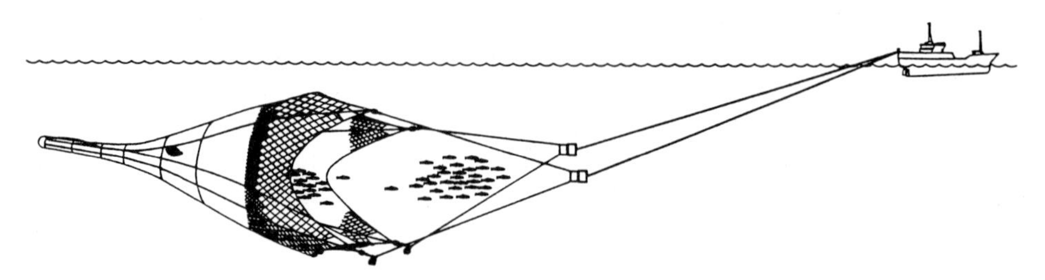

Fig. 2.2 Schematic drawing of a midwater trawl operation.

length, the towing speed and the vertical inclination of the doors. At a towing speed of about 6 ms^{-1} (3.5 knots), a warp length of about 500 m gives a fishing depth of about 200 m for a trawl with 30 m vertical opening, 1000 kg weights and about 3300 kg weight of the doors (Valdemarsen & Misund, 1995). To open the trawl vertically, there is a pair of weights (50–2000 kg) attached to the lower wings of the trawl to give it a downward pull. In most cases there are also floats or kites attached to the headline to give the trawl an upward pull. On single boat pelagic trawls, designed to catch small pelagic fish such as capelin (about 15 cm in length), there can be a pair of extra doors attached to the upper wings to give the trawl a proper opening. Such trawls are towed at low speed (about 3 ms^{-1}), and the lateral pull of just two doors will be too little to give the trawl the intended opening. According to the size, the vertical opening of a pelagic trawl varies by about an order of magnitude, from about 15 to 150 m. The horizontal opening is usually about equal to the vertical, and the total opening of pelagic trawls thus varies by about two orders of magnitude, from about 200 to 20 000 m^2.

Pelagic trawls are constructed according to specific drawings denoting the mesh size, twine, tapering and panel depth. Normally, the trawls are constructed of two or four panels that are joined in the selvedge or laced together. The size of the trawl is given as the circumference of the trawl opening. This is calculated as the number of meshes in the trawl opening (minus the number of meshes in the selvedge) multiplied by the stretched mesh size. The mesh size of pelagic trawls can be up to tens of metres in the front part, but decreases gradually to a few centimetres in the bag. The meshes in the front part of the trawl must only herd the target fish into the centre of the trawl opening (nevertheless, if these meshes are too large the target fish can escape, and the catching efficiency of the trawl decreases). In contrast, the meshes in the codend must be so small that it is physically impossible for the target fish to escape.

Pelagic trawling is conducted mainly on fish occurring in large shoals, extended aggregations and layers. The fish can be recorded by sonar or echo-sounder, and the trawl opening is normally monitored by a cable-connected net sonde or a trawl sonar. These instruments provide information on the opening of the trawl, and the presence of fish inside or outside the trawl opening. There are also acoustic sensors to monitor the door spread and the headline height and depth, and catch sensors that are activated when there is catch in the codend.

2.3.3 *Line fishing*

Large tunas, marlins and swordfish are also caught on hooks on longlines or with pole-and-line fishing. Longlining is conducted by Japanese, Korean and Taiwanese vessels in tropical and subtropical regions worldwide, and in total about 400 000 tonnes of tuna are caught by longlining (Bjordal & Løkkeborg, 1996). Bigeye tuna are fished almost exclusively by longlines, but substantial catches of yellowfin tuna and albacore are also taken.

Longlines for tuna are long ropes of synthetic fibres, with snoods of baited hooks attached up to 50 m apart (Bjordal & Løkkeborg, 1996). The snoods are often of monofilament and are attached to the mainline by a metal snap. To catch pelagic species, the longlines are set drifting in depths from the surface down to about 300 m. The pelagic longlines have marker buoys at each end, and are suspended by float and floatlines at regular intervals.

Pole-and-line is used to catch tuna shoaling near the surface. It is a traditional fishery around the Maldives area, which is now also developed in the western Pacific Ocean by Japanese fleets and on both sides of the Atlantic Ocean mainly by the Venezuelan fleet and the Senegalese, Côte d'Ivoire and French fleets. The pole can be operated manually or automatically. Attached to the pole there is a monofilament line with a barbless, shiny hook at the end, baited with a small fish or mounted with coloured filament forming a lure. Tuna that bite on the hook are immediately thrown onboard by swinging the hook. This can be a heavy job when catching large fish manually, and in some cases two fishermen operate a common hook so that they are able to handle large fish. A pole-and-line operation starts when the vessel has been manoeuvred gently into a shoal of tuna near the surface. To attract and keep the tuna near to the boat, live baitfish is thrown in the water regularly. Water spray from hoses at the rail may also help to attract fish.

2.4 Summary

In this chapter we have considered the major pelagic fisheries and the main fishing methods used to catch pelagic species. In particular we have focused on the annual variations in catches, which for most pelagic species are quite substantial but often unstable. We have also briefly presented the different gears commonly used and the fleets. One of the characteristics of most of the commercial fleets is their high mobility, from one country to another, when the resource is much depleted. For instance, the small pelagic fleet of purse seiners moved from California to Peru after the Californian sardine fishery collapse, then from Peru to Chile and to South Africa (some of these changes are reported by Glantz and Thompson, 1981). Similarly a large part of the purse seiner tuna fleet of the eastern Atlantic moved to the Indian Ocean at the beginning of the 1980s. Similar movements were observed between the eastern and western Pacific at the same period, mainly in reaction to a series of strong El Niño events.

In the next five chapters we will consider the major behaviour patterns of pelagic species that may affect fisheries and fish stock assessment. We start by habitat selection which can be quite variable for pelagic species that often live in the free water masses away from fixed landmarks.

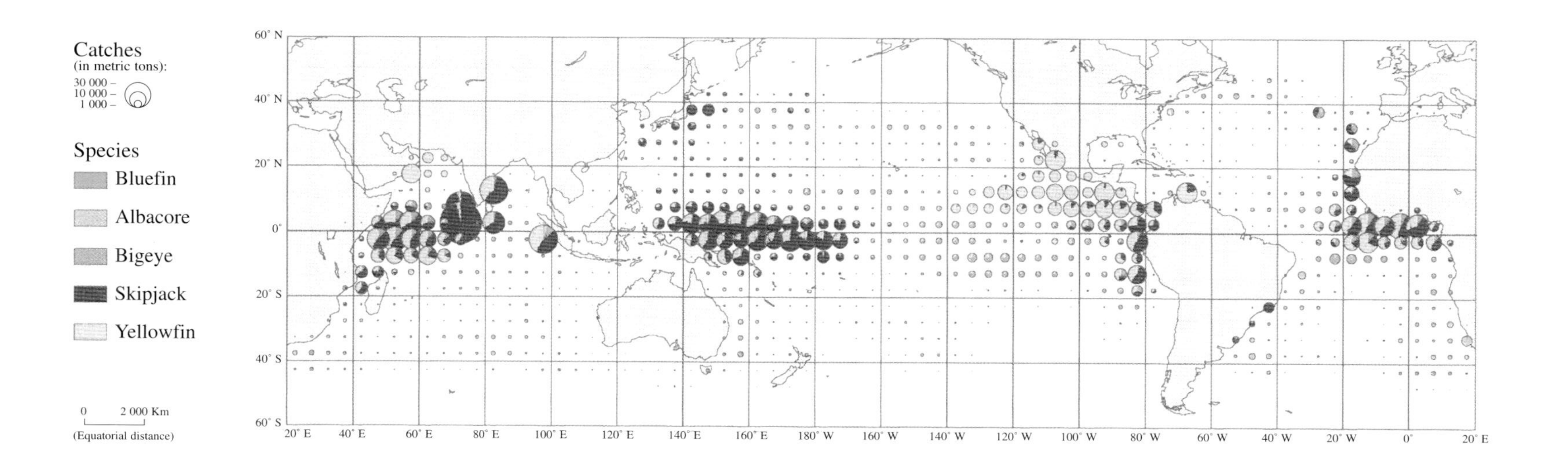

Plate 1 Average total catches of the tuna fisheries (purse-seine, pole-and-line, longline), 1989–1993.

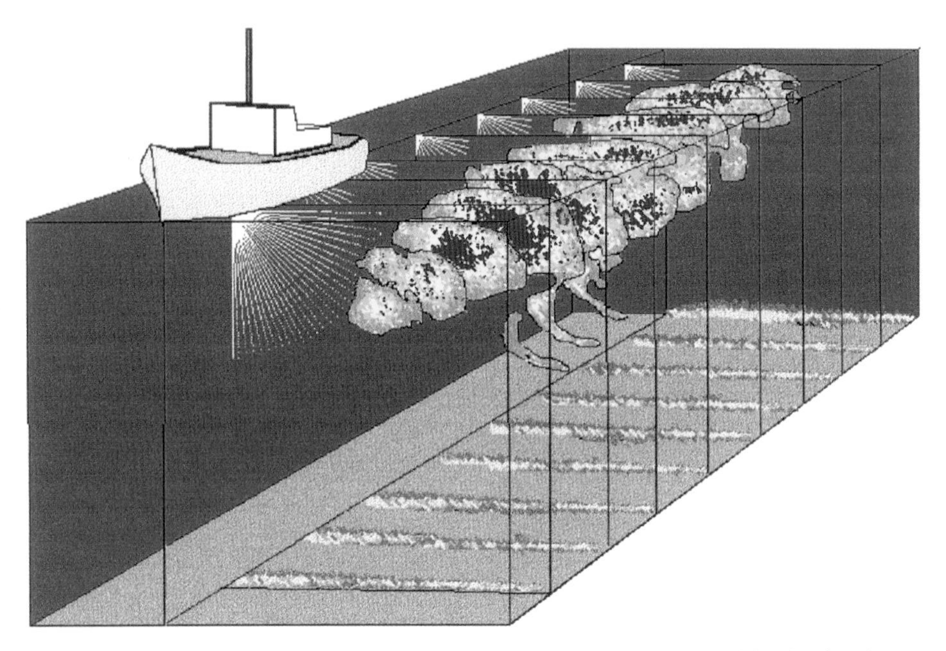

Plate 2 Three-dimensional reconstruction of a pelagic school of sardine (*Sardina pilchardus*) from images provided by a multi-beam sonar operated in side-scan mode. Note the heterogeneity in packing density and the irregular shape of the school (same cross sections of the school by a conventional single beam sonar would give a misleading image of two separated schools).

Chapter 3
Habitat Selection and Migration

3.1 Introduction

An animal's habitat is a complex of physical and biotic factors which determines or describes the place where the animal lives (Partridge, 1978). According to Fitzgerald and Wootton (1986), the optimal habitat is a place where an animal can maximise its lifetime production of offspring. In this book, 'habitat selection' will be used in a broader sense and refers both to the choice of a global environment favourable for the stock or at least the species and to a selection of a given micro-habitat at small time scale. Territoriality, which is irrelevant for most pelagic species in the wild, is excluded from our definition. Habitat selection is therefore an important behavioural function for adult fish especially in feeding, reaction to predator, agonistics, and sexual and parental behaviour (Huntingford, 1986). It has been also extensively studied in other vertebrates, particularly in birds (Rosenzweig, 1981, 1985).

For a number of economically important fishes, the selection of habitat varies with stages of the life history, because feeding and spawning often take place in different areas. Additionally, in order to maximise their fitness (physiological or physical factors), many fishes change habitat according to a circadian or seasonal rhythmicity, and sometimes from year to year. This results in fish migrations within and among areas (Harden Jones, 1968; McCleave *et al.*, 1982; Dingle, 1996). Baker (1978) broadly defines migration as 'the act of moving from a spatial unit to another', and mentions that 'spatial unit' has no restrictive overtones, unlike the term 'habitat'. Nevertheless, Dingle (1996) notes that most migrations are related to a change in habitat. In this book, migration, in the broad sense of the term (vertical and horizontal, short range and long range), refers to cyclical movement of a substantial part of the stock and is considered in this chapter with habitat selection.

Habitat selection and migration are of primary interest in stock assessment and management because they are key factors in the identification of stock units in relation to exploitation. It is necessary to know the distribution and time variability at different scales (circadian, seasonal, interannual) of the part of the species' population for which assessment data are collected, in relation to the fisheries that exploit it. This information will help to know how many fisheries are exploiting a given stock and conversely how many stocks a given fishery is exploiting. Moreover, habitat selection and migration govern the availability of the fish in both the horizontal and vertical dimension, and therefore the catch rates. The reasons for fluctuations in catch rates are of economic interest for fishermen, while they present a scientific interest for fishery biologists who use them as abundance indices in stock assessment models. Without this information, and an understanding of how it might change with environmental factors, the adequacy of assessments and related management advice is open to considerable doubt.

A naive observer of the marine environment could be surprised by the following discussion on habitat selection because, in contrast with the demersal realm, the pelagic environment looks uniform. In fact, numerous abiotic (temperature, oxygen, salinity, transparency, light intensity, current speed) or biotic factors (presence of conspecifics, prey or predators) characterise the pelagic environment. Fish are able to detect these characteristics and consequently react to their variation by horizontal or vertical displacement.

Habitat selection is presented here in three parts. First, the variation of habitat selection according to different time and space scales is reviewed, then the biotic factors which can be potential habitat cues are listed, and finally abiotic factors are considered.

3.2 Habitat selection according to different time and space scales

Changes in habitat selection may occur at different times (instantaneous, circadian, tidal, lunar, seasonal or interannual) and space scales. After Tinbergen (1963), Noakes (1992) applies the general terms 'ultimate' and 'proximate' to refer to different factors influencing behaviour. Ultimate refers to the final, long-term, evolutionary consequences of behaviour (function). Proximate refers to the immediate, short-term, physiological mechanisms of behaviour (causation). Instantaneous habitat selection is related to proximate cues and is expected to occur on a small spatial scale (micro habitat) and according to drastic changes in the environment. This is clearly the case of internal waves, which abruptly change the water mass characteristics, or predator arrival. This instantaneous selection and its relationship with the environment is usually easy to observe and to quantify. In contrast, ultimate relationships may be much more difficult to detect. For instance, two groups of individuals of the same species in the same population might differ in habitat selection relative to temperature.

If the survival of eggs and larvae is dependent on temperature for this species, there will be a difference in the reproductive success of the two groups and one of them will be favoured by natural selection. In this example, temperature is an ultimate factor that determines reproductive success and will usually act on large scales of time and space for pelagic species. In turn, once the selective pressure has been effective in selecting a particular behaviour, this behaviour may last for centuries, even when major change occurs in the environment. Carscadden *et al.* (1989) found that the only stock of capelin (*Mallotus villosus*) that was not an intertidal spawner was selecting areas of major deposition of gravel corresponding to ancestral beaches of the Wisconsin glaciation period. An opposite situation, but also interpreted as a behaviour-genetic case, is the discussion by Wyatt *et al.* (1986) of the unusual location of spawning grounds for sardine (*Sardina pilchardus*). The usual spawning grounds are coastal for this species, but a deep spawning ground is observed in an area that was coastal during the post-Pleistocene transgression. These two examples show that habitat selection during spawning can be the result of remote ultimate factors.

Circadian habitat selection is related mostly to light intensity and therefore to prey or predator behaviour, and is usually linked to a physiological and behavioural rhythm often in relation to the pineal organ (see Ali, 1992, for a review). A typical example of circadian habitat selection is the diel vertical migration of many fish species, which usually are found closer to the surface during the night than during the day during most of their life span (see Neilson & Perry, 1990, for review). Menhaden (*Brevoortia patronus*) move inshore and towards the surface at night, but are found offshore and close to the bottom during the day (Kemmerer, 1980). Most endogenous rhythms are synchronised by a natural cyclical phenomenon (such as light, temperature, tides), often termed 'zeitgeber' (Neilson & Perry, 1990). Inverted circadian vertical migration is sometimes observed, as in the case of the whitefish (*Coregonus lavaretus*), during the spawning season which is interpreted as the result of a selective pressure aimed to increase encounter probability among mature specimens and to avoid cannibalism on the eggs (Eckmann, 1991).

Lunar and semi-lunar rhythms in behaviour (especially migration) have been observed in different species like eels and salmon (Leatherland *et al.*, 1992), freshwater fish (Daget, 1954; Ghazaï *et al.*, 1991; Luecke & Wurtsbaugh, 1993). As far as we know, there is little documentation on this point related to marine pelagic species, except some variation of catch rate according to the lunar phase (e.g. Park, 1978; Anthony & Fogarty, 1985; Thomas and Schülein, 1988; Fréon *et al.*, 1993a) that will be analysed in Chapter 8. Nevertheless, it is likely that during the night the depth of the fish concentration in the water column and the level of aggregation depend on light intensity and therefore on the lunar phase.

Seasonal habitat selection is related to trophic and reproductive migrations and occurs at medium or large scale according to the species. Most pelagic species (both coastal and offshore) perform such seasonal migrations related to reproduction, feeding or wintering (Harden Jones, 1968; McCleave *et al.*, 1982; Cayré, 1990). The general triggering mechanism (zeitgeber) is usually thought to be the seasonal variation in daylength. Nevertheless, for most pelagic species, we do not clearly understand the key factors in the precise timing of migration and the mechanism of synchronisation between different schools belonging to the same group (cluster) of schools.

There are many well-documented examples of seasonal migration in coastal species derived from fish tagging or fisheries data analysis. In the Gulf of California, Hammann *et al.* (1991) indicated that the catch per unit of effort (CPUE) of sardine (*Sardinops sagax caeruleus*) was higher in autumn and that fish was pushed northward by the intrusion of tropical waters in the gulf during the spring. In northern Chile, sardines (*Sardinops sagax*) and anchoveta (*Engraulis ringens*) perform seasonal migration to the north from the cold season to the warm season. Nevertheless, this pattern of migration is altered by strong anomalies of sea surface temperature related to El Niño events as shown by Yáñez *et al.* (1995). The life cycle of the Atlantic menhaden (*Brevoortia tyrannus*) is well documented due to large scale tagging experiments. While the distribution of the young stage depends on salinity, large fish migrate northward along the coast in spring and stratify by age and size during summer, the large and oldest fish proceeding farthest north. A southward migration is observed in late autumn (Nicholson, 1978).

Hiramoto (1991) proposed a subdivision of the Pacific population of the Japanese sardine (*Sardinops melanostictus*) into two basic groups according to the amplitude of seasonal migration (and to the age of first reproduction). The coastal group lives in bays and coastal waters and displays limited coastal migration range. In contrast, the oceanic group displays extended seasonal migration to spawning grounds by late fall or early winter. It is interesting to note that the sardines that grow more slowly remain in the bays for a longer period of time.. In west Africa, adult gilt sardine (*Sardinella aurita*) migrate according to the upwelling season. The bulk of the northwestern stock is found in Mauritania during summertime because the trade winds are strong enough to generate an upwelling all year long only in this area, whereas during wintertime the stock is spread between Guinea and Mauritania due to the geographical extension of the upwelling (Boëly *et al.*, 1982).

Finally, in South Africa the anchovy stock (*Engraulis capensis*) is distributed over the whole western coast, from 28° S to 36° S, but there is a single main spawning area located on the Agulhas Bank, at the southernmost location. This spawning ground is located upstream of the strong Benguela current, a shelf-edge jet which transports eggs and larvae over the whole area of distribution, and spawning occurs in early summer when the current speed is still high (Shelton & Hutchings, 1982; Nelson & Hutchings, 1983; Armstrong & Thomas, 1989; Boyd *et al.*, 1992). In this last case, the habitat selection is obviously related to an ultimate factor because the spawning ground is not productive during the season of reproduction.

Such spatial and temporal changes in habitat selection are cyclical and therefore predictable. We will see in Chapters 8 and 9 how to take them into account in stock assessment methods. In contrast the interannual changes in habitat selection, which can also be termed emigration in opposition to migration (Heape, 1931), are not common in pelagic fish. Unlike migration, emigration does not involve return to the original habitat and is not cyclical. During drastic changes in population size, dramatic shifts in the distribution area of the stock may occur. Such changes may affect the annual migrations of pelagic fish stocks. A first example is the changes in migration behaviour of the Atlanto-Scandian herring in the late 1960s.

Due to a severe overfishing, both on the recruits and the Atlanto-Scandian adult stock of herring, the population declined drastically during the 1960s (Dragesund *et al.*, 1980). A stock which amounted to more than 10 million tonnes in the 1950s, was therefore reduced to only about 100 000 tonnes around 1970. Along with the decline in the stock size, the adult population changed the migration range and migration route during the feeding migrations in the Norwegian Sea.

Between World War I and II Iceland developed a substantial summer fishery on large herring that was feeding in the cold waters north of Iceland in summertime (Fig. 3.1). By recapture of tagged herring, Fridriksson and Aasen (1950) proved that this herring belonged to the same large stock as the herring that spawned on the west coast of Norway in winter. By use of a naval sonar onboard RV *G.O. Sars*, Devold (1953)

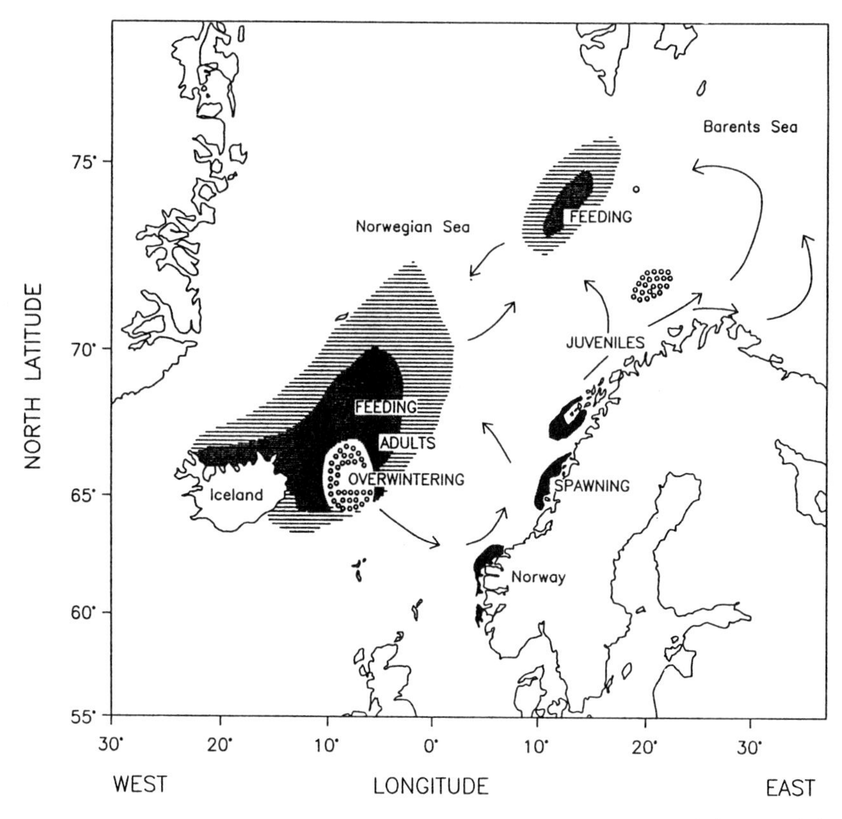

Fig. 3.1 Migration pattern of the Atlanto-Scandian herring (*Clupea harengus*) during a period of high stock level (redrawn from Bakken, 1983).

noted that herring concentrated in large aggregations in the deep waters east of Iceland in late autumn, and migrated to the spawning grounds at the coast of western Norway in December and January. This migration pattern with spawning at the coast of western Norway, feeding in summertime in the waters north of Iceland, concentration east of Iceland in late autumn, and migration back to western Norway for spawning in early winter, was maintained when the stock was large throughout the 1950s (Jakobsson, 1962, 1963; Devold, 1963, 1969; Østvedt, 1965; Jakobsson & Østvedt, 1996). As fishing mortality increased substantially throughout the 1960s the stock declined rapidly (Dragesund *et al.*, 1980), and in 1965 a change occurred in migration pattern (Jakobsson & Østvedt, 1996). Following the usual pattern of the last decades, the stock migrated out into the Norwegian Sea for feeding in the spring/early summer, but then the herring did not migrate across the East Icelandic current to the usual feeding grounds north of Iceland. That year the East Icelandic current, which brings cold polar water to the areas east of Iceland, was stronger than usual. The temperature in the areas east of Iceland in May 1965 was about 1°C lower than average for the last decade. This was the beginning of a cold period in the waters east and north of Iceland, and the herring did not enter these waters for feeding in the summertime any more. Due to the heavy fishing mortality in the 1960s, the stock in the late 1960s was only about 1% of the stock size in the 1950s.

In the middle of the 1960s, a subpopulation of Atlanto-Scandian herring was identified in the northern Norwegian Sea (Devold, 1968). This subpopulation spawned off northern Norway in winter and migrated to the Bjørnøya area for feeding in summertime. In autumn 1966, the whole subpopulation migrated to the east coast of Iceland and joined the rest of the stock (Jakobsson, 1968).

From about 1970 the herring was not observed to undertake the seasonal feeding migrations in the Norwegian Sea in summertime, but confined its distribution area to the coast of western Norway all year. As before, the herring spawned off western Norway in winter, but limited the feeding migration to the waters off north-western Norway, and entered fjords in north-western or northern Norway in autumn for hibernating (Røttingen, 1990). One component of the population overwintered in the Nordmøre fjords, another in the Lofoten fjords. The population was now named Norwegian spring spawning herring because its distribution area was totally confined to Norwegian waters. When the population recovered during the late 1980s and early 1990s, the herring made more extended feeding migrations in the Norwegian Sea in summertime. In autumn, the herring migrated to the coast of northern Norway for overwintering, and from 1988 the whole population has concentrated in the Ofoten/Tysfjord area from October to January (Røttingen, 1992).

In 1996 the stock of Norwegian spring spawning herring again reached a level of about 10 million tonnes (Anon., 1996), and in May 1996 the component of the stock feeding in the Norwegian Sea was estimated by conventional echo integration (Chapter 9) to be about 8 million tonnes or about 50 billion individuals (Misund *et al.*, 1997). The herring was then distributed over large areas in the Norwegian Sea. However, the westward migration still seemed limited by the East Icelandic current, as before the stock collapse in the 1960s, the westward migration seemed limited by the 2°C isotherm (Jakobsson & Østvedt, 1996; Misund *et al.*, 1997). The stock is expected to grow in the years to come (Anon., 1996), and a shift in the migration pattern and overwintering area may occur in the near future, especially if the temperature in the waters influenced by the East Icelandic current increases in early summer.

Other examples of interannual variation of habitat selection are:

- The shift in summer distribution of the adult herring of the North Sea in the 1980s, related to the previous example (Corten & van de Kamp, 1992)
- The dramatic change in both timing and route of the western mackerel (*Scombrus scombrus*) during its return southerly migration to Scotland and Ireland (Walsh & Martin, 1986; Anon., 1988)
- The excursion of capelin (*Mallotus villosus*) outside their normal range despite a low level of exploitation (Frank *et al.*, 1996)
- The large-scale changes in the distribution of the northern cod stock(s) (DeYoung & Rose, 1993; Hutchings & Myers, 1994; and other references in section 8.3.1 related to the shrinkage of the stock distribution)
- The decline of the sardine (*Sardina pilchardus*) production in the central fishery of Morocco related to a possible change in the migration pattern (Anon., 1996)

- The recent confinement of most of the northern Benguela pilchard stock to the Angolan area (Boyer *et al.*, 1995)
- The modifications of the horizontal and vertical distribution of the north-east Artic cod (Nakken & Michalsen, 1996)
- The changes in the distribution pattern of the blue whiting (*Micromesistius poutassou*) spawning stock in the north-east Atlantic (Monstad, 1990), etc.

In all these examples the influence of abiotic factors and/or biomass changes (related or not to exploitation) is mentioned. Seasonal migration seems regulated by the environment (e.g. Castonguay *et al.*, 1992; McCleave *et al.*, 1984; Walsh *et al.*, 1995) and possibly interannual variations in migration patterns are due to internannual changes in the abiotic factors. Nevertheless, another interpretation could be that migratory and sedentary populations are genetically distinct and that intensive fishing on the migratory fraction would select sedentary fish (Anon., 1996), a hypothesis that merits further investigations (see also section 7.3 on learning).

Let us see now in detail how biotic and abiotic factors can influence habitat selection at different time and space scales.

3.3 Influence of abiotic (physical) factors

Despite an abundant literature describing the water mass characteristics where different pelagic species live, or their tolerance under artificial conditions, it is not easy to define their precise preferendum (in the sense of active search for a specific water mass). Habitat selection is often the outcome of balanced conflicts between different determinants such as hereditary factors, learning, predation and optimal feeding. Also, the range of preferred experimental conditions changes according to the physiological state (hunger, reproductive stage, condition, etc.), age, season or previous acclimation, and is sometimes found outside the ecological range (Wootton, 1990). Moreover, in some instances, such as the temperature preferendum, the fish might react more to the gradient than to the absolute value of the parameter. This gradient can be spatial (e.g. Sharp & Dizon, 1978; Magnuson *et al.*, 1980, including discussion pp. 381–2; Cayré, 1990; Cayré & Marsac, 1993), or both temporal and spatial (Mendelssohn & Roy, 1986; Stretta, 1991).

As far as tuna species are concerned, Petit (1991) proposed a general hypothesis of opportunistic attraction of tunas by any kind of anomaly detectable by the perception organs of these animals: thermal structures but also structures related to oxygen or salinity, floating objects and bathymetric structures (shelf break, canyons, seamounts, islands). This hypothesis is difficult to validate at the moment, due to our poor knowledge of tuna physiology.

Finally, the different factors (biotic and abiotic) that characterise the habitat are often linked in such a way that it might be difficult to identify the key factor. For instance, in the coastal upwelling areas, a common habitat of pelagic species, it is well known that wind stress is responsible for raising cold and nutrient-rich water to the surface, resulting in high phytoplankton production. This makes it difficult to distinguish between the effects of temperature and plankton abundance. Moreover, it is obvious that in most cases there is not a single key factor, but a combination of factors which interact or impose a trade-off between conflicting demands.

3.3.1 *Temperature*

While most fishes are ectotherms, they are nevertheless able to thermoregulate by selecting appropriate water temperatures and avoiding those which are harmful. Reports of *in situ* mortality due only to low temperature are exceptional in pelagic fishes – in contrast to the less mobile demersal fish – and examples of such mortality of pelagic fish occur inside bays where circulation is limited. For instance, Esconomidis and Vogiatzis (1992) reported mass mortality of *Sardinella aurita* in Thessaloniki Bay (Greece) during an abrupt temperature fall in February 1991. Mortality due to high temperature is seldom observed and is always difficult to distinguish from the effect of the related decrease in dissolved oxygen (see section 3.3.3).

Populations of the same species inhabiting different thermal environments often exhibit differences in thermal physiology and behaviour, which presumably have a genetic basis (Reynolds & Casterlin, 1980). This seems also to be the case for most tuna species, despite their physiological thermoregulation (Sharp & Dizon, 1978). Nevertheless, the distribution of the southern bluefin, *Thunnus maccoyii*, is intriguing (Caton, 1991). This species exhibits a large tolerance to temperature variation, allowing large-scale migration from sub-tropical areas (south of Java) where reproduction occurs to circumpolar (11–13°C) areas south of 30°S. In contrast, the northern bluefin populations (*Thunnus thynnus*) of the Pacific Ocean and Atlantic Ocean display a narrower range of habitat temperature (Table 3.1).

A precise knowledge of habitat is of primary interest for tuna fisheries because tunas have an extremely wide distribution in the open oceans and perform rapid horizontal and vertical migrations; they are therefore difficult to locate. Studies in this field are now advanced, and temperature was identified early on as a key factor (Laevastu & Rosa, 1963). More recent studies show that even with a wide water

Table 3.1 Compilation of information relative to the effect of sea water temperature on the main commercial species of tuna: minimum lethal temperature, average temperature in the distribution area and temperature in areas of highest catches according to the gear.

Species	Min. lethal temperature (°C)	Average distribution (°C)	Highest catches/gear (°C) Surface	Deep longline
Yellowfin	14	18–31	24–30	19–22
Skipjack	15	17–30	20–29	nil
Bigeye	7	11–28	23–28	17–22
Albacore	9	11–25	16–19	17–21
Bluefin	5	10–28	17–20[a] 23–26[b]	13–15

[a] Temperatre areas
[b] Tropical and Mediterranean areas

temperature range where fish are distributed, there is a narrower range of optimal catches. From a review of the available literature (Grandperrin, 1975; Sund *et al.*, 1981; Collette and Nauen, 1983; Stretta and Petit, 1989; Holland *et al.*, 1990) we obtained the values presented in Table 3.1.

Table 3.1 distinguishes between surface fisheries (pole-and-line, purse seiners, trollers, surface longlines) and the deeper long-line fishery. Fish caught by surface gears are generally smaller than those caught by long-line, suggesting a length–dependent tolerance to cold temperature. This length–dependence relationship seems general in fish and can be interpreted as an influence of the ratio body-surface: body-length on the rate of caloric exchange with the water, or by the ratio gill surface: body-weight (see next section).

Most pelagic species are able to detect temperature variations as small as 0.1°C (Sund *et al.*, 1981) or less (Murray, 1971; Hoar & Randall, 1979), which allows them to orientate toward areas favourable to their metabolism, or more often to detect remote frontal areas where prey are usually more abundant. Comparisons between satellite images and the location of tuna catches confirm this fact (Fiedler & Bernard, 1987; Stretta, 1988). As far as small pelagic species are concerned, the combination of acoustic surveys and oceanographic observations indicated that often a higher concentration and patchiness in the distribution is found where frontal gradients are strongest, as in the case of the horse mackerel *Trachurus trachurus capensis* on the edge of the Agulhas Bank (Barange, 1994). Experimental observations on thermal preferendum or resistance of fish indicate the influence of thermal history and suggest that the response to temperature changes according to the season (e.g. Olla *et al.*, 1985).

The thermocline (which can be simply defined as a horizontal plane where the vertical gradient of temperature is at a maximum) often limits the vertical migration of surface pelagic species. For instance, young yellowfin tuna (*Thunnus albacares*) often stay above the thermocline, which explains a change of their availability to purse seining. The catch rate is on average greater when the thermocline is shallow (Fig. 3.2) and presents a sharp gradient (Green, 1967; Sharp, 1978a; González-Ramos, 1989). A physiological interpretation of the fence role of such a sharp gradient is provided by Brill (1997) who observed in tank a drastic decrease of the cardiac rhythm when the temperature decrease reached 10°C. This low cardiac rhythm certainly decreases the oxygen input.

Nevertheless, recent acoustic tagging experiments have shown that the thermocline is not an absolute fence. Fish can make short and repeated incursions below it (Carey & Robinson, 1981; Lévénez, 1982; Holland *et al.*, 1990), or remain within the area of maximum gradient of temperature (Cayré & Marsac, 1993). This behaviour has been related to the mechanism of thermoregulation of tunas (Sharp & Dizon, 1978; Cayré, 1985; Carey, 1992), but as far as young yellowfin (*Thunnus albacares*) and skipjack tuna (*Katsuwonus pelamis*) are concerned, the short duration of the stay was attributed to the low level of dissolved oxygen usually observed below the thermocline. Nevertheless, recent observations of such incursions have also been performed in areas where the gradients of temperature and oxygen around the thermocline were very low (Cayré & Chabanne, 1986). In the case of coastal pelagic species the vertical

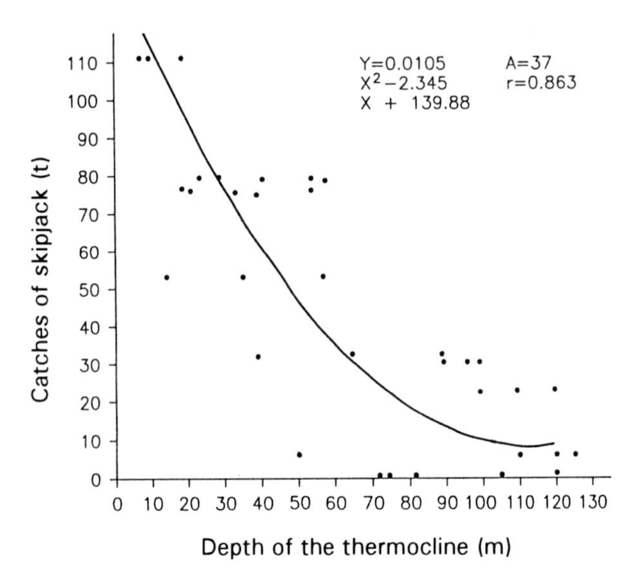

Fig. 3.2 Relationships between catches of skipjack tuna (*Katsuwonus pelamis*) and the depth of the thermocline in Gran Canaria (Canary Islands) in 1982 (redrawn from González-Ramos, 1989).

distribution of schools is difficult to interpret because it is difficult to distinguish between the effects of depth, temperature and distance to the bottom. Nevertheless some convincing examples are available, as for instance in the east coast of South Africa where the thermocline follows an oblique angle to the surface (Fig. 3.3, after Armstrong *et al.*, 1991).

Nonparametric regression methods, as the generalised additive models, can be used to distinguish the roles of different environmental factors, including the thermocline, the temperature at a given depth, the bottom depth, etc. (Swartzman *et al.*, 1994, 1995). In the Bering Sea, walleye pollock (*Theragra chalcogramma*) vertical distribution is related to the thermocline (Swartzman *et al.*, 1994). In deep water, adults remain below the thermocline when mid-water temperatures are below 0°C, while age-0 pollock predominate above it. The authors hypothesise that the thermocline serves as an effective barrier in the summer, separating age-0 pollock from potentially cannibalistic adults, due to different location of the prey of each group (euphausiids below the thermocline for adults and copepods in the upper water column for age-0). At the opposite end, O-group haddocks (*Melanogrammus aeglefinus*) are pelagic and distributed predominantly at the thermocline on Georges Bank (Neilson & Perry, 1990). The variability of results according to the species indicates that various determining factors (temperature, prey, distance to the bottom, etc.) can be involved in the process of habitat selection related to the thermocline, and therefore can affect the catchability.

Habitat selection during spawning is important for stock assessment because the bulk of the catch is often made before or during the reproductive season, when fish are concentrated and/or more vulnerable to the gears. For instance, Carscadden *et al.* (1989) found mature capelin (*Mallotus villosus*) in the north-west Atlantic mainly on

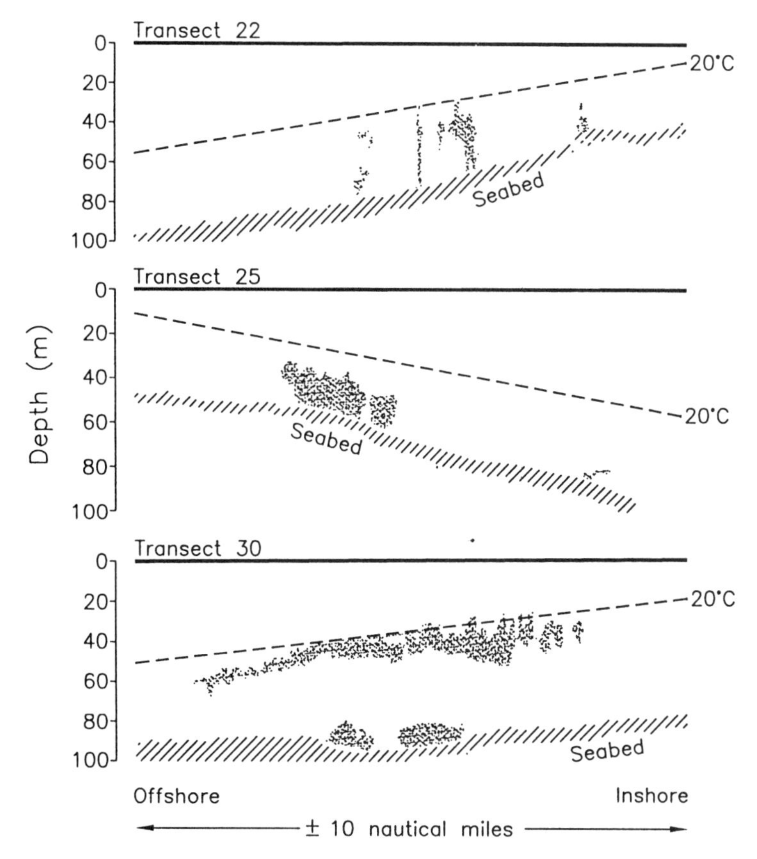

Fig. 3.3 Echograms from selected transects during the June 1990 acoustic survey on the East Coast of South Africa, showing pelagic fish targets in relation to the vertical temperature structure (redrawn from Armstrong *et al.*, 1991).

grounds where the bottom temperature was within the 2–4°C range. From a review of herring behaviour and spawning grounds in the Atlantic (*Clupea harengus harengus*) and in the Pacific (*Clupea harengus pallasi*), Haegele and Schweigert (1985) indicated that herring typically congregate near their spawning grounds for several weeks to months prior to spawning. While adults usually have a wide tolerance to temperature, the range of critical values is generally narrower for eggs and larvae, and may interact with salinity (Alderdice & Hourston, 1985). This is one reason why most pelagic species spawn in particular spatio-temporal windows, with temporal changes for a given species according to the latitude, as for instance in the case of different herring populations in the Atlantic and adjacent seas or in the Pacific Ocean (Haegele & Schweigert, 1985; Hay, 1985).

Among the other numerous reasons explaining the link between temperature (or temperature gradient) and spawning is a precise timing with optimal environmental factors, such as prey abundance, stability of the vertical column or low advection (Parrish *et al.*, 1983; Bakun & Parrish, 1990). Recently some authors suggested that this link could be mainly due to an 'obstinate reproductive strategy' which consists of searching for the environmental conditions of their own birth for spawning (Cury, 1994; Baras, 1996). This hypothesis is based on the generalisation of the natal homing behaviour, which is well documented in birds, sea turtles and salmon. These animals are known to be imprinted by environmental conditions a few hours or days after birth, and are able to come back to their native area for spawning, although imprinting and learning of smolts salmon during their seaward migration also plays an important role (Quinn *et al.*, 1989; Hansen & Jonsson, 1994). The possibility that natal homing in fish could greatly exceed the case of salmonids was first proposed by Blaxter and Holliday (1963) and then by Sharp (1978a). Later Hourston (1982) noticed that Canadian herring return to the spawning ground where they spawned for the first time in their life, and therefore possibly where they were born. The implication of imprinting on stock unity will be discussed in section 8.7.2.

The originality of the generalisation of the obstinate reproductive strategy hypothesis to pelagic species involves considering that these fishes are able to trigger their spawning from year to year in different spatio-temporal strata, taking into account only physical characteristics of the watermass (e.g. temperature). This promising hypothesis needs to be confirmed by experimental and *in situ* observations, and does not necessarily apply to all species (for instance Motos *et al.*, 1996 found that 1-year-old anchovy (*Engraulis encrasicolus*) do not share the same spawning grounds with older anchovy, which is difficult to conciliate with this theory). Interestingly, some simulations suggest that the obstinate strategy could be more efficient than the opportunistic strategy for avoiding population extinction (Le Page & Cury, 1996).

3.3.2 *Salinity*

Most of the main commercial stocks of marine pelagic fish develop their whole life-history in sea water where the salinity is generally over 32 psu. Nevertheless, there are remarkable exceptions in coastal pelagic fish. The menhaden (*Brevoortia tyrannus*) stocks, found along the south-east coast of the USA and secondarily in the Gulf of Mexico, spawn at sea but the juveniles migrate actively to brackish waters in the 'sounds' areas after an oceanic phase of 1.5 to 2 months, and stay there for several months before coming back to the continental shelf (Nelson *et al.*, 1977). Along the West African shelf, *Ethmalosa fimbriata* migrates far up rivers for spawning, in almost freshwater bodies where the salinity is over 4 psu (Charles-Dominique, 1982). In both these cases the adults are perfectly adapted to the high sea water salinity and can perform extensive migration (especially *B. tyrannus*), but their habitat selection is conditioned by the proximity of brackish waters. To a lesser extent, another African clupeoid, *Sardinella maderensis*, is related to coastal waters of lower salinity than a similar species of sardine, *S. aurita*, despite a large overlap in the distribution of those species (Boëly, 1980a). At the opposite end, another species of gilt sardine (*Sardinella marquesensis*) is very tolerant of salinity and is found in the range of 8.6 to 36.0 psu (Nakamura & Wilson, 1970), which indicates large variability of tolerance within the same genus.

The influence of salinity on habitat selection during spawning might be complex because the survival of larvae sometimes results from an interaction between several environmental variables (salinity, temperature, dissolved oxygen) and may also depend on the environmental past history of the larvae during hatching. Alderdice and Hourston (1985) studied the Pacific herring (*Clupea harengus pallasi*) and indicated a complex interaction between salinity and temperature on hatching. Moreover, they mentioned that survival of eggs on substrate, related to respiratory activity, appeared to be influenced by transport and perfusion velocity of interstitial water in an egg mass. They concluded that these factors, especially temperature and salinity, have a commanding influence on the reproductive cycle and thereby the distribution of the species (ultimate factor).

Salinity influence also seems important in the habitat selection of anchovies (*Engraulis encrasicolus*) as suggested by echo-survey observation (Massé *et al.*, 1995) and eggs collection (Motos *et al.*, 1996). The distribution of this species in the Bay of Biscay is related to the Gironde river plume (but also shelf edge fronts and oceanic eddies). From acoustic observations combined with commercial data on the mackerel, Walsh *et al.* (1995) studied the migration of the western mackerel (*Scombrus scombrus*) and supposed that its distribution was related to the salinity and the temperature, despite the low range of variability in the area. In such cases it is always difficult to distinguish the role of salinity by itself from the role of associated factors characteristic of the water mass (plankton composition, turbidity, temperature, etc.) due to the colinearity in their variability as shown by Reid *et al.* (1997).

Salinity does not seem directly to influence the distribution and behaviour of the most common commercial species of tunas within the range of their usual habitat (32 to 36 psu) (Blackburn, 1965; Dizon, 1977; Sund *et al.*, 1981; Stretta, 1988; Stretta & Petit, 1989). The geographical distribution of catches in the skipjack tuna surface fishery in the western Pacific is limited by the boundary of the 35 psu isohaline; this is not interpreted as the result of the salinity itself, but as the indicator of a convergence area (Donguy *et al.*, 1978).

3.3.3 *Dissolved oxygen*

Oxygen in the aquatic environment is provided by phytoplanktonic photosynthesis and, in the upper layer, by exchanges with the atmosphere. A number of factors affect oxygen solubility (temperature, salinity, exchange with atmosphere, turbulence, etc.). Of major importance is the decline of solubility of oxygen in water with increasing temperature (and salinity but not significantly as far as the marine environment is concerned). In sea water (e.g. 35 psu) at 5°C the saturation is around 10 $mg\,l^{-1}$ and falls to 6 $mg\,l^{-1}$ at 30°C. All these factors are responsible for a large variation of oxygen in the ocean, both in the horizontal and vertical distribution and also in time, which contrasts with the relative stability of oxygen concentration in the atmosphere. The oxygen demand of fish varies mainly according to activity, temperature, body size and the availability of oxygen. Among pelagic species, the Scombridae have a special status in that they are committed to continuous pelagic swimming because they cannot breathe by means other than by ram ventilation (except for Atlantic mackerel, *Scomber scombrus*), and consequently this continuous swimming increases their demand (Holeton, 1980; Holeton *et al.*, 1982).

Fish are able to rely on glycolytic fermentation or so called 'anaerobic metabolism' to cover high short-term energy requirements such as burst swimming in response to a predator or to fishing gear. According to Brett (1979), for periods of about 20 s they may expend energy at a rate about 100 times the basal rate. The maximum burst speed and the duration of the flight vary with temperature, and temperature is expected to influence the yield of trawl (Wardle, 1980; He, 1991; Smith & Page, 1996). Nevertheless, recent controlled experiments on acclimated Atlantic cod (*Gadus morhua*) and American plaice (*Hippoglossoides platessoides*) did not support this view and suggest thermal compensatory responses (Winger *et al.*, 1997).

Dizon (1977) found that the lethal value of dissolved oxygen for skipjack tuna is around 4 ppm (3 ml l^{-1}), in contrast to yellowfin tuna and especially bigeye tuna (*Thunnus obesus*) which is more adapted to deeper habitat. Sharp (1978a) estimated the lower oxygen tolerances for the main commercial tuna species according to their size, from energetic budget equations. The results ranged from 0.5 ml l^{-1} for a 50 cm bigeye tuna to 2.89 ml l^{-1} for a 75 cm fork length skipjack tuna and were related to the horizontal and vertical distribution of the species. Bushnell and Brill (1992) indicated that yellowfin tuna are sensitive (decreases of heart rate) to reduced oxygen values between 4.3 and 3.6 ml l^{-1}, despite the fact that lethal concentrations are usually estimated at 2.1 ml l^{-1}. However, mortality of pelagic fish due to hypoxic conditions is seldom observed because pelagic species are quickly able to change their habitat and/or behaviour in response to low oxygen levels. For instance, in the Gulf of Guinea, Cayré *et al.* (1988) indicated that dissolved oxygen in the upper layer (0–50 m) never reached the threshold of tolerance mentioned above for yellowfin and bigeye tuna. Nevertheless, Cayré and Marsac (1993) observed that tagged juvenile yellowfin tuna did not exceed the maximum vertical gradient of oxygen concentration (oxycline) and proposed a tentative modelling of the habitat selection of this species according to the vertical profiles of temperature and oxygen. Similarly, the occurrence of yellowfin and skipjack tuna off the west coast of South Africa during December–March is related to the seasonal occurrence of oxygen-depleted water observed at the boundary of the Agulhas current/return current system (Chapman, 1988).

The worldwide distribution of bigeye tuna catches (Fig. 2.1) indicates that this species is mainly caught in areas of depleted oxygen in the deep layers, below 150 m (Hanamoto, 1975; Cayré, 1987; Fonteneau, 1997), as if this species was taking advantage of its physiological capabilities for limiting competition with other tuna species. Nonetheless, bigeye tuna avoid the extremely low oxygen concentration (< 0.1 ml l^{-1}) observed in the eastern Pacific (Fonteneau, 1997) and in the northern Indian Ocean (Bay of Bengal and Arabian Sea, Isshiki *et al.*, 1997). As pointed out by Kramer (1987), 'although traditionally left to physiologists, breathing is also fundamentally a behavioural process; an ethological perspective immediately reveals many interesting problems from temporal organisation to motivation and sociality'.

There is a large variability in the avoidance reaction to low oxygen concentrations (Holeton, 1980), but a common behavioural response is an increase in general activity and movement towards the surface where higher oxygen concentrations are to be expected. Nevertheless, the immediate increase in activity, which seems contradictory to the expected reduction of energy expense, is difficult to interpret because it can also indicate attempted avoidance. Kramer (1987) distinguishes four categories of behavioural responses by fish to compensate for low oxygen availability:

(1) Changes in activity
(2) Increased use of air breathing (for species able to perform it)
(3) Increased use of aquatic surface respiration on surface film
(4) Vertical or horizontal habitat changes.

Some years demersal fish can colonise the pelagic ecosystem. In the summer of 1978, Peruvian hake (*Merluccius gayi peruanus*) became semi-pelagic and available to purse seiners, which caught 172 000 tonnes within two months. This change from a demersal habitat was related to changes in the oxygen and temperature distribution (Woznitza-Mendo & Espino, 1986).

For schooling fish the level of dissolved oxygen is more crucial especially for large schools with a high packing density (Chapter 4). There is a depletion of oxygen inside densely packed schools, especially in the rear part (Macfarland & Moss, 1967). When these schools live in a confined area, as in the case of hibernating herring schools in Norwegian fjords, the level of oxygen may decrease throughout the water body in the layer where the schools are living (Dommasnes *et al.*, 1994). Feeding activity of the Baltic herring (*Clupea harengus*) is highest close to the bottom in the shallow area, but occurs above the oxygen depleted bottom layer in the central basin of the Baltic (Koester, 1989). Off Peru, anchoveta (*Engraulis ringens*) schools change their maximal depth from 40 m in normal years to 20 m during El Niño events where there is an oxygen depletion (Mathisen, 1989). In areas of extremely intense upwelling (Yemen, Peru), during some peak upwelling seasons massive mortality of pelagic fish has been observed, attributed to the lack of oxygen in upwelled water (A.Y. Bakhdar, pers. comm.).

The body length-dependence of fish vertical and horizontal distribution is interpreted by Pauly (1981) as a combination of change in oxygen demand and gill surface:body weight ratio. The author submits that, in fishes, it is primarily oxygen, rather than food supply, which limits anabolism and growth performance because the gill area of fishes does not grow as fast as body weight (see also Longhurst & Pauly, 1987). Since oxygen demand increases with increasing temperature due to protein denaturation, large fish are less tolerant of high temperature than small ones and select their habitat accordingly. We think that this hypothesis is interesting to explain vertical dependance of fish distribution, in addition to the influence of the ratio body surface:body length. It also provides a valuable insight into the length-dependent migration distance (in addition to swimming performance) and to the length-dependent date emigration from warm to cold waters reported in this chapter. Nevertheless Pauly's (1997a) interpretation of the horizontal migration of pelagic species between Senegal and Mauritania by this theory does not explain why there is a sedentary fraction of the pelagic populations in Mauritania. The fraction of the migrating population and the timing of its migration to Senegal seems more related to the relaxation of the upwelling in Mauritania than to its intensity in Senegal (Fréon, 1986).

3.3.4 *Water transparency*

Except for a few strict phytoplankton feeders which filter only passively, most marine pelagic fish forage actively on prey with the help of vision (Guthrie & Muntz, 1993; sections 5.2.1 and 5.2.2). Aksnes and Giske (1993) derived and tested with experimental data a model for visual feeding by aquatic predators. It emphasised the predator's visual range according to its visual capability, surface light, water transparency, and size and contrast of the prey. Visual range increases almost linearly with increasing prey size and decreases non-linearly with increasing turbidity. Therefore water transparency is of primary importance in fish distribution (Nakamura, 1969; González-Ramos, 1989). However, as the topic has received little attention, we can only speculate on the relationship between water transparency and density. Possibly also this relationship is difficult to investigate because it is expected to be non-monotonic for particulate feeders, with a medium optimal value in water transparent enough to allow vision, but relatively turbid due to planktonic productivity.

Nevertheless, several coastal species are very tolerant to turbidity, like the Marquesan sardine (*Sardinella marquesensis*) which can be found either in white sand beaches of clear water or in bays of very turbid water (Nakamura & Wilson, 1970). In West Africa, S. *maderensis* is also observed in a wide range of turbidity, unlike *S. aurita,* which prefers clear water (Boëly, 1980a, b).

Some species seem to select habitat of optimal turbidity at the first stage of their ontological development. Uotani *et al.* (1994) performed experimental studies on the response and behaviour of postlarval Japanese anchovy 'shirasu' (*Engraulis japonica*) to differences in turbidity, prey density, and salinity. In an experiment on various combinations of conditions, anchovy showed a strong positive reaction only to turbidity. More than 90% of postlarvae moved from the tank section without turbidity to the section of turbid water of 5–10 ppm, while more than 75% moved to the section of 20–30 ppm turbidity. At much higher turbidity in the range of 40–50 ppm, shirasu were at first attracted into the turbid water, but soon escaped from it.

3.3.5 *Light intensity*

Clupeoids usually perform diel vertical migrations that seem mainly related to a light intensity preferendum but may be initiated by changes in brightness (Blaxter & Hunter, 1982). These migrations are probably related to an optimal combination of schooling ability, which depends largely on vision (section 4.8), decrease of predation risk and feeding behaviour. According to Hunter and Nicholl (1985), the threshold light intensity for schooling in northern anchovy (*Engraulis mordax*) estimated on 50 adults in the laboratory is 6×10^{-11} W cm^{-2} (2.6×10^{-4} meter candel). This was estimated to occur at a depth of 30 m on a starlit night and at 38 m during a full moon, when the chlorophyll concentration is 0.2 mg Chl-a m^{-3}. At 2.0 mg Chl-a m^{-3}, the threshold occurs at a depth of 8 m on a starlit night and at 20 m under full moon light. Sufficient light appears to exist at night within the upper 10 m for schooling to occur in most of the habitat of the anchovy.

In addition some species can use near-ultraviolet and polarisation vision to improve detection of targets. Shashar *et al.* (1995) demonstrated that polarisation vision increased by up to 82% detection range for transparent targets. A similar improvement is likely to exist for transparent organisms such as zooplankton, whose tissues depolarise light. Browman *et al.* (1994) tested the hypothesis that ultraviolet photoreception contributes to prey search in small juvenile rainbow trout (*Oncorhynchus mykiss*) and pumpkinseed sunfish (*Lepomis gibbosus*) while they are foraging on zooplankton, and they concluded that for both species, prey pursuit distances and angles were larger under full-spectrum illumination than under ultraviolet-absent illumination.

From a literature survey, James (1988) indicated that planktivores generally display diel feeding cycles, foraging either during the day or at twilight and by night. There is an agreement in the literature to classify *Clupea harengus* as a twilight forager despite the fact that – like most of the Clupeoid with the exception of menhaden – this species can perform both filter-feeding at night when exploiting dense patches of food and particulate-feeding during the day (Batty *et al.*, 1986). Data on the feeding periodicity of other clupeids are more conflicting, and probably result from the influence of other environmental and biological factors, the opportunism of some species and from sampling strategies.

As a consequence of the above-mentioned effect of light intensity on the schooling and feeding behaviour, the mean depth and the level of aggregation are expected to change, not only according to sunlight, but also to the moonlight. Such changes have been observed during acoustic survey of pelagic fish (primarily Bonneville ciscoes, *Prosopium gemmifer*) in three lakes of the USA. During the new moon, the fish were at the depths of 10-20 m, while at full moon, they were much closer to the bottom (Luecke & Wurtsbaugh, 1993).

In the Côte d'Ivoire purse seine fishery of small pelagics, Marchal (1993) found a clear negative correlation between the catch per unit of effort (CPUE) of *Sardinella maderensis* and an index of light intensity *in situ* (combining water transparency and surface light intensity). His interpretation was that fish were closer to the surface when light intensity in the water column was lower, and therefore more available for purse seiners.

From ultrasonic tagging experiments, several studies (e.g. Cayré & Chabanne, 1986; Holland *et al.*, 1990; Cayré, 1991; Marsac *et al.*, 1996) have confirmed that yellowfin tuna, skipjack and bigeye tuna swim in shallower water during night-time. In addition, Marsac *et al.* (1996) suggested that yellowfin habitat selection is influenced by the lunar phase. The fish swim in shallower water during nights of full moon; this probably relates to their feeding behaviour, which depends on light intensity. One of the best documented fisheries affected by circadian migration, probably in relation to light intensity, is the long-line fishery of swordfish (*Xiphias gladius*). This species is caught by deep long-line (200–400 m) during the day and by sub-surface long-line (0–50 m) during the night (Nakano & Bayliff (1992).

Neilson and Perry (1990) reviewed a large number of works (including their own) on vertical migration in relation to light intensity, and particularly the hypothesis that fish follow an isolume. They concluded that the role of light is questionable. Despite the fact that light obviously plays an important role in mediating diel vertical migrations, other factors might interact (predator avoidance, prey availability, depth of the thermocline, etc.).

3.3.6 *Current, turbulence and upwelling*

While temperature often sets limits on the geographical extent of species distribution, in the recent literature it appears to be less important than upwelling processes and locations promoting retention of larvae in this advective system. Such a retention area – often located near the coast – minimises turbulent mixing of the water column (Parrish *et al.*, 1981, 1983; Bakun & Parrish, 1991). Recent work on circulation indicates that retention can occur in certain areas even when a wind-driven upwelling is fully developed, due to special topographical features which favour double cells of circulation (Graham & Largier, 1997; Roy, 1998). Bakun (1996) defined the 'triad' of processes suitable for coastal pelagic fish reproductive habitat: enrichment (upwelling, mixing etc.), concentration (water column stability, convergence, frontal formation) and retention of ichthyoplankton within (or drift towards) appropriate habitat. In this triad, some factors are related to long-term habitat selection and others to the short term. Among the latter, the current and turbulence play a major role, and optimal values of wind during the reproduction period are expected to provide a suitable compromise for larval survival (Roy *et al.*, 1992).

The current in an upwelling area is often relatively strong near the surface (around 0.5 knot along shore) and can be higher in particular current systems such as the Kuroshio current or the Agulhas current, which often exceed 2 knots (Watanabe *et al.*, 1991; Boyd *et al.*, 1992). Fish usually avoid this area, probably to limit loss of energy through swimming against the current so as to remain in the upwelling area. The mean offshore velocity of the mixed layer in the Peruvian upwelling system varies from 0.09 to 0.3 knots according to the season (Bakun, 1985), and current speed along the coast in the Californian or Saharan upwelling system is estimated to be between 0.2 and 0.5 knots (Jacques & Tréguer, 1986). Despite the weakness of these speeds, they represent several km per day (1 knot $\approx$ 44 km day^{-1}). Moreover, high wind speed (over 15 m s^{-1}) generates high turbulence in the upper layer, which is probably uncomfortable for swimming but might be suitable for feeding up to a certain optimal level of turbulence (Littaye-Mariette, 1990). Unfortunately, references on the influence of turbulence seem only available for larvae behaviour (Rothschild & Osborn 1988; MacKenzie & Leggett 1991). This problem is difficult to investigate acoustically since high turbulence generates bubbles which limit detection in the surface layer.

Despite these negative effects of strong currents or high turbulence on fish behaviour, weak currents are known to play an important role in fish orientation or transport during migration (Harden Jones, 1968; McCleave *et al.*, 1982; Dingle, 1996). Blaxter and Hunter (1982) reported occasional passive transport of North Sea herring. From a long-range sonar survey, Revie *et al.* (1990) indicated that movements of sprat (*Sprattus sprattus*) schools over a six-day period agreed well with that expected from alternating tidal currents, i.e. schools were passively transported by the water mass. From scanning observations performed along the fishing grounds off the coast of Japan, Inoue and Arimoto (1989) noticed that the route of many pelagic species (flying fish, spotted mackerel, salmon) corresponded with the water flow. In South Africa, when anchovy spawners, *Engraulis capensis*, return to spawning grounds upstream of the Benguela current, they make use of inshore currents instead of swimming against the main current (Nelson & Hutchings, 1987). On the other hand, Walsh *et al.* (1995) found that the direction of migration of western mackerel (*Scombrus scombrus*) was counter to the direction of flow along the line of the shelf edge. The mean residual speed of this current was 17.1 cm s^{-1} in the core of the current and the migration speed varied from 13.0 to 25.9 cm s^{-1}.

Fish avoid migration against a strong current, but can use a weak counter-current to orientate during migration. The highest current speeds are observed in limited coastal areas where tidal movement of the water mass is amplified by the local topography (shallow water, capes, estuaries etc.). Even though some demersal species are able to select these tidal currents for migration (Arnold *et al.*, 1994), most pelagic species are simply transported if they are not able to compensate for the water flow. Nevertheless, in an extreme situation of strong tidal streams in the mouth of the Gulf of St Lawrence (Cabot Strait), where the current speed is up to 75 cm s^{-1}, Castonguay and Gilbert (1995) observed by acoustics a large variation in densities of the Atlantic mackerel (*Scomber scombrus*) according to the tidal cycle. They suggested that mackerel use selective tidal stream transport to enter the Gulf of St Lawrence, although they were not able to document the vertical migration through which such transport would occur.

Once the role of currents in migration is admitted, the following questions arise: how do fish detect current without visual (optomotor) contact with the ground, and how do they choose the appropriate current? These questions are still under debate, except for some species such as salmon and eels which are known to detect chemical cues associated with water masses. According to Dingle (1996), fish using tidal stream transport display what appear to be programmed vertical movements. It is likely that fish also use the 'trial and error' process, which supposes that they have the ability to orient and navigate. Dingle (1996) reviewed the different mechanisms of orientation and navigation presently known or strongly suspected in animals (chemical, visual and physical cues, including among others sun or stellar orientation and magnetic orientation). He concluded that, despite recent progress, these aspects are poorly known and require further investigation.

A commonly advocated physical cue in fish migration is the temperature gradient. Satellite images of infrared channels have shown the complexity of spatial structures of temperature distribution, especially in upwelling areas (front, eddies, filaments, etc.). Large-scale tuna movements have been modelled using daily sea surface temperature from satellite observation and artificial life techniques such as neural networks and genetic algorithm (Dagorn *et al.*, in press). A rule-based model exploiting knowledge on relationships between tuna and thermal gradients failed to reproduce known large-scale movements. In contrast individual based models using learning techniques and fine scale description of the oceanic environment (oceanic landscape) make it possible to mimic some large-scale movements of real fish in 15% of the cases. Other mechanisms of orientation are probably used by tuna during migration (Hunter *et al.*, 1986) but they have not been fully investigated (e.g. magnetic detection in yellowfin tuna; Walker *et al.*, 1982; Walker, 1984). Moreover, in some oceanic species like skipjack tuna and – to a certain extent – yellowfin tuna, it is not easy to distinguish opportunistic large-scale movement from regular seasonal migration (Hilborn & Sibert, 1986).

Tunas are also found in coastal upwelling areas, usually near the edge of the continental shelf where a thermal front is often observed between cold upwelled waters and warm offshore waters (e.g. Dufour & Stretta, 1973). They are also observed in the coastal domain near river plumes where haline fronts may be observed (Fiedler & Laurs, 1990). But in many instances they are located in specific offshore areas related to the large oceanic circulation systems like the thermal front of divergence zones, convergence zones and thermal ridges (Dufour & Stretta, 1973; Stretta & Petit, 1989). Although all thermal domes are productive, the link with the concentration of top predators is not obvious, as underlined by Marsac (1994) in the Indian Ocean, where the density of tuna catches along the 10° south latitude dome is lower than that of the other areas of the fishery.

Stretta and Petit underlined also the presence of tuna in oligotrophic areas for unknown reasons except when their migration route crosses these areas. A first possible reason why tuna stay in oligotrophic areas was advanced by Voituriez and Herbland (1982). They demonstrated that in the tropical Atlantic, the higher phytoplankton production was observed during the warm season at the top of the nitracline which defined a two-layer production system (the upper one is nutrient limited and the deeper one light limited), without major influence of the current. During the cold season, however, an upwelling takes place at the equator, bringing cold and nutrient-rich water to the surface. A second explanation is related to the high tuna catch rate in the western Pacific oligotrophic area ('warm pool'). Paradoxically, this area is one of the more productive for tuna fisheries targeting skipjack and yellowfin tuna (Fonteneau, 1997). This could be due to the combination of delays in the food web and the horizontal transport by current of the zooplankton and plankton feeders (Lehodey *et al.*, 1997).

3.3.7 *Depth of the fish in the water column*

The vertical distribution of fish in the water column is obviously related to the light intensity, which decreases dramatically according to depth (in transparent waters the light intensity at 100 m depth is only 1% of the surface light) and can seriously limit the vision and consequently the schooling capability (see section 4.8 for discussion). Nevertheless, we have seen that temperature also plays an important role in the vertical distribution. Moreover, pressure is certainly an important factor because it increases rapidly with depth (the change in pressure is around one bar – nearly 1 kg cm^{-2} – per 10 m depth interval) and so the volume of the swimbladder is greatly affected by changes in depth. Some species have an open swimbladder (physostomous) due to an anterior canal from the swim bladder to the oesophagus and a posterior canal from the swimbladder to the anus (Blaxter & Batty, 1990). They are able to release gas when coming up to the surface and cannot suffer damage due to gas expansion. Some of these species, like salmonids, secrete gas when they are in deep water to compensate for their decrease in buoyancy, but this process is slow (more than one day to compensate for a descent from the surface to 20 m) or nonexistent in other species like herring. Surprisingly, the value of acoustic target strength – which depends mainly on the swimbladder volume (Chapter 9) – does not vary as much as expected according to depth (Reynisson, 1993), which suggests other mechanisms of compensation.

Blaxter and Batty (1984) suggest that herring replenish the swimbladder by swallowing air at the surface, even to such an extent that an above atmospheric pressure is built up in the swimbladder (the big gulp hypothesis, Thorne & Thomas, 1990). However, opening of physoclist swimbladders of fish caught at shallow depths often indicates a substantial gas pressure in the swimbladder that is difficult to explain just by swallowing. In addition, Suuronen (1995, Suuronen *et al.* 1996b) did not find a significant difference in the survival of herring in cages when the fish had or did not have access to the surface. In wintertime this species is able to spend several months under the ice cover without access to the surface.

Rapid changes of depth may damage the fish, make it vulnerable to loss of depth control when rising to the surface, or force it to adopt a high-energy-consuming tilt angle when swimming to compensate for positive or negative buoyancy. For instance, overwintering Norwegian spring-spawning herring occupy deep water (50 to 400 m) in fjords and therefore are constantly negatively buoyant because they cannot refill their swimbladder. They compensate for their low buoyancy by constant swimming, which generates lift when the pectoral fins are used as spoilers, and they often adopt a 'rise and glide' swimming strategy (Huse & Ona, 1996).

Species having a closed swimbladder (physoclists) represent the majority of gadoids. They have a rete mirabile which is an organ for secretion/resorbtion of gas from the blood to the closed swimbladder. However, the buoyancy-regulating process is slow, and physostomous fishes are not believed to be neutrally buoyant through their often substantial diel vertical migrations (Blaxter & Tytler, 1978). Those pelagic fish species that are able to perform large vertical migrations (tunas and tuna-like species) have either no swimbladder or a small one. In the case of tunas, especially yellowfin and bigeye tuna, vertical migration seems not only related to temperature and oxygen concentration, but also to periods of possible fly-glide behaviour aimed at saving energy (Holland *et al.*, 1990; Marsac *et al.*, 1996).

3.3.8 *Bottom depth and the nature of the sea bed*

Many pelagic species display a depth-dependent distribution (as illustrated by the following examples) which must be taken into account by an appropriate stratification when estimating their abundance from fishing data or direct observation. On the Petite Côte of Senegal (South of Cap Vert) the two species of sardinella present a different depth-related distribution despite an overlapping of their distributions,

Sardinella maderensis being more coastal than *S. saurita (*Boëly *et al.,* 1982*)*. In Northern Chile, the Spanish sardine (*Sardinops sagax*) is usually found in deeper waters than the anchoveta (*Engraulis ringens*), which is located in coastal waters (Yañez *et al.*, 1993).

In addition to such depth-dependent distribution, some species display a depth-related length or age distribution. Usually younger fish are more coastal than adults. This is the case with both species of sardinella mentioned above, where young fish are found over 5–25 m grounds and adults over grounds deeper than 25 m, but also of other species from Senegal, such as horse mackerels (*Caranx rhonchus*, *Trachurus trecae*). Similarly, in South Africa the Agulhas bank horse mackerel (*Trachurus trachurus capensis*) are caught from 20 to 50 m depth when young and then progressively extend up to the shelf-break when adults (Barange, 1994; Kerstan, 1995). Similar examples are provided by Stevens *et al.* (1984) on *Trachurus declivis* and by Priede *et al.* (1995) on *Scomber scombrus*. Fishery management can benefit from this depth related length or age distribution by prohibiting fishing in nursery areas when applicable.

The shelf-break is the favourite habitat of many 'semi-pelagic' species, at least when adults. In addition to the Agulhas Bank horse mackerel, this is the case with the horse-mackerel (*Trachurus trecae*) in the Gulf of Guinea – contrary to its more coastal distribution in the Senegalese area (Boëly & Fréon, 1979) which suggests the influence of some factor other than bottom depth (probably temperature). In the same way, tuna concentrations are also found around islands (González-Ramos, 1992) and along the shelf break (Stretta & Petit, 1991). In addition, tuna fishermen concentrate their activity around seamounts which are known to attract tunas, as demonstrated by one of the pioneer works on ultrasonic tagging performed by Yuen (1970) on skipjack tuna around the Kaula Bank (Hawaii). This fishing tactic was developed in the central Pacific (Boehlert & Genin, 1987) and the eastern tropical Atlantic where young yellowfin, skipjack and bigeye tuna are caught in similar proportion (Petit *et al.*, 1989; Fonteneau, 1991). In the eastern tropical Atlantic, the mean total catch on 30 sensed seamounts between 1980 and 1987 was 2835 t per year, that is 95 t per seamount and per year, with a record of 1158 t for a single seamount (Fonteneau, 1991). More recently the Indian Ocean tuna fleet adopted the same tactic, but here yellowfin (53%) and skipjack (42%) tuna are more abundant than bigeye tuna (5%) and this proportion of species is different from the catch composition of sets performed on 'logs' (nearly any natural or artificial floating object above a few centimetres in size) (Hallier & Parajua, 1992a). The catches are performed mainly on a single seamount located near the equator line along the 56th meridian line, which has been exploited since 1984. The mean catch per year was 3800 t during the 1984–1990 period. In all these situations it is suspected that the relief generates a higher production. Genin and Boehlert (1985) found structures like Taylor's cells over the Minami-kasuga seamount, in the northwest Pacific, associated with a higher concentration of chlorophyll and zooplankton in layers deeper than 80 m.

Some pelagic species select habitat according to the nature of the sea bed, especially the demersal spawners. The different populations of Pacific herring (*Clupea harengus pallasi*), Atlantic herring (*Clupea harengus harengus*) and capelin (*Mallotus villosus*) select areas of major deposition of gravel, especially the intertidal-spawning populations, or shallow coastal areas covered with marine vegetation in the case of shelf spawners (Haegele & Schweigert, 1985; Carscadden *et al.*, 1989).

Pelagic fish may also relate their habitat to the depth of the sea bed, but the proximate reasons may not always be obvious. A striking feature is the distribution of most coastal pelagic species on the shelf platforms, even when the prey distribution is much wider. For instance, on both sides of the Atlantic (West Africa, South Africa, northern part of South America, North America), several acoustic or aerial surveys indicate that the limit of distribution of all pelagic species is the continental shelf break, even though this physical limit varies from 80 to 200 m depth according to the area. Remote-sensing measurements of chlorophyll in West Africa indicated that during the upwelling season, the planktonic production often extended to a distance from the shore double that of the width of the shelf (Dupouy & Demarcq, 1987).

We have mentioned that in upwelling areas the continental shelf often provides retention areas which are necessary to reproduction success, but this does not explain the adult distribution all year long. We can only speculate on the reasons for this habitat selection, but it could be related to the daily vertical migration of these species, which are often observed on the bottom during the day, as mentioned by Boëly *et al.* (1978) for the gilt sardine (*Sardinella aurita*) in Senegal. Nieland (1982) found in the stomach of this species a large proportion of mud and sediments that had a non-negligible energetic value. Moreover, the shelf may be used by this species for orientation during their migrations (Harden Jones, 1968; Kim *et al.*, 1993), a hypothesis that merits further exploration. Of course there are notable exceptions to this distribution pattern over the continental shelf, especially off Peru and Chile, where most of the small pelagic species (except anchoveta) are abundant up to 160 km from the coast although the continental shelf is often less than 30 km wide (Johannesson & Robles, 1977; Yáñez *et al.*, 1993).

Another important feature of habitat selection related to the sea bed is the frequent association of pelagic species (both coastal and offshore) with sudden breaks in the bottom depth. For instance, coastal pelagic schools are often observed on the top of submarine hills or big rocks, or conversely in escarpment areas of steep bottom gradients, submarine canyons and deepwater basins located close to shore. The latter examples are reported by Mais (1977) from the analysis of 38 acoustic surveys on the northern anchovies (*Engraulis mordax*) in the California current system. J.-P. Hallier (pers. comm.) reports that large yellowfin tuna are found above submarine canyons and trenches off Mauritania. Recently a surprising example of site homing and fidelity to underwater reef has been discovered by multiple acoustic tagging experiments and video camera observations on two schooling gadoides species (Glass *et al.*, 1992; Smith *et al.*, 1993; Sarno *et al.*, 1994). The studies were conducted in Loch Ewe, on the west coast of Scotland, on a 4471 m^2 reef located at 400 m offshore. During the first experiment, six saithe (*Pollachius virens*), length 35 to 43 cm, were tagged and tracked simultaneously by a hydrophone array over 1 to 21 days according to the individual. Each of the fish spent more than 68% of its time during the day on the reef while at night activity was restricted almost entirely to the reef ($> 90\%$). Moreover the fish displayed movements to another smaller reef during the day. Underwater television indicated that the tagged fish comprised part of a large schooling group of fish (Glass *et al.*, 1992; Smith *et al.*, 1993). During the second experiment, two saithe (results similar to the previous experiment) and two pollack (*P. pollachius*; 43 and 44 cm) were tracked simultaneously for 170 h. Pollack covered less than 50% of the reef, did not go to the second reef and were swimming more slowly than saithe during the day (but at the same speed at night). A large scale quantification of this kind of association between fish and bottom topography seems possible now with automatic classification of the sea bed (Reid, 1995). Most of the previous habitat selection according to bottom structure seems directly or indirectly related to the increase in current velocity and turbulence created by this structure. These hydrodynamic changes favour production and predation as mentioned earlier.

3.3.9 *Floating objects*

Many pelagic species (and some young stages of demersal species) are attracted by floating objects, especially offshore. This behaviour could be classified in the broad category of habitat selection. Indeed some species of small fish can spend a large part of their life under natural floating objects drifting offshore, but in many cases these are demersal species which have found a substitute habitat (Hunter & Mitchell, 1967). In pelagic species, especially tunas and tuna-like species, floating objects do not represent a temporary habitat because fishes associated with the object remain far from it (up to several miles). Even though the contact (visual or auditory) with the object can occur during a certain number of days, those pelagic species might spend several weeks without contact with the object. Therefore we have developed this specific behaviour in section 6.3 on 'associative behaviour'.

3.4 Influence of biotic factors

Most pelagic species are social animals living in groups (shoals, schools; see section 4.2 for distinction) and present a highly patchy distribution. Nevertheless, some internal (not reviewed here) and external regulating factors must limit their concentration in a single place. The external factors are the abiotic and biotic factors. Among biotic factors, we will see that different life history stages prey upon different species which do not necessarily have the same spatial distribution. Moreover, cannibalism of some species on their offspring is an additional reason for differential distribution of the different stages, either as the result of selective pressure (Foster *et al.*, 1988) or as the result of defensive behaviour of the offspring. Finally, we will see that predators might influence habitat selection.

3.4.1 *Conspecifics and the ideal free distribution*

The density-dependent habitat selection (DDHS) theory was first developed by ornithologists during the middle of the twentieth century. They observed differential relationships between change in local densities and changes in population abundance.

Later Fretwell and Lucas (1970) conceptualised the 'ideal free distribution' (IFD) which is a behavioural model explaining the DDHS. In the IFD theory and its numerous derivatives, individuals occupy initially habitats with the highest basic 'suitability', 'profitability' or 'fitness', but as realised suitability of these habitats declines due to increasing population density, other previously less suitable unoccupied habitats become equally attractive and are colonised. Criticisms of the IFD and common misconceptions are reviewed by Kennedy and Gray (1993) and Tregenza (1994), who insist that the original theory considers a situation in which food patches consist of continuously arriving resources. Since planktonic food is not evenly distributed, fish have to search for patches of plankton. Therefore Sutherland (1983) proposed a modified IFD model incorporating a level of interference between competitors. As a result the density in the best habitat is lower than expected from the original IFD theory. Parker and Sutherland (1986) contemplated the situation where competitors benefit from different capabilities, which logically leads to a segregation of the phenotypes according to the quality of the habitat. Other derivations of the IFD theory have been proposed and are reviewed by Gauthiez (1997). Bernstein *et al.* (1988) studied the case of limited food input and poor knowledge of the environment by the competitors. Bernstein *et al.* (1991) focused on the cost of displacements and the spatial structure of the environment.

The above-mentioned works are mainly theoretical or do not apply to fish. Several authors apply the IFD approach to demersal fish populations, including facultative schooling species, or at least discuss their results in the context of the IFD. In most cases, the results indicate that the less favourable habitats are occupied only when the stock abundance overpasses a certain threshold (e.g. Myers & Stokes, 1989; Crecco & Overholtz, 1990; Rose & Leggett, 1991; Swain & Wade, 1993; Swain & Sinclair, 1994; Gauthiez, 1997). These results contrast with those of Swain and Morin (1996) on a less mobile and gregarious demersal species, the American plaice (*Hippoglossoides platessoides*).

Another approach, involving diffusion models, is traditionally used by fishery scientists (see Mullen, 1989 for a review). Simple sets of equations are used, incorporating a 'diffusion coefficient' a^2 and a mean velocity vector v to characterise the movements of entire populations. Usually a^2 is assumed constant. Nevertheless, several scientists exploring tuna behaviour (e.g. Bayliff, 1984) reported that estimates of a^2 varied by one or two orders of magnitude for both yellowfin and skipjack tuna in the eastern Pacific. These results are interpreted as the consequence of permanent rich areas close to island and shallow banks. Mullen (1989) compared the results of simulation models where a^2 is either constant or proportional to the local abundance divided by the local carrying capacity of the ecosystem. As expected, models with variable a^2 simulated higher patchiness of fish distribution because these models allow fish to spend more time in good habitats. A more recent approach for modelling spatial heterogeneity due to fish behaviour is to use individual-based models (IBMs), which depict populations in which individuals follow specific rules of behaviour according to their environment (see review by Tyler & Rose, 1994).

MacCall (1990) extended the density-dependent habitat selection (DDHS) theory from discrete habitats to an entire population, i.e. a pelagic fish population. He reviewed some examples of expansion and contraction of population range or differential utilisation of marginal habitat with changes in population abundance in fishes, birds, insects, reptiles and mammals. The author developed a 'basin model' where habitat suitability is depicted graphically as increasing downward. Habitat is therefore described as a continuous geographic suitability topography having the appearance of an irregular basin (we will see in Chapter 8 how the population responds to climatic changes according to this model). The suitability of habitat is defined by MacCall (1990) according to different biotic and abiotic factors which favour the individual and population growth. The underlying mechanisms of density dependence rely on conventional population dynamics theory (cannibalism, starvation, individual growth), but implicitly the theory supposes that somewhere in the process it is the individual fish's decision to move from one habitat to another and the theory supposes that fish are able to detect gradients of suitability. The author does not try to explore this behavioural decision except on one occasion where a possible habitat selection according to detection of a food gradient is mentioned. He successfully applied his model to the northern anchovy (*Engraulis mordax*) population of the west coast of North America and concludes that although the basin model 'presents a grossly simplified and abstract view of population behaviour, the model conveys a strong intuitive image which allows holistic grasp of an entire suite of important considerations: abundance, distribution and geographic structure, movements and population growth'. Similar results of negative relationships between spawning biomass and spawning area have been observed in other stocks, as for instance the anchovy population of the Bay of Biscay (Uriarte *et al.*, 1996).

Notwithstanding the interest of the basin global approach, it seems that the IFD theory is not likely to be met in shoaling fish because individuals vary in their ability to compete or forage on food located by others (Wolf, 1987; Milinski, 1988). Aquarium and tank experiments suggest that in larger shoals fish find food faster, spend more time feeding despite predator threat because they are less timid and take advantage of vigilance from conspecifics, sample the habitat more effectively and are able to transfer information about feeding sites (Pitcher & Parrish, 1993). Pitcher (1997) proposed an alternative basin model for pelagic species in which the basin is 'flat-bottomed' in opposition to the irregular shape of MacCall's basin (Fig. 3.4). In this flat basin, fitness of individuals changes only at the lip of the basin, due to abrupt changes in the environment. This concept was developed mainly to prove that the shrinkage of the population area, also named 'range collapse', can be simulated without environmental determinants, but only with shoaling behavioural rules.

The two models produce contrasted spatial predictions of range collapse. In MacCall's basin, fish remain in specific areas corresponding to the best habitats, while in Pitcher's basin, fish concentrations are randomly located (which is not a common observation). The model proposed by Pitcher (1997) looks like a simple IBM structured on shoals as proxy for individuals. This approach seems promising but should benefit from the incorporation of more realistic spatial structures (variable size of the shoals, clusters) and deterministic hypotheses on school behaviour related to foraging, predation, reproduction and mutual attraction. Despite the pedagogical interest

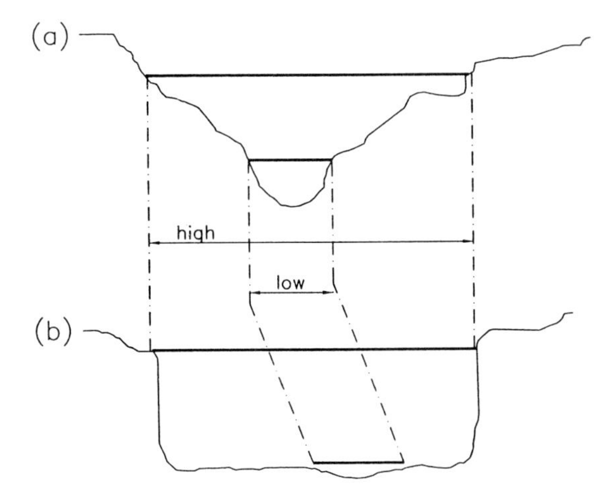

Fig. 3.4 Conceptual basis of two alternative 'range collapse' models. Vertical axis and profile of basins indicate relative habitat quality: deeper basin is better habitat. Spatial range of the fish stock at high and low abundance are indicated by arrows. (a) MacCall's basin model structured on habitat quality; (b) Pitcher's basin model (not structured) (redrawn from Pitcher, 1997).

of the 'flat-bottom basin' hypothesis, we think that a combination of irregular basin and schooling behaviour would better reflect the real world situation.

At a microscale, pelagic fishes usually school according to body size (Chapter 4), and on some occasions according to sex. For instance, the capelin (*Mallotus villosus*) of the east coast of Canada segregate in schools according to sex during their reproductive migration from offshore areas to coastal beaches in June and July. Males tend to school close to the beach and females congregate in schools in deeper water. When the females are ready to spawn, they move towards the beach through the male schools. These females are the main target of an intensive fishery (purse seine, trap and beach seines) to supply a lucrative Japanese capelin roe market (Nakashima *et al.*, 1989). Similarly the deep pelagic species named orange roughy (*Hoplostethus atlanticus*) presents a marked sex segregation and some evidence that males arrive on the spawning ground first (Pankhurst, 1988).

The role of pheromones in such habitat selection relative to conspecifics is likely, even though it seems poorly documented in marine species (Stacey, 1987) contrary to continental fish species (Myrberg, 1980) or insects. There is evidence that insects respond to aggregation pheromones (e.g. Brossut *et al.*, 1974; Rivault & Cloarec, 1992). As far as we know such aggregation pheromones have not been investigated in fish but could be a possible means of enhancing encounter frequency of dispersed fish or small schools with larger schools. We have found some references on sexual pheromones in pelagic species. From laboratory observation on captive Pacific herring (*Clupea harengus pallasi*), Stacey and Hourston (1982) suggested the existence of a pheromone in herring milt because spawning behaviour is rapidly initiated in ripe (ovulated and spermiated) herring of either sex following exposure to herring milt or to a filtrate of ripe herring testes. Later Sherwood *et al.* (1991) confirmed the existence of a pheromone-like spawning substance by males of this herring species and characterised its biochemical properties which are similar to those of polar steroids, prostaglandins, or their conjugated forms. A behavioural bioassay was successfully performed. The milt and testicular extracts retained their bioactivity during purification. The role of a male pheromone may be synchronisation of spawning in schools.

3.4.2 *Other species*

In pelagic species the influence of the presence of other species (except prey or predators) on habitat selection is not easy to interpret. Competition – defined by Mayr (1970) as being related to two species looking simultaneously for a basic and rare environmental need (e.g. food, space, shelter) – is certainly an important aspect, but as noted by Mayr, during the evolutionary process the speciation is supposed to largely limit the competition between species. If two species have exactly the same needs all through their life cycle they are not supposed to coexist ('competitive exclusion principle'; Hardin, 1960). Nevertheless, competition between two species can take place during a limited number of stages. Moreover, following Brown and Wilson (1956), Mayr suggested that strong competition is likely when two species have been in recent contact or when large environmental changes modify the previous dynamic of equilibrium between them. Despite the fact that such situations are likely to occur in upwelling ecosystems, Mayr said he did not know any valid example of competition or exclusion in planktivorous pelagic species. How species modify their habitat selection in the case of such competition is not well documented, nor is it clear whether it is an individual behaviour or a population behaviour.

The influence of competitors on the ideal free distribution (IFD) and its derivatives seems obvious but is not well documented. Different pelagic species often cohabit in the same area where they form ecological communities. The interaction among species of the community is not well known as far as habitat selection is concerned. Only some studies on freshwater fish in fluvarium are available. For instance, Allan (1986) shows that minnows (*Phoxinus phoxinus*) moved higher in the water and downstream in the presence of dace (*Leuciscus leuciscus*). Gudgeon (*Gobio gobio*) moved lower in the water and downstream in the presence of dace and of minnows. Rosenzweig (1981) stressed the importance of interspecific competition in the results of spatial modelling. According to the characteristics of the habitat, interspecific competition favours certain species over the others. As a consequence of this, the density is higher in the richest habitat than expected from the conventional IFD models. On a smaller scale, mixed-species schools are often observed (section 4.9) and in this case the IFD theory, and its limits, could apply if the two species are competing for the same prey. Similarly, inter-specific (and also intra-specific) competition for bait is reported for long-lines. The dominant individuals frighten and chase away the others from the baited hooks (review in Løkkeborg & Bjordal, 1992).

3.4.3 *Prey*

Let us first review the main types of prey in pelagic fish and the way they are detected. Most of the carangid species (*Caranx* spp., *Trachurus* spp.) are omnivorous when adults, preying on large zooplankton items (including ichthyoplankton) and small fishes (e.g. Brodeur, 1988). Most coastal clupeids are microphagous, preying on phytoplankton and/or zooplankton. From an extensive and critical review of the available literature, James (1988) stated that there are few true phytophagous fish in this group, contrary to the common belief. Most are omnivorous microphages, capable of particulate and filter feeding upon a wide range of particle sizes, but deriving the bulk of their energy from zooplankton. In northern anchovy (*Engraulis mordax*) for instance, the filtering mechanism acts over a 0.05 to 1.50 mm range of prey length, while selective particulate feeding occurs from 1.51 to 5.00 mm. Although phytoplankton is by number the main component of the stomach contents of this species, zooplankton prey are consistently the major source of carbon in the diet (Chiappa-Carrara & Gallardo-Cabello, 1993). Gibson and Ezzi (1992), who observed feeding behaviour of herring (*Clupea harengus*), estimated that the energy cost of filtering may be from 1.4 to 4.6 times higher than that of biting on the same zooplankton prey. Tunas and mackerels are carnivorous and feed on a large variety of items, from crustaceans to fish but also cephalopods, having a wide size distribution (Brock, 1985; Olson & Boggs, 1986; Stretta, 1988; Roger, 1993; Buckley & Miller, 1994). Despite a relationship between the maximum size of the prey and the fish size, large tunas still prey chiefly on small items (Roger, 1993). Large schooling semi-pelagic gadoids (cods, saithes, pollocks) are top predators that feed mainly on small fish, crustaceans and invertebrates.

How fish locate and orientate to their prey is a key question to understanding habitat selection. The range of vision in sea water is usually limited, especially for small prey. For instance, the visual field for prey detection of northern anchovy (*Engraulis mordax*) was estimated at 104 mm and the basal area of the vision cone at 125 cm^2 off the west coast of Baja California (Chiappa-Carrara & Gallardo-Cabello, 1993). Even in the favourable case of large oceanic predators preying on large prey in clear waters, the range of visual discrimination of two points 2.4 cm apart is around 4.4 m due to the limited penetration of light in sea water (Tamura *et al.*, 1972). Therefore, most pelagic fish detect their prey by vision at short range, even under low light condition (Guthrie & Muntz, 1993), while at large and medium range, chemical senses (smell and taste, two often overlapped senses) are recognised as the main factors (Atema,1980; Hara, 1993).

Atema (1980) distinguished three major situations as far as chemical stimuli distribution is concerned: nondirectional active space, gradient and trail (odour corridor). The nondirectional 'active space' is defined as any area where a chemical stimulus is present above a threshold enabling its detection, but not the location of the stimuli source owing to a too small gradient. This small gradient is due to several factors: irregular direction in the movement of the releasing source, turbulence of the environment or intermittent release of chemical stimuli (e.g. faeces). In contrast the gradient and trail (especially polarised trail) are special cases of active spaces because they are directional and therefore make easier the detection of the emitting source (e.g. carcass in decomposition laying on the bottom).

Nondirectional active space seems more common than gradient and trail in the pelagic marine environment, especially in turbulent upwelling areas. Therefore it is believed that the exact location of the source in a large three-dimensional active space cannot be determined from chemical cues alone but requires the use of other senses, in particular vision in the case of pelagic fishes. Olfactory stimulation appears to be a powerful means of alerting a fish to the presence of distant prey, exciting it and then keeping it motivated to pursue its search by vision. When tuna (little tuna, *Euthynnus affinis*, or yellowfin tuna, *Thunnus albacares*) detect chemical cues coming from prey extracts they react by increasing swimming speed, changing swimming pattern and head thrusts. Interestingly the water collected from a live prey school elicited predator responses (Magnuson, 1969; Atema, 1980; Atema *et al.*, 1980). Amino acids are a major component of tuna's prey odour image and the olfactory threshold of detection is about 10^{-11} m.

Experimentation on captive fish of different species indicates that amino acids (e.g. betaine, glycine, alanine) and related compounds present in faeces, mucus or skin products elicit both neurophysiological and behavioural responses (review in Atema, 1980). This capacity appears very early during the ontogeny. From experimental studies and a review of the available literature, Døving and Knutsen (1993) indicate that marine fish larvae use chemotaxis and respond to the concentration of metabolites emanating from planktonic prey for searching for, finding and remaining within clusters of food organisms.

One of the functions of diel vertical migration observed for most pelagic fish is to follow the vertical migration of their prey. In the eastern Bering Sea, 0-group walleye pollock (*Theragra chalcogramma*) modify their diel migration pattern according to prey availability. When prey are abundant, the pollock either do not migrate vertically or delay their migration for several hours (Bailey, 1989). Data on the feeding periodicity of coastal pelagic fish are scarce and conflicting, probably because most species are opportunistic foragers having flexible feeding cycles according to local conditions (James, 1988). Nevertheless, particulate feeding is often associated with peak feeding activity at dusk and dawn.

An interesting example of habitat selection related to prey is the recently investigated case of tuna concentrations found during summertime along the equator off Côte d'Ivoire and Liberia, which is not claimed to be a productive area. According to Marchal and Lebourges (1996), in this area tunas are feeding on a Photichthyidae species of the DSL, *Vinciguerria nimbaria*, which is surprisingly observed in the upper layer during daytime, providing excellent visual condition for tunas preying. At least during this period of the year, *Vinciguerria nimbaria* seems to have a reversed phototaxism for a reason not completely investigated but probably related to breeding. This species is therefore available to tunas in the upper layer during daytime, when the tunas can actively prey upon them owing to good visual conditions. An additional example is provided by Carey (1992), who followed by acoustic telemetry the location of a swordfish (*Xiphias gladius*) from 0200 to 0800 h. The fish was clearly following the descent of an acoustic sound-scattering layer of prey detected by a 50 kHz echosounder from one hour before sunrise (0445 h) to 0630 h (Fig. 3.5). Similar observations were performed on a 100 cm bigeye tuna in the open ocean in French Polynesia by Josse *et al.* (1998). During daytime, these authors also observed a smaller bigeye tuna (77 cm) swimming inside the DSL layer of maximum acoustic intensity. Josse *et al.* (1998) also tracked vertical movements of a 60 cm yellowfin tuna performing from time to time short incursions from the surface layer into the 100–200 m layer, and remaining precisely inside a concentration of prey for nearly one hour when detected (Fig. 3.6).

The relationship between the distribution of fish and their prey at a given time is

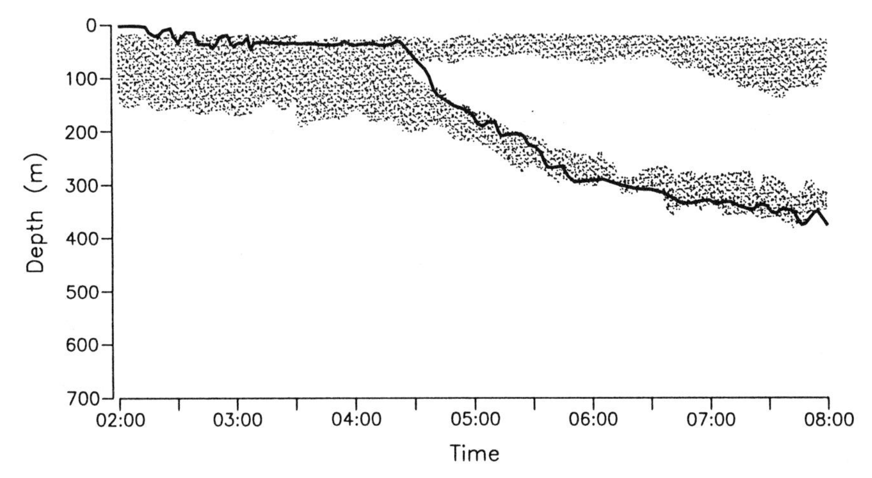

Fig. 3.5 Vertical movement (the black line) of an acoustically telemetered swordfish (*Xiphias gladius*) on Georges Bank in 1982, shown with respect to the movement of its presumed prey, as indicated by a 50-kHz echogram (shaded area). Dawn was at 0445 h (redrawn from Carey, 1992).

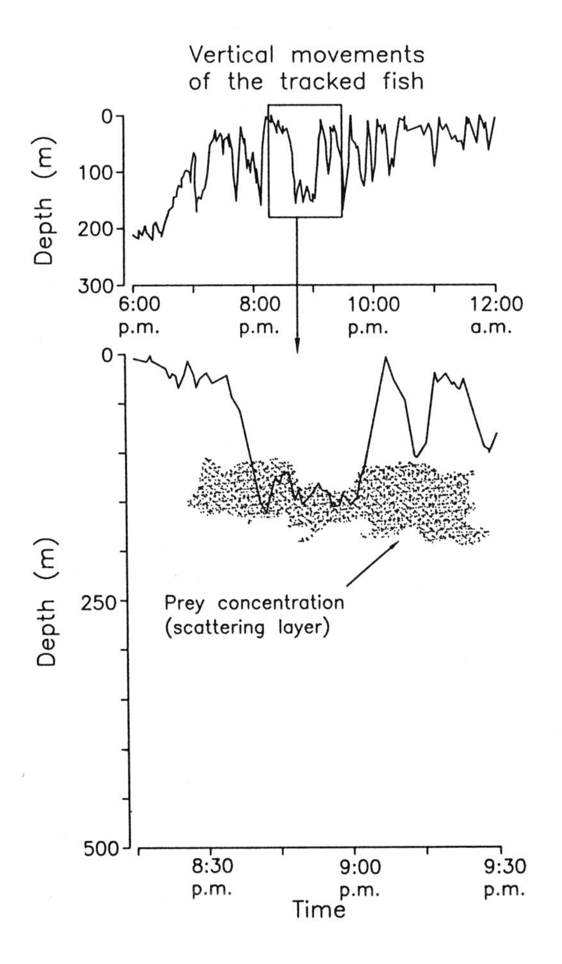

Fig. 3.6 Vertical movements (black line) of an acoustically telemetered yellowfin tuna (*Thunnus albacores*), 60 cm fork length, near Maupiti island (French Polynesia), shown with respect to the location of presumed prey concentration (dense scattering layer). (Redrawn from Figure 3 of Josse, E., Bach, P. and Dagorn, L. (1998) Tuna-prey relationships studied by simultaneous sonic tracking and acoustic surveys. *Hydrobiologia* (in press). With kind permission from Kluwer Academic Publishers.)

not easy to investigate because relevant parameters are not easy to collect simultaneously (except for phytoplankton measurement by satellite) and this relationship is not necessarily linear. Maravelias and Reid (1997) applied generalised additive models (GAMs) on prespawning herring distribution measured by acoustic and simultaneous recorded data of zooplankton, temperature and salinity in the northern North Sea. They found that the location of ocean fronts derived from oceanographic data was related to the herring distribution as in a previous study (Maravelias & Reid, 1995) and that this link was likely to be indirect and due to the availability of zooplankton prey in some water masses (Atlantic water).

Horizontal seasonal migrations related to prey selection are well documented (Harden Jones 1968; McCleave *et al.*, 1982; Dingle, 1996), but it seems that in some instances, interannual habitat selection might be related to prey abundance. Polovina (1996) suggested that changes that occurred during the previous decade in the proportion of northern bluefin tuna (*Thunnus thynnus*) migrating from the western Pacific to the eastern Pacific, were related to the decrease in prey abundance (of the Japanese sardine, *Sardinops melanosticta*). Similarly, Leroy and Binet (1986) speculated on the causes of the quasi-disappearance of this species in the North Sea and the Norwegian sea from the 1960s, and suggested the decrease of small pelagic prey abundance (especially herring), in addition to climatic changes.

In summary, prey abundance is an important factor of habitat selection variability both horizontally and vertically and at different scales of space and time (from minutes to years). Unfortunately the information on prey abundance is seldom available and therefore not taken into account in most of the assessment methods. The only attempts in this direction have been mainly applied to demersal ecosystems (Laevastu, 1990a, b) or have been limited to understanding the trophic food web in a steady-state situation (ECOPATH model, Polovina, 1984; Christensen & Pauly, 1992), which is a necessary but preliminary step. A dynamic ECOPATH model is investigated by Walters *et al.*, (1997) which is a promising approach despite the risk of overparameterisation in many insufficiently documented ecosystems.

3.4.4 *Predators*

Fish are able to detect and identify their predator at short distance by vision and at long distance by olfaction and sometimes by hearing. Predators can not only change the microdistribution of their prey, which usually react by increasing their packing density and performing a large repertoire of local evading manoeuvres including school splitting and avoidance, but can also cause a large decrease of abundance in the area due to long-distance flight by the prey. We did not find such evidence of long-distance flight in the literature on pelagic fish, but during acoustic surveys of *Sardinella aurita* in Venezuela, we observed that the fish disappeared when dolphins arrived in the area. Sunfish (*Lepomis macrochirus*) in the presence of predators move to a suboptimal habitat (Werner *et al.*, 1983). In this species an ontogenetic habitat shift is observed and Werner (1986) demonstrated that experimental habitat use is altered by the size-dependent risk of predation. In their review of the development of predator defences in fishes, Fuiman and Magurran (1994) give other examples of habitat shifts resulting from the trade-off between foraging and predation risk. In addition they underline the fact that niche shifts may also be mediated by the influence of the programme for morphological development on sensory or behavioural capabilities, especially when related to predator defences (camouflage, detection, evasion).

In the Bay of Biscay, the combination of pelagic trawl results and simultaneous school measurements by acoustics suggests that anchovy schools change their vertical habitat selection according to the occurrence of predators. When they are the dominant species, anchovies (*Engraulis encrasicolus*) are usually observed 16 m above the bottom, while they are observed 30 m above the bottom in the presence of horse mackerel (*Trachurus* spp.) in the bottom layer (Massé *et al.*, 1996). In this work the authors noticed that, surprisingly, this change in vertical distribution of the prey occurred despite the fact that the prey and the predator species had similar mean size (14 cm). We suggest that this could be due to predator recognition by prey, which is usually related to a limited number of key stimuli (see Chapter 7 for details). For instance we observed repeatedly the defensive behaviour of small sardine (*Sardina pilchardus*) schools during their encounter with schools of other species – including small predators – in the Carnon harbour (South of France) in November 1995. When sardines encountered large mullets (*Mugil* spp.), they did not react at all and the two schools crossed one over the other, but during encounters with small mackerels (*Scomber japonicus*) of similar body length to the sardine (around 12 cm), a strong avoidance reaction was displayed by the sardine school.

In large pelagic species, adult fish often prey on juveniles of their own species (section 6.3.1).

In addition to direct reaction to the presence of a predator, it is likely that pelagic fish can select their habitat according to the ability of predators to detect them, which depends on how conspicuous they are in this specific habitat. It is accepted that many predators take advantage of the reduced visual capabilities of prey to perform attacks during short twilight periods of dawn and dusk in tropical areas (Helfman, 1992; but see Parrish, 1992a for the importance of diurnal predation). Some predators have a better relative visual acuity than their prey during the crepuscular period because they possess larger (but fewer) cones in their retina (Munz & McFarland, 1973). According to McFarland *et al.* (1979), French and white grunts (*Haemulidae*) seem to reduce their level of conspicuousness during their dusk offshore migration by hugging the substrate in tight groups, which minimises the backlight effect. Another example of interaction between predator and habitat in freshwater fish is presented by Ekloev and Hamrin (1989).

The vertical distribution is likely to be affected by intensive bird predation, resulting in an increase of the mean school depth. We did not find references to this in marine habitats, except for describing the depth of predation by birds which can be important in some avian species (30 m on average for rhinoceros auklets; Burger *et al.*, 1993). We found some references to experimental work on juvenile chinook salmon which indicated such increase in depth in the presence of a bird model (Gregory, 1993) and similar experimental results in freshwater fish (Lonzarich & Quinn, 1995).

3.5 Conclusion

We have seen that the determination of habitat selection and migration is the result of many interactions at different space and time scales. Habitat selection depends on numerous biotic or abiotic factors which vary according to the species, the age and the physiological stage of the fish. These factors might be different according to the temporal and spatial scales, but they are sometimes conflicting in a given place at the same moment. Seasonal migrations over relatively large distances are usually related to reproduction and alimentation while habitat selection on other temporal and spatial scales depends on many other interacting factors. An example of such complexity is given by Swain and Kramer (1995) who studied interannual variation in temperature selection by cod in the southern Gulf of St. Lawrence from 1971 to 1991. They found that temperature selection was highly variable depending on year and age class. When density was high, old age classes tended to occupy colder waters and the authors suggested a trade-off between the density-dependent benefits of greater supplies in warm water and the density-dependent benefits of lower metabolic costs in the deeper water.

These facts can be linked to the classification of ethological programmes according to Mayr (1976). According to this classification, habitat selection programmes are 'open programmes' and the related phenotypic behaviour can be modified according to experience and to external constraints. This flexibility is an advantage because the different components which define an optimal habitat (especially predators, prey, physical needs) are highly variable. For instance, a strict mechanism of diel vertical migration in response to light intensity would in certain situations be a handicap for preying or predator avoidance. An additional difficulty encountered by students of habitat selection results from the fact that fishes living in unstable habitats may forever seek their environmental preferenda without achieving them, as a result of the complexity of the dynamics of habitat and behaviour (Neill & Gallaway, 1989). This situation could explain the difficulties in the studies of the distributional responses of fish to environment. The lack of dynamic models coupling environmental changes and fish movement dynamics could explain the high level of 'noise' in the deterministic modelling. Another interpretation of this 'noise' could be the mixture of different 'obstinated' strategies followed by different individuals belonging to the same stock (Cury, 1994). Finally, the depletion of prey by fish predation can also make it difficult to study the relationship between standing crop and fish density.

In other instances, the dynamics of the food chain process can be responsible for complex and time-lagged relationships between the effect of physical factors and biological factors responsible for habitat selection. Stretta (1991) proposed a dynamic model aimed at forecasting suitable tuna fishing areas according to the dynamics of sea surface temperature in the eastern Atlantic. The ideal thermal scenario leading to food availability for tuna is:

(1) Upwelling (cold water)
(2) Maturation of the water during four weeks (increase in surface temperature)
(3) Thermal stabilisation for two weeks.

In this situation it is likely that tunas are not able to take advantage of such a complex link between the temporal dynamic of a physical factor (temperature) and the location of the favourable habitat (presence of prey).

In this chapter we have seen that some biotic or abiotic factors responsible for habitat selection are easily predictable due to their circadian, seasonal or lunar periodicity, in contrast to other unpredictable factors (mainly the interannual variations and the biotic factors). The influence of all these factors on catch rates and population sampling is obvious. In contrast, it is not straightforward to take into account the influence of unpredictable factors for survey design. Therefore temporal or spatial pre-stratification is often useless in pelagic fish studies, contrary to the permanency of spatial structures in many benthic species and some demersal ones. Consequently the approaches by post-stratification (with all related difficulties in estimating a proper variance estimation) or better geostatistics are often used in pelagic fish stock assessment. Similar difficulties are encountered during the processing of commercial data aimed at computing abundance indices, with the additional problem of no control on the sampling strategy. From the experience accumulated during several generations, fishermen have learned how to take advantage of fish habitat selection to maximise their profits. We will detail in Chapter 8 how the different kinds of habitat selection and the consequent fishermen's strategy will generate error or biases in indirect methods of assessment and how easy or difficult it is to remove or limit these effects. We will do the same in Chapters 9 and 10 in relation to direct methods of abundance estimation and distribution. In the next chapter we move to the important matter of fish schooling, which is another fundamental behaviour pattern of many economically important fish stocks.

Chapter 4
Schooling Behaviour

4.1 Introduction

Fish can distribute and behave individually, they can aggregate in more or less social groups – for instance during feeding and spawning – and some species can swim in large, dense schools. These behaviour patterns determine to a large extent the volume occupied by the individuals in a stock (Pitcher 1980), and consequently have a significant influence on sampling.

Schooling behaviour is common among fishes. Shaw (1978) has estimated that about 25% of the approximately 20 000 teleost species are schoolers. Moreover, about 80% of all fish species exhibit a schooling phase in their life cycle (Burgess & Shaw 1979), and schooling as juveniles is especially prevalent. Even benthic species such as the naked goby (*Gobiosoma bosci*) that live among oyster shells and other hard substrates, school at the larval stage (Breitburg, 1989). Other aquatic organisms such as squids (Hurley, 1978), tadpoles (Wassersug *et al.*, 1981), and krill (Strand & Hamner 1990) may form aggregations that resemble fish schools. Even the behaviour of airborne bird flocks during migrations is a close analogy to fish schooling (Major & Dill 1978).

Schooling pelagic fishes, such as most clupeoids, scombroids and carangids, are the foundation of major worldwide fisheries and fishing industries. To optimise the long-term yield, fisheries on these stocks are regulated on the basis of biomass estimates from direct and indirect stock assessment methods. However, the schooling behaviour of these species puts important limitations on the applicability and accuracy of both indirect and direct stock assessment methods (Ulltang, 1980; Aglen, 1994).

In this chapter, the characteristic patterns of the schooling behaviour of fish are described, and features that differentiate schooling from other aggregative behaviours are emphasised. The function of schooling behaviour are considered, but most of the chapter will be devoted to illustrating how fish organise and maintain schooling. Finally, the senses involved during schooling, and communication between schooling individuals, are outlined.

4.2 School definitions

Since the first studies of fish schooling, there has been a debate about what the term 'school' really describes. Parr (1927) considered schools as fish herds, having an apparently permanent character and being a habitual, spatial relationship between individuals. The schools were claimed to exist on a diurnal basis, and to result from visual, mutual attraction and subsequent adjustment of direction to parallel swimming. Breder and Halpern (1946) simplified the definition by stating that schools are groups of fish that are equally orientated, regularly spaced, and swimming at the same speed. Distinguishing the schooling habit among species, Breder (1967) suggested the term 'obligate schoolers' for species swimming most of their lives in coherently polarised and permanent groups, while species forming such groups temporarily were called 'facultative schoolers'. Radakov (1973) considered a school just to be a group of fish swimming together. Shaw (1978) ended this group-emphasising tradition by arguing that groups of fish united by mutual attraction, and which may be either polarised or nonpolarised, should be considered as schools.

Illustrating that it is the individual's own decision to join, stay with, or leave a group of companions, Partridge (1982a) argued that schools should be characterised by independent measures of time spent schooling and degree of organisation. This view was developed further to the statement that a school is a group of three or more fish in which each member constantly adjusts its speed and direction to match those of the others (Partridge 1982b). Pitcher (1983) defined schooling simply as fish in polarised and synchronised swimming. The basic criterion of both these definitions is the element of organised motion accomplished by the participation of each individual in the school. A school is a functional unit.

4.3 Genetic basis of schooling

Like all other living creatures, fishes struggle to do what they have to do, i.e. avoid predators, feed so as to reach maturity, find a partner and reproduce so that their genes are passed to succeeding generations. Some species reproduce several times (iteroparity), others, such as the capelin (*Mallotus villosus*) reproduce just once (semelparity) and die afterwards. Tactics for predator avoidance also differ among species even if there can be striking similarities. At the juvenile stage most species perform schooling to enhance survival, but the pattern of schooling may differ among closely related species such as cod (*Gadus morhua*) and saithe (*Pollachius virens*) (Partridge *et al.*, 1980). Distinct patterns of behaviour unique to certain species contribute to reproductive isolation among species. As for all other living creatures, the behaviour of fishes is set within limits determined by their genetic constitution, which has been formed by natural selection through the evolution of their life histories.

If there is a certain degree of genetic variation within the population, natural selection may operate. In the case of fish behaviour, there are several intraspecific examples which illustrate how natural selection has been acting on genotypes so that the behaviour differs. Among the best known is the schooling behaviour of guppies (*Poecilia reticulata*) in various isolated streams in Trinidad (Seghers, 1974a). Schooling is well developed in populations living in rivers with a high density of piscivores (charachids and cichlids) taking all life stages of the guppy, is intermediately developed in populations living in streams with a high density of one fish species (*Rivulus hartii*) predating mainly on immature guppies, and is poorly developed or absent in populations living in streams with medium and low densities of this predator. The final evidence of a genetic basis for this intraspecific difference in behaviour was given by Magurran and Seghers (1990), who demonstrated that the differences in schooling behaviour among the various guppy populations were also present in newborn fish.

Similarly, European minnow (*Phoxinus phoxinus*) that live sympatrically with predatory pike (*Esox lucius*) organise more cohesive schools and perform better-integrated evasion tactics than fish from waters without the predator (Magurran, 1990). There was obviously a genetic, inheritable basis for this population difference, as Magurran (1990) reared the minnow populations studied in identical conditions in a laboratory without exposure to the predator. Moreover, the antipredator behaviour was modified most by early experience of a model predator in the population sympatric with pike, which indicates a genetic predisposition to respond to early experience that also varies intraspecifically.

Another way of revealing the genetic influence on fish behaviour is comparison of hybrids. In river populations of *Astyanax mexicanus*, schooling is prevalent, while it is not observed in blind cave populations. The inability to school can be caused by the loss of visual orientation, or a genetic change as a result of life in the caves with no selection for schooling. When crossing the fish from a cave and a river, it was revealed that despite good visual orientation, the hybrids showed a weaker tendency for schooling. This indicates that there exists a genetic reduction of schooling behaviour in the cave populations compared with the river populations. Similarly, genetic differences are expressed in different locomotor activity, fright reaction, feeding and sexual behaviour between cave and river populations of *Astyanax mexicanus* (Parzefall, 1986).

4.4 Ontogeny of schooling

Most fish larvae do not have fully developed sensory organs and locomotory systems when they are born; these develop gradually throughout the first critical stage of the life history. Due to the small size of most larvae (around 1 mm) and their weak locomotory ability, the viscous forces in the water make movement difficult. The hydrodynamic constraint put on objects moving in water by the viscous forces are expressed through the Reynolds number (dimensionless), which depends both on the size and on the speed of the object. At Reynolds numbers less than 10 the viscous forces are strong, at Reynolds numbers between 10 and 200 the viscous forces have an intermediate influence, and at Reynolds numbers over 200 they can be ignored. Fuiman and Webb (1988) estimated that newly hatched zebrafish (*Danio rerio*) have a Reynolds number less than 10 during ordinary swimming bouts, which means that the larvae are in fact moving in 'syrup', and reach a Reynolds number larger than 200 just for short periods during swimming bursts. Obviously, organisms living under such constraints have no possibility for synchronised, polarised movements in groups.

As the locomotory system develops, the larvae become capable of continuous swimming. The eyes which initially consist only of cones become more functional, and the number of rods necessary for perception of motion increases (Blaxter & Hunter, 1982). Most larvae hatch with free neuromasts, but when these become innervated, a moving larva becomes capable of detecting water movements generated by other nearby larvae. More pigmentation makes the initially transparent larvae more conspicuous and easier to keep sight of. Through these developments, the larvae may orientate in relation to conspecifics by monitoring their movements visually and the distance to them by the lateral line (Shaw, 1970).

As the sensory and locomotory organs of the larvae become more and more functional, the 'following response' develops. Shaw (1960, 1961) and Magurran (1986) describe how this response develops into schooling for silverside (*Menidia menidia*) and European minnow, respectively. The following response starts to be exhibited when the gradually developed larvae encounter conspecifics of the same life history stage. At first, encounters are followed by immediate withdrawals. Gradually, the larvae seem to be more mutually attracted and swim together as a pair, one leading, the other following somewhat behind and to one side. The duration of such following responses also increases gradually. Schooling develops as several pairs encounter and start swimming together as a group, at first for short periods, but regular schooling may rapidly be established.

The development of schooling behaviour is species dependent, and there is a clear relationship between the degree of functionality of the larvae at hatching and the time until schooling is fully established. Well-developed larvae of the ovoviviparous guppy in rivers on Trinidad are capable of fully developed schooling immediately after they hatch (Magurran & Seghers, 1990). This is very different from the poorly developed larvae of most marine pelagic species. The Atlantic silverside (*Menidia menidia*) is 10–12 mm in length or about 20 days old when polarised, synchronised swimming is established (Shaw, 1960; 1961). Other species that school before metamorphosis are northern anchovy (*Engraulis mordax*), which starts at lengths between 10 and 15 mm (Hunter & Coyne, 1982), and Atlantic menhaden when 22–25 mm in length (Blaxter & Hunter, 1982). Atlantic herring reach 35–40 mm and are at the onset of metamorphosis before schooling (Gallego & Heath, 1994). For European minnows, which emerge from eggs buried in gravel, schooling is fully developed about four weeks after the onset of free swimming (Magurran, 1986).

Schooling behaviour continues to develop after its initiation in the second half of the larval life stage. Van Olst and Hunter (1970) observed that for anchovy, jack mackerel (*Trachurus* sp.), and silversides (*Atherinops* sp.) the mean nearest neighbour distance and deviation in heading among fish in the school decreased from larvae near metamorphosis (20–40 mm long) to juveniles (60–80 mm long). The integrated schooling observed for adults is therefore a process that starts in the middle larval stage and continues through a substantial part of juvenile life.

4.5 Schooling and shoaling

At night-time, fish schools commonly disperse and extend in aggregations and layers (Shaw 1961; John 1964; Blaxter & Hunter 1982). Apart from being a social assembly of fish in one particular area, such groupings display no strict coordination among individuals. To distinguish them from schools, Pitcher (1983) proposes that such fish aggregations be termed 'shoals'. As with the flocking of birds, there are no implications for structure and function in fish shoals. By this heuristic definition, Pitcher (1983) argues that schooling is a special case of shoaling, but with strict criteria for synchronised and polarised motion.

Older literature often made no distinction between the behaviour patterns of fish schooling or shoaling in layers and aggregations (Radakov 1973; Shaw 1970), and according to Pitcher (1986) the term 'school' is still used in North America to describe both these behavioural patterns.

The long debate on how to define and distinguish schooling behaviour is a probable reason for the semantic confusion over how to describe the aggregative behaviour of fish. In addition, when observing fish at sea with low-resolution acoustic equipment, the degree of coordination in the movements within the fish recordings is nearly impossible to detect. In such situations, the distinction of the fish aggregative behaviour is made by considering the density and extent of the recordings. Experienced users of acoustic fish-detection equipment often term recordings of fish shoals of rather low density as 'slør' or layers which are easily differentiated from recordings of fish schools. When schools of large Norwegian spring-spawning herring disperse into shoals at night, the fish density in the recordings may drop by two orders of magnitude from an average of about 1 fish m^{-3} to 0.02 fish m^{-3} (Beltestad & Misund 1990). In other situations, for instance on the spawning grounds or when herring are hibernating in narrow fjords, there may be no large diurnal change in the extent and fish density of the aggregations, but the diurnal vertical migration may be substantial. Toresen (1991) measured densities of about 1 fish m^{-3}, even at night, of large herring in a fjord in northern Norway. Despite the high fish density, the degree of coordination among the individuals in such aggregations at night is not known. Similarly, the aggregative behaviour of *Sardinella aurita* in the Gulf of Cariaco, Venezuela, varied temporally and spatially, and dense schools were even detected at night. However, using criteria of fish density, it was still possible to group the recordings as schools, shoals (concentrations) and dispersed fish (Freon *et al.*, 1989). Shoal-to-school transitions in behaviour are absent for some species, and the sand lance (*Ammodytes* sp.) takes refuge in the sand when not schooling (Kuhlman and Karst 1967; Meyer *et al.*, 1979).

4.6 Functions of schooling

Schooling is regarded as an efficient way of conducting underwater movements, and is beneficial to the individual participants. This is achieved mainly through a higher probability of surviving predators, more effective feeding, hydrodynamic advantages and more precise migration when schooling than when solitary (Pitcher, 1986).

4.6.1 *Surviving predatory attack*

It is suggested that aggregating in groups decreases the probability of predation because of reduced probability of being detected (Olson, 1964; Breder, 1967; Cushing & Harden Jones, 1968). However, an advantage to a group member is present only if the consumption rate of a predator is less when feeding on groups, and not because of reduced probability of detection (Pitcher, 1986). Besides, many schooling fishes live in rather close proximity to their predators (Parrish, 1992b; Pitcher *et al.*, 1996). In addition, pelagic fish schools often occur close to the surface, and in such situations are more easily detected by predatory sea birds than are solitary fish.

Bertram (1978) and Foster and Treherne (1981) argue that the probability of being the victim during a predatory attack declines inversely with group size through a numerical dilution effect. Morgan and Godin (1985) show that such an effect operates in a laboratory study of white perch (*Morone americana*) predation on schooling killifish (*Fundulus diaphanus*). The perch attack rate per killifish decreased with school size by a slope no different than –1 on a double logarithmic plot, as predicted by the dilution effect (Fig. 4.1).

Because the dilution effect applies only to members of a group being attacked, Pitcher (1986) argues that simple dilution will not promote the evolution of grouping. If the attack rate of the predator increases in proportion to the size of the schools, the probability of being the victim is the same both for solitary and for schooling fish, and the dilution effect will be a fallacy. However, Morgan and Godin (1985) observed that the perch attack rate was independent of the killifish school size.

Pitcher (1986) and Turner and Pitcher (1986) suggest that the 'attack abatement' effect, in which predator search and attack dilution interact, can favour the evolution of grouping. In this case, a solitary fish will benefit by joining a group because even if the probability of detection is the same, the probability of being the victim of an attack decreases from 1.0 to an inverse proportion of the number of fish in the school.

An advantage of schooling for surviving attacks by predators is an earlier detection of approaching predators through the many eyes watching, and a repertoire of cooperative escape tactics (Godin, 1986; Pitcher, 1986). A stalking predator is often out-manoeuvred by the fountain effect, in which the schooling individuals maximise their speed relative to the attacker, avoid sideways, and reassemble behind it. When avoiding attacks from faster predators, the Trafalgar effect (Treherne & Foster, 1981) ensures that a flight reaction is transmitted through the school at a faster speed than the approach of the predator. The flight transmission speed through killifish schools was about twice the approach speed of a chasing perch model (Godin & Morgan, 1985).

Many predators can attack at a speed that is much faster than the burst speed of schooling prey. Still, such predators suffer a much lower capture success for schooling than for dispersed or solitary prey (Neill & Cullen 1974; Major 1978). A prey school will be a mass of moving targets within the visual field of the predator, which may thereby suffer a confusion effect arising from sensory channel overload or cognitive confusion (Pitcher, 1986). The predator may therefore hesitate and give the schooling prey more time to avoid, or it may perform imprecise attacks. To enhance the confusion effect, schools are often observed to perform a flash expansion in which the individuals respond to a striking predator by fleeing in all directions and rapidly reassembling after swimming 10 to 15 body lengths (Pitcher, 1979a; Pitcher & Wyche,

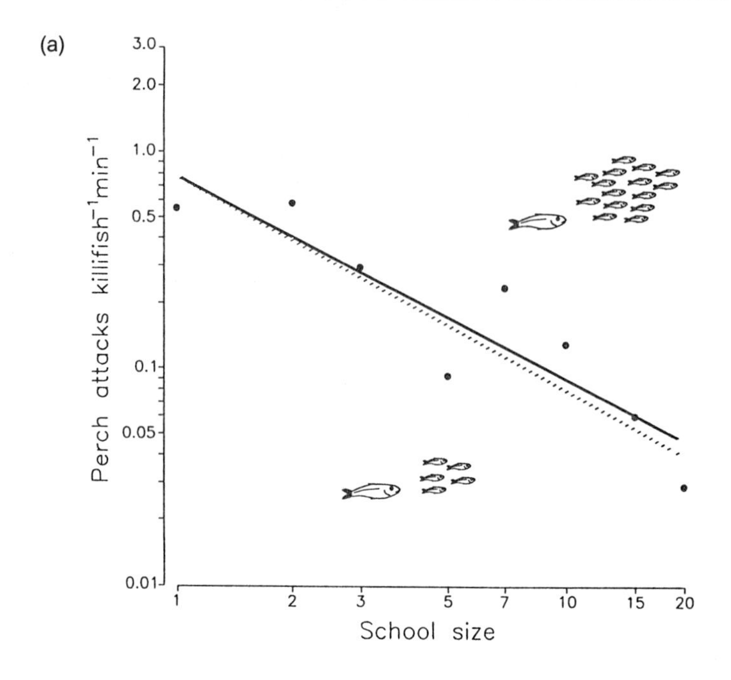

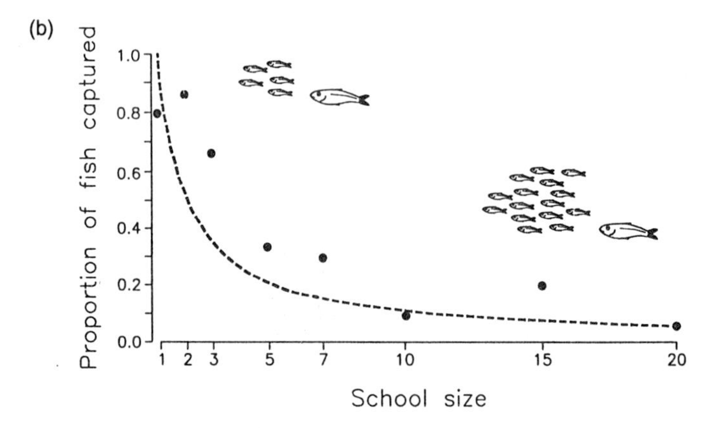

Fig. 4.1 The dilution effect. (a) Relationship between the rate of perch attacks per individual killifish and killifish school size (broken line = observed; solid line = predicted). (b) Proportion of killifish captured by perch from different school sizes (after Morgan & Godin, 1985).

1983). For a similar reason, European minnow (*Phoxinus phoxinus*) under threat may perform an individual skittering behaviour (Magurran & Pitcher 1987), which consists of rapid startle acceleration and subsequent rise in the water column. By performing this behaviour, an individual will have higher benefit than its schooling companions by confusing a predator that may have chosen that individual as a target.

Similar tendencies to gain higher individual benefits may be found when schoolers try to take refuge by packing in a dense ball. This behaviour may represent an example of Hamilton's (1971) selfish herd principle, in which individuals try to avoid the dangerous margins by seeking shelter in the central part of the group. However, when attacked by a black sea bass (*Centropristis striata*), Atlantic silversides (*Menidia menidia*) in central positions of the school suffered a higher attack rate than those in peripheral positions (Parrish, 1989a).

4.6.2 *Effective feeding*

When coming across patches of food, schools may break up into more loosely organised shoals because of individual activities needed to search for, capture and handle prey items. The distinction between schooling and shoaling during feeding may often be rather fluid, however. Fréon *et al.* (1992) observed that some parts of *Harengula* schools became disorganised to feed for short periods while the rest of the school continued its polarised and synchronised swimming. Anchovy maintain organised schooling during filter feeding on planktonic organisms (Koslow, 1981). Piscivorous fishes also often hunt and attack in schools.

A higher aggregated search rate results in faster food localisation for fish in larger groups (Pitcher *et al.*, 1982), and feeding sites are more effectively sampled (Pitcher & Magurran, 1983). Similarly, enhanced vigilance (Magurran *et al.*, 1985) and reduced timidity in larger groups enables the individuals to allocate more time to feeding, even when predators are present (Magurran & Pitcher, 1983; Morgan & Colgan, 1987).

Predator grouping may suppress the confusion effect when feeding on swarming prey; this happens through a reduced reliance on vigilance (Smith & Warburton, 1992). Similarly, when attacking schooling prey, schooling predators are more effective than solitary ones (Major, 1978). Probably schooling predators overcome the confusion effect by a better ability to split prey schools, creating stragglers that are easier to capture.

4.6.3 *Hydrodynamic advantages*

If schooling fish are swimming in the vortices left by the tailbeats of fish in front, they may gain hydrodynamic advantage (Weihs, 1973; 1975). This implies adopting a certain structure swimming in a diamond lattice, 0.4 body lengths away from lateral neighbours and 5 body lengths behind those in front. Such swimming was not observed when recording the structure of herring, saithe and cod schools cruising in an annular tank (Pitcher & Partridge, 1979).

Schooling fishes may also obtain hydrodynamic advantage through an improved thrust efficiency by swimming alongside each other, preferably at a distance of 0.3 body lengths. Tendencies to swim with neighbours of similar size in herring and mackerel schools (Pitcher *et al.*, 1982, 1985) and the alongside positioning of individuals within two-dimensional giant bluefin tuna schools (Partridge *et al.*, 1983) are indications that the lateral neighbour effect may operate when schooling. Abrahams and Colgan (1985) observed that shiner (*Notropis heterodon*) adopted a two-dimensional, hydrodynamically efficient school structure when swimming in an environment without predators, but reorganised to a three-dimensional structure that enhanced vigilance when a predator was present.

Breder (1976) argues that mucus, being continuously washed from fish in the front part of a school, will reduce the drag experienced by those following behind. To test this hypothesis, Breder (1976) added non-toxic polyox, an ethylene polymer of high molecular weight, to the water of one aquarium tank, but not to that of a control tank. He then observed menhaden (*Brevoortia patronus*) swimming in the two tanks, and found that fish in the treated water swam with twice the tail beat frequency and proportionally faster than those in the control. However, the experiment does not prove that the higher tail beat frequency was the result of the fish experiencing a lower drag. Parrish and Kroen (1988) measured the amount of mucus sloughed off individual silversides (*Menidia menidia*). Then they used polyox at different concentrations in the sea water to simulate the drag reduction due to the combined effect of solubilised mucus sloughed off a school of 10 000 silversides. The authors did not notice any changes in the tail beat frequency, even when the polyox concentration was two orders of magnitude higher than necessary to simulate the mucus effect.

4.6.4 *Migration*

Because the mean direction of a migrating group is likely to be more precise than each individual's choice (Larkin & Walton 1969), schools may also migrate more accurately. This effect will favour the formation of large schools, as are often observed when pelagic species like herring and capelin undertake long spawning migrations.

At the onset of migrations, there may be substantial individual differences in the underlying motivation within populations. For instance, the hunger level creating a motivation for feeding that initiates feeding migrations may vary substantially among conspecifics within an area. In such cases it is probable that schooling behaviour acts as a filtering mechanism so that only individuals having the same level of motivation participate in the migrating schools, while others continue doing something else. This filtering may be observed at the onset of the spawning migration of capelin in the Barents Sea. Both mature and immature fish are present in the same areas in the north-eastern Barents Sea, where they were feeding during the previous summer and autumn. In late autumn/early winter, most fish are distributed in layers that may extend over large areas. Within such layers, the maturing fish start organising dense schools that begin the long migration towards the coast of northern Norway for spawning. In this process, the schooling is a behavioural mechanism that separates the spawners from the immature population.

4.6.5 *Reproduction*

Schooling may be of adaptive significance for reproduction as a mechanism that facilitates mate-finding. This may be valid for pelagic fishes that often migrate over large distances and spend a relatively short time at the spawning ground. The spawning behaviour of herring seems to be of the promiscuous type where a large number of fish aggregate at the spawning ground and spawn simultaneously (Turner, 1986). Aneer *et al.* (1983) were not able to identify any pairing or courtship behaviour when observing a school of spawning Baltic herring, and describe the behaviour as 'a mass orgy of indiscriminate spawning with no apparent coordination between the two sexes'. However, many individuals were observed to spawn when swimming in parallel or alongside each other in mills, and it is possible that such behaviour represents pairwise spawning. Such spawning in schools may reduce the probability of predation, and during littoral spawning, as in the Baltic, herring may be exposed both to piscivorous fishes and to avian predators (Aneer *et al.*, 1983).

For many pelagic species that have the ability to school, the process of reproduction involves an element of competition among males. This may induce agonistic behaviour among the males when trying to find mates, when dismissing competitors that try to interfere with courtship behaviour, and when defending the spawning location. When such behaviour patterns start occurring, schools may rapidly disperse. During the spawning migration, schools of capelin tend to sort by sex. The males dominate in schools that first enter the spawning grounds near shore. There the schools more or less disperse as the males establish 'leks' (Turner 1986) where they start defending small territories and try to attract, court and mate with arriving females (Sætre & Gjøsæther, 1975). In a large tank in the Marineland Aquarium in Hawaii, Magnusson and Prescott (1966) observed that during spawning, Pacific bonito (*Sarda chiliensis*) formed pairs and conducted courtship behaviour within schools. The pair swim at a slightly faster speed than the rest of the school, with the female leading in a wobbling path and the male close behind. This courtship behaviour may end in circular swimming, temporarily separating the pair from the school, and during which the gametes may be released. The courtship behaviour ends if the pair is intercepted by one or several other males, and the pairing male tries to dismiss intruders by lateral threat display. If pair formation is maintained after the disturbance, courtship behaviour may be resumed. It is possible that such pair formation and courtship behaviour may lead to a gradual dispersal of schooling *in situ*. Mackerel seems not to organise into the usual large schools during the spawning season in the North Sea (Misund & Aglen, 1992).

4.6.6 *Learning*

Schooling may enhance social facilitation (Shaw, 1970). Carp trained to escape through a moving net were more successful when in a group than as isolates (Hunter & Wisby, 1964).

Soria *et al.* (1993) conditioned schooling thread herring (*Opisthonema oglinum*), by emitting a 500 Hz sound stimulus to which the fish were previously habituated during the hoisting of a net from the bottom of a tank. When mixed with naive fish, the conditioned fish seemed to transmit their experience throughout the inexperienced school by leading conditioned responses when exposed to the sound stimulus without the net being hoisted.

4.7 School size

When the school size increases, the advantage of grouping during feeding may be reduced due to increased competition for food among school members. Ranta and

Kaitala (1991) found that the number of strikes made by sticklebacks feeding on benthic prey increased with school size up to about 10 individuals and then levelled off. The advantage of schooling for avoiding predation probably increases asymptotically with school size. Pitcher (1986) therefore discusses the possibility of the existence of an optimal school size that has a maximum benefit/cost ratio. From a simple energy-balance model, Duffy and Wissel (1988) suggested that lower limits to school size are unlikely to be set by food but rather by predation, while upper limits depend on both food availability and school behaviour. When quantifying school size of herring, sprat and saithe in different geographic regions and seasons in the North Sea and in Norwegian fjords, Misund (1993b) observed a variation in school size by four orders of magnitude from about 100 to several million individuals in the same school (Fig. 4.2). In a majority of the different regions and seasons, the mode in the

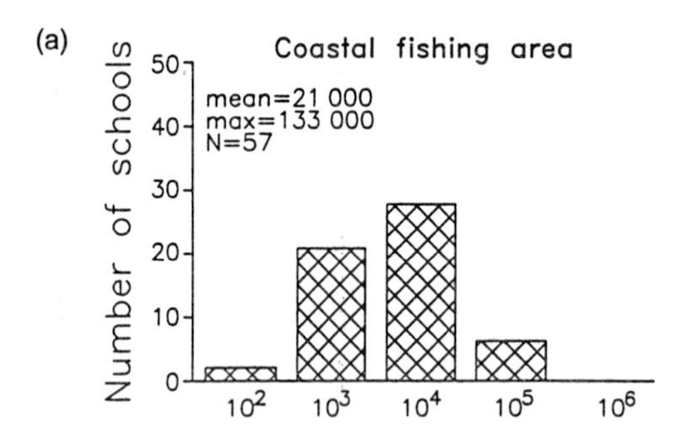

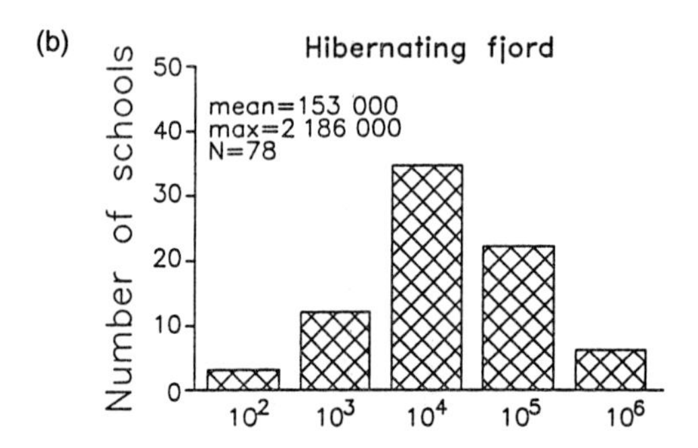

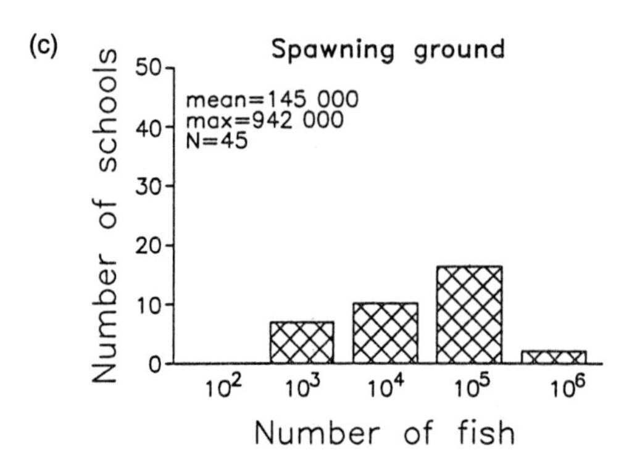

Fig. 4.2 School size distribution of Norwegian spring spawning herring (average length 33 cm) recorded by echo integration and sonar (a) at a coastal feeding area in Lofoten, northern Norway, in September, (b) in a hibernating fjord, northern Norway, in October, and (c) at the spawning ground off western Norway, in February (after Misund, 1993b).

school size distributions was in the order of 10 000 individuals. Variation in school size among regions and seasons probably reflects the fact that school size is influenced by the conflicting demands of avoiding predators and conducting feeding. In this context, it must be noted that the smallest herring schools were observed in the region where the predation pressure was most obvious, as schooling saithe were frequently observed to attack and split the herring schools.

In theory, the net benefits of schooling may be zero or even disadvantageous to participating individuals if the cost due to intraspecific competition increases with school size. However, on certain occasions such costs may be absent, as for instance when herring more or less stop feeding during hibernation and spawning migrations (Devold, 1969). In such situations, there will be a net advantage in joining schools of countless numbers. Rather than being examples of the 'tragedy of the common principle', huge schools such as those frequently observed in Norwegian fjords in autumn (Fig. 4.2) illustrate that on certain occasions the advantages of forming a school can still pay, even to millions of individuals.

Dramatic shifts in the motivational status of schooling fishes may be reflected through substantial changes in school size. Nøttestad *et al.* (1996) observed that the average size of herring schools at a coastal spawning site off south-western Norway varied by a factor of at least four. The herring entered the spawning site in large and dense immigrating schools that split into smaller but denser searching schools (Fig. 4.3). Spawning took place in a dense spawning layer from which small, short-lasting spawning schools erupted. After spawning, the herring aggregated in loose and dynamic feeding schools and left the spawning site in larger and rather unidirectional emigration schools.

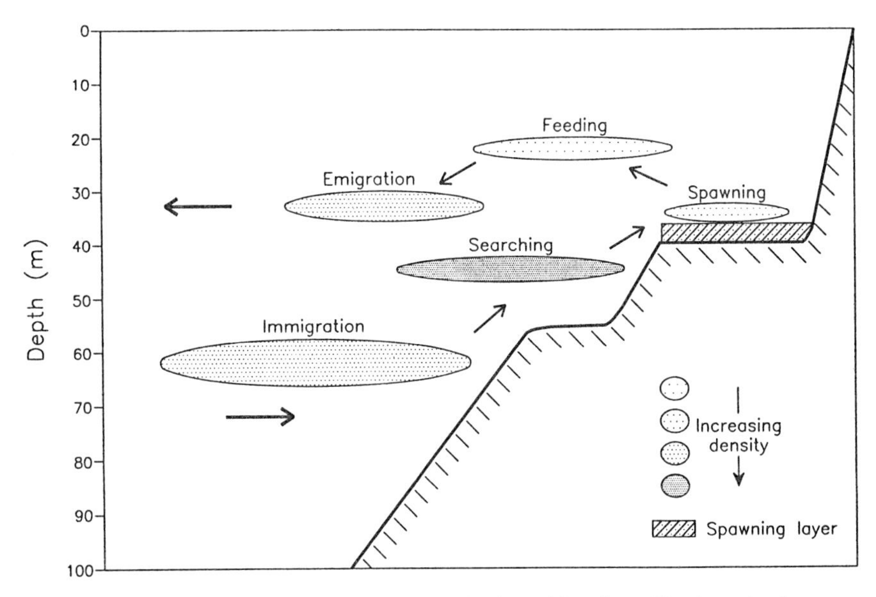

Fig. 4.3 Relative distribution of school size, density, depth, and location of herring schools at a coastal spawning site off southern Norway (after Nøttestad *et al.*, 1996).

The size of schools is not stable over time. Helfman (1984) observed that individuals in yellow perch schools frequently joined and left. Pitcher *et al.* (1996) recorded that herring schools on feeding migrations in the Norwegian Sea split and joined on average every 70 minutes. Hilborn (1991) estimated that 16–63% of individuals in schools of skipjack tuna (*Katsuwonus pelamis*) left the schools each day to join other schools. Based on tagging data, Bayliff (1988) estimated that individuals in a school of skipjack tuna became randomly mixed into other schools within 3–5 months. Anderson (1980) proposed a stochastic model for simulating the frequency distribution of fish school diameter observed acoustically. In this model the entrance rate of fish into a school is independent of the number of fish in the school, while the exit rates is proportional to this number. Wide distribution of school size is favoured by large entrance rates and a large amount of randomness to the schooling process. In contrast narrow distribution is simulated by large exit rates and low randomness. In conclusion to this section, further studies are required to better understand the distribution of school size of marine fish populations and its dynamics.

4.8 School organisation

Individuals in schools tend to take up positions that maximise the flow of information about the swimming movements of their neighbours (Partridge *et al.*, 1980). This results in a functional school structure which enables the individuals to perform complicated manoeuvres synchronously and in polarisation. In this section, the behavioural rules that fish apply when organising schools will be considered. The structural patterns that arise when fish swim in schools will be described through the parameters used to quantify school structure. Similarly, the variability of the school structure as well as the factors that affect it will be outlined. In addition, factors that tend to stabilise the school organisation are discussed. To describe how fish communicate when schooling, the senses involved are also considered. Vision and lateral line perception are recognised as the main sensory systems to maintain schooling, but the possibility that communication within schools can rely on hearing will also be discussed.

Measurement of fish school structure demanded development of special methodology. How knowledge of school organisation has been collected is therefore outlined here through a review of the development of methods applied for quantifying school structure. As most of this knowledge has emerged from studies conducted in laboratories, the focus here is on visual methods to quantify school structure in small tanks.

4.8.1 *Study methods*

Attempts to quantify the internal structure of fish schools were first conducted by Breder (1954), who measured the two-dimensional structure from photographs taken above a school cruising in a small tank. Keenleyside (1955) tried to take account of the three-dimensional school structure by estimating density within a small school visually. By using stereophotography or the shadow of each individual as a reference, Cullen *et al.* (1965) were able to find the coordinates of each individual of a small school of pilchard in the three-dimensional space. The shadow method was found to be most appropriate for such an investigation, the main argument against stereo-photography being that the accuracy was reduced too much for the necessary camera-to-object distance.

Measurements of internal school structure easily generate a substantial amount of data, and the results of the pilchard study were therefore based on coordinate analysis from just 11 photographs of the school. Hunter (1966) developed a procedure for quantification of two-dimensional school structure based on coordinate readings on motion picture frames, automatic digitising and recording, and computer calculation of the mean separation distance, mean nearest neighbour distance, and mean angular deviation for each frame.

Pitcher (1973) used the shadow method sideways to quantify the three-dimensional structure of schooling minnows held in a small tank. The school was photographed from the side against a mirror, and a calibration grid in front enabled the coordinates of each individual to be accurately calculated.

The above-mentioned methods were based on stationary cameras and small aquaria, a few metres wide only. This allowed testing of very small schools – fewer than 30 individuals could be used – and the study of Hunter (1966) was based on six jack mackerels only. Even if three or more fish swimming together in a polarised and synchronised pattern can be defined as a school (Partridge, 1982a), the degree of school organisation increases with the number of schooling companions (Partridge, 1980). A school of three minnows has a substantial element of leader-follower relations, while a school of six minnows has a more complicated organisation. Also, the small tanks put substantial constraints on the swimming movements, and behavioural changes induced by the tank edges can influence the school structure. Hunter (1966) observed an increase in mean angular deviation each time the jack mackerel school turned.

Partridge *et al.* (1980) used a 10 m annular gantry tank where fish could swim endlessly around without having to orientate towards edges. Schools of 20–30 cod, saithe and herring were conditioned to follow a red spotlit area on the tank floor projected at an angle from a rotating gantry (Fig. 4.4). The schools were filmed from the gantry while moving, and the conditioning gave the opportunity to regulate the swimming speed of the schools. The three-dimensional coordinates of each individual were calculated very accurately ($\pm$ 0.25 cm) using the shadow method. The data set collected was considerable; about 20 000 film frames were analysed and more than 1.2 million individual fish coordinates were calculated.

Underwater acoustics can be used to study internal structure and school shape, especially by using the smallest resolution of the sounder (e.g. one pulse × 0.5 m). The first step is to identify a school and many sounders now have an internal processor which does this automatically according to given thresholds of geometry, reflected energy and spatial continuity between echoes (e.g. Weill *et al.*, 1993). At this scale it is possible to obtain with a reasonable precision geometrical descriptors of the school

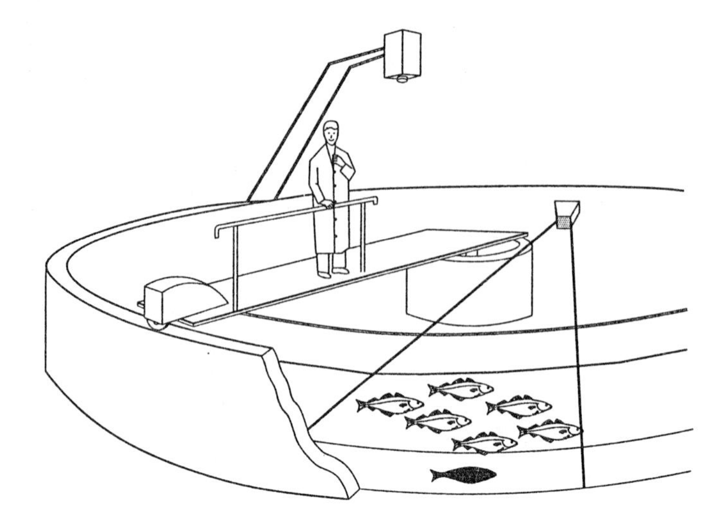

Fig. 4.4 Quantification of three-dimensional school structure using the shadow method in a circular tank with a rotating gantry (after Partridge, 1982b).

(height, width, perimeter, surface), estimates of the biomass, density and their variability, and the position of the school in the water column (distance to the bottom and to the surface). Another approach is to use image analysis algorithms which consider the smallest resolution as pixel (Reid & Simmonds, 1993).

Finally, an approach to evaluate hypotheses on school organisation is to perform simulation by object-oriented models. The interest of such an IBM model is to test if a minimum number of simple individual fish behavioural rules are able to generate the emergence of a realistic school behaviour. If the simulation is not able to reproduce the fish behaviour, it means that the proposed behaviour is not acceptable (or at least not alone). Nevertheless the emergence of a realistic school behaviour is not a sufficient condition to accept a model since similar results can be obtained with different behaviour rules and parameters. Pioneer work in this field has been carried out by Aoki (1982) and Huth and Wissel (1994).

4.8.2 *Minimum approach distance*

A fundamental behavioural rule for organising schools is that each individual maintains an empty space around itself. This became evident when the structure of a saithe school was compared with that of points generated at random with the same density and within a similar structure as the real school (Partridge *et al.* 1980). When plotting the cumulative frequency of neighbours at different distances related to the cube of the distance, it appears that the neighbours in the real school are distributed farther away and in a narrower interval than expected at random (Fig. 4.5). This indicates that each individual maintains a certain minimum approach distance. Consequently, if individuals come too close they quickly orientate away from each other. This has been observed in a school of jack mackerel in that the mean separation

(a)

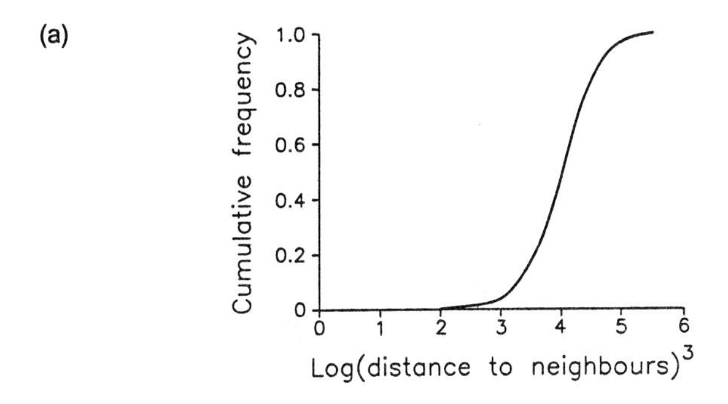

(b)

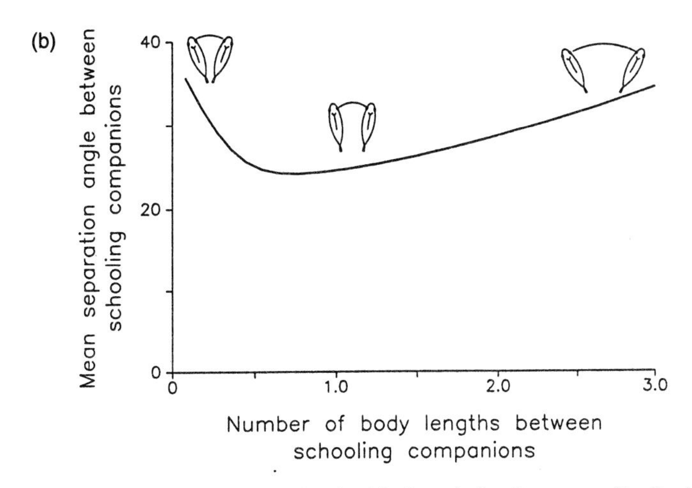

Fig. 4.5 Minimum approach distance in schools. (a) Cumulative frequency distribution of nearest neighbour distances in saithe school (after Partridge *et al.*, 1980); (b) mean separation angle between individuals in jack mackerel school (after Van Olst & Hunter, 1970).

angle between neighbouring individuals increased drastically if they came closer than 0.2 body lengths (Van Olst & Hunter 1970, Fig. 4.5).

Each species seems to have a specific minimum approach distance. In saithe schools, the individuals obey a minimum approach distance of about 0.3 body lengths (Partridge *et al.* 1980), while schooling pilchard have been observed as close to each other as 0.1 body length (Cullen *et al.*, 1965).

4.8.3 *Nearest neighbour distance*

The minimum approach distance rule results in the seemingly regular distance among schooling individuals. Typically, the average nearest neighbour distance is below one body length, but with a substantial variation and differences among species. Pilchard school at about 0.6 body lengths from their neighbours (Cullen *et al.*, 1965). The facultatively schooling minnow organises somewhat looser schools, as the nearest neighbours occur about 0.9 body lengths (SD = 0.4 body lengths) apart (Pitcher,

1973; Partridge, 1980). Cod will school at a smaller distance from their nearest neighbour than saithe, while in herring schools there is a greater distance among nearest companions (Partridge *et al.* 1980).

A cause of the substantial variation in nearest neighbour distance is the relative positions in space (Partridge *et al.*, 1980). Neighbours at the same level are generally farther apart than those above or below (Fig. 4.6). Herring swimming directly alongside are farther away than those in front and behind, a pattern also found for mackerel, jack mackerel, northern anchovy and silversides (Van Olst & Hunter, 1970). The gadoids, however, have their neighbours closer directly alongside than in front and behind (Fig. 4.6).

4.8.4 *Spatial distribution*

Schooling fishes tend to organise in certain spatial patterns at preferred positions relative to each other. Pilchard appears to school at diagonal positions relative to the nearest neighbour in both the horizontal and vertical planes (Cullen *et al.*,

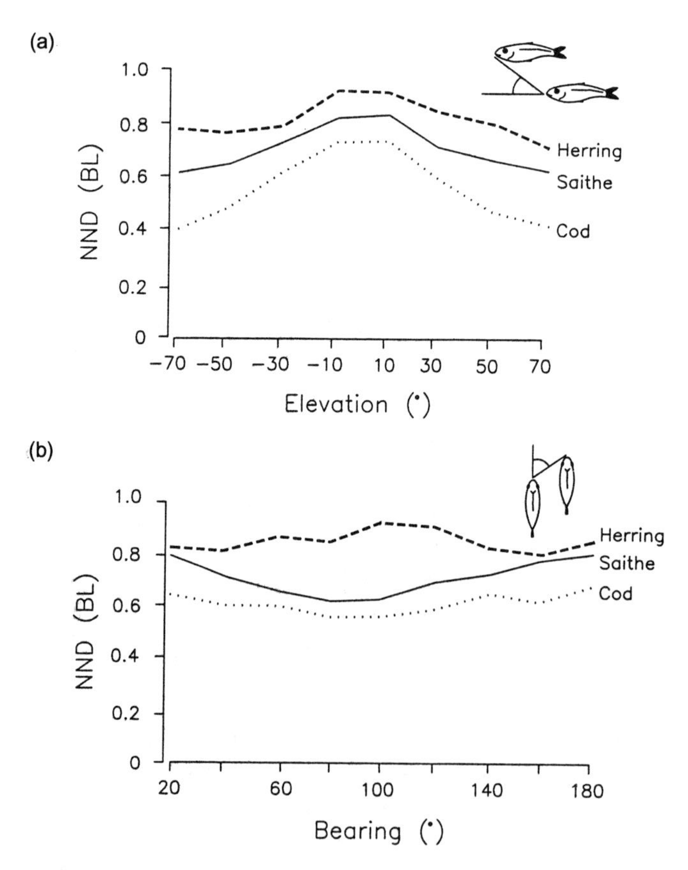

Fig. 4.6 Nearest neighbour distance (NND, in body lengths) as a function of (a) elevation and (b) bearing for cod, herring and saithe (after Partridge *et al.*, 1980).

1965, Fig. 4.7). Similarly, species like mackerel, jack mackerel and topsmelt have been observed to prefer nearest neighbours in diagonal positions horizontally, while schooling anchovy had their neighbours more directly to the front and rear (Van Olst & Hunter 1970). Minnows also tend to clump in preferred positions relative to the nearest neighbour, both horizontally (15°, 45°, 75°, 135° and 170°) and vertically (–45°,15°) (Pitcher, 1973). It is claimed that optimal packing (maximum number of equidistant neighbours) will be obtained in a tetradecahedral pattern (Pitcher, 1973). The schooling minnows tend to pack in a suboptimal tetrahedral pattern, while the diagonal positioning of pilchard resembles cubic lattice packing.

The idea that schooling individuals pack in a certain complicated geometric structure is appealing. Breder (1976) argued that the general school structure could not be regular cubic lattices because the individuals would then appear in regimented rows and colums, a structure 'so striking that the details of the regimentation would have been recorded long ago'. He argued that fish schools resemble more a rhombic lattice, which would give a denser packing because the number of equidistant neighbours increases from six, as in a cubic lattice, to twelve. If fishes in a school were

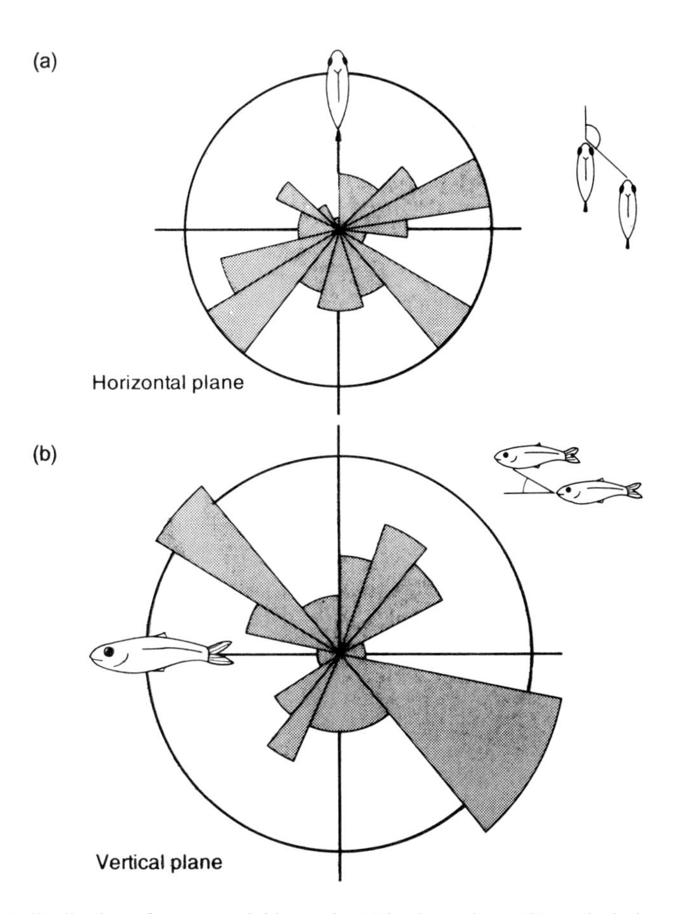

Fig. 4.7 Spatial distribution of nearest neighbours in (a) horizontal and (b) vertical planes in a pilchard school (after Cullen *et al.*, 1965).

packed in such a pattern, they would be swimming in each others' vortex trails at points which, according to the theoretical analysis of Weihs (1973), would give them a hydrodynamic benefit. Also, considerations and observations of school turning support packing in a rhombic structure. Schools seem to turn in clear sectors if their members are organised in a rhombic lattice, while if they are organised in a cubic lattice the turns of the school will occasionally be in sectors that cause individuals to collide.

If fish schools were organised in a regular lattice, the neighbours would have to be distributed at a uniform distance. However, the proportions between the first and second nearest neighbour in schools of herring, saithe and cod deviate from the one-to-one relationship found in regular lattices (Partridge *et al.* 1980). Fish schools consequently do not seem to be organised according to a regular geometric pattern, but there are still tendencies for the individuals to prefer specific positions relative to each other. Also, each species organises schools that have a characteristic spatial structure. Saithe and herring tend to school at different vertical positions from their neighbours, while this is not the case for cod (Partridge *et al.* 1980, Fig. 4.8). The

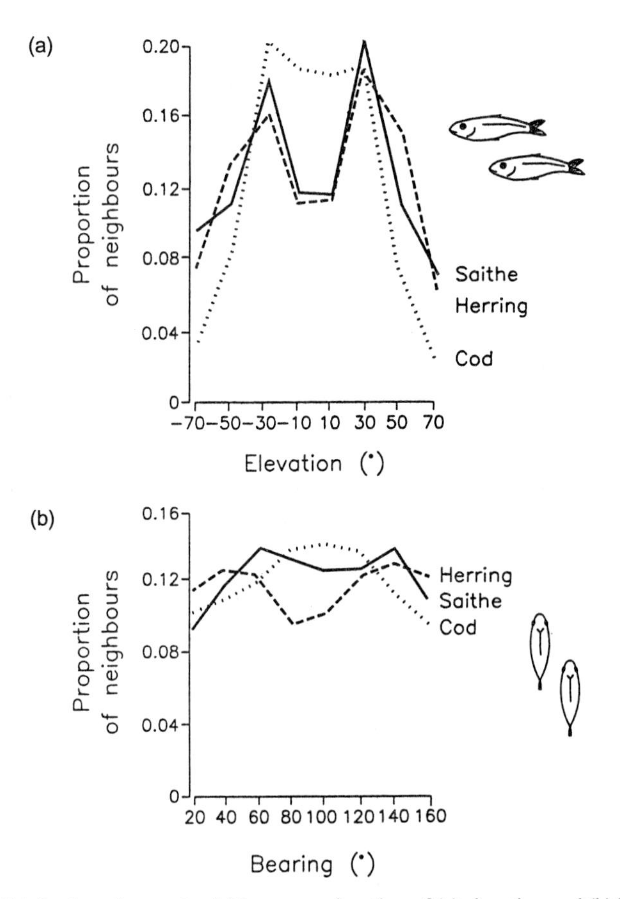

Fig. 4.8 Spatial distribution of nearest neighbours as a function of (a) elevation and (b) bearing in schools of cod, herring and saithe (after Partridge *et al.*, 1980).

gadoids are more likely to have their neighbours directly alongside, while herring tend to prefer neighbours at 45° and 135° (Fig. 4.8). Partridge *et al.* (1980) therefore concluded that individuals in schools position themselves so that they can most quickly respond to their neighbours, a statement that so far has ended the search for an undiscovered geometric packing structure of fish schools. They also argue that the spatial school structure reflects the sensory capabilities and manoeuvrability of each species.

The degree of structure in the school organisation is species dependent. The nearest neighbour proportions show that herring schools are more structured than those of saithe and cod. Partridge *et al.* (1980) claim that the structure mirrors the amount of time the fish spend in schools because the obligate schooler herring arrange more organised schools than facultative schoolers like the gadoids. The cod school structure is close to that expected at random, however, which is in agreement with the impression of cod as a weakly facultative schooler.

4.8.5 *Internal synchrony*

The position of individuals in schools is not fixed even if there are tendencies to a certain spatial structure within the school based on preferred angles and distances to the schooling companions. The individual swimming speed varies considerably in a school moving at constant speed, and the individual positions within the school change substantially (Partridge, 1981).

The synchrony of the moving school is maintained, however, as individuals match changes in swimming speed of their neighbours with a time lag of about 0.2 s (Hunter, 1969). The time lag is longer for neighbours changing speed in front and rear than for those directly alongside (Fig. 4.9). This is because the response to a neighbour changing velocity depends on perception of relative motion. The number of degrees through which the neighbour changes position will be greater and easier to detect alongside than in front and rear. In the narrow binocular field directly in front, the perception of changes in relative motion is increased, and consequently the lag is reduced. The result of the relative motion perception is that the fish shows maximum correlation to the movements of neighbours directly alongside and in front (Partridge, 1981). In response to fright stimuli, sticklebacks (*Gasterosteus aculeatus*) minimise the approach time to a conspecific according to body orientation and initial distance (Krause & Tegeder, 1994). This means that the stickleback is able to take into account the time necessary for turning and for swimming toward the conspecific which is faster to join.

The individual fish matches the swimming speed to its nearest neighbour better than to the second and third nearest neighbour (Partridge, 1981). However, a certain correlation to the speed of the second nearest neighbour at short lags indicates that the individual fish does orientate to more than one neighbour. Similarly, the correlations of heading to neighbours in a school are not very high, and no better to the nearest neighbour than to the second or third nearest neighbour (Partridge, 1981). This implies that individuals adjust their heading not only according to that of the nearest neighbour, but also on the basis of headings of a number of surrounding fish.

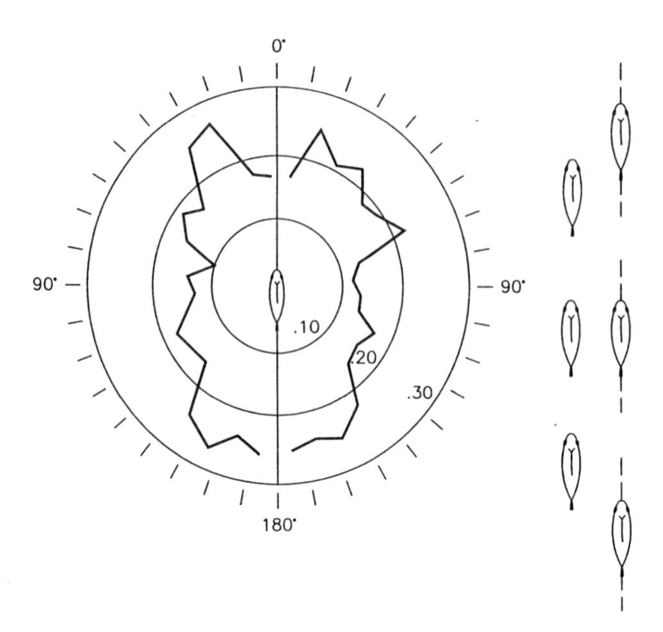

Fig. 4.9 Reaction latency of schooling jack mackerel to swimming behaviour change of nearest neighbour related to bearing. Full circles represent 0.1, 0.2 and 0.3 s latency, respectively (after Hunter, 1969).

These findings were used in simulation by object-oriented models by Huth and Wissel (1993, 1994) to test the 'front priority rule'. This rule means that only the neighbours in front of a fish are recognised by this fish. They averaged the influence of a maximum of four front neighbours on the swimming behaviour of a given fish (repulsion at short distance, parallel orientation at intermediate distance, attraction at long distance). The authors compared successfully the results of their simulations (frequency of nearest neighbour distance, index of polarisation, time spent by a fish at the top of the school) to experimental data found in the literature. They showed that a leader fish is unnecessary for simulation of realistic school behaviour (merging of two schools, splitting, exploitation of a food path, higher efficiency of the school to migration according to a gradient of better abiotic condition compared to a single fish). Nevertheless, the Huth and Wissel model fails to simulate the formation process of schools, as underlined by Reuter and Breckling (1994) who proposed an alternative IBM in which a fish is influenced by all visible neighbours. Huth and Wissel (1994) did not agree with this model and instead suggested a modification to their own model by incorporating a dynamic range of sight which would allow a longer range of sight for fish not surrounded by neighbours. This controversy illustrates the need for a combination of field observations, laboratory experiments (e.g. distance of perception by vision and lateral line) and simulations in this branch of science as in many others.

Formation of subgroups can also affect school synchrony. Subgroups can be quite stable over time, and members of one subgroup show high correlations to members of the same group, but low or negative correlations to members of other groups (Partridge, 1981). For the school to exist as a unit, the movements of adjacent subgroups must be linked even if they are not quite synchronous (Pitcher, 1973).

4.8.6 *Individual preferences and differences*

There are individual preferences for positions within a school. These may induce size segregation within schools because of each individual's preference to swim among neighbours of similar size, as observed for herring and mackerel when schooling in a small net cage (Pitcher *et al.*, 1985; Fig. 4.10). The size preference may indicate a functional aspect of schooling in that individuals swimming beside neighbours of similar size may gain a hydrodynamic advantage (Weihs, 1973). Individual position preferences that are not size dependent may also exist within schools. This was observed in a saithe school in which individuals varied in length by a factor of 2.5 (Partridge, 1981). Magurran *et al.* (1994) observed schooling preferences for familiar fish in small schools (12–15 fish) of guppy (*Poecilia reticulata*). The existence of such a preference in schools of several thousand individuals, a common size for marine pelagic fish, is questionable.

Individual differences in ability to school are also quite plausible. Herring diseased by the fungus *Ichthyophonus hoferi* are not able to participate in normal schools and are less able to avoid trawls (Kvalsvik & Skagen, 1995). During predator-evasion manoeuvres, individuals frequently become isolated from the rest of the school. To

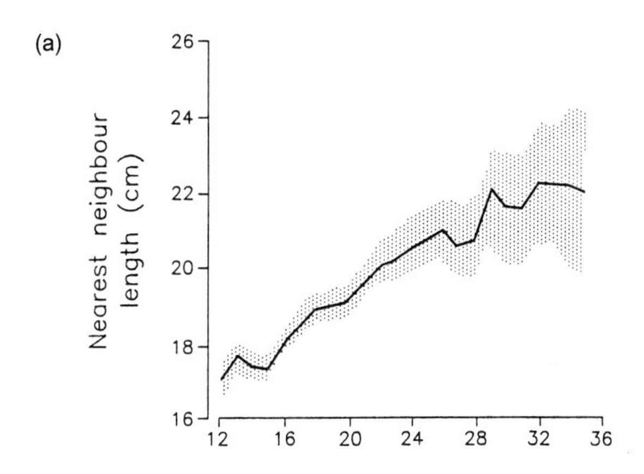

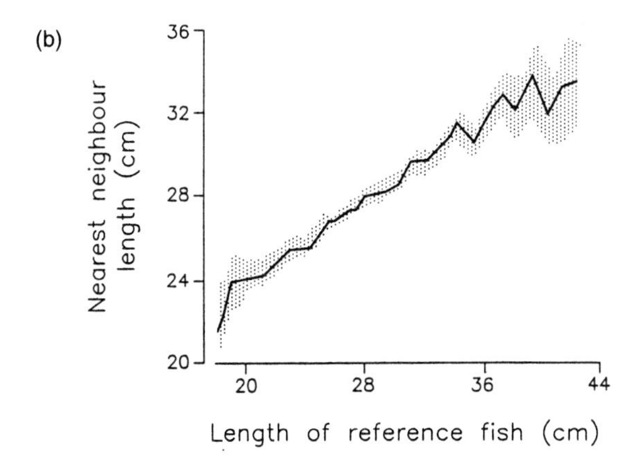

Fig. 4.10 Distribution of nearest neighbour size in (a) herring and (b) mackerel schools swimming in net cages submerged in a Scottish sea loch (after Pitcher *et al.*, 1985).

keep within the functional school is fundamental as isolates or stragglers are more easily predated (Major, 1978). Within the functional school, however, peripheral individuals may in fact be safer than those sited more centrally if the predator performs striking attacks (Parrish, 1989a). Individual differences in schooling ability among those that keep within the functional school are therefore probably not easy to detect but can be simulated by object-oriented models (Romey, 1996). Of five saithe examined within a school containing a total of 16 individuals, there was no consistent difference in nearest neighbour distance or degree to which the five individuals matched either the velocity or the heading of their nearest neighbours (Partridge, 1981).

4.8.7 *Packing density*

The school volume is generally proportional to the number of individuals and the cube of the average body length (Pitcher & Partridge, 1979). This means that the packing density of schools in number of individuals per unit volume decreases in inverse proportion to the length of the fish. The mean volume per fish is larger in saithe schools than in herring schools. This indicates that even if the nearest neighbour distance is larger in herring schools, the overall structure is more compact and dense than that of the saithe schools. The density in the herring school observed by Pitcher and Partridge (1979) is comparable to that observed for minnows and pilchard when schooling in small tanks, but free-swimming herring schools in nature seem to school at a density which is an order of magnitude lower (Radakov, 1973; Serebrov, 1974; Misund, 1993b; Fig. 4.11). This discrepancy has caused speculation on the reliability of school structure studies on fish confined in artificial environments in small tanks, but also on the accuracy of the observation methods used in the field.

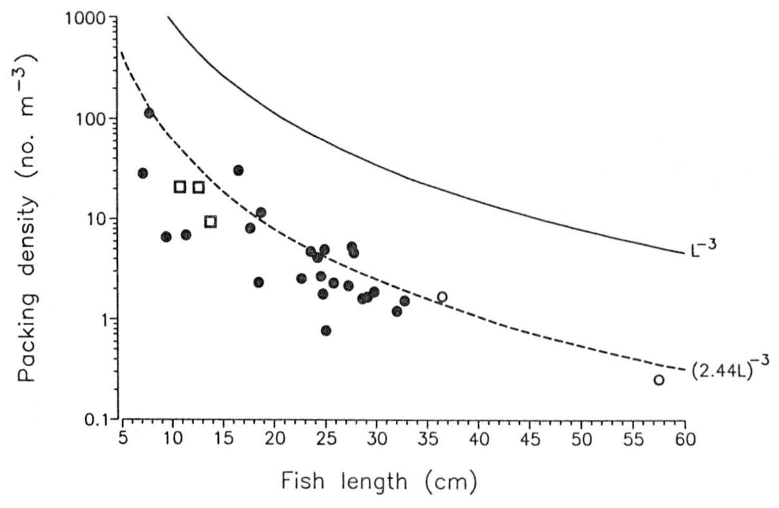

Fig. 4.11 Packing density of schools related to fish size (after Misund, 1993b). ○ = Saithe; ● = herring; □ = sprat.

4.8.8 *Packing density structure*

When observing herring and sprat schools with a high resolution sonar, Cushing (1977) observed that the packing structure within the schools was rather heterogeneous. This has been confirmed by measurements of free-swimming schools using photography and high-resolution echo integration, which shows that the packing density distribution in capelin and clupeoid schools varies considerably (Serebrov, 1984; Fréon *et al.*, 1992; Misund, 1993b). Regions of high density are usually found within the schools, and even empty vacuoles have been recorded. In capelin and herring schools, the packing in the densest regions is comparable to that recorded in laboratories (Serebrov, 1984; Misund, 1993b). Similar high-density regions have been recorded by photographing within anchovy schools in the Pacific (Graves, 1977). Misund and Floen (1993) observed by repeated echo integration that there were large variations in internal packing density among schools, but that a single school seemed to maintain a certain packing density structure for time intervals of up to about 10 min (Fig. 4.12).

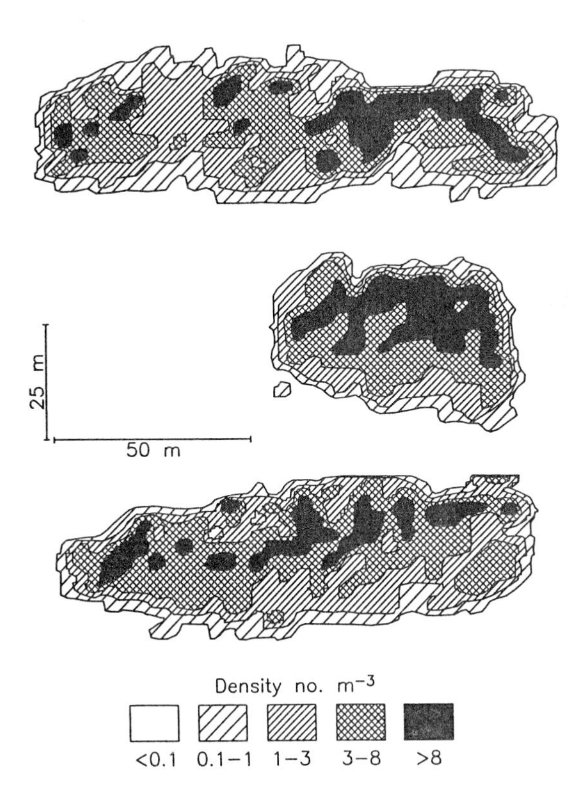

Fig. 4.12 Packing density structure of a herring school as recorded during three successive transects by a small research vessel (after Misund & Floen, 1993).

To explain how such large internal variations in packing density may occur within a school, Misund (1993b) proposed the 'moving mass dynamic' hypothesis. This is based on the great dynamic of individual positions within schools (Partridge, 1981) which results in rapid changes of neighbours and small differences in swimming speed among individuals. School members matching changes in speed and direction of their nearest neighbours within short time lags (Hunter, 1969; Partridge, 1981) may cause short-term variation in speed and level of arousal among different regions of a moving mass of individuals. If individuals pack more densely at greater speeds or level of arousal (Pitcher & Partridge, 1979; Partridge *et al.*, 1980), this could cause variation in packing density between regions of the school. As relatively small changes in nearest neighbour distances may create large changes in number of fish per unit volume, such variation may be especially pronounced in free-swimming schools with a much lower overall density compared with more compact 'aquaria' schools. The internal speed and density variations may be especially apparent when large schools change direction, come across patches of food or respond to predators. High-density regions or empty lacunas may thereby be the result of the moving mass dynamic within the schools.

Fréon *et al.* (1992) explain the internal variations in packing density of schools by a compressing/stretching mechanism. According to this hypothesis, the interfish distance decreases, especially at the periphery, when the school is confronted by danger (Fig. 4.13). When the stress is very strong, the interfish distance decreases rapidly to a minimum and all the vacuoles in the near part of the school collapse quickly. The reaction may be rapidly transmitted throughout the school, but due to a certain attenuation in the propagation, several repeated stimuli may be necessary to compress the school as a whole. From such a dense packing, the school stretches out as individual exploratory behaviour starts taking place, and the interfish distance may reach the upper limit. Then the stretching/tearing phenomenon occurs: a given individual must choose which fish to join in order to maintain this maximum stretching distance within the normal range, and small vacuoles start appearing (Fig. 4.14). The individual following the 'disrupting' fish then faces the same problem, but with a greater intensity because the possibility of keeping an equal distance from neighbours means a greater withdrawal from each of them. As a consequence the vacuole enlarges. Fréon *et al.* (1992) observed that such internal vacuoles do not move with the school, but rather that the fish move around the vacuoles as a river flows around rocks.

4.8.9 *School shape*

Adopting a special external school shape may reduce the probability of detection by predators. Several authors (Breder, 1959, 1976; Cushing & Harden Jones, 1968; Radakov, 1973) claim that this is achieved with a spherical shape in which the area-to-volume ratio is minimised. Partridge and Pitcher (1980) argue that a disc shape will better reduce the chance of being discovered.

In general, the horizontal dimensions of schools are greater than the vertical. This gives a discoid shape that differs to a certain extent among species. Oshihimo (1996) found that the horizontal extent was about five times the vertical extent in anchovy (*Engraulis japonicus*) schools that were distributed in shallow water (< 50 m depth) in the China Sea. The shape is seldom constant, but changes continuously as individuals perform various manoeuvres. Fréon *et al.* (1992) observed from an aircraft that a *Harengula clupeola* school in a shallow bay at Martinique varied from an amoeboid, unstructured form to a densely packed, egg-shaped type (Fig. 4.15). During a one hour period the surface area of the school varied from 145–522 m^2. Some species, like

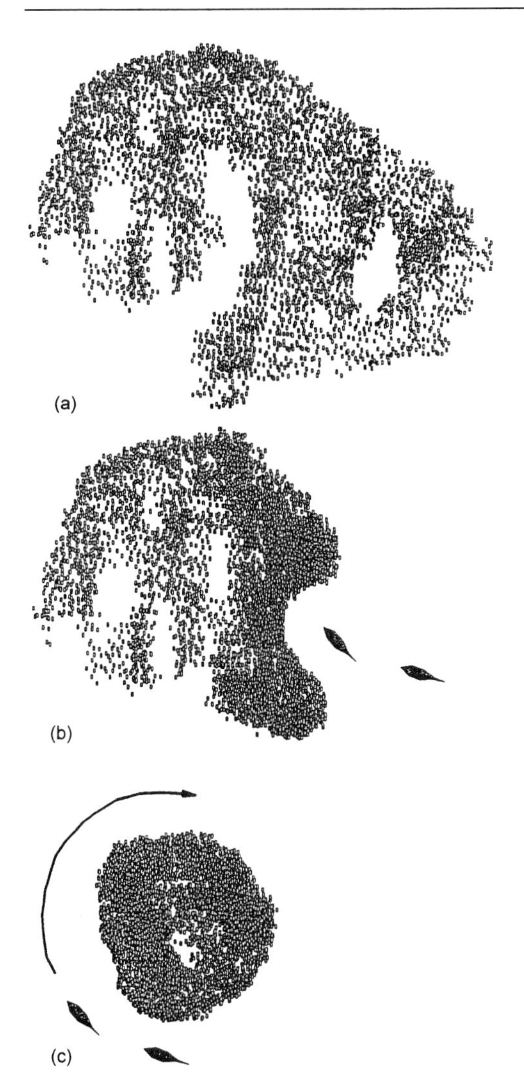

Fig. 4.13 Compression of a school during a predator evasion reaction (after Fréon *et al.*, 1992).

herring, form schools in which the shape varies less than schools of other species (Partridge *et al.* 1980).

By aerial photographs of daytime schools or night-time school bioluminescence, Squire (1978) found that most schools of northern anchovy (*Engraulis mordax*) were shaped like rods and ovals. The average length-to-width ratio of the schools was 2.1:1 during the day and 2.5:1 at night. Misund *et al.* (1995) found that about 70% of the herring schools in the North Sea had a compact appearance (circular, oval or square), about 20% were more stretched (shaped like a rod or parabola), and about 10% were amorphous. Hara (1985) observed that the shape of Japanese sardine (*Sardinops melanosticta*) schools was related to size. Elongated or crescent-like schools were 10 m in width and 100–200 m in length, while oval schools were at most 20–30 m across.

The shape of herring schools is dependent on swimming depth (Misund, 1993b). Midwater schools are spherical while schools close to the surface and bottom are

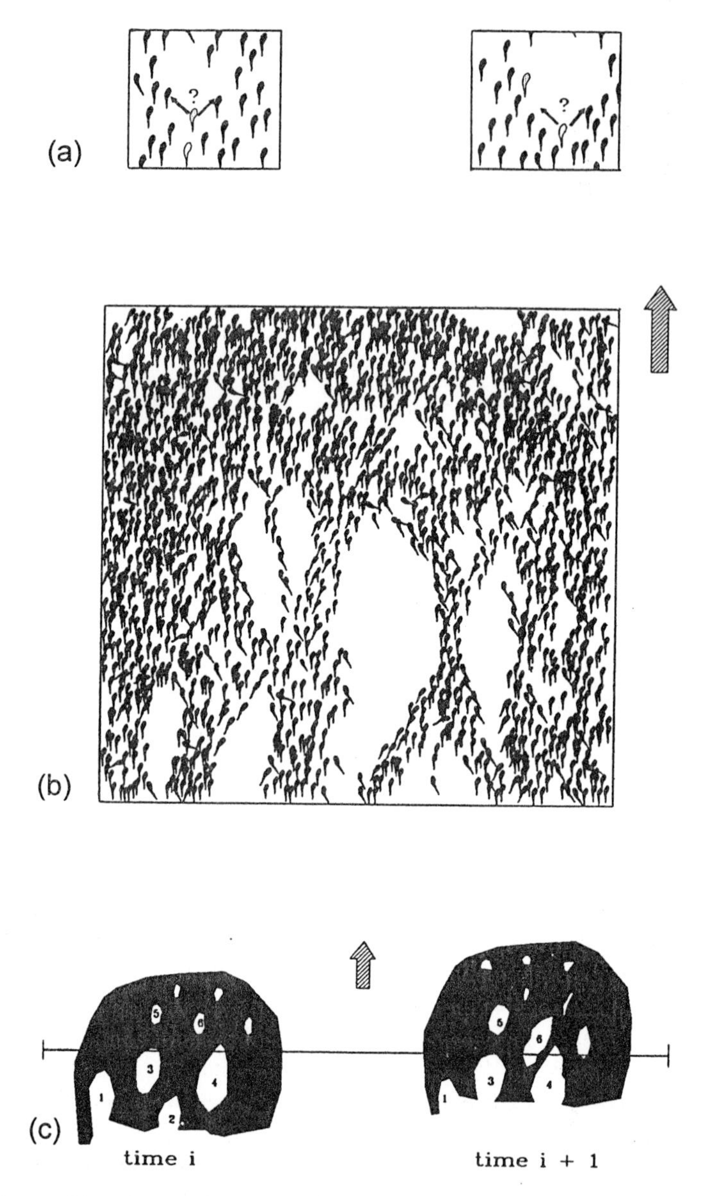

Fig. 4.14 The stretching of a school through formation of internal lacunas (after Fréon *et al.*, 1992).

discoidal. This may indicate that the spherical shape is primary according to the detection-minimising hypothesis, but that other functional or ecological aspects result in more flattened schools close to the surface or sea bottom.

4.8.10 *Factors affecting school structure*

A school is a functional unit in which each individual participates to its own benefit. In principle, everything that affects the individual's fitness may also affect the school

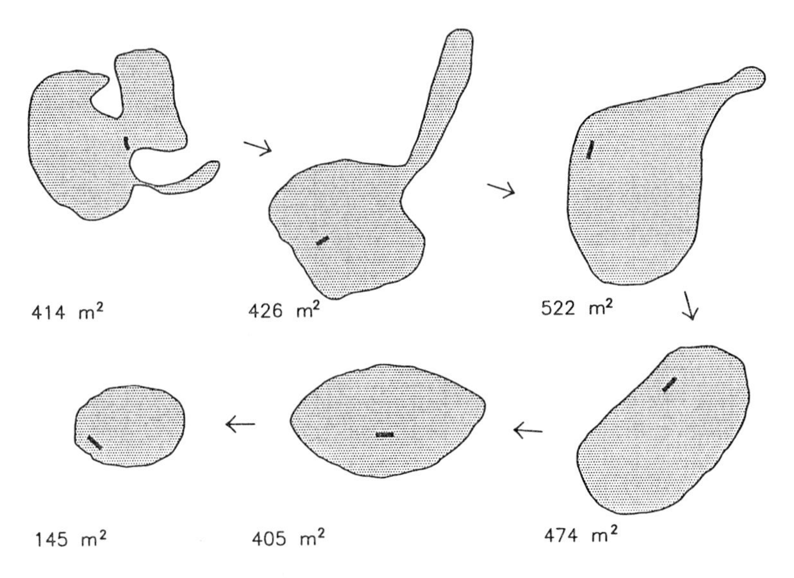

Fig. 4.15 Variation of horizontal area of a *Harengula clupeola* school during one hour as recorded from an aircraft. The small black bar represents the size of an observing scuba diver (after Fréon *et al.*, 1992).

structure. Most schooling fishes live in close contact with their predators (Major, 1977), and the main function of a school is not to minimise detection probability but rather to reduce predation probability after detection (Pitcher, 1986). This reduction is achieved when the schooling individuals cooperate in complicated predator-evasion manoeuvres such as fountain, ball packing or flash expansion in which the school structure changes dramatically within tenths of a second. Fréon *et al.* (1993b) observed that a bonito lure that approached *Harengula* schools induced mainly local changes in school structure, whereas the changes due to attacks from real bonitos were dramatic (Fig. 4.16), and could even lead to splitting of the school.

Schooling fishes find food faster and can spend more time feeding in the vicinity of predators due to increased vigilance and reduced timidity (Pitcher, 1986). During feeding, the individuals may behave more or less independently and thereby break up the functional school to a more loosely organised shoal. However, if necessary due to a severe predator threat, individuals in visual contact can rapidly reorganise the school. In free-swimming *Sardinella* schools, Fréon *et al.* (1992) often observed that one part of the school broke up the structure to feed, while the rest of the school continued polarised and synchronised swimming.

Among the proximate factors affecting the functional school structure is hunger. Individuals deprived of food tend towards an individual food-search behaviour and thereby loosen the school structure (Morgan, 1988; Robinson & Pitcher, 1989a,b; Fig. 4.17). A similar explanation has been given for the phenomenon that younger individuals organise looser school structures than adults because the younger and fast-growing stages have relatively higher food requirements (Van Olst & Hunter 1970).

The benefits of schooling increase asymptotically as the number of participants

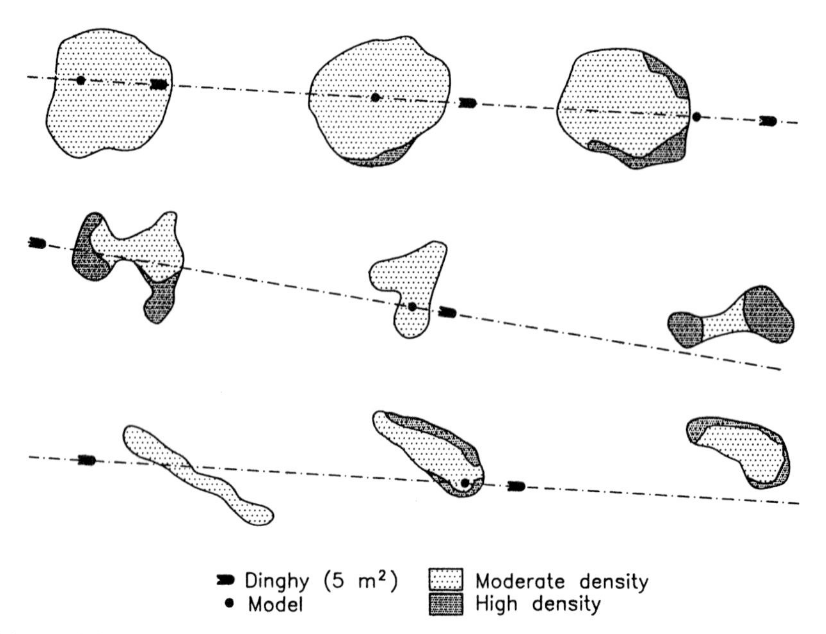

Fig. 4.16 Aerial observations of *Harengula clupeola* school reactions during three passes of a dinghy towing a model predator above the same school (after Fréon *et al.*, 1993b).

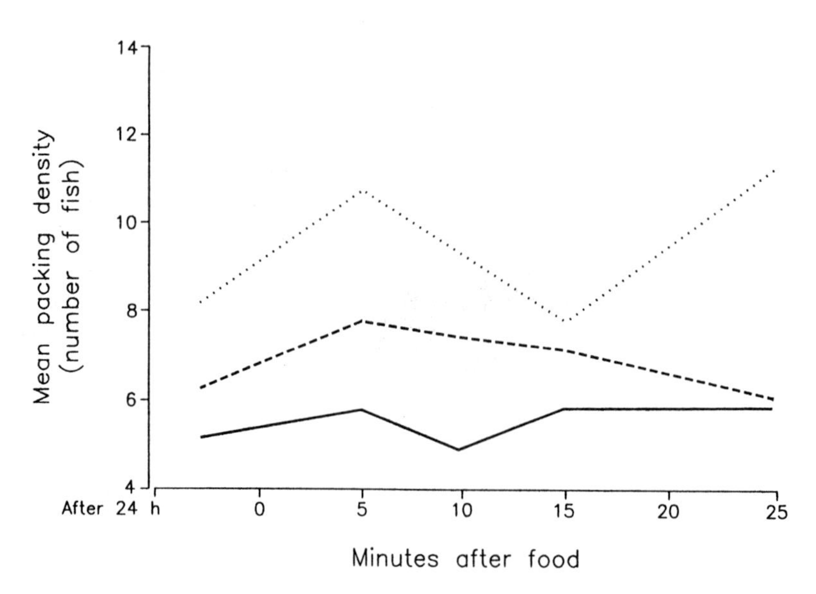

Fig. 4.17 Packing density of a herring school as a function of diet ration (after Robinson & Pitcher 1989a). (. . . .) 130 g diet; (- - -) 40 g diet; (—) 10 g diet.

goes up. Individual benefits will therefore be highest when school size increases from low numbers. This is reflected as a tendency towards more compact packing as the individuals rely more on the movements of their neighbours when the size of small schools increases (Partridge *et al.* 1980). Ross and Backman (1992) found that sub-yearling American shad (*Alosa sapidissima*) spent more time schooling the larger the group, and that schooling in larger groups enhanced vigilance.

At faster swimming speed, individuals also pack more densely (Partridge *et al.* 1980; Partridge 1981). Individuals gain a better monitoring of each other's movements when the interfish distance is reduced, and this may be necessary for keeping up the functional school structure at faster speed. A lower internal synchrony, with a relatively wider range of individual speed variation and change of positions, indicates that the internal school structure is more difficult to maintain when swimming faster (Partridge, 1981).

A source of substantial variation in internal school structure is the formation of subgroups. Such groups have been observed in saithe schools containing more than 10 individuals (Partridge, 1981), and in small schools of minnows (Pitcher, 1973) and herring (Pitcher & Partridge, 1979). Relatively independent movements of such clusters of individuals can open up empty spaces and cause large variation in school volume.

Physical environmental factors also to a certain extent affect the structure of schools. This probably occurs indirectly as the schooling individuals have preferences for environmental conditions such as current, temperature, light level and oxygen content, within certain limits. This probably affects just the external organisation of schools. As the schooling individuals may prefer to swim within definite depth intervals due to such preferences, this will set the limits for the vertical extent of the schools. Off Peru most anchoveta (*Engraulis ringens*) are found between 20 m down and a vertical boundary at 40 m depth where there is an onshore undercurrent with an oxygen content less than 2 ml l^{-1} (Villanueva, 1970). During the 1977 El Niño, this vertical boundary occurred at about 20 m depth and forced the schools closer to the surface, where they were exposed to an offshore current and high temperatures (Mathisen, 1989).

It is more doubtful whether physical environmental conditions directly affect the internal structure of schools. In schools of migrating *Mugil cephalus*, MacFarland and Moss (1967) observed that the amount of dissolved oxygen declined towards the rear of the schools where surface rolling, the highest packing densities, and tendencies towards loosening of the school structure by leaving of small subgroups, occurred. The higher densities at the rear were probably not an effect, but rather the cause, of the lower dissolved oxygen content. The schools' structural changes were therefore more likely the result of predator-prey interactions taking place at the rear, because migrating *Mugil cephălus* suffer from heavy predation (Petersen, 1976).

Koltes (1984) observed seasonal variation in the packing density of an Atlantic silverside (*Menidia menidia*) school (Fig. 4.18). The school was loose, but active throughout spring, less active in a compact mill in summer, dense in the autumn, and mostly had a random organisation of individuals in winter. The packing density of the school was not correlated with temperature or photoperiod, but there was some

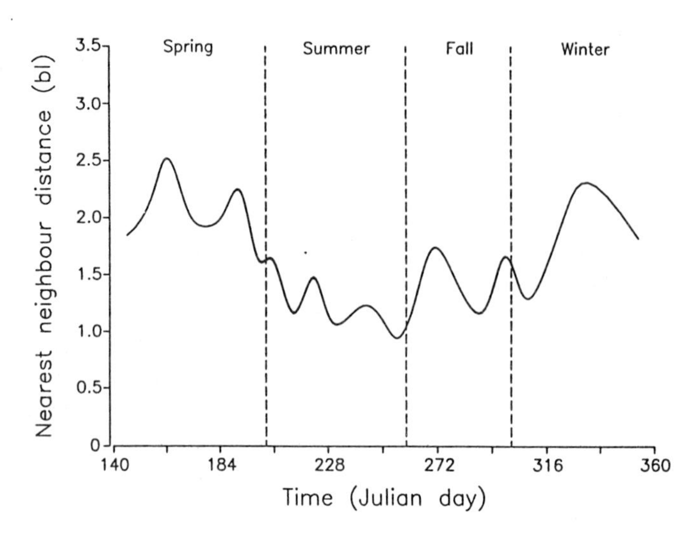

Fig. 4.18 Seasonal variation in packing density of an Atlantic silverside school (after Koltes, 1984).

indication of a lunar cycle. Koltes (1984) argued that the seasonal variation in the packing density of the school correlated with annual spring and autumn migrations within New England estuaries. Similarly, Hergenrader and Hasler (1968) observed that yellow perch (*Perca flavescens*) formed dense schools in summer, but loose aggregations in winter.

The numerous factors affecting school structure and shape of a given species make it difficult to automatically allocate echo recordings to different species. Due to the plasticity of school geometry and internal structure, the algorithms used for school identification have limited performance. These algorithms are able to identify only some species with an acceptable error, though their performance improves if they are calibrated for each survey when used in neural networks, or if they have input from sophisticated wideband frequency equipment which has been tested only on caged fish (Haralabous & Georgakarakos, 1996; Scalabrin *et al.*, 1996; Simmonds *et al.*, 1996).

4.8.11 *Factors selecting for homogeneity, structure and synchrony*

After having considered all these proximate factors affecting school organisation, we may reasonably ask if there is any school structure at all. From their detailed investigation of the three-dimensional structure of herring, cod and saithe schools, Partridge *et al.* (1980) claimed that the school structure is present in a statistical sense only. Similary, Pitcher (1986) concludes that homogeneity and synchrony have been overemphasised in schooling. Both arguments are in clear contrast to the group-selectionistic view of schools as regularly spaced, synchronously behaving units of one species of equal size. The variations quantified in relative positions, speed, interfish distance and spatial distribution were too great to support arguments of schools being organised according to regular geometric lattices, however. There is also clear evidence of interindividual tensions in schools such as the skittering behaviour of minnows under predator threats (Pitcher, 1986). Such selfish behaviour indicates that individuals try to maximise the advantage of schooling at the expense of others. The only remains of a homogeneous, synchronised behaving geometric organisation is that schooling individuals maintain a minimum of free space around themselves, prefer certain positions relative to each other, and match changes in the swimming movements of their neighbours within short time lags.

However, there is doubt as to whether the experimentally based conclusion is consistent with our image of schooling. Laboratory observations have been made exclusively in small tanks and net enclosures, some rectangular, others circular. In all set-ups, however, the observed schools have been constantly turning. Therefore the quantified structure is skewed. In addition, the coordinates of neighbours on both sides of an individual have been pooled. Even if there were no significant differences among coordinates from the right and left sides of an individual, the total variance is likely to have increased, thereby creating an impression of a more random school structure than is real.

In most studies of schooling, the nearest neighbour distance has been used as a measure of structure. Whether this parameter is appropriate for this purpose is doubtful, however. Due to the internal dynamic, individuals may rapidly change neighbours. Two fish that are nearest neighbours in one videotape frame may be involved in different nearest neighbour relationships in the next frame. If the interfish distance varies within preferred limits, a neighbour's status as the nearest, second nearest etc. may alternate constantly. The nearest-to-second-nearest-neighbour proportion may therefore give the impression that the school structure is more occasional than in reality.

In many circumstances it is the characteristic homogeneous, synchronised and structured organisation that makes schooling advantageous. During migrations, the route may be more precise as the path of the school will be the average of each individual's preference. On such occasions, individuals may gain an additional hydrodynamic advantage if they swim with neighbours of similar size and in certain positions relative to each other. Such preference of similar-sized neighbours, as has been shown to exist in herring and mackerel schools (Pitcher *et al.*, 1985), therefore selects for homogeneity in schools. Similarly, large size variation may cause splitting of schools into separate size groups due to asymmetric pay-offs in intraspecies competition during feeding (Pitcher *et al.*, 1986), and size-assortative schooling in the presence of predators (Ranta *et al.*, 1992). These factors select for homogeneity in size of the individuals participating in a school, as has been shown to exist in capelin schools in the Barents Sea (Gjøsæter & Korsbrekke, 1990) and gilt sardines off Senegal (Fréon, 1984).

In large pelagic schools, not all fish can be expected to have consumed equal amounts of food and the same occurs in large schools of migrating gadoids (DeBlois & Rose, 1996). Hungry schooling fish tend towards individual food-searching behaviour, which may create tensions among the individuals, and this may eventually break up the school structure into subschools of more equally fed individuals (Robinson & Pitcher 1989b). This indicates that the homogeneity of schools may apply also to the motivational state of each individual participating.

During predator threats, the factors resulting in homogeneous, synchronised schools may be especially pronounced. In mixed-species schools, one species may be more conspicuous than the other, and sorting by species may reduce the probability of one species being more susceptible to predation than the other. Similarly, there may be species differences in organisation and performance of predator-evasion tactics (see following section).

To be effective, tactics to evade attacking predators rely on the cooperation of each individual. The point of such tactics is performance of organised spatial structures with precisely synchronised swimming movements. This increases the confusion of the predator and enhances the survival probability of the individuals. Those individuals performing best have the highest probability of surviving, those losing position are most likely to be predated. The selection forces for homogeneous, synchronised and structured schooling are therefore quite strong and obviously are evolutionarily stable.

4.9 Mixed-species schools

Mixed-species schools or shoals seem common in demersal and semi-demersal fish communities, especially in coral reef ecosystems in the tropics (e.g. Ehrlich & Ehrlich, 1973; Alevizon, 1976). In contrast, fewer *in situ* observations of mixed-species pelagic schools are reported. Hobson (1963) observed the association of flatiron herring (*Harengula thrissina*) and juvenile anchovetas (*Cetengraulis mysticetus*) in the same school in the Gulf of California. Radovich (1979) observed large schools of northern anchovy (*Engraulix mordax*) surrounded by small groups of Californian sardines (*Sardinops sagax*). This low number of direct observations seems due to the poor transparency of the water in areas where several of the large commercial pelagic stocks are living (upwelling or river discharge areas). One of the best documented examples of *in situ* observation of mixed-species aggregation is provided by Parrish (1989b) in the relatively transparent waters of Bermuda (see later in this section).

Purse seine-fishery data may be used to identify the presence of such mixed-species schools of commercial species. Each set of a purse seiner is usually performed on a school or a dense shoal which is totally or partially caught. The analyses of temperate fisheries data suggest that mixed-species schools are seldom caught, or that the association between species lasts only a short time and in particular areas. For instance, herring and mackerel can mix in schools for a short period during summer in the North Sea. In the Bay of Biscay, mixed schools of anchovy and sprat or anchovy and horse-mackerel are often observed, even though their proportion is difficult to estimate using a pelagic trawl (Massé *et al.*, 1996). Nevertheless, a single species is often largely dominant in number and represents the bulk of the school. This contrasts with tropical or sub-tropical areas, where mixed-species schools are frequent.

In the Senegalese fishery, for instance, a database of 15 419 sets performed from 1969 to 1987 shows that most of the schools of gilt sardine are a mixture of two species (*Sardinella aurita* and *S. maderensis*) or an association with horse mackerel (mainly *Caranx rhonchus*) or mackerel (*Scomber japonicus*) (Table 4.1). The proportion of

Table 4.1 Half Burt table showing the co-occurrence of the two species of gilt sardine in a single set, according to their body length group (empirical commercial categories by increasing size from I to VIII), from 1969 to 1987. The lower line represents the mixing ratio of the two species, regardless of group size and other species, and the diagonal line of numbers corresponds to non-mixed schools (from Fréon, 1991).

Species	Size	*Sardinella aurita*								*Sardinella maderensis*						
		I	II	III	IV	V	VI	VII	VIII	I	II	III	IV	V	VI	VII
S. aurita	I	88														
	II	0	86													
	III	0	4	476												
	IV	0	1	4	2608											
	V	1	2	0	5	3314										
	VI	0	0	0	0	5	305									
	VII	0	0	1	8	2	0	3258								
	VIII	0	0	0	0	0	0	0	136							
S. maderensis	I	42	5	10	9	6	0	0	0	89						
	II	0	45	13	45	2	0	2	0	0	113					
	III	36	11	210	164	175	4	3	3	4	3	641				
	IV	2	14	108	1746	1487	50	132	44	2	0	7	4724			
	V	0	2	9	110	1083	77	375	26	0	0	5	10	2103		
	VI	0	0	0	5	34	63	7	0	0	0	5	6	11	121	
	VII	0	0	0	6	4	0	99	0	0	0	0	0	0	0	120
Mixing ratio %		91	90	74	80	84	64	19	54	81	95	94	76	80	90	91

mixed schools varies according to season and to the threshold retained for the definition of mixed-species schools (Fig. 4.19). In most cases, one of the species is largely dominant. The different species in a mixed school would usually have similar body lengths, and in general, species of similar body shape school together (Fréon, 1984). The comparison of length ranges in individual mixed schools of gilt sardines and all schools caught in the small fishing area shows that the association between species is related first to similarities in body length and shape and then to species groups (Table 4.2). Nevertheless, the usual dominance in number of a single species in mixed schools strongly suggests that fish of the secondary species join a school of a different species when they are not abundant in a large enough number to form a school on their own. This supposition is supported by the low abundance of the secondary species during the season of its common association with other species in the area.

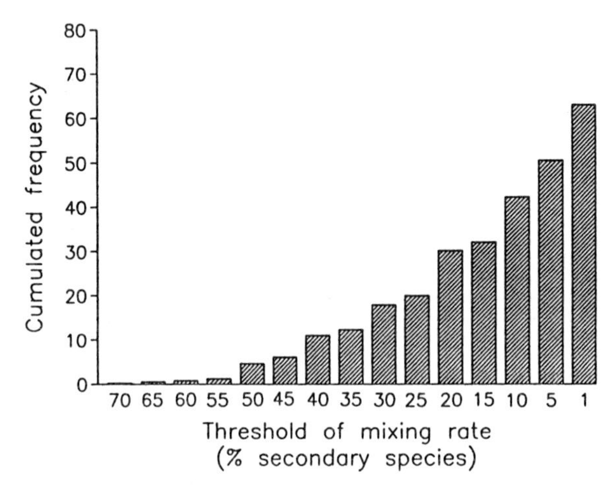

Fig. 4.19 Cumulative frequency of the occurrence of mixed schools in the Senegalese commercial fishery according to the threshold of mixing rate chosen (expressed as a percentage of the secondary species).

Similarly, tuna are known to seldom school only with conspecifics. Skipjack tuna (*Katsuwonus pelamis*) are mainly caught by purse seine or pole and line in association with other tunas in the Eastern Atlantic (Table 4.3; Cayré, 1985) and in the Pacific as well (Brock, 1954). Tuna are frequently associated with drifting floating objects ('logs' or drifting artificial objects set by fishermen), or with fixed, permanent structures (anchored FADs, seamounts, islands, etc.). In a given area, the assemblage of mixed-species schools in this various type of object is different, and it also differs from the assemblage in non-associated schools (Hallier & Parajua, 1992a). Another particularity of tunas is their frequent association with schools of dolphins, especially in the eastern Pacific Ocean (Chapter 6). This association between fish and mammals can be considered as a mixed school in so far as there is no predator-prey relationship between the two groups, the two species being of similar size.

The arrangement of the different small pelagic species inside the school is not well documented. Nevertheless Parrish (1989b) observed layering in a heterospecific assemblage of four species of morphologically and ecologically similar fish: two

Table 4.2 Comparison between length ranges in individual mixed schools of gilt sardine *Sardinella aurita* and *Sardinella maderensis* and all schools caught in the small fishing area (15 n.mi^2) from May to November 1978 in Senegal (from Fréon, 1984).

	Mixed schools *S. aurita* + *S. maderensis*			All schools *S. aurita*		*S. maderensis*	
Month	Mean	Length range	n	Length range	n	Length range	n
5	27.2	8	319	22	3131	14	942
	28.1	7	202				
	22.8	11	327				
	23.9	12	213				
6	24.7	9	214	22	2470	20	2754
	23.4	11	269				
	21.8	11	496				
	24.0	8	256				
	23.0	9	306				
7	23.4	12	386	11	1168	16	3184
	22.4	8	355				
	22.9	8	261				
	22.0	7	350				
	23.3	8	216				
8	23.4	9	218	23	1215	15	3424
	23.2	10	327				
	20.6	10	366				
9	22.9	10	234	11	1667	22	4029
	21.1	10	256				
	20.5	12	333				
	20.0	13	396				
	20.4	11	366				
	22.9	11	327				
10	21.9	7	256	19	2119	19	3396
	21.7	9	270				
	21.6	8	297				
	20.6	8	265				
	20.0	8	380				
11	21.1	8	362	24	1989	18	2384
	21.1	9	276				
	19.8	15	402				
	18.4	14	374				

Table 4.3 Frequency of skipjack tuna (*Katsuwonus pelamis*) monospecies and mixed schools (with other tuna species) in the east Atlantic (from Cayré, 1985).

	Monospecific skipjack schools	Mixed skipjack schools
Purse seine	10.1%	89.9%
Pole and line	11.5%	88.5%

clupeids (*Jenkensia lamprotaenia* and *Harengula humeralis*), an engraulid (*Anchoa choerostoma*) and an atherinid (*Allanetta harringtonensis*). Up to seven fish-types were defined according to the species and the size (except for A. *harringtonensis* which had a single fish-type). Usually two to four fish-types were observed in a single aggregation. This assemblage formed shoals or schools, and it was possible to identify three distinct layers inside. The surface layer was occupied by *A. harringtonensis*, the middle one by *J. lamprotaenia* (juveniles and adults) as well as juveniles of *A. choerostoma* and *H. humeralis*, and the lowest layer occupied by adults of *A. choerostoma* and *H. humeralis.* In general, smaller fish were found in higher positions than large ones. The big tunas associated with dolphins are travelling under them at the same speed (Anon., 1992). Aggregations of tuna under floating objects also stratify according to body size and species: the smallest fish are near the surface (skipjack tuna, juveniles of other tuna species) and larger fish are found below, sometimes as deep as 200 m, and therefore there is no continuity between the two related schools (Anon, 1992; Josse, 1992). Therefore layering according to the size and species seems frequent in mixed-species schools (other examples are reported by Parrish, 1989b).

Other types of spatial segregation between species, besides layering, have been described. Hobson's (1963, 1968) *in situ* observations indicate that the anchovetas were forming well defined sub-units, often located in the centre of the school, surrounded by herring. During her observation of mixed-species layers, Parrish (1989b) noticed that juveniles of *H. humeralis* were usually observed in small discrete schools single. Fréon (1988) described tropical, mixed clupeoid schools inside large enclosures in coastal waters of Venezuela where three species occupied different positions: the bulk of the school consisted of *Sardinella aurita* while *Opisthonema oglinum* were observed in the rear part and *Harengula jaguana* partially covered the upper part of the school.

Some of the advantages of monospecific schools remain obviously the same for mixed schools: effective food detection (when the diet is the same), hydrodynamic advantages (when existing), and some aspects of survival to predators such as dilution effect or early detection. However, the confusion effect may decrease when the different species in a school are so different in colour or shape that they can be distinguished by the predator, especially if one species is present in a low proportion. Hobson (1963, 1968) reported that in a school consisting of 90% herring and 10% anchovetas, the latter suffered a selective predation by pompano (*Trachinotus rhodopus*) in a ratio of 6 to 1 although located in the centre of the school. This author suggested that anchovetas were more conspicuous because of their flashing gill-covers. Coral reef species abandon mixed group when threatened, leaving sooner if a group has fewer conspecific members, but continue schooling if enough of their own species are present (Wolf, 1985). Allan and Pitcher (1986) observed that mixed shoals of cyprinids actively segregated when under threat of predation (by a pike model) by a tendency to join conspecifics. Similar findings exist for other animals, especially birds (Powell, 1974; Bertram, 1978; Caraco, 1980). This tactic could be selected for two reasons: better coordination with conspecific neighbours and therefore more effective evasion manoeuvres, and enhancing the predator's confusion by increasing locally the uniformity of appearance of the school (Allan & Pitcher, 1986). Concerning foraging, when the diet of each species is different, the disadvantage of overall competition decreases. In the case of tuna associated with dolphin, we can also speculate on the advantage gain by the two species in terms of prey detection (Chapter 6).

Finally, although more investigations are required on cost-benefit trade-offs of mixed-species schools, our present knowledge on their behaviour supports the view that pelagic fish constantly assess the shifting costs and benefits of schooling or shoaling, and this strongly influences their decision making as stressed by Lima and Dill (1990) and Pitcher and Parrish (1993).

4.10 Spatial distribution (clustering)

It is a common strategy for purse seine skippers to search near other vessels that have located and made sets on large schools, because if there is one large school in the area, others are usually nearby. This feature of spatial distribution of schools is known from purse seine fisheries in both the northern and southern hemispheres, and thus probably represents a general behaviour pattern of spatial school distribution.

Although there have been few scientific investigations on the spatial distribution of fish schools in the sea, at present this is a field of considerable interest, the more so because it has direct relevance to the design of abundance estimation by scientific surveys (Fréon *et al.*, 1989; Simmonds *et al.*, 1992). Therefore patchiness in fish distribution is commonly studied by the geostatistical approach applied to data from a scientific survey performed for a short time in order to limit temporal effects (e.g. acoustic or aerial survey). Autocorrelation distance, distance to the closest neighbour (MacLennan & MacKenzie, 1988), or the range of the variogram are indicators of the cluster size. Some examples of current range values for pelagic stock are 10 n.mi. for the pelagic stock of the Catalan Sea dominated by sardine (Bahri, 1995) and 7 n.mi. for the spring-spawning herring off the Norwegian coast (Petitgas, 1993). Higher values can be found for gadoid stocks, such as the Alaskan walley pollock (*Theragra chalcogramma*) whose autocorrelation distance in density is 30 n.mi. (Sullivan, 1991). Note that the previous values have been obtained from the analysis of the total biomass (dispersed fish + schools) per elementary sample, whereas only the biomass in school is available for many fishing gears.

In contrast, in her study on the pelagic stock of the Bay of Biscay, dominated by sardine and anchovies, Patty (1996) studied the range of variograms of the number of schools per n.mi during daytime. She found a value of 8 n.mi., that is close to other results. These clusters of schools are often multispecific (Massé *et al.*, 1996). Petitgas and Lévénez (1996) studied a time series of acoustic recordings of schools off Senegal. They reported that the schools occurred in clusters, and that the dimensions and number of clusters relative to the area where they were found stayed relatively constant with varying total biomass (range of the variogram close to 5 n.mi.). However, in the areas in which fish were present, the occurrence of dense schools varied according to the total biomass.

Tuna schools are also clustered and the cluster size has been estimated to around 40 n.mi. by Fonteneau (1985) using georeferenced data of purse seiner catches on a small temporal and spatial scale. Similarly small scombroids like the western mackerel (*Scombrus scombrus*) display a patchy distribution, in large clusters of about 40 n.mi, and the clusters are separated by distances of up to 50 n.mi. during migration (Walsh *et al.*, 1995).

The reason for school clustering remains largely speculative. It is probably related to the patchiness of the environment (Schneider, 1989; Swartzman, 1997), but social behaviour is also likely to play an important role which is presently not well understood.

Finally, the spatial structure of pelagic fish species (one or several mixed species) can be qualified by 'fractal' distribution (Fig. 4.20): in a wide area of the species distribution there are different discrete populations and within each population several clusters of schools can be found. Within the school the spatial distribution can also be very heterogeneous with several nuclei or cores.

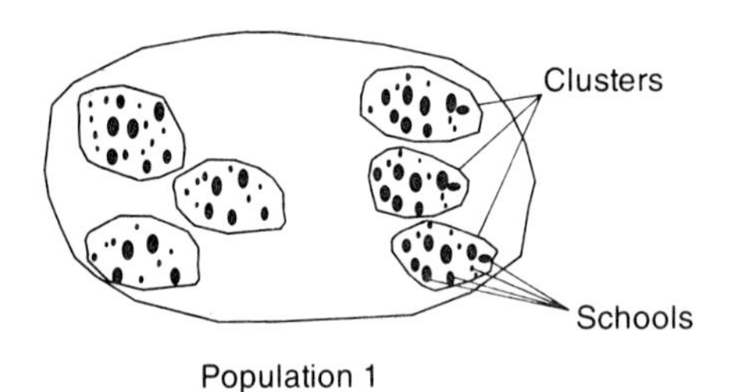

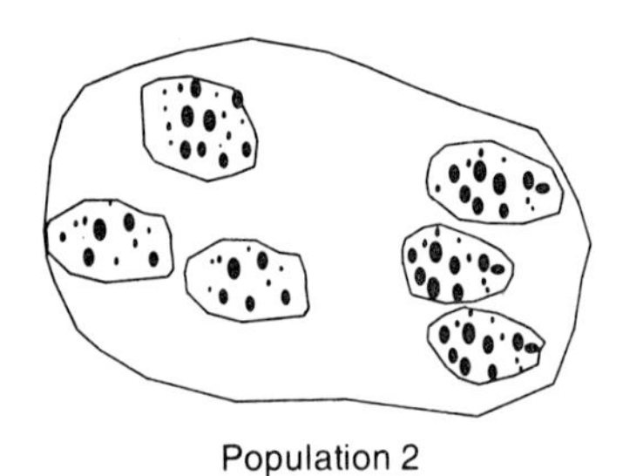

Fig. 4.20 Simplified diagram showing the 'fractal' distribution of pelagic fish species.

4.11 Communication

Fish aggregate in schools to increase their individual fitness. Organisation of the school unit does not fulfil the rules for packing in rigid geometrical patterns, but there are tendencies to preferred positions relative to other individuals. This probably indicates that each individual tries to maximise the information flow from its neighbours, so as to keep up with the movements of the functional school. The particular school structure is therefore organised to enhance interfish communication. In fact, during predator threats, schooling blackchin shiners reorganise from a flat, hydrodynamically effective positioning to a vertically extended structure in which the visual field of each individual is improved (Abrahams & Colgan, 1985).

Efficient interfish communication is a fundamental factor when individuals perform complicated predator-avoidance manoeuvres with remarkable synchrony and precision without colliding.

4.11.1 *Vision*

Visual sensing is fundamental to the formation of schools. A common behavioural feature of pelagic schooling species is that the schools break up into looser-organised shoals at night. This dispersal is thought to arise mainly because the individuals lose visual contact with each other in darkness. Experiments have shown that school structure breaks up at light intensities around 10^{-8} µE s^{-1} m^{-2} for jack mackerel (Hunter, 1968) and at 10^{-7} µE s^{-1} m^{-2} for Atlantic mackerel and *Astyanax* (John, 1964; Glass *et al.*, 1986; Fig. 4.21). Other authors claim higher values for other species (about 10^{-6} µE s^{-1} m^{-2} for northern anchovy, Hunter & Nicholl, 1985) but comparison of light intensity thresholds for schooling is difficult due to differences in experimental set-up and measurement procedures.

Position holding and movement monitoring within functional schools are also

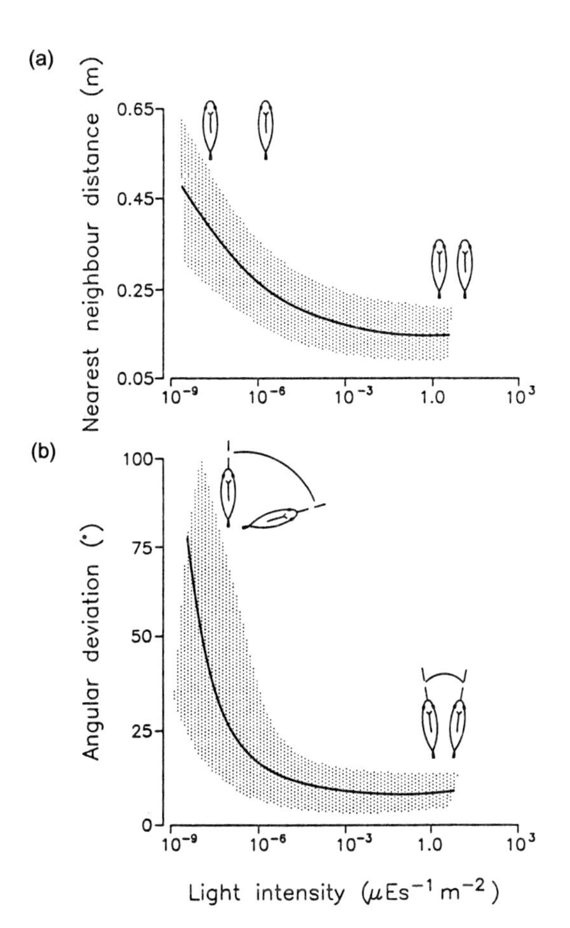

Fig. 4.21 (a) Nearest neighbour distance and (b) angular deviation of schooling mackerel related to light intensity. Curve = average. Shaded region = ± standard deviation (after Glass *et al.*, 1986).

based partly on visual sensing. Fish deprived of other sensory information by having the afferent lateral line nerves cut or by being placed behind transparent barriers are still able to perform schooling behaviour, but the position preferences relative to other fish are changed. Such fish have to rely purely on visual cues for their positioning and therefore tend to take up positions directly alongside and nearer their neighbours to better monitor changes in swimming movements (Cahn 1972; Pitcher 1979b; Partridge & Pitcher, 1980; Fig. 4.22). Determination of heading to individuals alongside is more difficult than to those in front. In a school of lateral-sectioned saithe this results in an increase in angular deviation from 2.5° as for a normal school to 10.4°. Partridge & Pitcher (1980) therefore claim that vision is of primary importance for maintaining interfish position.

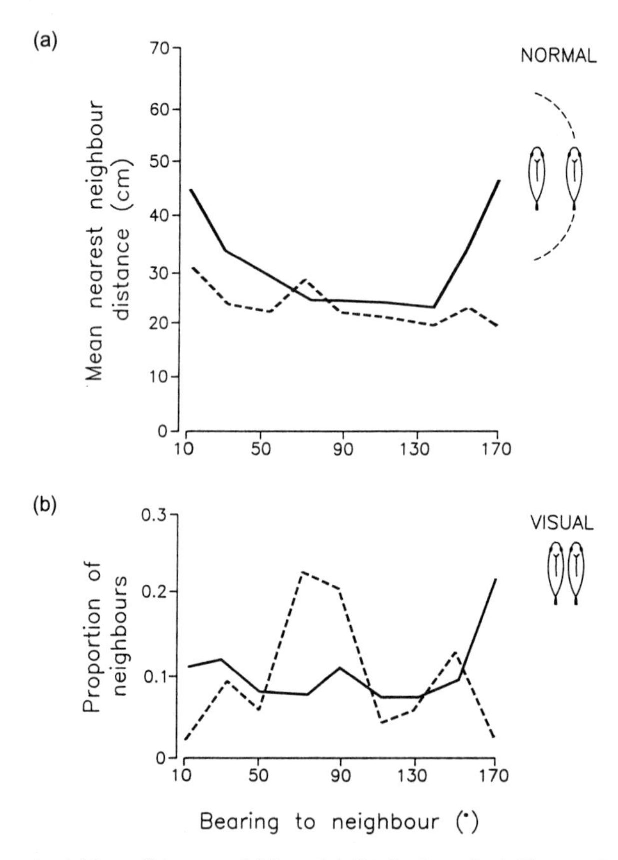

Fig. 4.22 (a) Nearest neighbour distance and (b) spatial distribution of neighbours related to direction of bearing of lateral line-sectioned saithe. (—) Normal nearest neighbour. (- - -) Visual orienting nearest neighbour (lateral line sectioned) (after Partridge & Pitcher, 1980).

Social communication in schools is mainly visual. For predator-evasion tactics to be efficient, individuals in the school that do not see the predator must perceive and react to the change in behaviour of their neighbours. As in the Trafalgar effect, this may occur faster than the approach of the predator itself. Minnows that could see a group of conspecifics, but not a predator model, stopped feeding and started hiding when their conspecifics reacted to the approaching predator by inspection behaviour and skittering (Magurran & Higham 1988). Likewise, isolated juvenile chum salmon (*Oncorhynchus keta*) stopped feeding and remained motionless when exposed to alarmed conspecifics (Ryer & Olla, 1991). When some school members come across food patches and start feeding, neighbouring individuals will observe the changed behaviour and join the foragers (Pitcher, 1986).

4.11.2 *Lateral line*

Swimming fishes generate low-frequency water displacements or net flow that can be sensed by neighbouring fish through the lateral line system. During schooling, gathering information on the swimming movements of neighbours through the lateral line system is fundamental for maintaining a normal school structure. Pitcher *et al.* (1976) showed that blinded saithe were able to school, but with slighly changed positional preferences. Quite surprisingly, blinded fish school at greater nearest neighbour distance (Partridge & Pitcher 1980; Fig. 4.23). Such fish tend to take up normal bearings, however, but slightly above their nearest neighbours. Blinded fish

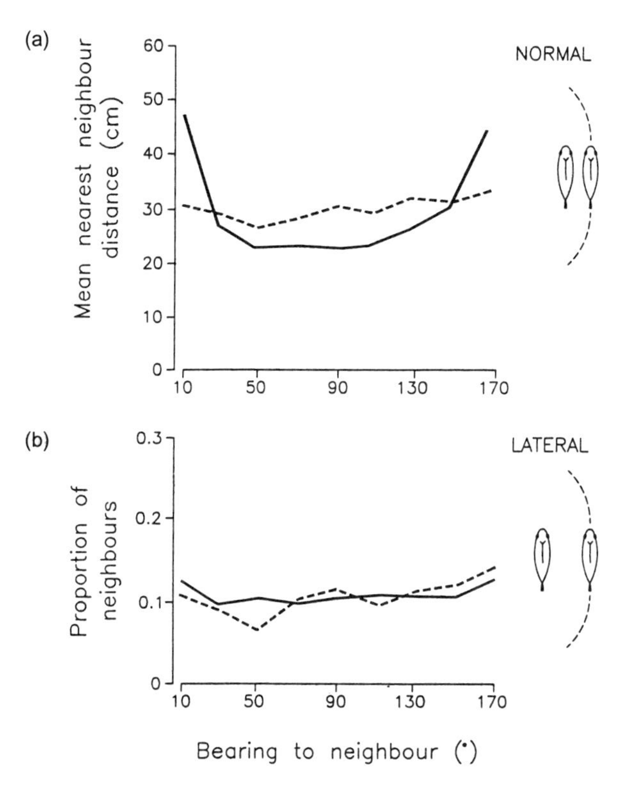

Fig. 4.23 (a) Nearest neighbour distance and (b) spatial distribution of neighbours related to direction of bearing of blinded saithe. (—) normal nearest neighbour. (- - -) lateral line orientating nearest neighbour (blinded) (after Partridge & Pitcher, 1980).

are also able to match short-term changes in swimming movements as well as normal fish, which led Partridge and Pitcher (1980) to conclude that lateral line sensing is particularly important for monitoring changes in swimming speed and direction of neighbours. This is especially important during quick predator-evasion manoeuvres. Fish with severed lateral lines occasionally collide during startle responses (Partridge & Pitcher 1980), whereas normal fish never do (Blaxter *et al.*, 1981). However, even if maintainance of school structure is possible by lateral line sensing alone, formation of schools probably is not.

4.11.3 *Hearing*

During swimming, the movements of fish are preceded by transient pressure pulses (Gray & Denton 1991). The pulses have their maximum intensity at low frequencies and within the hearing range of fish. The pulses decay rapidly with distance from the source, for whiting by the power of –1.5, for herring by the power of –2.5. This indicates that the pulses are nearfield effects, but their sound intensities are above the detection level of fish at a distance below about 0.5 m. The spatial distribution and polarity differ for species like whiting and herring. Whether the fast pressure pulses play a role in interfish communication during schooling has not been tested. However, there are certain factors that indicate that this is likely, even if saithe that were both blinded and lateralis sectioned were not able to school (Partridge & Pitcher 1981).

The pressure pulses are detectable at interfish distances observed during normal schooling (below one body length). For herring, the quadri-lobed emission pattern of the pressure pulses coincides with the diagonal position preferences in that the individuals tend to take up bearings in which there are minima in the pressure fields from the neighbour (Fig. 4.24). In these positions the individuals can best monitor changes in swimming speed and direction of the neighbour on the basis of changes in the pressure fields. If hearing plays a role complementary to the other senses during schooling, the possibility exists that schools can be maintained perfectly in the absence of the dominant visual sense. In fact, herring have frequently been observed to school at substantial depth even in darkness at night-time (Devold, 1969; Misund, 1990).

4.12 Schooling, modern fishing and natural selection

Fish schools are easier to locate in the sea than fish swimming individually. This applies for visual detection when fish schools are breaking the surface, and when the schools generate bioluminescence close to the surface at night. Similarly, for fish finding by acoustic instruments, the probability of detection is much higher for schools than for single fish (MacLennan & Simmonds, 1992).

When schooling, pelagic species often aggregate in great numbers in a unit that occupies a limited volume. This greatly increases the potential for catching a large amount of fish in a short time, and has led to the development of surrounding nets and large towed nets. In particular, the combination of large purse seines or pelagic trawls, hydraulic handling equipment and acoustic fish detection instruments has proved to be a very effective way of fishing. By these methods, several of the biggest

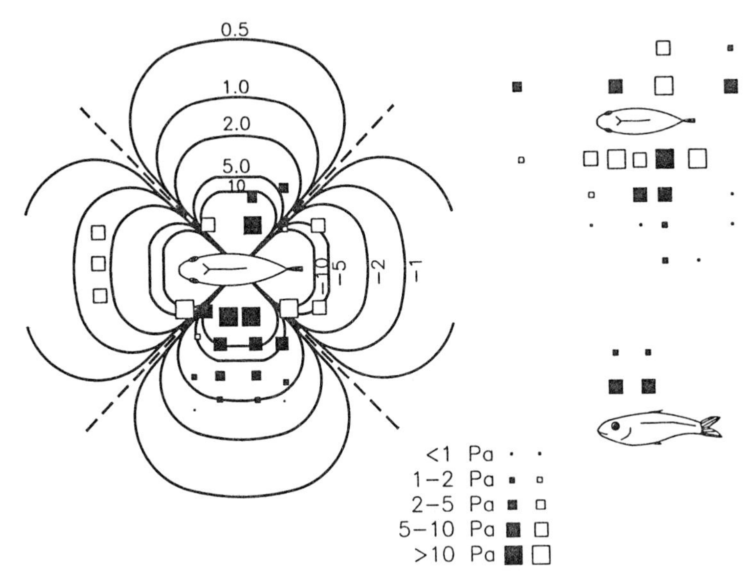

Fig. 4.24 Distribution of low-frequency pressure pulses around swimming herring. Filled squares = compression; open squares = decompression (after Gray & Denton, 1991).

clupeoid stocks in the world have been fished down to very low levels. This has been the fate of the Hokkaido-Sakhalin herring, the Downs herring, the Atlantic herring, the Pacific sardine, the Japanese sardine, the South African sardine and the Peruvian anchoveta, and has caused the loss of a total fish production in the order of 10 million tonnes annually (Murphy, 1980).

The life history parameters of the clupeoid stocks can indicate a critical stock-recruitment depensation (Ulltang, 1980). This means that if the stock of spawners falls below a certain critical value, it will not be able to produce the number of offspring necessary to sustain the population. Theoretically, clupeoid fish populations may thereby go extinct. Bjørndal and Conrad (1987) showed that the closure of the fishery at the end of the 1977 season saved the North Sea herring from possible extinction.

From a conservation-orientated view, we may therefore argue that schooling of fishes is a disadvantage because this makes such species more vulnerable to capture. When many pelagic species reach a fishable size, the mortality due to fishing is often much higher than the natural mortality. The argument may therefore be put forward that the advantages of schooling gained in predator avoidance may be lost through the higher probability of being fished.

We may therefore speculate on how the modern fishery affects the schooling behaviour of fishes. Because the fishery may be the main cause of mortality when many species reach a certain size, does the fishery thereby select against schooling? Does the fishery exhibit a selection force that in the long run mediates a genetically determined change in the schooling behaviour of fish? Could this result, for instance, in enhanced vigilance and avoidance of fishing vessels, and smaller schools that are more difficult to capture? Certainly there is no evidence to support such speculations, but there are some indications that may support them.

The capture success for Norwegian purse seiners when fishing herring in the North Sea was about 63% for the 1985 and 1986 seasons, and at about the same level as experienced when the purse seine fishery started in this area in 1963 (Misund, 1986). At that time the biomass was about three times higher than in the 1985 and 1986 seasons, but since then the stock has been fished down to a critically low level (Burd, 1990). The equipment used for fishing in the 1980s was considerably improved from that being used in 1963. The purse seines have nearly doubled in size, the vessels are much bigger, the hydraulic gear-handling equipment is more powerful, and the acoustic fish detection instruments are far more sophisticated. Modern sonars not only detect schools but also have a substantial tactical improvement in that they now are able to track the movements of both vessels and fish schools (Bodholdt & Olsen, 1977; Bodholdt, 1982). Still, there does not seem to have been any improvement in the capture success for schools in the Norwegian summer fishery for herring in the North Sea in this period. We may therefore reasonably speculate whether there has been a change in behaviour of the schools so that they have become more difficult to catch, and whether this change is genetically determined. The possibility that the change may have been caused by learning is less probable, as the fishery was closed from 1977 to 1984 while the stock recovered.

However, even if the fish exhibit the highest mortality when many pelagic species reach a certain size, it is still the forces of natural selection that exert the main impact on the younger life stages. The natural death rate is tremendous in the first larval stages, probably mostly due to predation, but also probably in part due to starvation if the larvae hatch in areas with little food or are brought out of such areas by advection. When the larvae start schooling, the natural death rate levels off (Moksness & Øiestad, 1987). The adaptive value of schooling is therefore significant, and there will be strong selective forces for schooling in the youngest stages of the life history. Most schooling pelagic species are also preyed upon in later life stages, and this, together with the other functional advantages of schooling, will select for schooling throughout the life history of these species.

On certain occasions the predation pressure may be so great that large stocks of small pelagic species may be reduced to a fraction of their usual size. In the Barents Sea, large stocks of young cod and herring exerted such heavy predation pressure upon the capelin that the stock declined from about 2.5 million tonnes in 1984 to a few hundred thousand tonnes in 1986 (Hamre, 1991). Other stocks of small pelagic species may undergo similar or even more dramatic stock size fluctuations because of environmental events such as the inflow of warm water (El Niño) that affects the stock size of the Peruvian anchoveta (Pauly & Palomares, 1989). However, this does not exclude the possibility that in the long run, heavy fishing mortality may modify the behaviour of schools.

4.13 Conclusion

In this chapter we have considered probably the most characteristic and admirable behaviour of pelagic fish: their schooling. The development of the understanding of fish schooling behaviour is reviewed in a historical perspective, leading up to the present definition of schooling as fish in polarised and synchronised swimming. There is a genetic basis for schooling, and fish from populations that live sympatrically with predators seem to be better schoolers than fish from less exposed populations. The schooling behaviour develops when the locomotory and sensory systems become functional in the latter part of the fish larvae stage.

The main function of fish schooling is to survive predatory attack, enhance feeding, gain hydrodynamic advantages during swimming, contribute to more precise migration, and facilitate reproduction and learning. Fish organise schools that become structured by maintaining a minimum approach distance and tendencies to preferred nearest neighbour distances and spatial positions relative to neighbouring individuals. It is now possible to mimic schooling behaviour by IBM. The packing density in terms of number of individuals decreases in inverse proportion to the cube of the length of the fish, and the density in free-swimming schools is usually an order of magnitude lower than in schools in captivity. However, the packing density structure of free-swimming schools is rather heterogeneous, and regions of high density have been recorded. The horizontal dimensions of schools are generally larger than the vertical, which gives schools a discoid appearance. This is the common shape of schools near surface or bottom, while midwater schools tend to be more spherical. However, the shape of schools varies considerably and is affected by predators, foraging, hunger, swimming speed, formation of sub-groups, and physical environmental factors. Mixed-species schools are common in some areas, and schools tend to occur in clusters.

To keep the high synchrony within schools, fish communicate through vision, lateral line and possibly hearing. Vision is essential for school formation, but a blinded fish can still school through lateral line sensing.

Chapter 5
Avoidance

Avoidance is a general term for behavioural patterns fish perform to move away from stimulation that elicits fright or is sensed as unpleasant. During fisheries and stock assessment, avoidance can be of significance if the vessel, the monitoring instruments or the sampling gear generate stimuli to which fish react in a way that causes the measured density to be lower than the actual value, or the recorded species composition and distribution of size groups to be different from the undisturbed ones. With regard to avoidance behaviour of pelagic fish, which are mainly caught by large purse seines or pelagic trawls and directly assessed by the hydroacoustic method, we will concentrate on avoidance elicited by sounds from vessel and gear. This is simply because avoidance reactions during pelagic fishing or hydroacoustic surveys often occur within the range of sound detection, but usually beyond the range of visual detection of vessel and gear. Despite this, vision plays a major role in an acoustic survey when the fish is close to the vessel (Gerlotto & Fréon, 1990). This occurs for surface fish and in shallow water. Moreover visual sensing may be a determining factor in the capture of pelagic fish, because the final result of fish capture by purse seines and pelagic trawls may be determined by near-field avoidance reactions to the gear (Misund, 1994; Wardle, 1993).

5.1 Avoidance of sounds from vessel and gear

5.1.1 *Ambient noise in the sea*

People often describe the sea as a silent world. Few sounds revealing underwater activity can be detected at the surface, and even during diving few sounds are detected by the human ear. The reason is that the water–air interface is an effective barrier for propagating sounds, and the human ear has evolved to detect sounds in air, not in water. If a wideband hydrophone is put in the sea and connected to a loudspeaker, the underwater environment appears far from silent. Adding a frequency analyser to the circuit would reveal loud sounds beyond the frequency range of sound detection by the human ear.

The sea is always in motion. Surface winds set up waves that wash the shores and create oscillating bubbles and spray as the surface breaks and air is hit into the water. The wind also generates turbulence and pressure fluctuations due to wave and current motion. The sounds produced by these sources are usually dominant underwater. In addition, seismic events generate vibrations that are readily transmitted as sound waves in water, and that can be detected hundreds of kilometres away from the source. Then there are the marine animals which may produce sound to communicate or as a result of their motions. Man-made activities may also contribute, and ocean traffic may constitute a substantial part of the underwater sound in many regions. More locally and regionally, geological mapping by use of explosives to generate seismic waves may generate substantial sound underwater.

According to Wenz (1962), the various sources produce a composite ambient sound spectrum that is characterised by a dominant low-frequency region up to about 100 Hz. The low-frequency component falls off rather steeply by about −10 dB per octave. Turbulence-pressure fluctuations due to eddies and long waves constitute the major natural source of these sounds. However, depending on time and location, sounds from seismic sources such as earthquakes and underwater explosions may dominate parts of the spectrum below 100 Hz. In the region 100–10 000 Hz the spectrum level drops by about −6 dB per octave, and the ambient level in this region is about 5 dB higher in shallow water than in the deep. The sources of the sound in this region are oscillating and collapsing bubbles (cavitation) and spray from surface agitation. Most of the sounds produced by ships are within this frequency region, with maximum energy output between 50 and 200 Hz. In addition, many sounds produced by marine animals, such as the whale barks, the crackling potpourri of shrimps and the chorus of croakers are within this frequency region. Above 20 kHz, thermal noise from molecular agitation starts appearing. The echolocation sounds of the porpoises are also of high frequency.

As the natural sound sources underwater are the result of wind-driven movements, the sound level in the region below about 20 kHz may vary substantially. In shallow water, the wind may cause variations in the low-frequency component by about 20 dB. Then there may be additional regional variation depending on the bottom depth, degree of shelter and proximity to seismic sources. However, even during calm weather with a mirror-like surface, there may still be substantial sound in the low-frequency region underwater due to the turbulence-pressure fluctuations generated by eddies and long waves. In the frequency region from 100 to 10 000 Hz the wind may cause variations in the ambient sound level by about 40 dB.

5.1.2 *Noise from vessels*

Modern fishing and fisheries research vessels are steel-hulled and are powered by diesel engines that drive the propeller. A running diesel engine generates substantial sound due to the cylinder firing and friction created by the revolving engine parts, and resulting vibrations in the engine block. These sounds are really an unwanted output from the engine system and are therefore termed noise (Mitson, 1989). The noises produced are so loud that being in the engine room of an operating vessel for long periods without ear protection may be damaging to human hearing. The noises are also readily transmitted through the steel structure of the vessels, and the noise level may be irritatingly high in the cabins and working rooms of many older vessels. Legislation in many countries now sets limits to the maximum noise level in the various rooms and sections of vessels, and modern vessels are therefore built with precautions that reduce internal noise substantially.

The steel hull transmits some of the engine noise underwater. The hull may also vibrate due to the hydrodynamic forces that act as it moves through the sea. In addition, the hull is affected by the propeller shaft which sets up vibrations and generates friction, and by the propeller itself at a frequency equal to the blade rate (rotations per minute multiplied by the number of propeller blades and divided by 60). A revolving propeller also sets up strong negative pressures in the water that passes the propeller blades. This creates small bubbles that rupture and thereby generate a loud, sharp sound, the cavitation noise (Urick, 1983). The sound from thousands of collapsing bubbles merges into a continuous and characteristic 'hiss' from the propeller. The main cavitation mechanisms are negative pressure set up from front to rear along the propeller blade surface, and the vortex stream of bubbles caused by the tip of the revolving propeller blades (Mitson, 1992).

Underwater, the noise from the engines, the propeller shaft and the blade rate of the propeller is transmitted by the hull mainly as discrete line frequencies. The frequency range in which these sources may be recognised will depend on the number of revolutions, but the shaft rate will usually lie from below 1 up to a few Hz, the blade rate from a few to about 10 Hz, and the engine firing rate from below 10 up to over 100 Hz (Fig. 5.1). The line frequencies are superimposed on a more continuous

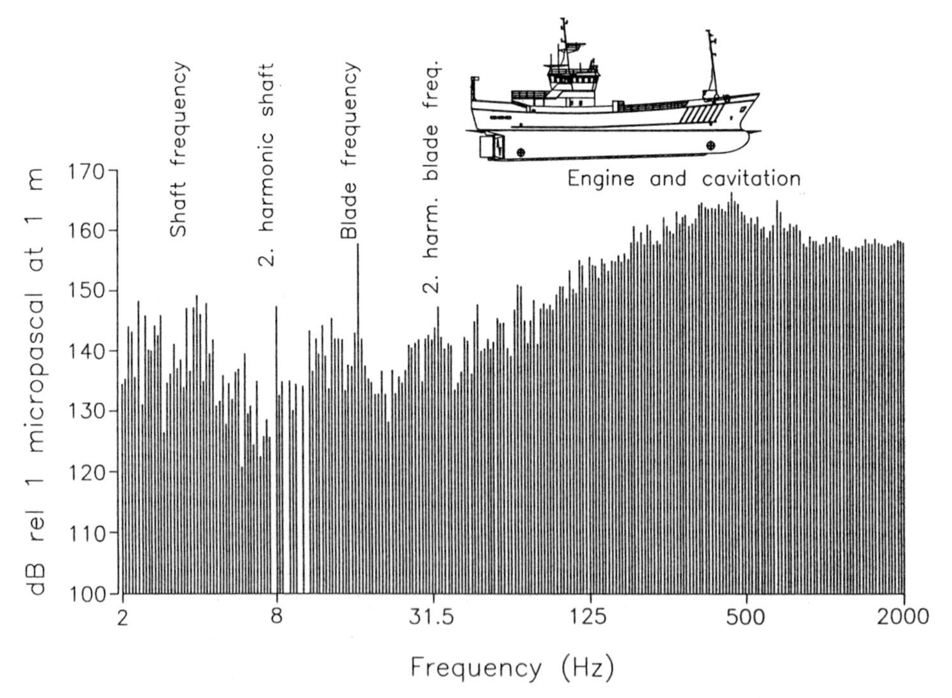

Fig. 5.1 Distribution of sound intensity recorded at 10 m depth, analysed in 1/24 octave and calculated to source level at starboard side from the Norwegian purse seiner MV *Ligrunn* (769 GRT, engine 2250 HP) when cruising at 10 knots.

spectrum generated by the propeller cavitation (Urick, 1983). This is best demonstrated if the vessel-generated noise is illustrated in a spectrum that presents measurements carried out with a narrow, 1 Hz bandwidth. In such graphs the line frequencies will appear as distinct spikes that rise above the more continuous spectrum. Harmonics to loud engine- or propeller-generated spikes may then be recognised at higher frequencies.

However, it is difficult to get a clear overall picture of a vessel's noise signature from narrow band spectrum graphs. This can be gained by using a third-octave analysis where the energy is averaged over 1/3 of each doubling of the frequency, and which permits graphing of the whole vessel noise signature over the whole frequency range up to several hundred kilohertz. It is more difficult to distinguish line frequencies as the higher the frequency interval in which they appear, the more they will be smoothed out, but they will contribute to a general rise in the actual frequency intervals. The maximum energy output from the cavitation noise is in the frequency range from about 50 Hz up to about 1000 Hz.

The vessel-generated sound follows the physical laws for transmission of sound underwater. The noise energy therefore decays due to geometrical spread, which causes a decibel reduction in the level equal to 20 multiplied by the logarithm of the range from the source. Shadowing by the hull and the bubbles in the propeller wake generate a certain directivity in the transmitted sound (Urick, 1983). In front and aft of the vessel there will be lower noise intensity than to the sides (Misund *et al.*, 1996). In a horizontal plane close to the surface, the minima will occur within 20° in the fore-and-aft direction (Fig. 5.2). There will also be a similar directivity in the vertical plane. The noise intensity may also differ between the port and starboard sides of the vessel, depending on the direction of the revolution of the propeller and where noisy auxiliaries are positioned.

There are many other sources of variation in the underwater noise radiated from operating vessels. Naturally, the noise depends on the construction, power output and mounting of the engine and the auxiliaries, the size and construction of the vessel, the size, number of blades and their design, and the pitch of the propeller. Therefore, different vessels will have different noise signatures. Comparison of some European fisheries research vessels when running at 11 knots revealed an overall variation of as much as about 30 dB between the most silent and the most noisy vessel (Garnier *et al.*, 1992). Even rather similar vessels may have substantial differences in the noise signature. Comparison of three Norwegian factory trawlers built after the same sheer draught, that were equipped with engines of equal power output but with slightly different propeller units, revealed substantial differences over the whole frequency range (Engås *et al.*, 1995; Fig. 5.3). Modern vessels are often equipped with axial generators for production of electricity that require constant revolution of the main engine. To adjust the speed, such vessels are often equipped with a controllable-pitch propeller that can be considered as an almost infinitely variable noise generator (Mitson, 1992).

At the end of the 1960s, some concern was aroused in the Icelandic and Norwegian herring purse seine fleet that the underwater noise generated by the vessels had a significant influence on the catch rates (Olsen, 1971). A series of noise measurements

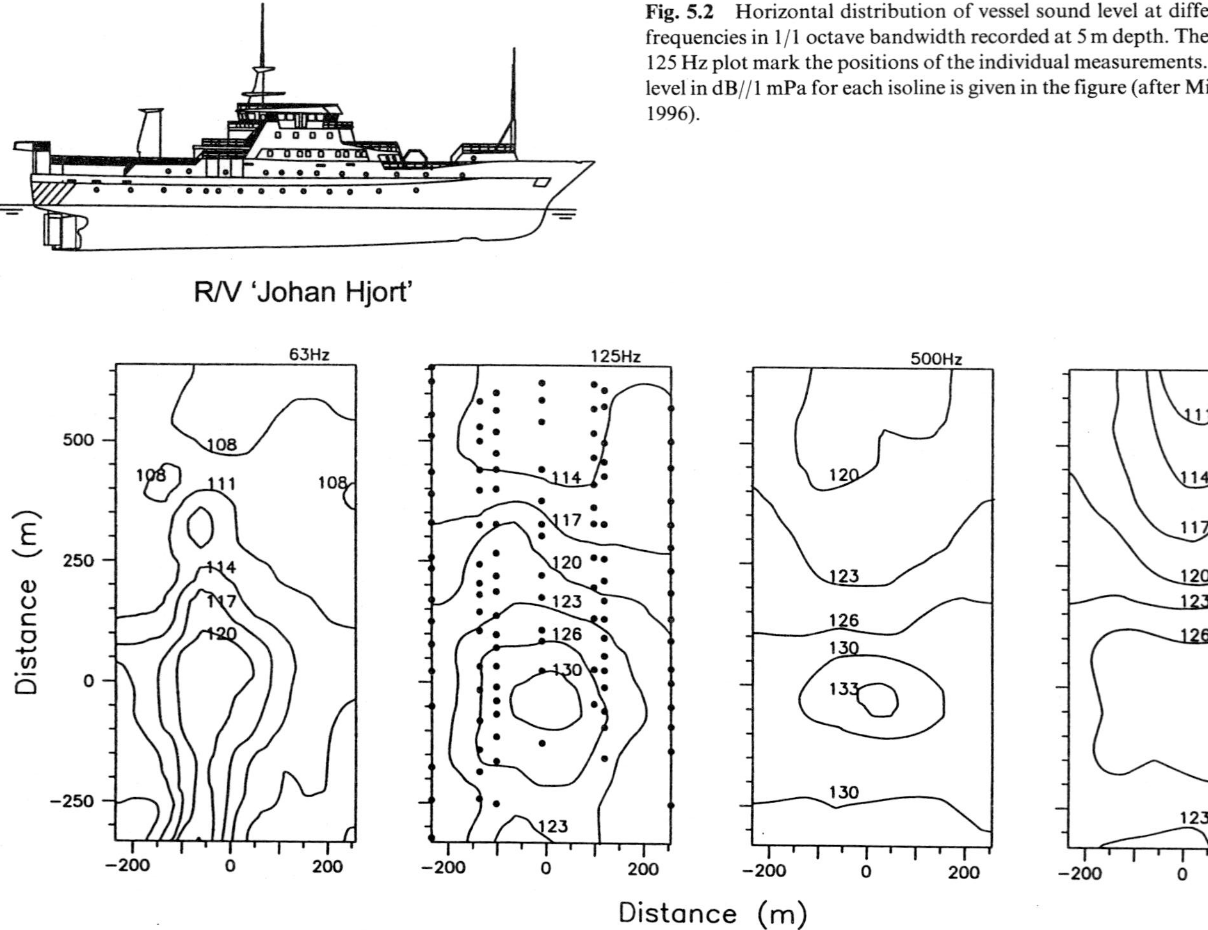

Fig. 5.2 Horizontal distribution of vessel sound level at different centre frequencies in 1/1 octave bandwidth recorded at 5 m depth. The dots in the 125 Hz plot mark the positions of the individual measurements. The sound level in dB//1 mPa for each isoline is given in the figure (after Misund *et al.*, 1996).

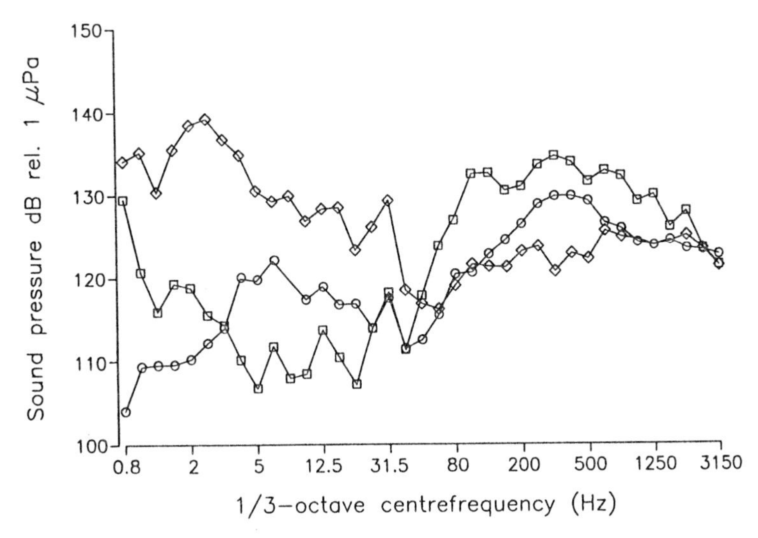

Fig. 5.3 Frequency distribution at maximal sound intensity in 1/3 octave bandwidth during passage of trawlers A, B, and C (after Engås *et al.*, 1995). □ = Vessel A; ○ = vessel B; ◇ = vessel C.

confirmed that operating purse seine vessels could be very noisy (Gjestland, 1968). In some cases, extreme sound sources such as strongly cavitating propellers were diagnosed and the necessary action to reduce the noise was prescribed. There was also some concern that the vessel-generated noise might have significant influence on the catch rates in other fisheries and for other fishing methods (Chapman & Hawkins, 1969; FAO, 1970).

A problem that has received some attention, especially on fisheries research vessels, is that the vessel noise can interfere with echo sounders and sonars working at about 10 kHz and higher, and can be a limiting factor in the fish detection capability of these instruments. However, most vessels are rather silent at such high frequencies.

Throughout the 1970s and 1980s, little effort was given to reducing the underwater noise emitted from fishing and fisheries research vessels. The last generation of Norwegian factory trawlers, delivered between 1987 and 1989, was therefore built without any special precautions to reduce underwater noise. Some of these vessels experience lower catch rates than expected and this is claimed to be the result of strong fish avoidance reactions (Engås *et al.*, 1995). Special attention was paid to one of the vessels which was claimed to be very noisy and to fish significantly less than the other trawlers. Indeed, a rumour arose that the other vessels had developed a tactic to take up a position somewhat behind and slightly to the side of the noisy vessel when trawling. This was claimed to give higher catches because some of the fish in the path of the noisy vessel were scared to the side and into the path of the trawlers following.

Likewise, series of fisheries research vessels have been built without any other precautions for the noise radiated underwater than that it must not interfere too much with the working frequencies of the acoustic fish recording instruments. An exception to this trend was the English RV *Ernest Holt* driven by diesel-electric engines mounted in an acoustic enclosure to reduce the radiated noise (FAO, 1970). On the Polish RV

Prof. Siedilecki, antinoise and antivibration precautions such as an elastically supported propulsion system made the vessel more silent than conventionally built research vessels (Ojak, 1988).

In some of the more recently delivered fisheries research vessels substantial efforts have been made to reduce the noise radiated underwater. The English RV *Corystes* delivered in 1988 is so far the most silent among the operating European fisheries research vessels (Garnier *et al.*, 1992). A maximum noise level radiated underwater was specified for this vessel when contracted, and the vessels was therefore designed, constructed and engined to be silent (Kay *et al.*, 1991). Control measurements of underwater noise during test trials revealed unacceptably loud noise düe to the alternating current speed control, and a solution that eliminated this noise source was therefore mounted (Mitson, 1989). It was also detected that polyamide bearings of the rudder flap produced a loud squeal above 1 kHz each time the rudder moved (Mitson, 1992), and therefore had to be replaced by bearings of another material.

For the Japanese *Kaiyo-Maru*, which was delivered in 1991, there were also specified certain criteria for maximum radiated noise underwater (Furusawa & Takao, 1992). A series of noise-reducing actions were therefore taken during building, and selection and installation of the engines and propulsion systems. Among others, the vessel is equipped with two engine systems: one diesel-engined propulsion system for ordinary operation, and one silent electric propulsion system to be used during quantitative echo integration. Underwater noise measurements revealed that the vessel was more silent than specified. To avoid unwanted effects of vessel-generated noise, future fishing and fisheries research vessels should be constructed and equipped so that the radiated noise underwater is minimised. Such objectives were applied during the new building of the French *Thalassa* and the Scottish *Scottia*.

5.1.3 *Fish hearing*

The hearing ability of fish has been mapped through experiments to find the frequency range of sounds that fish can detect, and whether fish can detect the direction and distance to the sound source. Several studies have also been conducted to reveal whether fish can discriminate frequencies, amplitudes and temporal structure of sound (Schwarz, 1985). Many experiments have been conducted in small tanks and aquaria, often constructed and arranged so as to reduce the noise from the surroundings (Hawkins, 1981). Special tanks have also been constructed for standing waves and with the possibility of varying the pressure and particle motion of the sound independently (Hawkins & MacLennan, 1976). To avoid unpredictable effects from the tank walls, some experiments have been conducted in midwater in sea lochs and fjords, well away from reflecting boundaries (Chapman, 1970; Chapman and Hawkins, 1974; Schuijf, 1975).

The hearing organ

Fish detect sounds through the otolith organs which are embedded in a pair of labyrinths located in the skull. The labyrinths are complicated structures of canals, sacs and ducts filled with an endolymph fluid. There are three semicircular canals orientated at right angles to each other. At the base of each canal there is an ampulla containing sensory hair cells embedded in a cupula of jelly. Through these structures, fish sense spatial acceleration because the endolymph will lag behind movements of the canals and bend the cupula and thereby stimulate the hair cells.

Each labyrinth has three otoliths, the utriculus, sacculus and lagena. The otoliths are located in sacs connected to each other and the semicircular canals, and they are orientated at different axes in space. The stone-like otoliths are made of calcium carbonate and inorganic salts covered by jelly, and lie on a thin membrane that protects a macula of numerous hair cells. The membrane connects to both structures and holds them in the same position relative to each other. The hair cells are made of a bundle of apical cilia with an eccentrically positioned kinocilium and more than 40 stereocilia (Rogers *et al.*, 1988). There are numerous hair cells in the sensory epithelia under each otolith, and the hair cells are orientated in groups which have the kinocilium in the same direction as the ciliary bundle.

Each hair cell is directionally sensitive. Maximum depolarisation receptor potential is generated when the ciliary bundle of the hair cells bends directly in the direction of the kinocilium. Bending in the opposite direction causes a hyperpolarisation receptor potential, while bending in any other direction induces a receptor potential that is proportional to the cosine of the direction of stimulation (Rogers *et al.*, 1988).

The otoliths are three times denser than water and the flesh of the fish, and thus have different acoustic properties. Imposed sound waves will cause movements of the fish body at the same amplitude and phase as in the surrounding water. The denser otoliths, however, will move at smaller amplitudes than the rest of the body and in a different phase. The otoliths and the sensory epithelia with the hair cells will therefore move relative to each other as they are connected only by the thin otolith membrane. The cilia of the hair cells will bend proportionally to the particle displacements from the relative motions between the otoliths and the sensory epithelia. As the hair cells bend they will produce physiological responses that are transmitted through the nervous system to the brain, where they are interpreted and analysed as sound. Thus the signals from the hair cells contain information on the amplitude, the frequency and the direction of the sound stimuli. Besides the function of sound detection, the otoliths are also gravistatic receptors, giving the necessary signals for determining spatial orientation in most fishes (Platt & Popper, 1981).

The principle for sensing the particle motion component of imposed sound waves is regarded as the direct method of sound detection. Many fishes also have the possibility of sensing sounds through indirect detection of the pressure component of imposed sound fields. This is possible through the swimbladder. Due to the high compressibility of gas, sound pressure fluctuations will set up proportional motions of the swimbladder wall (Sand & Enger, 1973; Sand & Hawkins, 1973). These motions reradiate some of the imposed sound energy, and the reflected sound waves have an enhanced particle-displacement-to-pressure ratio. The reflected sound waves will set up proportional movements of the otoliths relative to the sensory epithelia, thereby producing the necessary physiological signals for sound detection. This indirect method of sound detection seems most effective for frequencies above a few hundred Hz (Fay & Popper, 1974, 1975).

Detection of direction and distance to sound sources

Implicit from the structure of the sensory epithelia of the hearing organs, fish are able to detect the direction of a sound source. By observing the direction of avoidance responses of caged herring, Olsen (1969, 1976) found that herring could sense the direction of sound sources to within at least 45°, a result confirmed by Sorokin (1989). Using reward conditioning, Schuijf (1974) observed that the wrasse (*Labrus bergylta*) located sound sources to within + 20°. Similar accuracy in the ability to detect sound direction has been documented through both cardiac and reward conditioning experiments for cod (Chapman & Johnston, 1974; Schuijf & Siemelink, 1974; Schuijf, 1975; Hawkins & Sand, 1977).

Even a single sound wave in the sensitive frequency range is enough for the herring to receive sufficient information on amplitude and direction of the sound source, and thereby elicit directional avoidance responses (Blaxter *et al.*, 1981). The polarity (i.e. whether the sound wave starts with a compression or a decompression) does not matter. This indicates that fish have a precise sensory system for spatial hearing that rapidly determines the direction of the sound source. Although a proper, general explanation of how such a system functions is still lacking, promising suggestions have been put forward.

For fish without a swim bladder or any other sound-reflecting organ, the particle motion of an impinging sound wave will make the hair cells of the sensory macula bend proportionally to and fro in the direction of the sound. This may indicate that simple vector determination of the direction of stimulation will give the fish the direction of the sound source. The problem, however, is that sounds from directly opposing sources will cause identical movements of the hair cells, and thereby produce the same nervous signals. This gives rise to 180° ambiguity in simple vector determination of the sound direction.

For fish having a swimbladder, an impinging sound wave will reach the otoliths both directly and indirectly as a reradiation from the swimbladder. The indirect sound wave will contain no information on the direction of the sound source as its direction is only determined by the geometry of the swimbladder-labyrinth system. However, the direct and indirect sound waves will be out of phase. Moreover, sound waves from diametrically opposite sources will produce unequal differences in phase between the direct and indirect stimuli. Schuijf (1975, 1976) therefore suggested that a system for directional hearing could operate by resolving this phase difference in addition to vector determination of the direct input.

There is experimental support for such a mechanism. Cod (*Gadus moruha*) have been shown to be able to discriminate sounds from sources positioned in front and behind (Schuijf & Buwalda, 1975) and this was confirmed by Berg (1985) on whiting (*Odontogadus merlangus*), catfish (*Ictalurus nebulosus*) and cod. The phase difference was shown to be essential for this ability, as the fish interpreted the sound to come from a diametrically opposite source when there was an artificial inversion of this difference. Buwalda *et al.* (1983) showed that cod could discriminate sounds coming from opposite sources also in the vertical and transversal planes. Experiments with manipulating the phase relationships in standing wave fields confirmed that the phase difference of the direct and indirect stimuli was necessary for detection of the direction of the sound source. Rogers *et al.* (1988) put forward a complete hypothesis for fish hearing, including a numerical model for the directional determination based on vector and phase analysis in the central nervous system of the signals from the sensory macula. The direction of a sound source could be determined either by the principle of performed beams in which the sound is located to be within a definite direction interval, or by an algorithm for detecting the incident angle of the sound directly.

The movements of the hair cells set up by an impinging sound wave will be the vectorial sum of the direct and indirect stimuli (Fig. 5.4). Munck and Schellart (1987) calculated that the particle displacement in waves reradiated from a prolate spheroidal air bubble which resembles a swimbladder will be elliptical orbits. By calculating the resulting displacements of the hair cells from the direct and indirect stimuli vectors, Schellart and Munck (1987) proposed that directional hearing could operate through simple vector analysis of the resulting elliptical displacement orbits of the hair cells. This model had difficulty in explaining how fish could separate sources at 0° and 180° which gave rather similar orbits. The model could also provide fish with directional hearing with just one ear, even though Schuijf (1975) had shown that two intact labyrinths were necessary for this ability. Based on experiments with rainbow trout (*Oncorhynchus mykiss*), Schellart and Buwalda (1990) proposed a revised version of the orbit analysis model by incorporating an angle of 90° of the indirect stimuli vectors between the two otholiths in the calculation of the displacement of the hair cells. This gave different displacement orbits for two functional ears for sounds coming from all different directions (Fig. 5.5).

However, despite this promising hypothesis for directional hearing of fish having a swimbladder, it still remains to be explained how this is done in fish without a

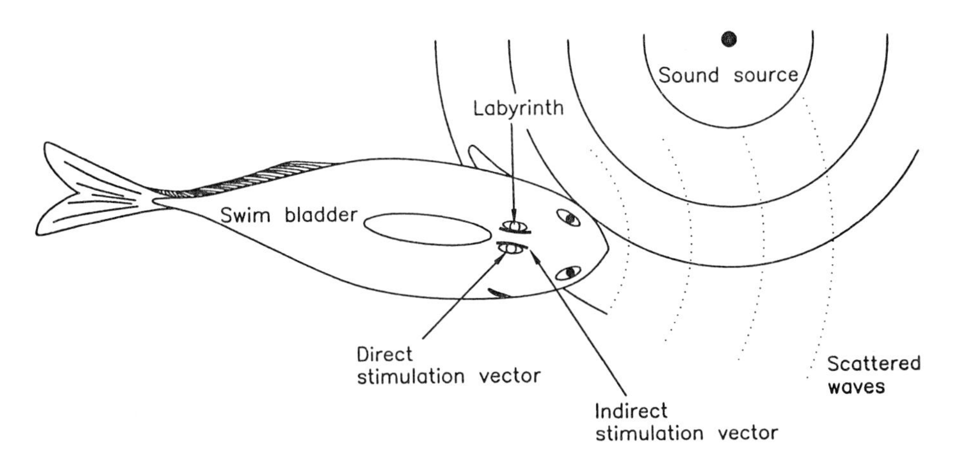

Fig. 5.4 Direct and indirect stimulation of the inner ear by sound waves originating from the sound source and scattered by the swimbladder, respectively (after Schellart & Buwalda, 1990).

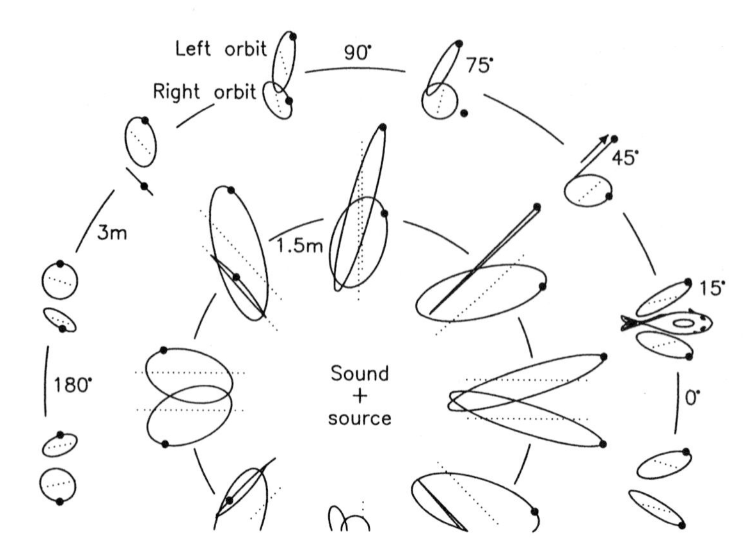

Fig. 5.5 Orbits of the utricular maculae of a 30 cm cod with a 90° difference in the indirect vector acting on each otolith. The upper and lower ellipses represent the left and right utricular orbit, respectively. Each orbit pair gives the position of the swim bladder of the fish relative to the source position. The dashed line within each orbit represents the direct input. Left-right mirrored fish positions give the same result (after Schellart & Buwalda, 1990).

swimbladder. Like sharks that have been shown to orientate towards low-frequency sound sources (Myrberg *et al.*, 1976), it is probable that such fish have directional hearing. As a substitute for the lack of an organ that transforms sound pressure to particle displacement, Schuijf (1975) proposed that fish without a swimbladder, and sharks, could utilise the sounds reradiated from the surface or the sea bottom as indirect stimuli.

As with fish having a swimbladder, the threshold for directional hearing for those lacking this organ lies above the absolute threshold for sound detection. In fish with a swimbladder, indirect stimuli will be stronger than direct stimuli, while the opposite will be the case for fish without a swimbladder. For cod, which has a swimbladder, Schuijf (1975) found that the threshold for directional hearing was about 6 dB above the absolute hearing threshold.

The phase relationship between the indirect sound stimuli transformed from sound pressure and the direct particle motion stimuli is dependent on range from the source. Analysis of this phase relationship could therefore provide fish with perception of distance to the sound source. The distance dependence is most pronounced close to the source. Based on experiments to determine the minimum phase difference cod could detect, Buwalda *et al.* (1983) proposed that cod could distinguish distances in categories of 0–1, 1–2, 2–5 and more than 5 m. Indeed, Schuijf and Hawkins (1983) found that cod were able to distinguish between sound sources placed 1.3, 4.5 and 7.7 m away. Fish are therefore probably able to perceive the distance to sound sources at close range.

Frequency range of fish hearing

The frequency range that fish can sense is illustrated through audiograms (Fig. 5.6), which show the relationship between the minimum detectable sound level and frequency. Such audiograms have been established for a number of species ranging from herring and small goldfish (Enger, 1967), pollack (*Pollachius virens*; Chapman, 1973) and eastern mackerel (*Scomber japonicus*; Sorokin, 1987) to large yellowfin tuna (*Thunnus albacares*; Iversen, 1969). The general trend is that fish are most sensitive to frequencies up to a few kilohertz, and seem incapable of detecting frequencies above 10 kHz. This means that fish are most sensitive to sounds in the frequency region where fishing and fisheries research vessels have the maximum sound energy output. On the other hand, fish are not able to detect the working frequencies of most echosounders and sonars (from 24 kHz to about 200 kHz).

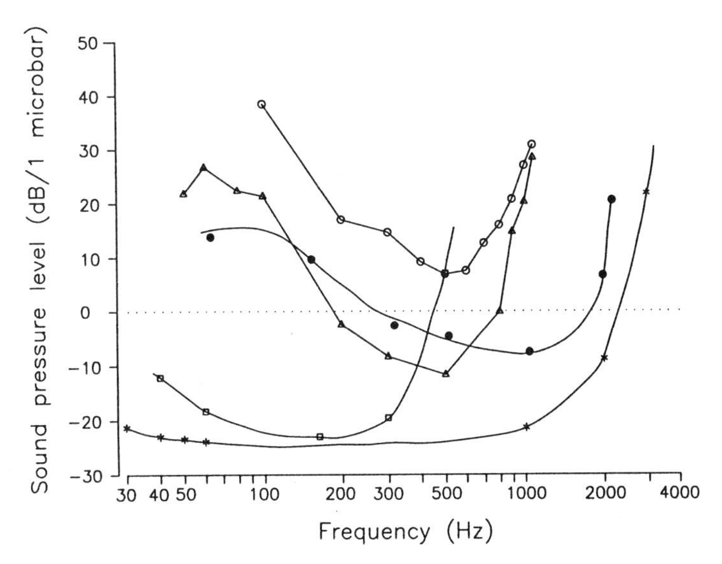

Fig. 5.6 Audiograms of pelagic fishes: herring (after Enger, 1967), eastern mackerel, *Scomber japonicus* (after Sorokin, 1987), pollack, *Pollachius pollachius* (after Chapman, 1973), little tuna, *Euthynnus affinis*, and yellowfin tuna, *Thunnus albacares* (after Iversen, 1969). ○ = Little tuna; △ = yellowfin tuna; ● = eastern mackerel; □ = pollack; * = herring.

Differences in the sensitive frequency range among species are linked to differences in their anatomy. Fish having a linkage between the swimbladder and the inner ear, the otolith organs, can sense sounds both at lower intensities and over a wider frequency range than fish without a swimbladder or those having an isolated one. Hearing ability is therefore well developed in the ostariophysan fishes, like the catfish (*Ictalurus nebulosus*) and the cyprinids that have a bony structure (the Weberian ossicles) between the swimbladder and the skull where the inner ear is embedded (Hawkins, 1986). In the cypriniform fishes there is also a close connection between the two organs, and in the Holocentridae the hearing ability seems proportional to the degree of linkage between the swimbladder and the inner ear (Hawkins, 1986).

Among pelagic species, the herring seems to have a very good hearing ability, being able to sense frequencies of up to about 4 kHz (Enger, 1967). For cod that have an isolated swimbladder, hearing ceases above 470 Hz (Chapman & Hawkins, 1973). This is the case also for other gadoids such as haddock (*Melanogrammus aeglefinus*), ling (*Molva molva*), and pollack (Chapman, 1973; Fig. 5.6).

The herring has a direct coupling by a pair of pro-otic bullae connecting gas-filled diverticula extending from the swimbladder to the perilymph of the otholith organ (Allen *et al.*, 1976). A thin membrane under tension separates the swimbladder gas and the perilymph in the pro-otic bullae. When the herring changes depth, the compressible swimbladder will change volume and gas will pass through the diverticula to equalise the pressure. This probably ensures that the pressure in the gas of the otic bullae is equal to that in the perilymph at all depths and maintains a constant tension on the membrane (Blaxter, 1985). Imposed sound waves will be amplified by the swimbladder and will set up vibrations in the membrane which are thereby transmitted to the perilymph and the inner ear. In herring there is also contact between the lateral line and the inner ear through a membrane on each side of the skull that separates the perilymph and the endolymph of the lateral line system.

Older audiograms indicated that the sensitivity to sounds decreased towards the very low frequencies. However, more recent experiments have shown that species like cod (Sand & Karlsen, 1986), plaice (Karlsen, 1992a), perch (Karlsen, 1992b) and Atlantic salmon (Knudsen *et al.*, 1992) can detect sounds in the infrasound region (< 10 Hz) at about the same low intensities as for higher frequencies (Fig. 5.7). As generating very-low-frequency sound underwater is rather difficult, these experiments were conducted in a specially constructed, standing-wave tube with the fish placed in

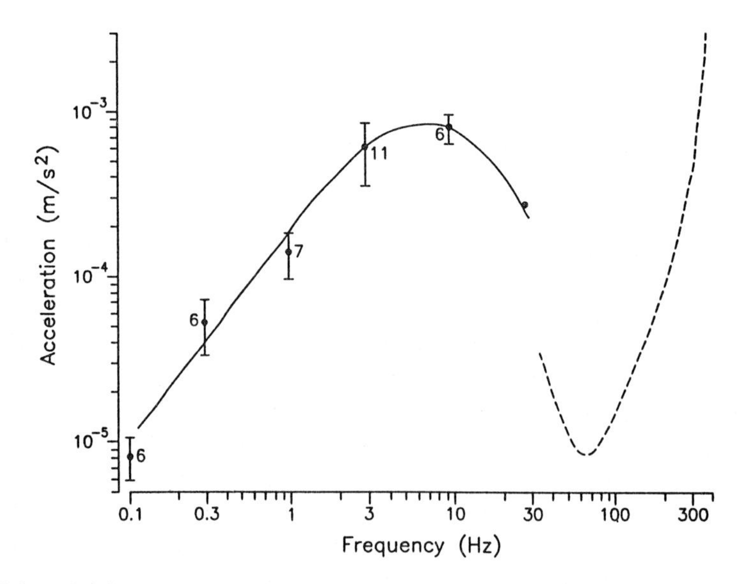

Fig. 5.7 Infrasound threshold for cod for frequencies < 30 Hz, and with the motion sensitivity of the otolith organs for frequencies > 30 Hz recalculated from Chapman and Hawkins (1973) (after Sand & Karlsen, 1986).

a central plastic cage. The sound waves were generated by vibrator-driven pistons at each end of the tube.

The ambient noise level in the sea is a limiting factor for the hearing of fish, especially for those having a swimbladder (Hawkins, 1986). Buerkle (1968) showed that the hearing threshold for cod could vary by nearly 20 dB depending on the background noise level in the experimental tank, and Chapman (1973) established a direct relationship between the background noise level and the minimum sensitivity level of pollack (Fig. 5.8). This ambient noise dependence induces difficulty in comparing audiograms established both for a single species and among species. The audiograms established for goldfish differ by up to about 40 dB in the most sensitive frequency region, probably due in part to varying ambient noise levels in the different experiments (Hawkins, 1973).

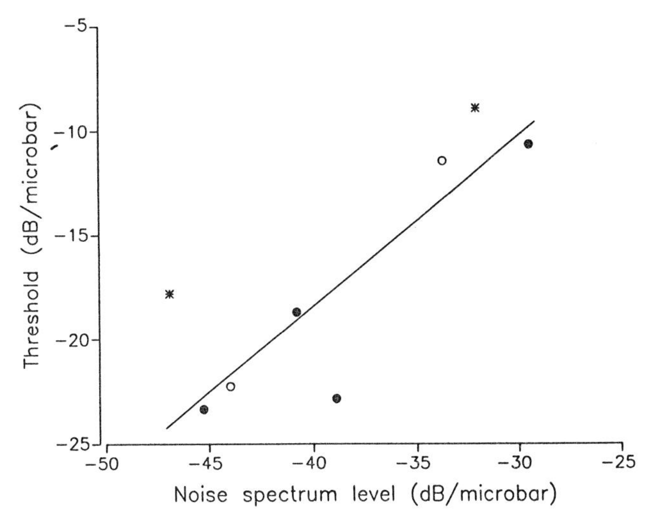

Fig. 5.8 Relationship between thresholds obtained from pollack and ambient noise level for three frequencies (after Chapman, 1973). ○ = 110 Hz; ● = 160 Hz; * = 310 Hz.

Another reason for caution when comparing various audiograms is that different techniques may have been used to measure the sensitivity of the fish. In most experiments, the fish have been conditioned to respond to detectable sounds by obtaining reward (food) or punishment (electric shock) in association with the stimuli. A much-used and objective technique is cardiac conditioning, where a clearly detectable sound stimulus is followed by an electric shock. After repeated treatment, the fish will develop the cardiac response, which consists of a slowing of the heart rate upon detection of sound stimuli. Some experiments have also been based on electrophysiological recordings of the functioning of the hearing organs. For instance, as herring may be a difficult species to keep alive and condition in controllable laboratory conditions, Enger (1967) established a tentative audiogram for this species by measuring electric potentials in the brain of anaesthetised individuals. It is therefore to be hoped that a proper audiogram for live herring with full sensory capacity will be established in future experiments, as observations on the swimming behaviour during survey and fishing operations of this economically important species are accumulating.

There are now indications that clupeoids in fresh water are able to detect sounds two orders of magnitude higher in frequency than older literature indicates. Nestler *et al.* (1992) observed that blueback herring (*Alosa aestivalis*) reacted more strongly to high-frequency sounds between 110 and 140 kHz than to low-frequency sounds between 0.1 and 1 kHz. Dunning *et al.* (1992) also found that the alewife (*Alosa pseudoharengus*) responded to high-frequency sounds. Nestler *et al.* (1992) speculate that high-frequency sound may be detected through the auditory bullae system, and that the detection of high-frequency sounds may have evolved as a predator warning of the high-frequency echolocation clicks of whales.

Discrimination of frequencies, amplitude and temporal structure of sound

Fish can distinguish between frequencies of sounds. As with sound intensity, there are clear species differences in this ability (Hawkins, 1986). Fay (1970) found that a greater separation interval was required, the higher the frequencies to be distinguished. As a means for discriminating frequencies, fish seem able to filter out and tune in on particular frequencies. This enables detection of weak tones in the presence of a high ambient noise lèvel. However, sounds may be masked by high noise that occurs at about the same frequency band (Buerkle 1969), and the higher the frequency of the sound, the wider the frequency band of the noise that can mask it (Hawkins & Chapman, 1975).

Some fishes can produce and use sounds to communicate, for instance to find mates, during courtship or agonistic behaviour, but as far as we know this kind of communication does not exist for pelagic species. The sounds produced are not very complex in frequency structure but can be modulated with a complex repetition rate and varying intensity. To be able to receive information through such sounds fish hearing has adapted to resolve the temporal structure of the sound, and goldfish seem able to distinguish pulses of short duration better than humans (Fay, 1970). Similar adaptations may exist for discrimination of sounds differing in amplitude, and haddock are able to separate 50 Hz tones differing by only 1.3 dB in intensity (Hawkins, 1986).

5.1.4 *Fish reactions to noise*

Fishermen have observed that the approach of noisy vessels might scare the fish. This was especially noted during purse seine fisheries for herring in the 1960s (Olsen, 1971). The reactions could often be observed during sudden changes of the vessel manoeuvring which caused simultaneous changes in the vessel noise. This was therefore taken as a clear indication that the avoidance reactions were elicited by vessel noise.

Similarly, comparison of audiograms of gadoids and noise spectra emitted from fisheries research and fishing vessels revealed that fish were most sensitive to sounds in the low-frequency region where the vessel noise was most intense (Olsen, 1967;

Hering, 1968; Chapman & Hawkins, 1969; Freytag & Karger, 1969). Based on an audiogram established by food rewarding of cod and noise measurements of the Norwegian research vessels RV *G.O. Sars* and RV *Johan Hjort*, Olsen (1967) suggested that the cod could hear the two vessels at distances of up to at least about 80 m. Hawkins and Chapman (1969) established another audiogram for cod by cardiac conditioning which indicated that the fish could be sensitive to noise from vessels several miles away. The discrepancy between these two estimates is most probably due to different ambient noise levels during the fish hearing experiments. This factor was taken into account by Buerkle (1974), who measured the noise from a trawler at the same location during summer, and in winter when the ambient noise level was about 10 dB higher. Comparison of these measurements with previously established audiograms for cod under different noise conditions (Buerkle, 1968, 1969), indicated that the fish could hear the trawler at 3.2 km away in summer and at 2.5 km in winter. Buerkle (1977) recorded higher gillnet catches of cod in the presence of a trawler, which indicated that the cod reacted to sound of the nearby trawler with increased swimming activity and thereby higher probability of being caught by the gillnet.

The Olsen model

By generalising a substantial number of fish reactions to an approaching survey vessel, Olsen *et al.* (1983a) proposed a vessel avoidance model in which the low-frequency vessel noise is supposed to be the eliciting stimulus (Fig. 5.9). At close range to the vessel, it is also assumed that the hydrodynamic pressure wave produced by the hull when moving through the water may have some effect. The model assumes that the vessel noise can be sensed by detection of the particle motion component of sound via the reradiated indirect stimuli from the pressure-converting swimbladder. Avoidance reactions are supposed to be triggered mainly by a momentary increase in the sound pressure, but the actual noise level from the vessel is also of importance. The sound pressure gradient will therefore be a relevant measure of the stimulus that elicits the avoidance. To formulate the model mathematically, let P_f be the pressure amplitude of the vessel noise as perceived by a fish at a distance R from the noise source. The vessel noise pressure in the position of the fish is related to the pressure at the noise source (P_o) by:

$$P_f = P_o/R \qquad (5.1)$$

If the vessel is running at a speed V and the fish is at an angle a from the horizontal in a plane coordinate system with the noise source in the centre, the pressure gradient of the vessel noise is expressed by:

$$dPf/dR = -P_o/R^2 * \sin a * (V * \cos a) \qquad (5.2)$$

The term (sin a) is an approximation for the assumed directivity characteristics of the noise source, and (V * cos a) is the component of the vessel speed in the direction of the fish. To reduce the strength of the imposed noise stimuli, the fish may react to the approaching vessel by trying to swim away from it. This will alter the vessel noise gradient by:

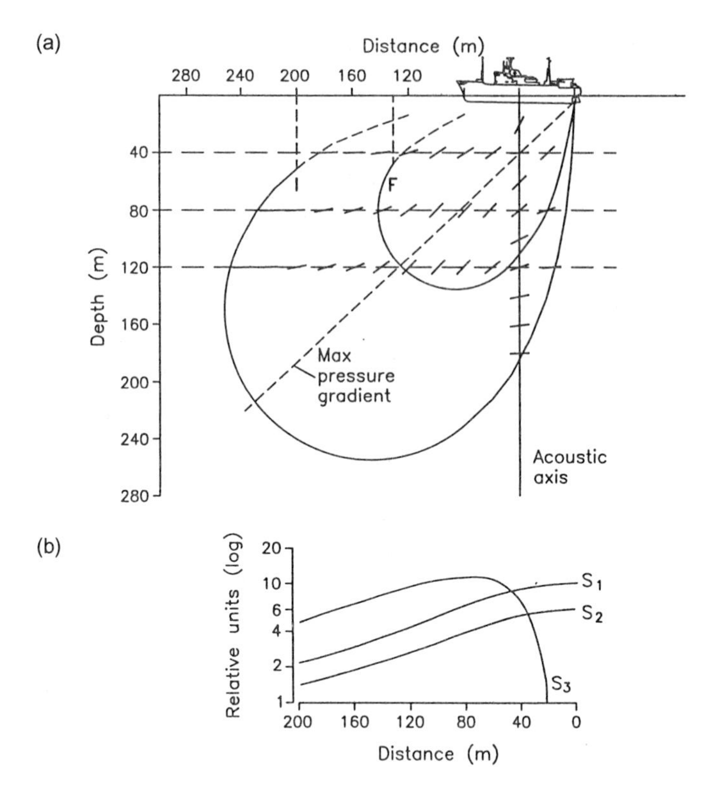

Fig. 5.9 The Olsen model. (a) Vertical distribution of pressure gradients from RV *G.O. Sars* at 12 knots. (b) Propagation of different stimuli. S_1 = pressure amplitude; S_2 = pressure gradient; S_3 = deviation from direct approach.

$$dP_f/dR = -P_o/R^2 * \sin a * \{(V * \cos a) - [V_f * \cos (a - b)]\} \quad (5.3)$$

The term $[V_f * \cos (a - b)]$ is an expression for the swimming speed component V_f of the fish away from the vessel where b is the tilt angle of the fish.

The connection between the sound pressure and the sound pressure gradient as related to the distance to approaching vessels is illustrated in Fig. 5.9. The sound pressure increases gradually as the vessel approaches, but reaches an asymptote just before the vessel passes over. This leads to a sharp drop in the sound pressure gradient just before the vessel passes. Due to the directivity of the vessel noise, the fall in the sound pressure gradient occurs further in front of the vessel the greater the depth. Figure 5.9 also illustrates that fish that adjust their behaviour to minimise the noise stimuli directly approaching can reduce the imposed noise pressure substantially.

The pressure wave (P_h) induced by the movement of the hull may be sensed by fish close to the surface. The nature of this stimulus is looked upon as originating from a Bernoulli source, and is proportional to the length (L) and speed (V) of the vessel:

$$P_h = L^2 * V^2/R^2 \quad (5.4)$$

This indicates that the pressure gradient in this noise source will decrease in a manner inversely proportional to the cube of the depth. This is also the case for the nearfield wave stimuli from the propeller where the particle motion component (d_P) is given by:

$$d_P = (1/R + 1/R^2) \quad (5.5)$$

According to this model of the vessel noise stimuli, fish may minimise the imposed noise stimuli by performing avoidance swimming directly away from the approaching noise source. Fish that react according to this principle will increase their swimming speed and swim gradually more downward as the vessel approaches. If fish come so close to the vessel that they detect the nearfield stimulus from the hull or the propeller itself, panic reactions with fleeing in all directions may be elicited. The avoidance swimming will cease when the fish are out of the near field and start perceiving that the avoidance swimming outruns the approach of the noise source.

Playback experiments

To study the isolated effect of vessel noise on fish behaviour, experiments have been conducted by playback of vessel noise recordings to fish in pens. Schwarz and Greer (1984) recorded sounds from seiners (size 18–20 m), trollers (size 15 m), and gill-netters (size 12 m). The sound pressure level was highest for the bigger vessels, which also generated most of the sound energy at lower frequencies than the smaller vessels. The sounds were projected at about 30 to 40 dB above the ambient noise level to groups of about 500 Pacific herring (*Clupea harengus pallasi*) shoaling in a small net pen. To most sounds from the different vessel categories, the herring showed directional avoidance responses away from the sound projector. The responses occurred as changes from shoaling activities such as feeding to organisation in structured schooling and swimming in the lower part of the pen.

Occasionally, the herring performed startle responses with quick swimming movements for a short time when exposed to the vessel noise playbacks. The responses were most pronounced to the noise from the bigger vessels, but were also significant for smaller vessels on accelerated approach, or triads of synthesised sounds, contrary to natural sounds (rain on the water, predator noise or conspecific noise). There was an inverse relationship between the number of groups that responded to a noise category and the number of groups that habituated to that category. This indicates that the more effective the sound from a certain vessel category in eliciting avoidance responses, the lower the probability that the herring were habituated to that sound.

A similar playback experiment was set up by Engås *et al.* (1995) to study fish reactions to sounds from a Norwegian factory trawler that was claimed to be noisy and therefore fishing less than the sister vessels (vessel A, Fig. 5.3). Groups of about 20 cod or herring were placed in a small net pen (3 × 3 × 3 m) and exposed to playback of sound recordings of the trawler through an underwater loudspeaker positioned at 2 m depth and 11 m away from the pen. At that distance the fish were held outside the near field of the lowest frequencies transmitted of about 20 Hz (Hawkins, 1986). The fish were exposed to the original sounds recorded when the vessel was towing a bottom trawl, emitted at an intensity of at most about 40 dB above the ambient noise level at the experimental site.

Both cod and herring reacted when exposed to the vessel noise. The cod packed together, polarised and swam slowly down to the bottom of the pen, up along the net walls, and down to the bottom again. Prior to the playbacks, the herring were mostly shoaling slowly just above the bottom of the pen. When exposed to the sound, the herring began strict schooling and swam in a diagonal path along the bottom of the pen. The reactions of both cod and herring were interpreted as distinct avoidance. No alarm responses – as seen when the cod fled at burst speed to the bottom of the pen on being approached closely by a small skiff running at full speed – were observed during the playback sequences. The reason is probably the need for combined visual and auditive stimuli to elicit alarm responses.

To study whether there were frequency-dependent effects, the fish were also exposed to sequences filtered to contain the original sound in frequency regions to which the cod, according to Chapman and Hawkins (1973), has maximum sensitivity (60–300 Hz) and decreasing sensitivity (20–60 Hz and 300–3000 Hz). The herring seem to be maximally sensitive to sounds from 50 Hz to about 1200 Hz (Enger, 1967), and should thereby have rather similar sensitivity to the three frequency categories. Despite the filtering, the sound sequences contained substantial sound outside untouched frequency regions. The maximum sound level during the projections of these sequences was from about 20 to 30 dB above the ambient noise level.

In the most sensitive frequency region for both species (60–300 Hz), the sound level for the sequences of the 20–60 Hz category reached about 118 dB re 1 mPa. This is about 10 dB above the hearing threshold for cod at an ambient noise level of 90 dB re 1 mPa when extrapolating from Chapman and Hawkins (1973). However, both cod and herring hardly reacted to the 20–60 Hz sequences, but performed distinct reactions to the other sequences, which reached 5 to 10 dB higher in the most sensitive frequency region. This indicates that the vessel noise level in the most sensitive frequency region is the main determinant for eliciting avoidance reactions.

Because fish are capable of discriminating sounds differing in temporal structure, sequences were also presented in which the temporal variation in the original vessel sound had been smoothed. These were obtained by editing the original recordings by an automatic gain-controlled amplifier and subsequent triangle amplification. The fish were exposed to both full and frequency-filtered sequences that were smoothed.

The fish started to react to the smoothed sequences at the same time as to the original sequences, which indicated that sound components that elicit the avoidance responses were present also in the smoothed versions. However, both cod and herring clearly judged the smoothed sequences to be different from the original ones as the frequency of responses decreased and the reactions ceased earlier. This indicates that the temporal structure of vessel noise may have significance for the avoidance behaviour of fish. Similarly, Schwarz and Greer (1984) found that herring reacted more to pulsed than to continuous sound.

Analysis of catch data

The catching efficiency of vessels depends on a number of factors, and acceptable connections between the catch results and the sound characteristics of vessels are therefore scarce. Engås *et al.* (1995) found that a Norwegian factory trawler, claimed to fish less than the others due to heavy noise, actually caught about the same per unit of effort and even more cod (Table 5.1) than some of the more silent sister vessels (Fig. 5.3). For saithe, however, the noisy vessel caught significantly less than the others (Table 5.1). At least this indicates that the heavy noise of this vessel may have a negative effect on the catching efficiency of saithe.

Table 5.1 Catch per unit effort (CPUE) in tonnes per hour of the noisy vessel A compared with the sister vessels B and C. The comparisons are made from catch data when the trawlers fished with bottom trawls in the same area at the same time in the North Sea and the Barents Sea in 1988 (after Engås *et al.* 1995).

Species	Vessel A CPUE	Vessel B CPUE	n	Vessel A CPUE	Vessel C CPUE	n
Cod	0.72	0.45	41*	0.33	0.49	37
Saithe	0.24	0.38	26*	0.16	0.28	31*

n = number of comparisons
* = significant paired *t*-test at $p < 0.01$

Erickson (1979) found that trolling vessels that generated distinct, high-intensity spikes above 1500 Hz in the frequency spectrum caught significantly less albacore tuna than those vessels generating a more even frequency spectrum. Foote (1979) proposed that this result could alternatively be explained by attraction to the vessels generating a smooth sound spectrum because these vessels also generated the highest intensity at the lowest frequencies. However, Bercy and Bordeau (1987a) have also shown that high-intensity spikes in the frequency spectrum of vessel noise can affect the catch of tuna. They categorised the sound spectrum from pole-and-line vessels fishing for tuna in the Indian Ocean into those with a smooth spectrum, those with medium-intensity spikes, and those with high-intensity spikes ranging 3–5 dB over the average spectrum (Fig. 5.10). The catch results of the vessels were then grouped according to the respective sound spectrum category. This revealed a rather evident connection between the catching efficiency and the sounds emitted by the vessel. The less spiky the frequency spectrum, the better the catching efficiency. However, the amount of data is rather limited, and it is not known whether there are other characteristics of these vessels such as speed, rigging and handling equipment, instrumentation, age etc. that could categorise the vessels in a similar manner and thereby also contribute to the observed differences in catching efficiency.

Nicholson *et al.* (1992) conducted a fishing experiment with RV *Corystes* to test the effect of a spike at about 300 Hz which ranged high above the spectrum level. The spike was due to an alternating current speed control of the diesel-electric engines. It was possible to switch this speed control and thereby the spike on and off, and this was conducted during alternating trawl hauls in the North Sea. However, analysis of

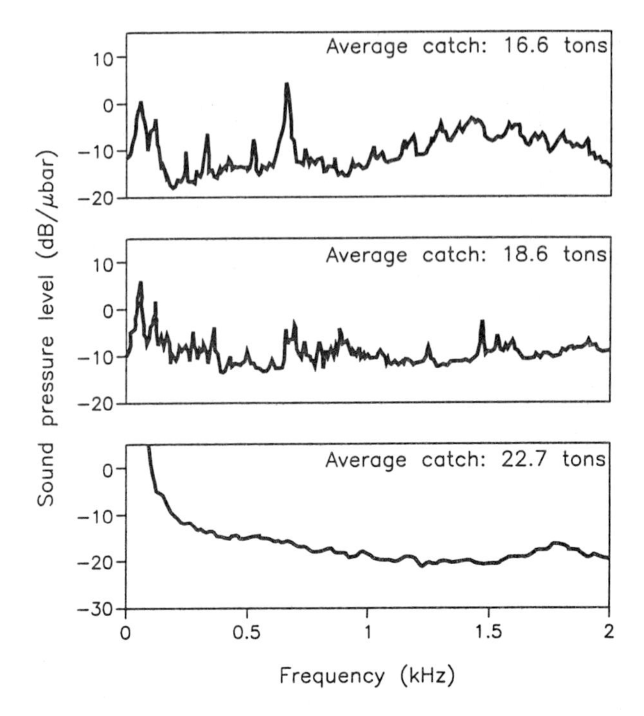

Fig. 5.10 Characteristic sound spectra and average catch rates of trolling vessels in the French tuna fleet (after Bercy & Bordeau, 1987a).

the catch data did not reveal any systematic influence of the high-intensity spike either on the species composition or on the size of the catches.

5.2 Visually elicited avoidance

As described in Chapter 4, vision is a fundamentally sensory modality for pelagic fish behaviour, and school formation, social communication, feeding, and predator avoidance rely on visual sensing of conspecifics, prey and predators. However, vision is a sensory modality for near-field orientation because of the high absorption of light waves in the sea. Avoidance reactions of pelagic fish that are initiated at some distance from vessel or fishing gear are therefore not elicited by visual stimuli, but rather by sounds from vessel and gear as described in section 5.1. Nevertheless, visual sensing of approaching vessels and fishing gears such as purse seines or trawls at close range may probably initiate even stronger avoidance reactions than those elicited by distant sounds (Gerlotto & Fréon, 1990).

5.2.1 *Light in the sea*

Light is rapidly attenuated in water by the water molecules, suspended matter and objects present. In clear sea water, light can reach down to more than 100 m. Then the sea has a blue colour because maximum penetration is for wavelengths between 400 and 500 nano-meter (nm). When the sea contains high concentrations of phytoplankton near the surface, most light is absorbed above 25 m, and maximum penetration is for wavelengths between 500 and 600 nm, giving the sea a more greenish colour. The amount and spectral quality of the light sensed by fish will depend on the line of sight involved (Guthrie, 1986). Maximum visibility underwater is only about 40 m in clear sea water under the most favourable light conditions (Tyler, 1969). In the presence of high concentrations of phytoplankton or other suspended matter, the visibility is often reduce to 10 m or less.

5.2.2 *Fish vision*

Pelagic fishes have paired image forming eyes which conform to the vertebrate type (Guthrie, 1986). Lens focusing is done by muscular control of the distance to the receptor cells, and not by changing the shape of the lens as in mammals. The fish eye has a weak pupillary function only. The inverted retina contains rod pigments with peak absorbency usually within the range 480–530 nm, and middle-or long-wave sensitive cones that probably assist discrimination of brightness contrast. According to a classification by Levine *et al.* (1980), fish living near the surface have double cones that absorb in the range 460–570 nm with offset single cones absorbing at wavelength near 415 nm. Nonetheless, near-ultraviolet and polarised light photosensitivity have been described in several species of fishes, both freshwater and marine taxa (Loew & McFarland, 1990). For midwater species, the pigment range is shifted upwards by about 40 nm as an adjustment to the longer wavelengths present midwater.

The fish eye sees a field of vision of about 180°, and fish have binocular vision where the two fields overlap in a sector of 20°–30° to either side in front. There is a corresponding blind zone of about 20°–30° to each side of the fish tail. Therefore, when escaping a predator, a fishing gear or a vessel, fish perform a zigzag swimming or display a circular movement around the detected danger ('fountain effect') in order to maintain visual contact with it (Hall *et al.*, 1986). To each side fish have a relatively large field of monocular vision of about 120°–140° where the ability for distance perception is limited. When seen with one eye, it is difficult to distinguish a 7 m mesh of 23 mm twine at 40 m range from 1.7 m mesh of 6 mm rope at 10 m because both objects give a 10° image (Wardle, 1993). This is probably the reason why large meshes of thick ropes are effective in herding fish in the mouth of large midwater trawls.

The visual acuity of the fish eye determines the maximum sighting distance of objects, and is dependent on density of cones in the retina and the size of the object (Zhang & Arimoto, 1993). On the basis of the cone density in the eye of large walley pollock (*Theragra chalcogramma*), Zang and Arimoto (1993) estimated a minimum separation angle of 0.17°, and calculated the maximum sighting distance according to:

$$D = l/\alpha$$

where l = object size and α is the minimum separable angle in radians. According to this, a 4 cm object can be recognised at a distance of 13 m by large pollock. Smaller fish have a larger separable angle and thus shorter distance recognition. Visual acuity reduces with decreasing light intensity, and visual range is also influenced by the contrast of the target. Anthony (1981) found that cod had about the same ability for contrast discrimination as man, and observed a reduction of reaction distance with reduced contrast.

5.2.3 *Avoidance reactions elicited by visual stimuli of vessel and gear*

During pelagic fishing, visually based avoidance reactions probably depend on fishing gear, time of day, species, and motivational state of the species. When purse seining mature, shoaling herring on the spawning grounds off western Norway in winter darkness there is normally little avoidance, and the capture success exceeds 90% of the sets (Misund, 1993c). In contrast, when purse seining mature, schooling herring in the North Sea in summer daylight, the herring schools often avoid the net and escape out under the groundline or through the opening before the net is closed in about 35% of the sets. Similarly, in the North Sea mackerel fishery in autumn daylight, the capture success is only about 65% due to avoidance reactions and escapement out of the net (Misund, 1993c). In both the herring and mackerel fisheries in the North Sea, the avoidance of the gear and subsequent escapement occur during the last part of the pursing when the fish are in close contact with the lower part of the net, and are probably frightened or herded out by the sight of the net (Misund, 1993c) in the same way as in front of trawls (Mohr, 1971; Wardle, 1993). Similar day-and-night differences in behaviour towards purse seines as described here for mature herring in winter darkness and summer daylight are also reported for Pacific tuna (Scott & Flittner, 1972).

As described above for purse seining, pelagic fish are normally easier to catch by pelagic trawls when shoaling at night-time than when schooling in daytime. This indicates that visual sensing in daytime enables precise location of the gear and enhances coordinated escape reactions. On the basis of underwater observations, Wardle (1993) describes a fish behaviour model to trawl gear where the sound of the vessel causes the fish to swim deeper, eventually towards the bottom. However, the sound of vessel and gear mainly causes curiosity, and the first reaction generally occurs when the trawl boards can be seen. The fish then avoid the trawl boards by swimming so that the trawl boards are kept at an angle of 155° or at the rear edge of monocular vision. Fish that come between the boards are guided by the bridles (if the boards touch the bottom there will also be a guiding sand cloud) towards the trawl mouth, where the fish turn and attempt to swim forward at the same speed as the trawl. This optomotor response (Harden Jones, 1963) to try to hold the same relative position to the visual stimulus of the moving gear, lasts until the fish are exhausted.

The swimming ability of fish depends on size and species (Viedeler & Wardle, 1991); small fish are rapidly exhausted and turn and drop back towards the codend. Larger fish can swim for a long time in the trawl mouth, and when trawling on fast swimming fish such as mackerel or sardinella, the speed must exceed 4 knots for effective capture.

Observations of fish behaviour when pelagic trawling indicate a strong effect of avoidance reactions elicited by sounds of both vessel and gear. Misund and Aglen (1992) observed that midwater herring schools in the North Sea avoided downwards to the bottom when approached by a survey vessel towing a pelagic sampling trawl with about 15 m vertical opening in daytime. Behind the vessel the schools seem soon to be influenced by stimuli (possibly low frequency vibrations or visual detection) from the warps and the gear, and were recorded to swim in the same direction as the approaching gear. When at close range, the sight of the gear obviously elicited strong avoidance reactions, and the schools tended to compress towards the bottom. To obtain catches, the trawl had to be lowered close to the bottom at the location of the schools. Similar strong avoidance reactions have been reported by Mohr (1969) when pelagic trawling on spawning migrating herring schools in the Norwegian Sea in the 1960s. In front of the vessel the schools performed strong downwards avoidance, dived 50–100 m, frequently changed swimming direction, and even split up. The herring seemed to be visually aware of the approaching net at a distance of 30–50 m, swam in front of it for some time, and then avoided downwards (Fig. 5.11).

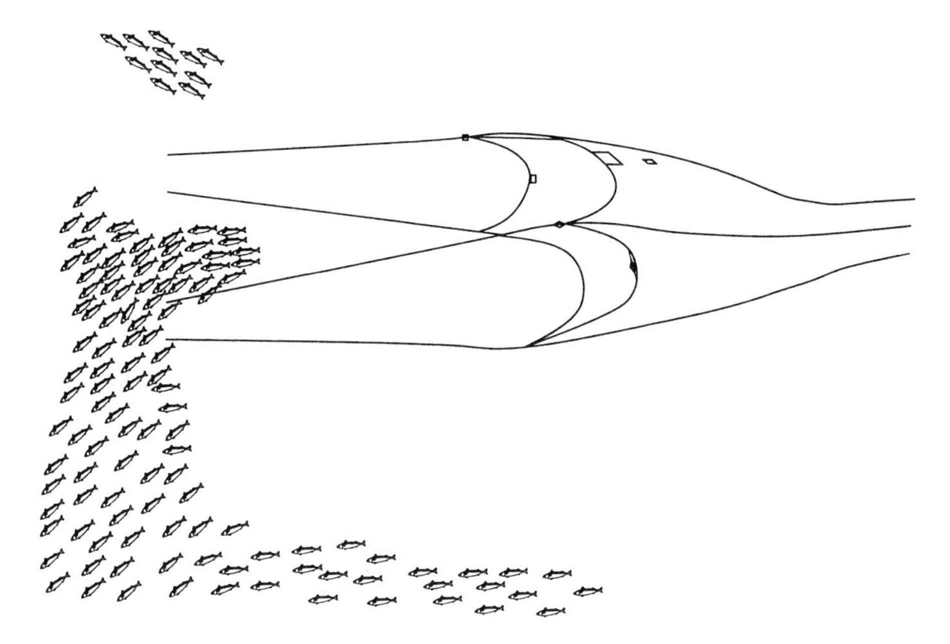

Fig. 5.11 Reaction of migrating herring schools to pelagic trawl (after Mohr, 1969).

At low light levels there is normally no ordered reaction pattern to approaching nets (Glass & Wardle, 1989; Walsh & Hickey, 1993). Inoue *et al.* (1993) observed that walleye pollock were mostly inactive when captured by trawl at depths of 150–300 m at low light levels and low sea temperature north of Japan in autumn. However, when pelagic trawling on dispersed herring concentrations off east Iceland in autumn in the 1960s, Mohr (1969) observed a general downward flight to the approaching gear, and the fish seemed to concentrate in the upper half of the trawl mouth (Fig. 5.12). Dense concentrations of herring were often present under the trawl, and an effective tech-

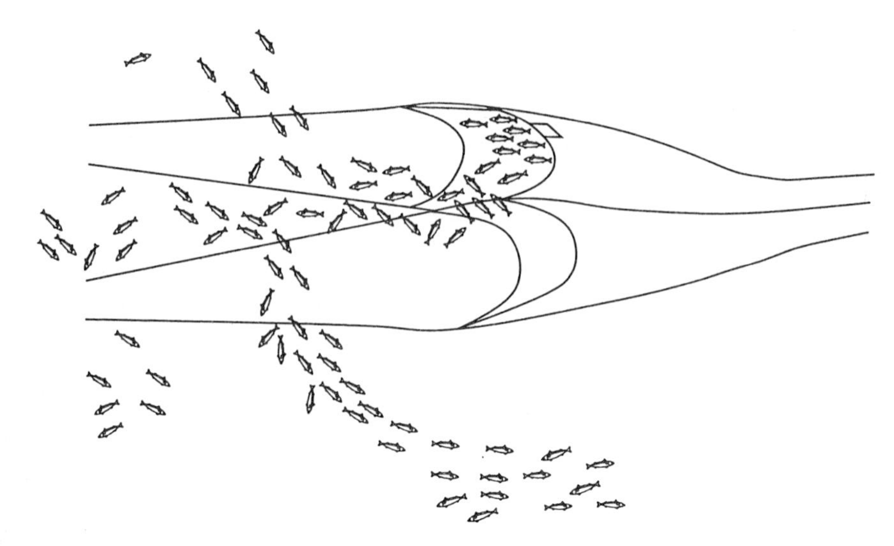

Fig. 5.12 Reaction of dispersed herring to pelagic trawl (after Mohr, 1969).

nique was therefore to allow the net to sink slowly when towing through an aggregation.

5.3 Conclusion

In this chapter we have defined avoidance as a general term for behaviours that fish perform to move away from stimulation that elicits fright or is sensed as unpleasant. We have focused on avoidance elicited by sounds and visual stimuli from vessel and gear. Avoidance reactions at long range during pelagic fishing and hydroacoustic surveys are probably caused by sounds, while visual sensing of the vessel and fishing gear at close range may be a determining factor during gear surveys and fish capture.

The ambient noise in the sea, produced by sea motion as waves, currents, turbulence, bubbles, animal communication etc., has a composite frequency spectrum dominated by low-frequency noise. Noise from vessels generated by the engines and propeller cavitation also has the main energy in the low frequency region (10–1000 Hz). The line frequencies generated by the engines are superimposed on a more continuous spectrum generated by propeller cavitation. In the 1970s and 1980s, little effort was given to reducing the underwater noise emitted from fishing and fisheries research vessels, but on more recently delivered fisheries research vessels actions have been taken to reduce the emitted noise.

Fish detect sounds through the otolith organs embedded in a pair of labyrinths in the skull. The three calcium carbonate otoliths (utriculus, sacculus and lagena) lie upon a thin membrane that protects a macula of numerous directionally sensitive hair cells. The otoliths are three times denser than the surrounding flesh of the fish, and will thus move at smaller amplitudes than the rest of the fish body when imposed on by sound waves. The relative movements of the otoliths and the hair cells cause physiological responses that are analysed as sound and enable discrimination of its intensity, frequency, and direction and distance to the sound source. The swim-bladder enhances sound detection through amplification. Fish audiograms show that most species are sensitive to low frequency sounds of up to a few kilohertz.

Field observations, playback experiments, and analysis of catch data show that fish react to vessel noise, and the Olsen model is a conceptual model of avoidance behaviour elicited by vessel sound stimuli.

Light is rapidly attenuated in the sea, and maximum underwater visibility is about 40 m. Maximum penetration is for wavelengths between 400–500 m giving the sea a blue colour. Pelagic fish have paired image forming eyes with muscular control of the distance between the lens and the receptors cells – rods and cones in the inverted retina. The field of vision of the fish eye is about 180°, and fish have binocular vision where the two fields overlap in a sector of 20°–30° to either side in front. The visual acuity that determines the maximum sighting distance of objects is dependent on the density of cones in the retina, the size of the object, light intensity and contrast.

Visually elicited avoidance reactions have been observed during purse seining and pelagic trawling in daylight. Such reactions often enable pelagic fish to escape capture.

Chapter 6
Attraction and Association

6.1 Attraction to light

Fishing with light was a major technique for coastal pelagic fish and tuna up to the middle of the twentieth century. Many pelagic fisheries around the world are still based on aggregating fish by use of artificial light (Chapter 2). Pole-and-line tuna fisheries usually fish the bait with light.

Schools of pelagic species usually disperse at dusk and become distributed in loose concentrations throughout the night. The density in such concentrations is often too low for feasible fishing. However, most pelagic schooling species can be attracted to areas illuminated by artificial light at night. The source of the illumination may be simple, such as a wooden fire, a burning chemical, oil or gas, or it may be an electric lamp. Normally, the light source hangs above the surface in the rigging of a vessel, but there are electric lamps that can be immersed to 200 m depth.

A recent fishing technique combining fishing with light and association with a floating object (see next section) was implemented in the middle of the 1980s by the pole-and-line fishery based in Dakar (Senegal) when it operated near Mauritania. This 'log-boat' tactic consists of using a drifting fishing boat as an artificial log to concentrate a huge school (1 to 8 days) and then to exploit it for several months, replacing one vessel by another when necessary, usually during the night (Fonteneau & Diouf, 1994). The boat usually stops its engine during the night and drifts, using powerful lights (25 kW). Usually two pole-and-line boats take turns on the same associated school, but sometimes a third, non-fishing vessel is used as a 'buffer' between rotations. When the vessel shifts to unsuitable areas, the skipper is able to move it slowly using the engine without losing contact with the school. According to the skippers, a permanent vessel speed of 1 to 3 knots is typical. Therefore this technique seems a hybrid technique which combines attraction by light and association. The expression 'offshore sea ranching' proposed by Fonteneau and Diouf is partly justified by the fact that fishermen provide additional food and light in order to increase the productivity, as in some ranches (e.g. battery chicken). This fishing tactic is very efficient for catching the three dominant commercial species of tuna in this area (yellowfin, skipjack and bigeye) plus a bycatch species, the black skipjack tuna (*Euthynnus alleteratus*). Young fish dominate the catches, with an average weight of 2.4 kg for skipjack, 6.8 kg for yellowfin and 10.4 kg for bigeye tuna. The average daily catch rate is 5.6 t per fishing day (1988–1991) and 10 to 20 t per day are commonly recorded; days without catch are extremely rare, unlike the conventional pole-and-line technique (Fonteneau & Diouf, 1994). The 'log-boat' tactic is also used by the Spanish tuna fleet based on the Canary Islands from the beginning of the 1990s (Delgado de Molina *et al.*, 1995).

The behaviour of pelagic species towards artificial light is to a certain extent species dependent, but varies within species according to age and physiological or environmental conditions (Ben-Yami, 1976). Some species are positively phototaxic and are attracted to artificial light, others are negatively phototaxic and avoid artificial light. Small herring are generally positively phototaxic while large herring may be negatively phototaxic (Ben-Yami, 1976). According to Dragesund (1957), the behaviour of herring during illumination by surface lights can be generalised in four behaviour patterns:

(1) The density of the herring layer increases, and the fish swim deeper
(2) The herring disappear from the illumination area
(3) The herring rise quickly towards the light source, and then either disappear or swim deeper and denser
(4) The herring swim denser and rise slowly towards the light source.

Light is rather quickly absorbed in water, and the distance of light penetration depends on light intensity, water turbidity, and colour. The radius of attraction of kilka (Caspian sprat, *Clupeonella* spp.) in the pump fishery in the Caspian Sea increased asymptotically from 35 m in response to a 15 W lamp to 72 m to 9 kW lamps (Ben-Yami, 1976). An electric surface lamp of 1.5 kW affected the herring within a radius of about 50–60 m (Dragesund, 1957). Using a 3.5 kW lamp underwater at 5 m depth, Beltestad and Misund (1988b) observed that large herring (24–38 cm) concentrated in a region at 50–110 m depth.

In water with high turbidity due to algae or planktonic organisms or inorganic particles, light penetration may be very limited. In the turbid waters in the Finnish archipelago, Tshernij (1988) measured that the light intensity at 6 m depth was only 0.03% of that at the surface. Still, the Baltic herring (*Clupea harengus membras* L.) aggregated in shoals up to 7 m in horizontal diameter and 10 m in vertical extent below a light source made of three 20 W underwater lamps. From comparative experimental trials, Ona and Beltestad (1986) concluded that saithe seem more attracted to blue than to red light. The aggregation behaviour of herring was similar towards red, blue and white light sources (Beltestad & Misund, 1988b). Kilka (*Clupeonella* spp.) in the Caspian Sea and saury (*Cololabis* spp.) in the Pacific seem most attracted to yellow light (Ben-Yami, 1976). Other factors such as water current, surface waves, moonlight, presence of food or predators may also have strong influence on the aggregation behaviour of pelagic species towards artificial light (Ben-Yami, 1976; Mohan & Kunhikoya, 1985).

The vertical distribution towards a light source probably depends on the initial natural vertical distribution (Beltestad & Misund, 1988). In one season, large herring were distributed from 0 to 50 m naturally, and then concentrated both at the surface at 45–100 m to the side and a 20–70 m below the source when illuminated by artificial light (Fig. 6.1). During another season in the same Norwegian fjord, herring were concentrated at 50–100 m depth when illuminated by a 3.5 kW light source at 5 m

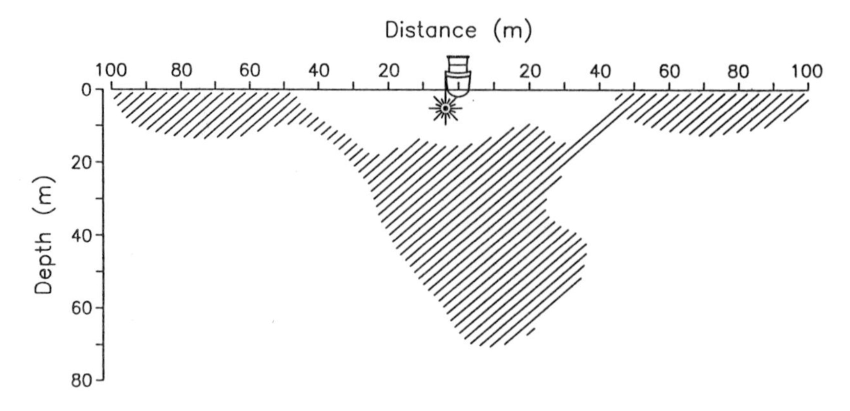

Fig. 6.1 Distribution of herring (*Clupea harengus*) in a vertical section around the light sources as recorded by SIMRAD FS3300 during trials by underwater light in a Norwegian fjord (redrawn after Beltestad and Misund, 1988a).

depth; this was also the vertical distribution range when not influenced by artificial light. At such depths, the herring were still unavailable to coastal purse seiners with shallow nets. A solution for such occasions was to lower a lamp with an upward reflector under the herring concentration. When it was switched on, the herring rose rapidly to the surface at a speed of 20 m/min.

A substantial part of the light from a source above the surface will be reflected at the surface and will not penetrate the water. Such surface reflection will be avoided if the light source can function underwater. To test if underwater light is more efficient than surface light, Beltestad and Misund (1988) compared the attraction of large herring (24–38 cm) to 3.5 kW surface (lamp 1 m above surface) or underwater light (lamp at 5 m depth). The abundance of herring aggregated to the two light sources within one hour was similar, but the herring seemed to aggregate somewhat faster to the underwater light than to the surface light (Fig. 6.2a). The vertical distribution of herring when attracted to the underwater light was about twice as deep as when attracted to the surface light. This indicated that the light penetrated much deeper when the source was underwater than at the surface. When the light intensity was gradually reduced after one hour, the herring schooled more densely and rose towards the light source (Fig. 6.2b).

There are several hypotheses why most pelagic schooling species are attracted to artificial light. Fish perceiving the artificial light may become curious and search the illuminated area to explore it. Since planktonic organisms may be concentrated within the illuminated region, the fish may stay in the area because of enhanced feeding opportunities. In the above-mentioned case of tunas associated with a pole-and-line vessel for several days and aggregated by powerful light (commonly 25 000 watts) at night, J.-P. Hallier (pers. comm.) reported active feeding in the illuminated area. Another hypothesis is that the artificial illumination enables the visual contact necessary for establishing schooling behaviour, and when coming into the illuminated region, fish find it advantageous to join schooling. Possibly, pelagic species combine schooling and feeding in the artificially illuminated region as observed for sardinella

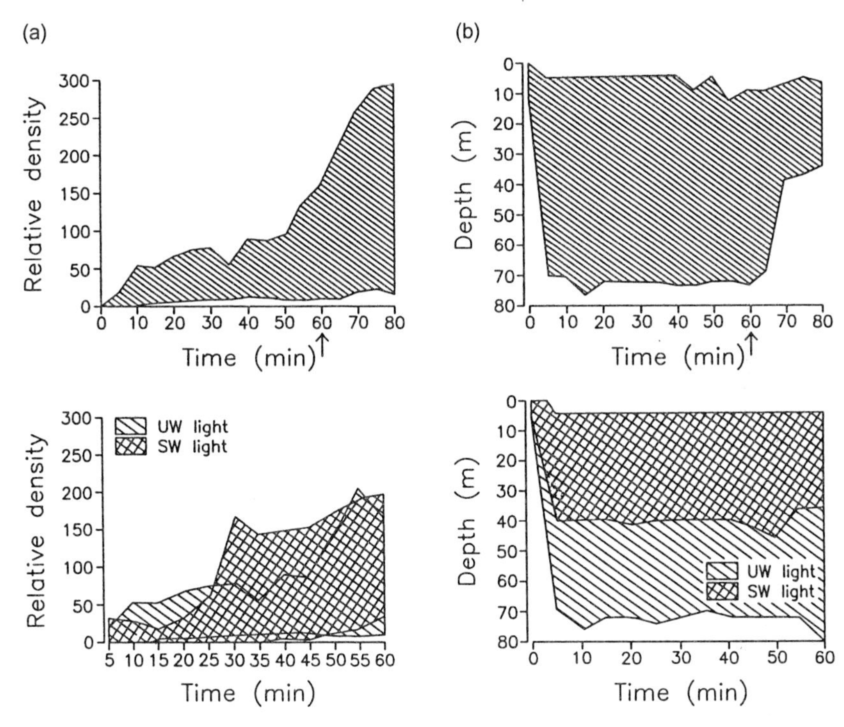

Fig. 6.2 Interval of variation for reflected echo energy (a) and depth distribution (b) of herring (*Clupea harengus*) during comparing trials by surface (SW) and underwater (UW) light in a Norwegian fjord after starting to reduce light intensity (redrawn after Beltestad and Misund, 1988b).

(*Sardinella aurita*) in the eastern Mediterranean (Ben Yami, 1976), or for tuna species in the eastern Atlantic (Hallier, pers. comm.).

6.2 Attraction to bait

Several gears (handline, longline, pole-and-line, some type of trap) use bait for attracting the fish. Nevertheless, as far as pelagic species are concerned, use of bait is limited to tunas and tuna-like fishes which are caught by different types of line in tropical regions worldwide (Chapter 2).

It was usually accepted that skipjack, yellowfin and bigeye tuna rest at night (review in Cayré *et al.*, 1988) and feed visually during daytime, which suggested that the visual stimulus of the bait is a major determinant for the attraction of tunas to baited hooks. Nevertheless, the previous results were based on stomach content studies of fish caught by surface fisheries (pole-and-line or purse seiner). Recent findings (see section 6.3.1 and Josse *et al.*, 1998) suggest that nocturnal foraging might also occur, at least in some areas, at certain depth and for some species. In the French Polynesia EEZ (Exclusive Economic Zone), acoustic surveys are simultaneously deployed with sonic trackings to observe the biological environment of the tracked tuna. A yellowfin tuna (*Thunnus albacares*) tagged by an ultrasonic device and tracked during 21 hours around Maupiti Island remained in a non-migrant scattering layer at 45 n.mi. during daytime and 30 n.mi. during night-time where its behaviour suggested foraging (ECOTAP project, 1995).

Chemical stimuli, which are related to the soaking time and the type of bait, could also play an important role as for most other fish caught by bottom long-lines (Løkkeborg & Johannessen, 1992; Løkkeborg, 1994). The influence of chemical stimuli and their diffusion according to the current explains why the catch rate is increased when the line is set perpendicularly to the current and upstream of fish location (Bjordal & Løkkeborg, 1997). Chemical stimuli are likely to explain why Sivasubramaniam (1961) found a negative relationship between soaking time and catch of tuna, in the Japanese long-line fisheries. In addition, the size of the bait might play a significant role in species and size selectivity, but is not directly related to attraction (Løkkeborg & Bjordal, 1992). From laboratory experiments, Johannessen *et al.* (1993) found that haddock (*Melanogrammus aeglefinus*) bit only a part of the big bait, leaving the hook outside the mouth, and would then jerk or rush to tear the bait apart.

To our knowledge, little is known about tuna baiting behaviour except in the case of swordfish (*Xiphias gladius*) longlining, which is performed at night-time (e.g. Berkeley & Waugh, 1989) and recently incorporated light sticks from place to place along the line, thus combining bait attraction and light attraction (Mejuto & De la Serna, 1995). Nonetheless, recent studies have been performed in the Pacific around Hawaii and Tahiti islands on other tuna species. These studies are based on longlines equipped with depth and time recorders, including recorders of the baiting time on hooks (hook timers). Preliminary results obtained around Tahiti (ECOTAP, 1995) indicate that the depth of a given longline depends on the strength of the current, among other parameters, as expected from previous theoretical models (Yoshihara, 1954), but here with a gain of precision (84% of variance explained). It is therefore important to take current into account, since the proportion of the target species in the catch is related to the depth of the hooks. Bigeye tune (*Thunnus obesus*) is caught in deeper – and therefore colder and darker – waters than albacore (*Thunnus alalunga*) and yellowfin (*Thunnus albacares*). The time of baiting seems also to vary according to the species. Albacore bite during daytime as well as at night, with a maximum at dawn and around 0010 h. Bigeye tuna was not caught at night, but only four longline sets were performed at night-time and therefore it cannot be firmly concluded that this species does not bite at night. Nevertheless, in the Hawaii-based longline fishery, most of the longline sets targeting tuna and swordfish are performed during the night and bigeye tuna is the dominant species (Boggs, 1997).

An interesting aspect of the biting behaviour is related to the moment of baiting according to the fishing operations. Fish bite not only when the longline is completely set, but also during the setting and hauling operations. Nearly one third of the bites are observed during these two operations, mainly during setting for yellowfin and albacore, and mainly during hauling for bigeye tuna (ECOTAP, 1995). Around Hawaii, hook timers indicated that 32% of the striped marlin (*Tetrapturus audux*), 21% of the spearfish (*T. angustirostris*) and 12% of the bigeye tune (*Thunnus obesus*) were caught on sinking or rising hooks (Boggs, 1992). The increase of biting during fishing operation suggests that bait movement is an important stimulus, implying that this movement favours visual attraction (but an increase of odour diffusion cannot be eliminated).

The case of pole-and-line fisheries is particular in the sense that fish schools are not directly attracted by the bait but are first detected by the skippers (through visual observations of birds, logs or directly of the school); then the fish is maintained close to the boat by feeding it with live bait (usually small coastal pelagic fishes). The hook of the pole-and-line can be baited with a fish, especially at the beginning of the fishing period, or with artificial lures which must be in movement to attract the fish. Nevertheless, the success of the catch depends strongly on the quantity and quality of the available bait. The effectiveness of baiting is related to the species used for baiting and to the targeted tuna species (Hallier & Kulbicki, 1985; Mohan & Kunhikoya, 1985; Argue *et al.*, 1987). From a field study in New Caledonia and a review of the pioneering work of Yuen (1959, 1969) and Strasburg (1961), Conand *et al.* (1988) classified small fish from the lagoon of New Caledonia according to their potential efficiency during baiting operations of a pole-and-line fishery. He suggested that efficiency depends on the species' attractiveness for tuna, which is related to its taste and smell, but also depends on its silvered aspect and its behaviour. The most interesting species are those that tend to remain and swim actively at the surface, near the boat, rather than those that dive or escape rapidly.

To some extent attraction by lure can be related to bait attraction. Lures are mainly used in recreational fisheries, in pole-and-line fishery in association with bait fish (Ben Yami, 1995), but also in some professional small-scale trolling fisheries around FADs (fish aggregating devices) or in open waters (Drew, 1988).

In conclusion, bait attraction seems to be based both on olfactory stimuli and on visual stimuli. Movement represents an important aspect for artificial lures but also for natural bait, alive or dead. The combination of attraction by light and by bait seems promising.

6.3 Associative behaviour

6.3.1 *Association with floating objects*

Overview of association

The attraction of tuna, tuna-like species and many other pelagic or bottom species to natural floating objects (logs, mammals, etc.) or artificial ones (often termed FADs: fish aggregating devices) is well known. Tuna fishermen have taken advantage of this association for many years, especially in Asian countries (e.g. Uda, 1933; von Brandt, 1984). But there are many other species that display such association. Let us overview this strange phenomenon of association before focusing on tunas.

Hunter and Mitchell (1967) made 70 purse seining collections around natural and artificial floating objects in Pacific waters offshore of Costa Rica. They found that 12 families of fishes (Lobotidae, Carangidae, Coryphaenidae, Mullidae, Kyphosidae,

Pomacentridae, Scombridae, Blenniidae, Stromateidae, Mugilidae, Polynemidae and Balistidae) and 32 species were represented in the collections, dominated by the carangids (nine species), particularly *Caranx cavallus* and *Selar crumenophtalmus*. Parin and Fedoryako (1992) have gathered a large amount of information collected on board Soviet research vessels in all the oceans since the mid 1950s. They indicate that the total number of fish species associated with floating seaweeds and flotsam varies from 186 (60 families) in neritic to 81 (27 families) in oceanic areas. The species assemblages are similar between flotsam (logs) and seaweeds. Similar results were obtained in the Atlantic and Indian Oceans by Stretta *et al.* (1996).

Sharks and barracuda are also attracted by floating objects, as reported by Stéquert and Marsac (1989) in the Indian Ocean, by Au (1991) and Hall (1992a) in the eastern Pacific and by Batalyants (1992) in the eastern Atlantic, who mentioned different species of Carcharinidae, Sphyrnidae and Sphyraenidae.

Flyingfishes are attracted by floating objects, but unlike tunas, the determinism is clearly identified: these species are searching for any floating substrate for spawning (Oxenford *et al.*, 1993). In the Caribbean Sea, especially around Barbados, fishermen use coconut branches to attract flyingfish and catch them with dip nets or gill nets (Gomes *et al.*, 1994) and a similar technique is used around Sao Tome island in the eastern Atlantic (F.X. Bard, pers comm.).

The attraction by objects in the water might also exist for some coastal small pelagic species in some areas, but as far as we know, unlike tunas or flyingfishes, coastal pelagic species attracted by floating objects are not commercially exploited, with the notable exception of the Java Sea fishery. This traditional small-scale fishery uses floating or submerged bamboo, named 'rumpons', to attract different species of small pelagic fishes (mainly: *Decapterus russelli, D. macrosoma, Rastrelliger kanagurta, Sardinella* spp., *Amblygaster sirm* and *Selar crumenophtalmus*) (Potier & Sadhotomo, 1995). Some of these species have an oceanic stage of their life, but others do not. This is the case of *Sardinella gibosa* and *Sardinella lemuru*, despite the fact that the same genus is not known to display such an associative behaviour in other areas where large stocks are exploited (e.g. *Sardinella aurita* and *Sardinella maderensis* in the Atlantic). Here also the mechanism of attraction is not well known (except from 1987, when fishermen started to combine the use of rumpons with light). Similarly, several small-scale fisheries based on islands take advantage of the association of small or medium-size pelagic fish to floating objects, but as far as we know it concerns oceanic species.

In coastal waters near Sydney, Druce and Kingsford (1995) used natural (algae) and experimental drifting objects. Drift algae attracted mainly pigmented juveniles, but also larvae (ambassids and gerreids) and a few adult fishes (mullids, pomacentrids, teraponids). Experimental drifting objects attracted mainly juvenile carangids, sphyraenids, mullids, mugilids and larval ambassids, sillaginids, sparids and gerreids, but many larval species demonstrated no affinity for floating objects, as for instance clupeids and atherinids.

The colonisation of the floating object is usually very fast, often less than one week, as reported in the works mentioned above. In the eastern Atlantic, Bard *et al.* (1985) indicated that tuna schools were attracted a few days after setting a drifting FAD, and that a FAD set in the vicinity of a school (probably skipjack tuna) was immediately 'adopted' by the fish. In the western Indian Ocean, Yu (1992) performed short time-scale observation of the colonisation of drifting FADs by visual and hydroacoustic methods and concluded that species of Aluteridae, Carangidae and Serranidae arrived 25–35 hours after setting and concentrated in the immediate vicinity of the FAD (5–20 m), while predators (mainly Coryphaenidae, Sphyraenidae, Scombridae and Carcharrhinidae) were observed regularly within a longer range (50–100 m). After this period, a stabilisation of the specific composition within 50 m of the FAD was observed, followed by a mere increase of its biomass. Usually the tuna schools arrived 30–50 hours after setting the FAD: first small schools of juvenile yellowfin tuna, followed by small schools of skipjack tuna, then other juveniles or adults of those two species.

Midwater artificial structures which resembled a pup tent (Fig. 6.3) were positioned off Panama City in the Gulf of Mexico and were effective in attracting up to 25 tonnes of small pelagic mixed school: round scad (*Decapterus punctatus*), Spanish sardine (*Sardinella aurita*) and scaled sardine (*Harengula pensacolae*). Larger pelagic species were also attracted during this experiment (*Seriola* spp., *Elegatis bipinnulata* and *Caranx chrysos*) but were not observed to feed on the smallest ones (Klima & Wickham, 1971). The authors observed that attraction to structures relies on the individual fish or school passing within the visual range of the structure or of fish already associated with it. Wickham *et al.* (1973) concluded that structures function only to concentrate fish already present in the vicinity. The association with structures seems rather transient as individuals and schools have been observed to move towards and away from structures rather frequently (Klima & Wickham, 1971).

There seem to be certain species-dependent behaviour patterns when associating with structures. In the above-mentioned observations of Hunter and Mitchell (1967) in offshore Pacific waters of Costa Rica, juvenile fishes were more abundant and located closer (vertically and horizontally) to the floating object than adult fishes. These juveniles usually stratify vertically according to the species and the whole distribution resembles a cone, the apex of which is at the underside of the floating object. In the study of Klima and Wickham (1971) off Florida, the baitfish group tended to maintain position upcurrent from the structure and in the upper half of the water column, while the jacks stayed close to and at the same depth or slightly below the structure (Fig. 6.3). The baitfish group seemed to behave as mixed species schools. The baitfish were frequently observed feeding while associated with the structures, but not the jacks. Presencc of wounded individuals of both groups indicated predation attempts by great barracuda (*Spyranea barracuda*) and sharks observed in the vicinity of the structures. However, there are indications that prey fish are able to take advantage of the structure for predator avoidance and sheltering. Little tunny and king mackerel immediately launched successful attacks on baitfish when a structure was removed, while such attacks were seldom successful when the structure was present (Wickham *et al.*, 1973).

The characteristics of floating objects that facilitate attraction have been studied by several authors in the search for the reason for the associative behaviour. Hunter and Mitchell (1968) evaluated different kinds of artificial floating objects in the same area.

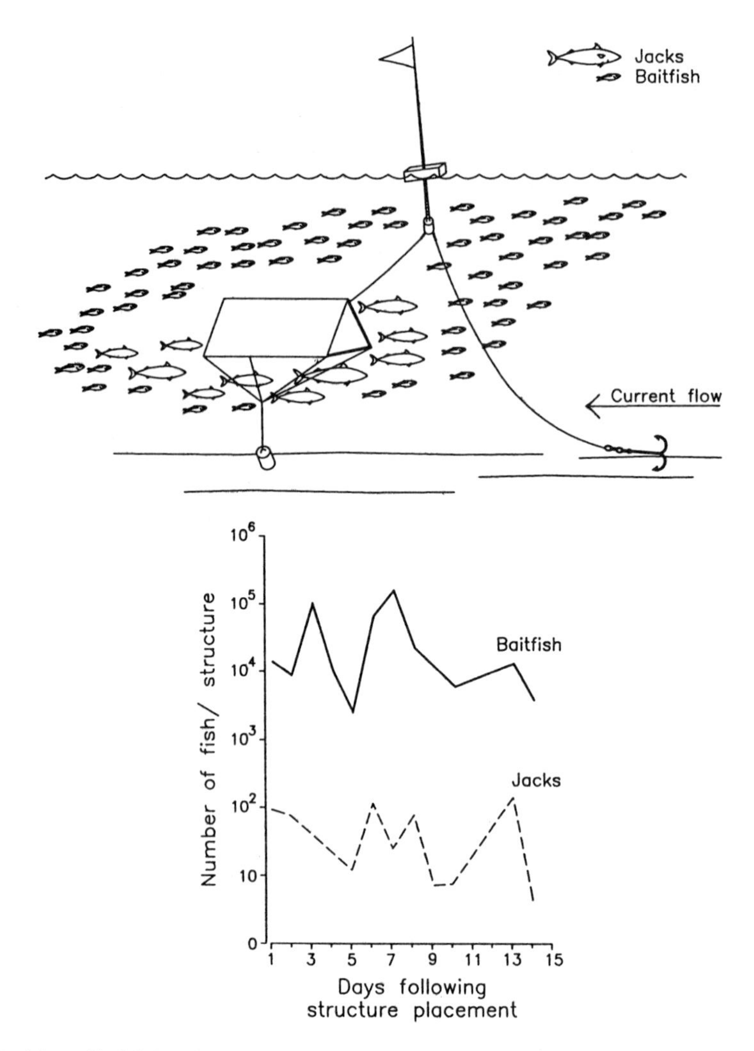

Fig. 6.3 Position of baitfish and jacks in relation to submerged structure set off Panama City (redrawn from Figures 2 and 6 from Klima, E.F. and Wickham, D.A. (1971) Attraction of coastal pelagic fishes with artificial structures. *Transactions of the American Fisheries Society* **100**, 86–99).

They found that surface objects attracted more fish than submerged ones of the same size. Moreover, they indicated that a plastic sheet, bent at a 60° angle at its midline so that it resembled a small tent which floated with both sides submerged, was more effective than horizontal floating sheets. Unfortunately the authors did not mention if the differences observed in fish density were mainly due to small fish, solitary predators or tunas and such experiments were not repeated.

During the Inter-American Tropical Tuna Commission (IATTC) symposium, Hall *et al.* (1992b) presented observations on 5518 floating objects, made on board commercial tuna purse seiners in the eastern Pacific, of which 2793 resulted in a set. From this database they studied the influence of the characteristics of the floating object on the concentration of tuna. Their conclusions were that the size of the object is not an important factor for medium-sized objects (1 to 6 m for the longest dimension) but that very small and large objects are less attractive (see also Akishige *et al.*, 1996, on this point). The results are similar when surface or volume are considered. It seems that the aggregation of several objects is more attractive than a single object (this last result is in agreement with the above-mentioned experimental results of Wickham *et al.* (1973)). Artificial objects seem more attractive, especially for skipjack tuna. The percentage of the surface of the object covered by epibionts did not appear to play a role in its attractiveness, nor the time at sea as already mentioned by Bard *et al.* (1985). The percentage of the submerged volume seems to play a role for yellowfin, but it should be confirmed by experimental observation. In Hall *et al.*'s study, the colour does not seem to play a major role, except that black seems to be preferred (but there were not enough objects of this colour in the database to validate the observed difference). Nevertheless, in the recent study of tuna fisheries performed in the Atlantic and Indian oceans, Stretta *et al.* (1996) reported that more than half of the floating objects associated with tuna were brown and that black objects represented around 20% on a total of 792 observations.

Association of tunas with floating objects

During the last three decades, industrial tuna purse seining has increasingly used a strategy which consists of localising and searching around biotic and abiotic objects at the surface. The reason is that tuna tend to aggregate underneath and are less active and therefore easier to catch. In the western Indian Ocean, for instance, the frequency of unsuccessful sets between 1984 and 1990 was only 8% on 'logs', against 53% on non-associated schools (Hallier & Parajua, 1992b). Similar results have been obtained for the French and Spanish fleets operating in the Atlantic and Indian oceans in 1995 (Stretta & Slepouka, 1986; Stretta *et al.*, 1996). Catches from a few hundred kilos up to several hundred tonnes of tuna are taken under logs or floating timber, kelp, branches etc. (Greenblatt, 1979). Association with mammals is also frequent and will be developed in section 6.3.2, even though this type of association could depend on the same behaviour as association with floating objects.

Fishermen have discovered that the aggregative behaviour of tuna also applies to non-natural objects at sea, and in many countries it has become very common for fishermen to construct and deposit artificial fish aggregating devices (FADs) to enhance aggregation of tuna (Marsac & Stéquert, 1986; Josse, 1992). Those FADs can be either anchored or not, and we will see many differences between drifting FADS – the most commonly used by commercial fisheries – and anchored ones. Curiously, costly anchored FADs were first used, usually near islands for artisanal trolling fisheries, during the 1980s. In French Polynesia for instance, FADs have been anchored since 1981 and have completely modified the exploitation pattern in the area (Josse, 1992) and increased by 33% the number of fish caught per fishing day. During the 1990s, purse seiners used drifting FADs which are identified, easy to locate (radar reflector, radio beacons; Fig. 6.4) and frequently re-collected on board

Fig. 6.4 A common drifting floating aggregative device (FAD) used in the purse seine tuna fisheries (modified after Bard *et al.*, 1985).

and displaced (Hallier, 1995). Some FADs are equipped with an echo-sounder which automatically transmits data to the fishing vessel by radio. In the Spanish fishery of the Atlantic, supply vessels are used only to build drifting FADs on board, release them and alert the purse seiners when the tuna concentration is estimated to be high enough for a commercial catch (Ariz *et al.*, 1992).

Fishing under artificial floating objects is traditional in the Philippines fishery where 95% of the catches are performed under 'payaos' (Barut, 1992). At the beginning of the 1990s, the proportion of catches related to floating objects was over 50% of the total catch in the western Indian Ocean purse seine fishery (Hallier & Parajua, 1992b), in the western Pacific (Hampton & Bailey, 1992; Suzuki, 1992) and in the south of the Caribbean Sea (Gaertner & Medina-Gaertner, 1992). Log fishery accounted for only 15% in the eastern tropical Atlantic at the beginning of the 1990s (Ariz *et al.*, 1992) but according to Stretta *et al.* (1996), in 1995 the Spanish and French purse seiner fleets performed around 80% of their catches under floating objects (90% of these objects were artificial) in the Atlantic Ocean and more than 90% in the Indian Ocean (50% artificial objects). In the eastern Pacific, this proportion is only known for large purse seiners (26% at the end of the 1980s) but is probably larger for the small ones (Hall *et al.*, 1992a).

Due to this increase of tuna commercial catches under natural or artificial floating objects, the Inter-American Tropical Tuna Commission (IATTC) convoked in 1992 an international workshop on the floating objects and tunas (Hall, 1992b). The situation is contrasted according to the different species and oceans (Gaertner & Medina-Gaertner, 1992). Skipjack tuna are more often associated with floating objects than are other commercial tuna species (yellowfin, bigeye), as shown in Fig. 6.5 (Fonteneau, 1992; Hall *et al.*, 1992a; Hallier & Parajua, 1992b). Yellowfin dominate the association with large mammals and therefore are more abundant in the landings of the Venezuelan fishery which target such association (probably due to the lack of logs). The tuna catches under floating objects are substantial all around the world, but more important in the Indian Ocean, West Pacific and Philippines area (where anchored devices locally named 'payaos' are commonly used). The association of tuna with natural 'logs' is most common in tropical coastal waters off mangrove regions or off large rivers that supply the ocean surface with floating trees, branches etc. (Stretta & Slepoukha, 1986; Ariz *et al.*, 1992; Hall *et al.*, 1992b).

At the beginning of the industrial tuna fisheries, floating objects were mainly parts of trees transported by rivers to the sea, and secondarily dead mammals. The conventional 'log' fishing was therefore predominantly coastal. But nowadays the importance of tree residuals (logs, large branches) used as a help for fishing is lower than in the past. Tree residuals represent less than half the total number of floating objects found by fishermen (47% in the eastern Pacific for instance). Most of the objects result from human activity – boards, parts of fishing gears (net, floats, ropes), housing, etc. – and artificial logs are launched by fishermen. As a consequence, 'log' fishing is now conducted in productive offshore regions, in the eastern Pacific (Hall *et al.*, 1992a), in the eastern Atlantic (Ariz *et al.*, 1992), and in the western Indian Ocean (Hallier & Parajua, 1992b).

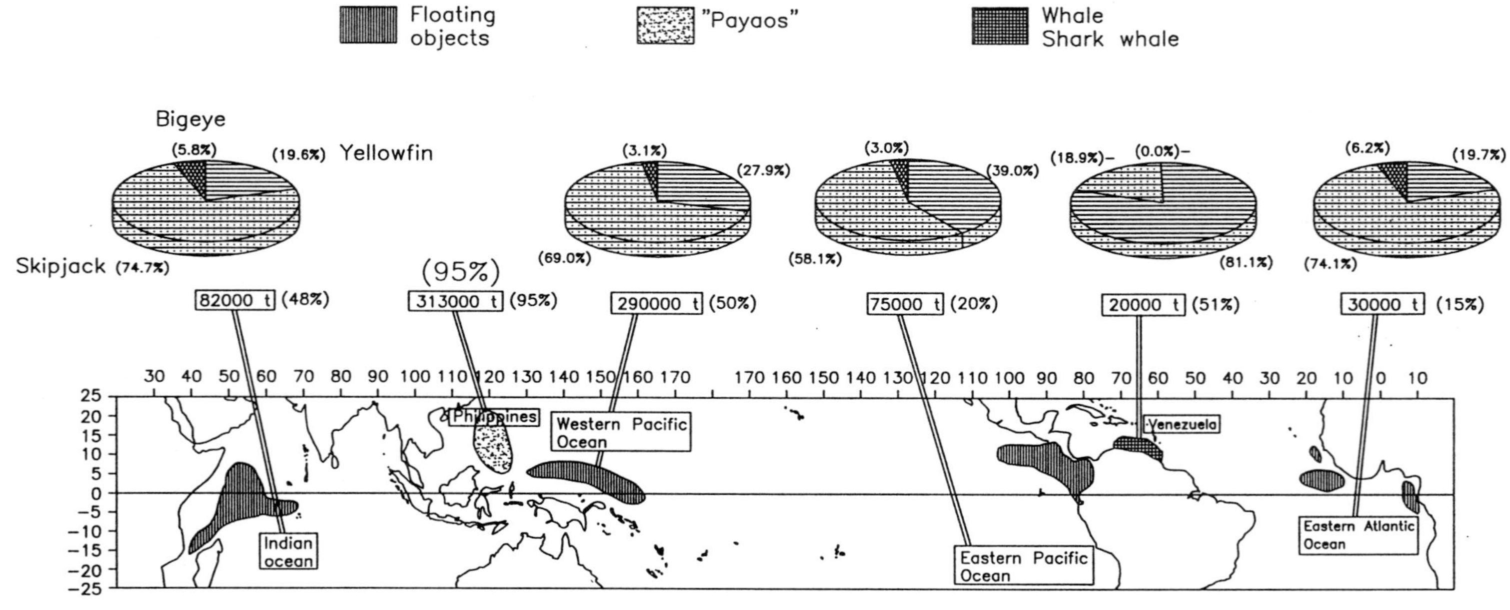

Fig. 6.5 Synopsis of tuna catches associated with floating objects in 1990: main fishing areas, total catches per fisheries and species composition (modified after Fonteneau, 1992).

The range of body length of fishes in catches associated with floating objects is similar to those of unassociated fishes in the case of skipjack tuna, but yellowfin and, to a lesser extent, bigeye tuna, are usually smaller. In many fisheries (Fonteneau, 1992; Hall *et al.*, 1992a; Hallier & Parajua, 1992b; Stretta *et al.*, 1996), a bimodal distribution of length frequencies is observed both in associated and unassociated schools, but the relative size of each mode is different, with a higher proportion of small fishes in the case of associated schools of yellowfin and bigeye tuna (Fig. 6.6). This difference in length frequency distribution is a major problem for stock management due to the growing percentage of tunas caught under floating objects, the smallest being discarded dead. The scientific community working on tuna is at present targeting this problem. The fishermen themselves decided to prohibit tuna catches under floating objects in the eastern Atlantic Ocean for a period of three months beginning 1 November 1997, controlled by observers onboard and encouraged by the International Commission for the Conservation of Atlantic Tunas (ICCAT).

Despite the fact that the mean individual weight is lower in associated fishes than in set schools, the school weight is usually higher. This is especially true in the eastern Atlantic and in the eastern Pacific oceans, where the mean catch per set of log-schools is more than double that of non-associated schools (Fonteneau, 1992; Hall, 1992a).

Cillaurren (1994) analysed a large amount of data on experimental fishing around five offshore FADs anchored in the Pacific off Efate island (Vanuatu) during the daylight hours. The results, expressed in catch per hour trolling, suggest interactions between time, space and body-length effects. A decrease in tuna abundance with increasing distance from the FADs was observed within the 0–1600 m range for the two main tuna species: yellowfin and skipjack (3796 and 3442 individuals respectively). Altogether, 77% of the fish were caught within 200 m of the FAD, 20% between 200 and 500 m and 3% beyond 500 m, but the yellowfin tuna tended to be located nearer to the FADs than the skipjack tuna.

The daily fluctuations of the yield clearly varied according to the distance from the FADs. No catches were observed during the afternoon over 300 m from the FADs and after 0010 h over 500 m. This suggested that the radius of distribution around the FAD decreases from sunrise to sunset (nevertheless note that the temporal distribution of fishing effort without catch is not known). The best catches of yellowfin within a 100 m range were observed around midday and secondarily around sunrise, whereas for skipjack tuna the yield decreased from one hour before sunrise to sunset, with a secondary peak three hours prior to sunset. According to Cillaurren (1994), this variability in the yield within the daylight period could be due both to daily migration and to feeding rhythms. Body length frequency distributions indicate that small specimens (30–40 cm) are found at all ranges, but large ones, especially yellowfin, are found mainly over 300 m from the FAD. Moreover, the proportion of smaller fish increases throughout the day and the author suggested that this indicates that the smaller fish progressively take the place left by the departure of larger fish, in agreement with the model proposed by Hallier (1986) for log-schools in the western Pacific. All these results are fruitful for the management of FAD fisheries, but their

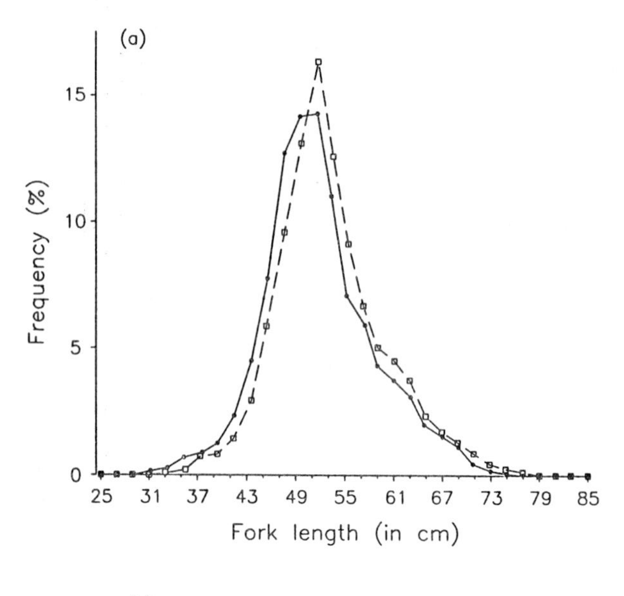

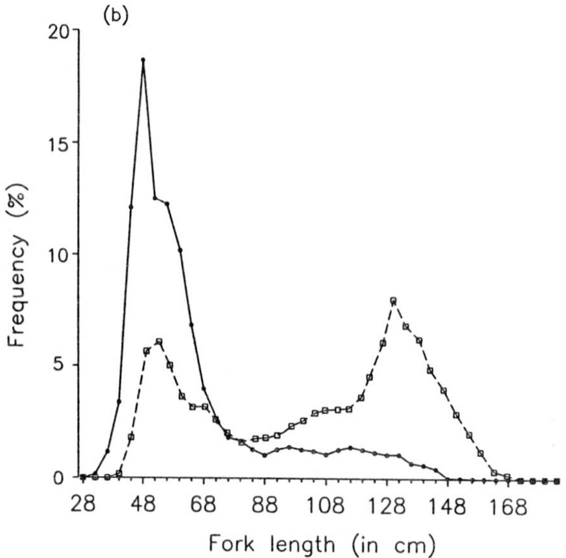

Fig. 6.6 Size frequency distribution of (a) skipjack tuna (*Katsuwonus pelamis*), (b) yellowfin tuna (*Thunnus albacores*) and (c) bigeye tuna (*Thunnus obesus*) in the French purse seine fishery of the western Indian Ocean from 1984 to 1990 (redrawn after Hallier and Parajua, 1992a). For all figures: ● = log set; □ = school set.

interpretation for the reason for fish aggregation is limited due to the unknown effect of time-dependent catchability and fisherman behaviour.

The turnover of fishes under a floating object seems relatively high, at least when fish are removed by exploitation, because series of successful sets are often observed under the same object every day for a few days (Cayré & Marsac, 1991). In the Indian Ocean, there is a slow decrease in the catches only due to a lower abundance of

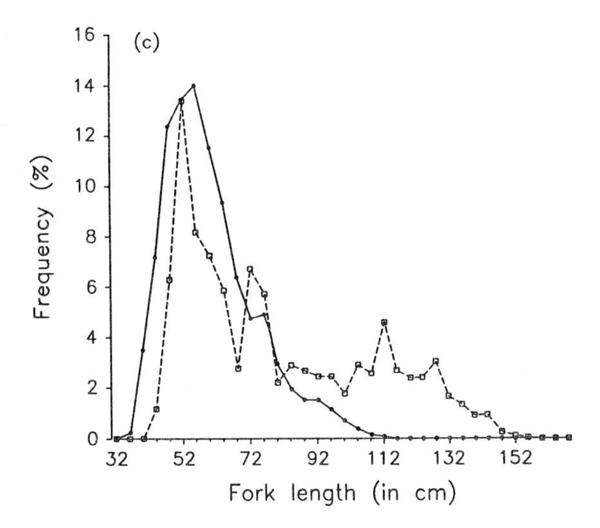

Fig. 6.6 *Contd.*

skipjack tuna, while the catch of yellowfin remains constant (Fig. 6.7), which indicates a steady recruitment of fish under the log (Hallier & Parajua, 1992c). In the eastern Pacific, Hall and Garcia (1992) observed a similar decrease in repeated sets on the same object and a lower decrease of yellowfin abundance compared with skipjack tuna. In the tropical western and central Pacific, the catches decreased markedly between the first and the second set but then remained similar or increased in subsequent sets (Suzuki, 1992), which was probably due to the higher proportion of skipjack tuna compared with the Indian Ocean. Despite the high turnover suggested by these results, one must bear in mind that they have been obtained on a limited number of days (commonly seven) and probably depend on the local abundance. Longer experiments show that once the local abundance is depleted, the catch per set of all species decreases markedly (Farman, 1985; Ianelli, 1987). It seems that the removal of fish around the floating object makes possible the arrival of new individuals, possibly responding to an optimal distribution around a favourable habitat, as stated in the ideal free distribution. Interestingly, F. Marsac (pers. comm.) noted that the removal of young fish by purse seining results in the arrival of older ones (which are usually found in deeper layers) as if they could then occupy the space made free by this removal. Nevertheless, in unexploited concentrations it is not certain that a real turnover occurs and possibly the same fish remain associated with a given FAD (or set of FADs) for several days.

The fidelity of tuna to anchored FADs is now better known. Hunter and Mitchell (1968) tagged fish that remained near moored objects from 8 days for *Labotes pacificus* (33 cm) to over 32 days for *Sectator ocyurus* (8–16 cm). Estimates of residency were 11–14 days for untagged yellowfin tuna and 9 days for black skipjack tuna (*Euthynnus lineatus*). Recently, Klimley and Holloway (1996a) tagged 14 yellowfin tuna with ultrasonic transmitters and attached detectors on five FADs around Oahu Island (Hawaii, USA). The experiment was seven months old at the time of pub-

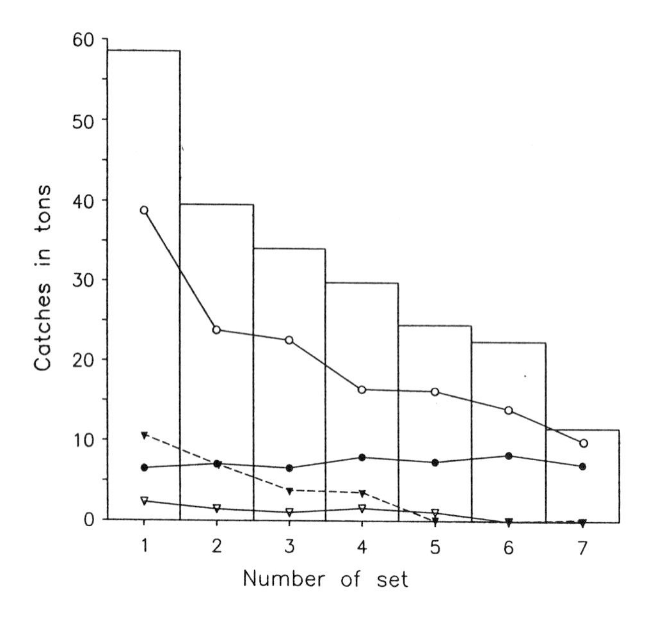

Fig. 6.7 Average catch by species (including mixture of species) and total average catch according to the number of repeated sets on the same floating object in the French purse seine fishery of the western Indian Ocean from 1984 to 1991 (redrawn after Hallier and Parajua, 1992b). ○ = Skipjack; ● = yellowfin; ▽ = bigeye; ▼ = mixture; bars = total.

lication and is continuing at the time of writing. They observed that tunas returned from one to ten times to the same FAD at which they were tagged over periods of 1–18 days, usually in two large periods centred on sunrise (0200–0900) and sunset (1500–2200). Moreover, the authors noticed on several occasions that two tunas arrived at the FAD less than a minute apart after absences of many days, indicating that yellowfin tuna can remain together within a single school on the above time scale of several months.

A more detailed study on a group of five among fourteen tagged fish (Klimley & Holloway, 1996b) suggests that there is a fidelity of tuna to a given FAD (tunas are faithful to a FAD and rarely visit the other ones located at less than 6 miles). In addition, the time of visiting the FAD seems different according to the schools. It is possible that separate schools have separate feeding routes or, conversely, the majority of schools visit the same waypoint at the same times. The authors did not comment on the fact that most of the tagged yellowfin remained in the Hawaii area during the seven months of observation, which is partially in contradiction to the common belief of seasonal migration of this species (Suzuki, 1994), but can be explained by low seasonal variation of surface temperature in this area.

From an experiment with ordinary dart tags, Kleiber and Hampton (1994) found evidence of the effect of FADs and of islands on the movements of skipjack tuna. They fitted a model of fish movements to the tag data and estimated that the presence of up to four or five FADs per 50 km^2 can reduce the propensity of skipjack tuna to leave that area by approximately 50%. Possibly part of the tuna population (at least for yellowfin and skipjack tuna) is sedentary and remains all year round associated with FADs or with other landmarks such as islands, underwater reefs or seamounts (Holland & Kajiura, 1997).

The associative behaviour of tuna towards logs is believed to be nocturnal because most purse seine sets on tuna shoals associated with logs are conducted before sunrise (Fig. 6.8; Hall *et al.*, 1992a; Hallier & Parajua, 1992b; Hampton & Bailey, 1992). Purse seiners stay close to logs during the night and make a set early next morning if tuna are detected underneath the log. It is believed the association breaks or is not so close during the morning when the tuna shoal starts searching for food. Still, logs are investigated by purse seiners throughout the day, and about 17% of successful sets are performed in the Indian Ocean from 0800 to 1800 h (Fig. 6.8). Interestingly Hampton and Bailey (1992) found a second peak of sets at dusk, which suggests that fish get closer to the floating object when visual conditions are still good. All these facts indicate that the association is not strictly nocturnal, and possibly that there are variations according to the areas (but we do not know if this is due to differences in fishing strategies, fishing equipment, environment or fish behaviour). In the next

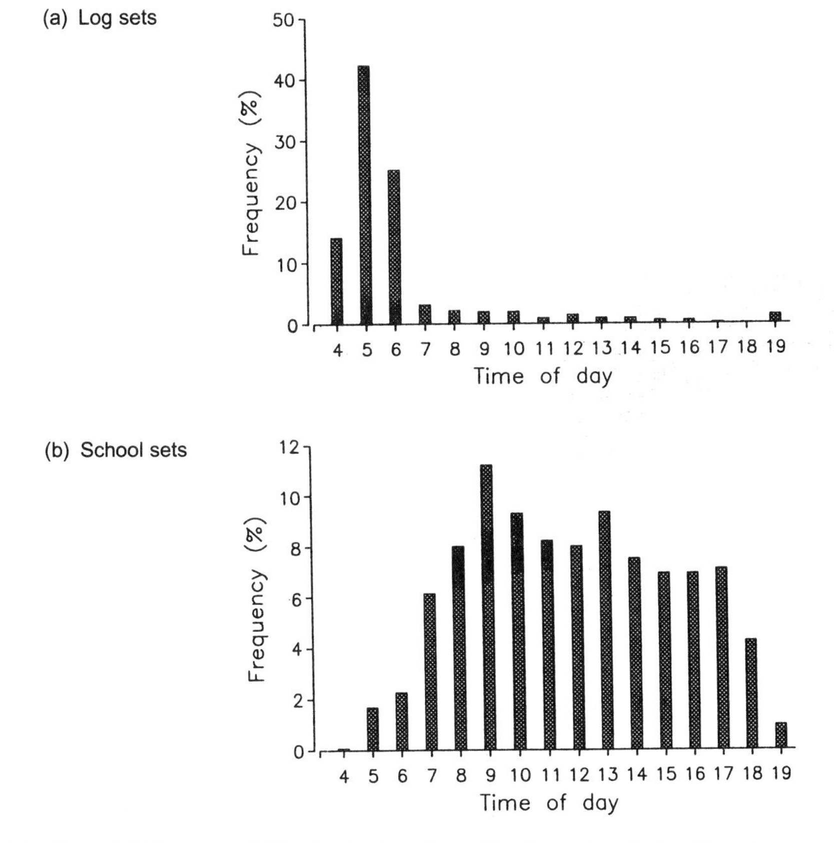

Fig. 6.8 Time of (a) log sets and (b) school sets performed by the western Indian Ocean by purse seiners from 1981 to 1989 (redrawn after Hallier and Parajua, 1992a).

section we will see that ultrasonic and acoustic surveys around anchored FADs give contrasting results, often opposed to those related to log-associations.

Different attempts at modelling the influence on tuna aggregation of population dynamics have been proposed during the last two decades of the twentieth century (see section 8.3). These models are sensitive to the estimates of the radius of attraction of FADs. From ultrasonic tagging experiments performed around anchored FADs, Cayré and Chabanne (1986) estimated this radius at 6 n.mi., Hilborn and Medley (1989) at 4.3 n.mi., Holland *et al.* (1990) at 5 nmi., Cayré (1991) at 7 n.mi. and Marsac and Cayré (1998) at 5 n.mi. As far as we know, there is unfortunately no radius estimate available in the case of drifting objects, nor on the fishes' fidelity to the object. Hall and Garcia (1992) speculated on different possible options for visits to one or several objects by tuna and concluded that there most likely is an association with any object encountered in the evening by the tuna instead of fidelity to one or several objects as described above for anchored FADs.

The puzzling determination of association

The underlying mechanism of association is controversial. We review the different hypotheses on this topic, from the oldest to the recent ones. Hunter and Mitchell (1967) analysed the 'concentration of food supply hypothesis' proposed by previous workers (e.g. Kojima, 1956), which argued that small fish, zooplankton and sessile biota are more abundant on or near the floating objects, but they did not find strong evidence to support it and proposed two other suggestions. The first, 'the schooling companion hypothesis', which was first proposed by Atz (1953), is based on the observation that many schooling fish associate frequently with animals other than their own species. Atz suggested that for schooling fish an aggregating companion could represent merely a simple point of reference for optical fixation. We will name the second hypothesis 'substitute environment hypothesis', which is mainly based on a generalisation of the observation of Carlisle *et al.* (1964) that artificial reefs established in sandy locations rapidly attract groups of demersal fishes that would not otherwise inhabit these areas (nowadays many fisheries in the world take advantage of this behaviour). Similarly for species not adapted to a pelagic life, and others undergoing a change from pelagic to other modes of existence, a floating object may function as a substitute.

As did Hunter and Mitchell (1967), Gooding and Magnuson (1967) discarded the 'concentration of food supply hypothesis' and proposed the 'cleaning station hypothesis' for some of the fishes, based on observations of rough triggerfish (*Canthidermis maculatus*) consuming parasites on conspecific and on other species. But finally, on the base of their *in situ* observation from a raft with an observation chamber, Gooding and Magnuson (1967) preferred the 'shelter from predator' hypothesis (but surprisingly they did not observe tuna during their two drifts which lasted over a week in the central Pacific). This hypothesis had already been proposed by Suyehiro (1952) and Soemarto (1960) and received later support from the experiment of Helfman (1981), who made visual observations of a black and white cylinder (12 cm high, 6.5 cm in diameter) below the surface, by snorkling under a shelter. He noted that a shaded observer could see a sunlit target at more than 2.5 times the distance at which a sunlit observer can see a shaded target. This is due to the reduced background light and veiling brightness of the shaded object. Therefore Helfman (1981) concluded that a fish hovering in the shade is better able to see approaching objects and is simultaneously more difficult to see. Nevertheless, it is likely that this conclusion does not apply for silvered fish, as their coloration is an adaptation to diminish the contrast with the surface, but this could explain the observation of Hunter and Mitchell (1967), who stated that unlike silvery fishes, dark-coloured fishes swim close to floating objects. More recently, Rountree (1989) compared the effects of structure size on fish abundance around coastal midwater FADs located in 14 m of water off South Carolina. The structures attracted mostly round scad (*Decapterus punctatus*) and exhibited a significant linear FAD size effect. This suggests that the FADs can be used as shelter by such small bait fishes.

Klima and Wickham (1971) proposed that floating objects and underwater structures provide spatial references around which fish orientate in the otherwise unstructured pelagic environment, but this hypothesis is difficult to demonstrate and is not valid for objects quickly drifted by the wind. Hunter and Mitchell (1967) marked ten adults of *Canthidermis maculatus* and released them separately at 7.5, 15 and 30.5 m from their original log. One hour and 30 minutes later, three out of four released at 7.5 m and three out of four released at 15 m had returned. Neither of the two fishes released at 30.5 m returned to their log. Despite the low number of observations in this *in situ* experiment (which would merit being repeated with other species), it seems that this species of fish returned to the log when it was within visual range.

In contrast, experiments on ultrasonic tagged tunas clearly established the existence of a homing behaviour related to anchored FADs. These observations of tagged fish were performed on yellowfin and bigeye tuna (Holland *et al.*, 1990), on skipjack and yellowfin tuna (Cayré & Chabanne, 1986; Cayré, 1991; Marsac *et al.*, 1996) or on yellowfin tuna only (Brill *et al.*, 1996). They indicate that the fish tended to remain tightly associated with the FADs only during the day, but most of them were able to swim as far as 10 n.mi. away from the FAD during the night and return directly to it in the morning. These experiments indicate also that the fish can locate the FAD and come directly back to it. For instance, Marsac *et al.* (1996) reported that a yellowfin tuna covered the 7 n.m.i. distance between two FADs in less than 2 hours. Nevertheless, it is not clear whether the fish are able to detect the FAD at long distance (by its noise signature or by olfaction, for instance) or whether they navigate after memorising its geographical position as suggested by the authors (see Chapter 7, on learning capabilities in fish). Even though some daytime excursions were also observed during tagging experiments around anchored FADs, the predominance of night-time excursions contrasts with the log-set fishing data analysis already reported, which mentioned that catches around logs occur chiefly around sunrise. Acoustic surveys performed around FADs in French Polynesia gave conflicting results in terms of daily variability of abundance (Depoutot, 1987; Josse, 1992), partially due to the difficulty of echo identification. Despite many uncertainties, it is obvious that fish behaviour associated with anchored FADs differs from the behaviour of fish associated with drifting objects.

It is noteworthy that in some areas where natural floating objects are not abundant, some tuna species were seldom caught before the use of artificial floating objects. This is observed in the eastern Atlantic tuna fishery, south of the Equator, where dispersed skipjack tuna are concentrated by artificial floating objects and caught in substantial quantities (Fonteneau, 1992). In contrast, A. Fonteneau (pers. comm.) indicates that artificial FADs do not give the expected result in the Canary current area where natural logs are few. Nevertheless, we have seen in the previous section that the 'log-boat' tactic was used successfully in this area. According to A. Fonteneau, artificial FADs are not efficient in this area because they drift too rapidly due to the strength of the Canary current and/or to the trade-wind in the case of partially submerged objects. This movement could result in a loss of geographical reference for the fish. The 'log-boat', however, is able to remain in a given area by use of its engine. The biomass associated with the baitboat is usually large, around 500 t according to skippers' estimates. It is reduced by only about 50% at the end of the fishing season, despite the fact that total catches might exceed 500 t. In addition to a permanent recruitment near the log-boat, there are indirect evidences of a high turnover of the fish associated with it, for instance a large day-to-day variability in the proportions of the three species in the catches.

As tentative explanation for the association between fish and floating objects, Parin and Fedoryako (1992) proposed a classification of the animals associated with floating objects. They distinguished three ecological communities related to floating objects. The two first communities are directly related to the floating object: the 'intranatant' animals (invertebrates, fishes) live inside or less than half a metre from the object, whereas the 'extranatants' are located between half a metre and two metres from the object but may come closer when they need shelter. The fauna of these two first communities originates mainly from the coast or the continental shelf and simply follows the drifting object, or is made up of juvenile pelagic fishes. In a typical log consisting of a 4 m tree trunk, the average biomass of these two communities is very low (less than 1 t) compared with the third community of 'circumatants' which are mainly young or adult predators temporarily associated with the floating object and swimming at a greater distance, generally on the leeward side of the floating object and often out of the field of vision, up to 5–7 n.mi. The biomass of this third community of predators generally exceeds 10 t (usually in the order of 10–100 t). Therefore it is obvious that the concentration of prey around the floating object is not sufficient to sustain tunas, which need to eat approximately 5% of their weight each day (Olson & Boggs, 1986), despite the fact that they are able to starve for several days.

Nevertheless, the reason for the associative behaviour is still unclear for tunas, especially its link with feeding behaviour on remote prey. Batalyants (1992) agrees that the 'concentration of food supply hypothesis' explained above is not relevant for tunas, but could be a good one for sharks and large dorado (*Coryphaena hippurus*) which may have small tunas in their stomach, even though there is no evidence of feeding under the floating objects. For tunas this author proposes the 'comfortability stipulation hypothesis', which means that tunas rest near floating objects for regenerating energy when satiated, even though he does not exclude the possibility of a hungry tuna resting after an unsuccessful search for prey. According to Batalyants' hypothesis, feeding of associated tunas is supposed to occur during the night on prey migrating to the surface layers (which could fit in with previously mentioned swimming behaviour observed with acoustic tagging, but not with most of the stomach content studies performed on anchored FADs, e.g. Buckley & Miller, 1994) and around noon, and resting occurs early in the morning and during the afternoon. This hypothesis is based on stomach content analysis, but the reason why tunas choose floating objects to rest instead of any other place is not clearly explained.

Buckley and Miller (1994) did not find any significant difference in diet between yellowfin tuna associated and unassociated with FADs, and reinterpreted contradictory results in the literature (e.g. Brock, 1985) in order to defend the similarity of feeding behaviour. They suggest that tunas are not feeding at night but may prey on vertically migrating mesopelagic organisms during crepuscular periods. Nevertheless, the night-time excursions of yellowfin away from FADs (or island reefs) are interpreted by Holland *et al.* (1990) as a foraging behaviour targeting on squids and shrimp, and Brock (1985) found that 84% of the yellowfin in the FAD fish community feed almost exclusively (92% by volume) on deep-dwelling oplophorid shrimp, against only 4% of the non-FAD-associated yellowfin. Brock interpreted this shift in feeding habits as the result of the high competition for food under the FADs due to concentration of tunas and other predators. In the case of seamounts, Yuen (1970) had a similar interpretation of skipjack tuna nocturnal excursions, which were related to feeding and exploration of the surroundings.

The seven former hypotheses are:

(1) Concentration of food supply hypothesis (Kojima, 1956 or earlier)
(2) Schooling companion hypothesis (Hunter & Mitchell, 1967)
(3) Substitute environment hypothesis (Hunter & Mitchell, 1967)
(4) Cleaning station hypothesis (Gooding & Magnuson, 1967)
(5) Shelter from predator hypothesis (Suyehiro, 1952; Soemarto, 1960; Gooding & Magnuson, 1967)
(6) Spatial reference hypothesis (Klima & Wickham, 1971)
(7) Comfortability stipulation hypothesis (Batalyants, 1992).

Two new hypotheses are:

(1) Generic-log hypothesis (and related hypotheses) (Hall, 1992a)
(2) Meeting point hypothesis (Soria & Dagorn, 1992; Dagorn, 1994; and this book).

Hall (1992a) proposed a long series of hypotheses in order to explain the migratory cycle of yellowfin in the eastern Pacific Ocean. Among those hypotheses, four complementary ones ('indicator-log', 'generic-log', 'follow-the-local' and 'dolphins-are-simply-fast-logs') are related to associative behaviour, but can be regrouped under the name of the second one, 'the generic-log' hypothesis. The author states first that the association of tunas with floating objects develops in response to the tunas' need to remain within a biologically rich water mass when they are not searching for prey, especially during the night. Floating objects are often indicators of such water masses and are moving with them or are trapped in vertical convection cells named

Langmuir cells (Fedoryako, 1982) or in convergence areas. This hypothesis fits with the second of the three requirements necessary in a habitat suitable for reproduction in pelagic fishes according to Bakun (1996):

(1) Enrichment (upwelling, mixing, etc.)
(2) Concentration processes (water-column stability, convergence, frontal formation)
(3) Retention of ichthyoplankton within an appropriate habitat.

Possibly the association with floating objects is the result of an evolutionary process in tunas which could facilitate their detection of, and remaining in, concentration areas. When they associated with anchored FADs they could simply have been misled (but see below for other interpretation related to anchored FADs). The 'generic-log' hypothesis itself is simply a generalisation. It relates association with any floating object (natural, FADs, boat) to association with animals of other species (dolphins, whales), dead or alive. Tunas associate with anything on the surface, floating or swimming, because it is an indicator, the only constraint being that the object does not move faster than the tunas can swim. This would explain why only tunas can associate with fast-swimming dolphins (see next section). Let us argue now on our preferred hypothesis, the 'meeting point hypothesis'.

The meeting point hypothesis

We name as the 'meeting point hypothesis' the possible enhancement of fish aggregations by floating objects as a process aimed to favour the encounter rate between small schools or isolated individuals. The idea was first suggested during a workshop on fish aggregation in Montpellier, France (Soria & Dagorn, 1992), along with another hypothesis related to a possible effect of schooling reinforcement under floating objects that we will not develop here. Later, the meeting point hypothesis was successfully simulated by artificial life (Dagorn, 1994), and we will try to develop it further here.

First the meeting point hypothesis considered the facilitation of aggregation of dispersed fish around the object which would act as a meeting point. This facilitation can occur for fish dispersing during the day or the first part of night, and resuming schooling during the second part of the night around the floating object (at least in the case of non-anchored objects). Second, this hypothesis also considered the possibility for small schools, which might be under an 'optimal size' (Sibly, 1983) to take advantage of a meeting point for gathering and might therefore benefit from the numerous advantages of schooling (sections 4.6 and 4.7). This could explain the direct observations reported by Yu (1992), who noticed that small schools appeared before large schools around floating objects. Moreover, it can explain the 'snowballing effect' observed around the floating objects: biomass increases with time and we have mentioned that schools under floating objects are larger than free-swimming schools (Fonteneau, 1992) despite the fact that the frequency of repeated sets on floating objects certainly contributes to the decrease in the expected biomass under natural conditions. This snowballing effect has been successfully simulated by Dagorn (1994).

In fact Dagorn did not simulate specifically FADs' effect on aggregation but more generally the attraction of any detectable anomaly in the tuna's field of perception, following that in Petit (1991). We feel that association with FADs is not comparable to attraction by other anomalies, but this point does not interfere with the interpretation of the results obtained by Dagorn (1994). The meeting point hypothesis (and others) can also explain why artificial drifting objects (including the 'log-boats') are so successful in areas where natural logs are few. As in the generic-log hypothesis, evolutionary theory could explain the selection of this behavioural trait.

In addition, Dagorn (1994) supposes that the smaller length of individuals (especially yellowfin) caught under floating objects compared with those from free-swimming schools could be explained by the 'critical school biomass' expressed in weight of school. If this critical school biomass is exceeded, the individuals in the school do not look for other conspecifics, they have no further interest in searching for a floating object, nor in remaining around it, and they leave it. Since the critical school biomass requires fewer individuals to be reached for large fish (for instance a 50 tonne school requires 25 000 individuals of 2 kg whereas the same biomass is reached with only 500 individuals of 100 kg), this could explain why a majority of small fish are caught under floating objects.

We think that another factor related to predation could explain the smaller length of individuals associated with floating objects compared with those in free-swimming schools. Small tunas can suffer attacks from different predators such as mammals – particularly killer whales (*Orcinus orca*) – or other tunas, including billfish (*Istiophoridae*) and conspecifics (Tsukagoshi, 1983; Kikawa & Nishikawa, 1986; Zavala-camin, 1986; Reiner & Lacerda, 1989; Nitta & Henderson 1993), whereas large tunas are too big and swim too fast for most top predators. Despite the lack of behavioural observation on schools of small tunas suffering attack, we can suppose that they react similarly to small pelagic fish, that is by a large repertoire of manoeuvres, including school splitting (Pitcher & Parrish, 1993). In addition, the cohesion of a school certainly decreases with the number of individuals under different circumstances (attack by predators but also feeding, migration, night dispersion). Therefore, for a given school biomass, the conservation of school integrity is probably more difficult for a large school of small tunas than for a small school of large ones. Once more this could explain a more frequent need to use a meeting point for small tunas than for large ones.

A remaining question could be the reason for the variability of differences in length according to the species (similar lengths for skipjack and bigeye tuna when associated or not with floating objects, but smaller lengths for yellowfins associated with floating objects compared with free-swimming schools). The reason could be a difference in habitat selection (mainly depth and oxygen; see section 3.3) between the two large tuna species when adult: large bigeye tuna live deeper and are less available to surface gears than large yellowfins. Skipjack tuna is a smaller surface species, in which maximum fork length seldom exceeds 70 cm in commercial landings, and it suffers predation during most of its life span (Zavala-camin, 1986). Nevertheless, an interesting exception in the size distribution of this species is the western Pacific, where fish over 70 cm are caught in substantial numbers (Hampton & Bailey, 1992). A detailed comparison between length frequency distributions made on a large number of fish caught under logs and length frequency distributions from non-associated schools in this area indicates that these large skipjack tuna are found only in non-associated schools and are nearly totally absent under logs (Fig. 6.9). Therefore the association with logs in the surface layer could be more size dependent than species dependent.

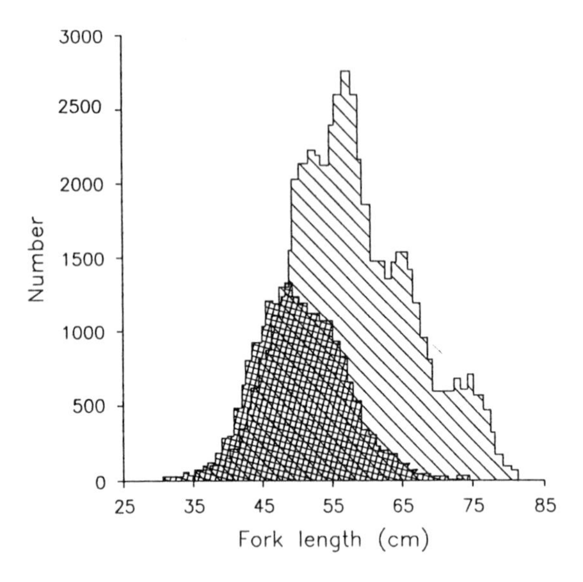

Fig. 6.9 Size frequency distribution of skipjack tuna (*Katsuwonus pelamis*) from school sets and log sets by US purse seiners in the western Pacific from 1988 to 1991 (redrawn after Hampton and Bailey, 1992). School sets (▧), n = 48 148; log sets (▩), n = 20 172.

In the case of isolated fish, as in a small school, the probability of encountering conspecifics by simple random swimming in the wide habitat of tunas is very low, taking into account the limited range of vision, especially during the night and at dawn. If we assume that tuna are not able to maintain schooling all night long, it is likely that they resume schooling before sunrise, and possibly drifting objects can enhance this aggregation. This difficulty of encountering conspecifics can be compared to the problem of some solitary terrestrial animals during the mating season, especially when they are in a situation where vision is limited. They have developed sophisticated cues in order to detect the mate at long distance. In nocturnal solitary insects, like the bombyx moth, the male is able to detect very low concentration of pheromone emitted by the female at more than 10 km (Fabre, 1930). In pelagic fish, aggregation pheromones have not been extensively investigated but could be an interesting candidate to explain the social behaviour (section 3.3). In the forest, for instance, the bell of the hart is a low frequency sound that can be heard at several kilometres by the females. Similarly, the low-frequency components in the song of humpback whales can be used in long-distance location of conspecifics. In tunas, if we imagine that when encountering a floating object a fish remains around it for a certain period of time before searching for another floating object (or a conspecific), this mechanism would not necessarily increase the probability of encounter with other conspecifics. But if the floating objects are easier to detect than conspecifics, and neither too numerous nor too greatly dispersed, then there is certainly a gain to this behaviour. Usually floating objects are less numerous than tunas and are clustered by physical processes. They are probably easier to detect by vision than conspecifics because, like most pelagic fishes, tuna species are countershaded. Floating objects, however, are seldom silvered, and it seems that dark objects are more attractive.

Auditory detection of floating objects is probably also possible, especially for medium or large objects that can amplify the noise of slapping on the surface in a choppy sea. Anchored FADs probably also emit sounds from mooring-line vibrations, and drifting pole-and-line vessels used as FADs are not silent because their secondary engine is working permanently. Olfactory detection is also possible, due either to the object itself (wood, synthetic material) or to the epifauna. Nevertheless, when a tuna patrols in the vicinity of an FAD it can be either upcurrent (e.g. Holland *et al.*, 1990) or on the leeward side (e.g. Parin & Fedoryako, 1992; Cillaurren, 1994, but only in the case of yellowfin tuna), and only this last location is believed to favour detection by olfaction. Moreover, we have already reported from the literature evidence for the possibility that tuna come back to the same anchored FAD from several miles upstream. This suggests an orientation capability not related to olfaction (e.g. geomagnetic field; section 3.3.6).

If tuna are to resume schooling before sunrise (as do coastal pelagic fish, e.g. Fréon *et al.*, 1996), they need to be close to the floating object to meet at this time. As with the effect of light attraction, the school is probably easier to capture at this particular moment. This could explain why tuna fishermen perform their sets mainly around dawn, despite the fact that schools could remain at a few miles from the floating object during daytime. It could also explain the larger catch per set if we suppose that the whole school is caught under a floating object whereas part of it can escape in other circumstances (but note that the meeting point hypothesis can explain higher school biomass under floating objects if dispersive processes are more important than aggregative processes away from the floating object's influence). Some contradictions found in the literature on the times of day that are believed to correspond to closer association between the fish and the floating object could also be partially explained by the circadian variation of the distance between the school and the object. Association around anchored FADs seems mainly diurnal, whereas it seems nocturnal for logs. But contrary to ultrasonic observations around FADs, log data come from fishermen and depend on the range of exploration around the object and on the definition of daytime and night-time. Possibly catches made not exactly under logs are reported as non-associated school catches. In addition, the importance of twilight transition periods could be underestimated, especially at dawn.

Nevertheless, it is likely that differences between log-associated and anchored FAD-associated tuna behavioural mechanisms do exist. They could be due to the fact that tuna can more readily locate an anchored FAD than a log, not only because it is fixed but due to its size (always greater than the one metre length limit found by Hall *et al.*, 1992b), together with its vertical extension due to the mooring, its probably sound emission and in many cases its light signal (also present on some drifting FADs). It seems that association with FADs can be related to association with seamounts, underwater reefs and islands, contrary to association with drifting objects. In all these situations, the permanence of the structure and its close distance to a coast or island allows for the development of learning capabilities related for instance to orientation to usual coastal foraging areas, as suggested by Yuen (1970), Holland *et al.* (1990) or Klimley and Holloway (1996a). These interpretations on anchored FAD-associated tuna are compatible with the need for a meeting point. Similar *in situ* observation of searching strategy involving site fidelity, spatial memory and expectation have been observed by Noda *et al.* (1994). In addition Warburton (1990) studied the use of local landmarks by fish (see section 7.3 for details). Some authors speculated on other specific mechanisms of tuna association with anchored FADs, but their proposals are not very convincing[1]. An additional difference between a drifting object and an anchored FAD is that fish associated with an anchored FAD might have to swim against the current, and this can be energy consuming in certain situations, contrary to the case of log-associated fish (except if emerged structures allow for fast wind drifting).

The meeting point hypothesis is not incompatible with some of the others. It presents some obvious relationships with the spatial reference hypothesis, at least in the case of anchored FADs, but at a larger range than stated by Klima and Wickham (1971). The comfortability stipulation hypothesis (Batalyants, 1992) could be complementary to our hypothesis since the need for tunas to rest can be combined with the need to wait for conspecifics at a meeting point. Even if association with a log would not increase the probability of encounter but just keep it equal, it would still present the enormous advantage of saving energy by limiting displacements. The mean travelling speed of tuna has been estimated by different authors (Carey & Olson, 1982; Cayré & Chabanne, 1986; Holland *et al.*, 1990; Brill *et al.*, 1996; Marsac & Cayré, 1998) and is usually around 2.5 n.mi. per hour. The generic-log hypothesis, and more specifically the indicator-log sub-hypothesis (Hall, 1992a), specifies a possibility for tunas to gather in an area where prey concentration is likely to be high, and to follow this concentration by remaining around drifting objects (logs or drifting FADs). This is not in contradiction with the idea of a meeting point, but the generic-log hypothesis does not apply to anchored FADs – except if we imagine that tunas are misled by these artificially fixed objects – nor for logs drifting mainly by the wind, which do not necessarily fit with the concentration of pelagic prey.

6.3.2 *Association with other species*

In the eastern Pacific Ocean, surface yellowfin schools are often observed associated with dolphin schools, and a major part of the catches are taken by aiming the fishing directly at dolphin schools (Perrin, 1969; Hall 1992b). In this portion of the ocean there is a rather shallow thermocline (< 100 m) that probably facilitates the

[1] Holland *et al.* (1990) suggested that by orienting to FADs the fish could maintain a fixed position in the current flow and may feed on current-borne prey, similarly to trout holding station in a river. But the results presented in their paper do not support this hypothesis because during the day fish are found moving permanently and are located out of the visual range of the FAD (> 500 m), an observation also made by Marsac and Cayré (1998). Buckley and Miller (1994) suppose that at night tunas avoid a fixed point of reference where nocturnal predators may easily find them.

encounter and preserves the stability of the association (Hall *et al.*, 1992a). The species of dolphin most frequently involved is the spotted dolphin (*Stenella attenuata*). Other tuna species (skipjack, bigeye, black skipjack *Euthynnus lineatus* and frigate tuna *Auxis* spp.) and dolphin species (mainly *Stenella longirostris* and *Delphinus delphis*) are also involved, but much less frequently than the association between yellowfin and spotted dolphin (Perrin, 1968; Allen, 1985; Au & Pitman, 1988; Hall, 1992a). Usually only large tunas associate with dolphins and the reason for the association of tunas with dolphins is still uncertain. Stuntz (1981) and Hall (1992a) summarise the main hypotheses in the early literature:

(1) In the 'contagious distribution hypothesis', both species are large piscivores that feed upon concentrations of similar food at the same place; this hypothesis does not explain the fact that the two species are swimming so closely to each other.
(2) In the 'innate schooling behaviour hypothesis', schooling is interpreted as a cover-seeking habit that evolved in a pelagic habitat where objects that provide shelter are rare, which would explain the occurrence of mixed schools; this hypothesis does not match the present interpretation of schooling behaviour, unlike the following one.
(3) The 'predator avoidance hypothesis' is based on evidence that predators are more easily detected and confused by schooling prey; this is obvious for small pelagic species, including juvenile tunas, but in the case of tuna–dolphin associations the large species involved have few predators.
(4) The 'food finding' hypothesis suggests that tunas and dolphins travelling together can prey more efficiently by complementary sensory modes (visual and olfactory for the tunas, sonar and visual for the dolphins) and by cooperative hunting. There are no *in situ* visual observations to support this hypothesis and indirect observations provide contrasting results. From simultaneous acoustic tracking of tunas and dolphins, Scott *et al.* (1995) concluded that their association often breaks due to opposite diel vertical migrations. A study of the comparison of feeding behaviour of yellowfin tuna and dolphins (Olson & Galván-Magaña, 1995) indicates the different feeding periods and prey. But two years later and with more data the same authors found a substantial overlapping in the diet of these two animals (Galván-Magaña & Olson, 1997).

From the end of the 1980s to the beginning of the 1990s, fishermen were under pressure from environmentalists due to the killing of dolphins during tuna fishing in the eastern Pacific purse seining fishery. The cooperation of fishermen and the implementation of systems permitting dolphin escapement have now resulted in a very low mortality of the mammals (Chapter 2).

The reason why yellowfin are found associated with dolphins in only the eastern Pacific Ocean remains unexplained. In the European purse seine fisheries of the Atlantic and Indian oceans, despite the presence of these cetaceans, the mean number of dolphins circled by net in 1995 was very low: 0.1 in the Atlantic on a sample of 360 sets, and 0.05 in the Indian ocean on a sample of 432 sets. The mortality is even lower because most of the dolphins escape or are released before the end of the circling process (Stretta *et al.*, 1996). We speculate on the possible role of the marked thermocline and especially oxycline in the eastern Pacific (Fonteneau, 1997), which could limit deep excursions of dolphins. In other oceans those deep excursions could be possible only for dolphins and not for yellowfin, which have a narrower preferendum of temperature and oxygen.

Marine birds are currently observed near surface tuna schools because they feed on the same prey as tuna or on prey associated with tuna prey (Au, 1991) and the presence of birds is used to spot the tuna schools. Fishermen use powerful binoculars and high-resolution radars to detect flocks of birds (Hervé *et al.*, 1991). Interestingly, birds are more often seen above non-associated schools than above schools associated with floating objects. Stretta *et al.* (1996) estimated that birds were observed in 68% of the set schools caught in the eastern Atlantic and 64% of those caught in the Indian Ocean. In contrast, birds were present above only 23% of the schools associated with floating objects in the Atlantic and 32% in the Indian Ocean. Despite the likely underestimation of the proportion of surface schools not surrounded by birds and therefore not detected by fishermen, these observations reinforce the idea that the reason for the association of tuna with floating objects is not related to prey.

In some areas, catches of tuna are also taken in association with larger marine animals such as the whale shark and whales (Stretta & Slepoukha, 1986; Fonteneau & Marcille, 1988). In the tropical west Atlantic, a study during 1987–1991 indicated that more than half the tuna catches (dominated by yellowfin) were made under whales (22%) or whale sharks (32%), against less than 1% under logs, while the rest of the catch was made on non-associated schools (Gaertner & Medina-Gaertner, 1992). A negative trend with time in the proportion of schools associated with whales is observed, but no information is available to relate this decrease to a possible decrease of whale abundance. The main whale species present in the Caribbean are *Baleanoptera edeni*, *Megaptera novaeangliae*, *Ziphius cavirostris* and *Physeter macrocephalus*. The main whale shark species is *Rhincodon typus*. This kind of association is scarce in the other oceans (Fonteneau, 1992; Stretta *et al.*, 1996). Here we also lack a valid explanation for the reason for this association. Nevertheless, contrary to the tuna–dolphin association, which is very dynamic, the association with shark whales could be considered similar to the floating object association discussed in the previous section, especially in the case of dead whales. Similar to observations reported on floating objects, repeated sets on carcasses during a week or even more indicate a slow decrease in the catch (Bard *et al.*, 1985; Cayré *et al.*, 1988). Here also the hypotheses of meeting points or water-mass indicators, among others, can apply.

In the west Pacific the association of tunas with whales is less frequent. From an analysis of the South Pacific Commission database, Hampton and Bailey (1992) estimated a 2% occurrence of this type of association in the purse seine fishery, despite large variation according to the flag of the tuna fleet. But due to the huge landing of the international fishery, this represents more than 40 000 tonnes. The authors identified three distinct association types. The first type is tuna aggregating and feeding with sei whales (*Balaenoptera borealis*) and, to a lesser extent, minke whales (*B. acutorostrata*) on ocean anchovy (*Stolephorus punctifer*). This is not considered a real association but a co-occurrence of species feeding on the same prey for a short period of time. The second is schools aggregating around floating carcasses of sperm whales (*Physeter catodon*), which is similar to log association. The third type of association, considered as intermediate between the two previous ones, consists of schools associated with the slow-moving whale shark (*Rhincodon typus*). In this last case there is also a common prey (ocean anchovy), but the association can be maintained for some time.

6.3.3 *Summary of associative behaviour*

Many pelagic species display an associative behaviour with items located at the surface or subsurface, either drifting ('logs', drifting FADs) or anchored (anchored FADs), either object or animals (dolphins, sharks, whales, either dead or alive), either natural or artificial. Nevertheless, only some coastal pelagic species display this behaviour, especially towards anchored FADs, and most of these coastal species have an oceanic stage of life. In contrast tunas are champions of all types of associations, but the underlying behavioural mechanism(s) are not well understood despite the large number of descriptive works produced during the last 10 years. It is not certain that a single behavioural trait can account for the different kinds of association, especially for the three main groups:

(1) Drifting objects or animals
(2) Anchored FADs
(3) Fast-moving mammals.

At least in the case of tuna, it seems that these associations depend on genetic behavioural trait(s), probably related to the schooling behaviour of this species.

Among the numerous hypotheses proposed to explain this behaviour in tunas, we think that the two most credible ones are the 'generic-log' hypothesis (Hall, 1992a), which permits the fish to detect and follow prey concentration areas by following a drifting object, and the 'meeting point hypothesis' (Soria & Dagorn, 1992; Dagorn, 1994; and this book), which could favour schooling behaviour and save waste of energy in searching for conspecifics. Both hypotheses agree with the fact that the only factors which seem to influence the attractiveness of floating objects are related to the possibility of detecting them (minimum size, submerged area, dark colour, noise). Nevertheless, further studies are required to understand the difference in tuna behaviour according to their association with logs or anchored FADs. Possibly the additional complexity and variability of association with fixed items (anchored FADs, islands, seamounts) can be partially explained by the possibility of learning how to use a reference point, which does not exist with drifting objects. This could explain differences in the results of studies aimed at studying school fidelity in tunas.

From the frequency of rare allele in Pacific yellowfin and Pacific skipjack tuna, Sharp (1978b) suggested long-term fidelity to school of these two species of tuna. External tag experiments performed by Bayliff (1988) out of anchored FADs (but maybe around logs) indicate that after three to five months at liberty, tagged and untagged skipjack tuna were randomly mixed. In contrast, Klimley and Holloway (1996b) demonstrated that yellowfins tagged around FADs are able to remain in the same school for at least seven months. It seems advisable to recommend ultrasonic tagging experiments based around drifting objects (natural logs or drifting FADs) in order to better understand the differences in tuna behaviour according to whether the object is fixed or not.

The association with mammals, especially with dolphins, is also puzzling. Possibly it is related to the floating object association as suggested by Hall (1992a), but it seems more easily explained by the 'food-finding' hypothesis (Stuntz, 1981) based on the complementary sensory modes (visual for the tunas, sonar and visual for the dolphins) in prey detection. For other species, especially bait fish, young stages of demersal fish and non-schooling top predators, other hypotheses might be valuable, especially the 'concentration of food supply hypothesis' (Kojima, 1956, or earlier) and the 'shelter from predator hypothesis' (Gooding & Magnuson, 1967).

Chapter 7
Learning

7.1 Introduction

Early ethologists believed that animal behaviour was mainly instinctive or pre-programmed and that experience did not play a major role, which was mainly true for insect behaviour (Fabre, 1930). In contrast, psychologists claimed that experience and learning were the major determinants. Even in the case of fish, it is now recognised that neither of the two opposite viewpoints is entirely correct. Animal behaviour in general and learning in particular are the result of a complex interaction of genotype, physiology and experience.

Different types of learning can be identified. The following ones are commonly identified:

- Habituation (persistent waning of a response to a repeated stimulus, without reinforcement)
- Social facilitation (an individual learns from its conspecifics)
- Conditioning, including classical conditioning or reflex conditioning (typically Pavlov's salivary responses of dog) and operant conditioning (association of the animal's behaviour with the consequences of that behaviour)
- Imprinting (fast and irreversible acquisition; in fish biology, the most important class of imprinting is locality imprinting by larvae or juveniles, which are then able to identify their place of birth during the natal homing process)
- Avoidance learning (also named aversive conditioning).

The classification of these types of learning varies with different authors (e.g. Guyomarc'h, 1980; Eibl-Eibesfeldt, 1984; Lorentz, 1984; Drickamer & Vessey, 1992; Soria, 1994).

In this chapter we will review the most important categories of learning regarding spatial dynamics and stock assessment of pelagic fish. Unfortunately, most of this work has been performed on coral reef species or freshwater species. Nevertheless, it is likely that most of the results can apply to other species and, as far as possible, we will give several references in order to show the possible generalisation of the authors' conclusions.

7.2 Learning in fish predation

In fish, the antipredator influence of learning has retained the attention of scientists for the last two decades of the twentieth century. In their review of learning in fish predation, Csányi and Dóka (1993) distinguish four classes of interactions between prey and predator fish: individual level, group level, among prey level and learning by predator (small section, not reported here). These authors also present aquarium experiments designed to determine the key stimuli in learning and hypotheses on interactive learning. We try here to summarise their work and incorporate some other new references. Finally, we give some thoughts on the comparison between learning in avoiding predators and in avoiding fishing gear.

7.2.1 *Individual level*

The individual level involves recognition and inspection of the predator, including discrimination between a hungry and a satiated one, and finally defence responses. We will review these different aspects and terminate this section by some considerations on the mechanisms of learning and on trade-off and learning.

Fish, like most vertebrates, are able to recognise their natural predators without prior experience and respond appropriately on their first exposure to them, which suggests a strong genetic component (Pitcher & Parrish, 1993). Studies on the ontogeny of antipredator behaviour indicate that ability to recognize predators appears in paradise fish larvae (*Macropodus opercularis*) between 15 and 20 days old and is under broad genetic control (Miklósi *et al.*, 1995). This genetic control was already evidenced on the Trinidadian guppy (*Poecilia reticulata*): local variation in predation pressure significantly affects the avoidance, schooling behaviour and courtship behaviour of this species, even on laboratory-reared populations, as shown by Seghers (1974b), Magurran and Seghers (1990, 1991) and Magurran *et al.* (1993). These results were also confirmed by observations on sticklebacks (*Gasterosteus aculeatus*) (Giles & Huntingford, 1984).

Despite the strong genetic component in both predator recognition and inspection, Csányi (1985) and Csányi *et al.* (1989) have observed in 'naive' paradise fish (*Macropodus opercularis*) a decrease in duration of inspection in front of another fish species, the goldfish *Carassius auratus* (here, and in the rest of the book, 'naive' means a fish that has no experience regarding a given stimulus). This decrease was still significant if the second encounter took place three months later (Fig. 7.1). If the olfactory nerves of the paradise fish are cut, the inspection duration does not decrease as fast during repeated encounters as in the controls (Fig. 7.2 from Miklósi & Csányi, 1989). This suggested that the memory formed during the encounters contains not only visual but also olfactory information (also confirmed by Gerlai (1993) in experiments on the same species and by Magurran (1989) on the European minnow, *Phoxinus phoxinus*). Moreover, paradise fishes which experienced repeated attacks by a predator diminished their inspection and finally tried to 'escape' by constantly swimming perpendicular to the glass wall of the aquarium (Fig. 7.3). Even though this manoeuvre was unsuccessful, it suggests some learning process expressed as an attempt to avoid the next encounter (Csányi, 1985). Similarly Chivers and Smith (1994a, 1994b) found no response of predator-naive fathead minnows (*Pimephales promelas*) to chemical stimuli from northern pike (*Esox lucius*) while wild-caught fish of the same age and size did respond. Fourteen days were necessary for minnows from

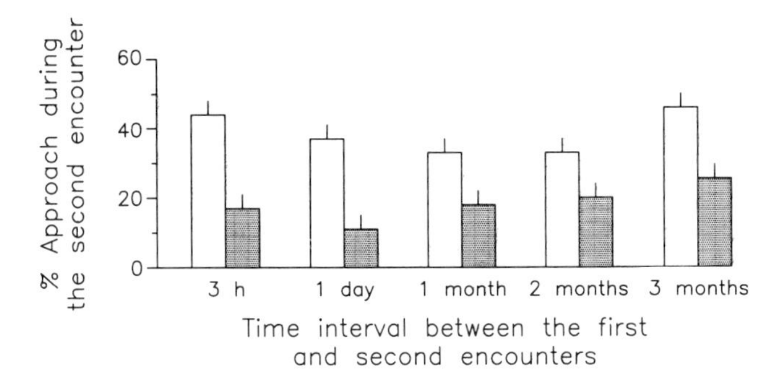

Fig. 7.1 The percentage of time spent in exploration during a second encounter of paradise fish (*Macropodus opercularis*) with a goldfish (*Carassius auratus*) at various intervals after the first encounter. Bars represent means of observation within a group (+ SE). (Redrawn from Csányi *et al.*, 1989.) □ = Control; ▩ = treated.

a pike-free pond to acquire recognition of pike odour (Chivers & Smith, 1995). These results suggest that previous experience is necessary for chemical predator recognition. Nevertheless, the experiment also demonstrated that reinforcement was not required for predator recognition to be retained.

In schooling species, Magurran and Pitcher (1987) also found some evidence of learning in the so-called 'predator inspection visit', which is the observation of a predator by an individual fish leaving the school for a few seconds to approach a predator and return to the school (Pitcher *et al.*, 1986b). Magurran and Pitcher (1987) found inspection of pike by minnows ceased after a successful attack on another fish, which strongly suggests some learning during the process. Huntingford and Coulter (1989) found habituation of predator inspection in sticklebacks. Habituation to predator models was observed by Welch and Colgan (1990).

Csányi and Dóka (1993) found little information on the role of learning in the discrimination between hungry and satiated predators by paradise fish. In contrast

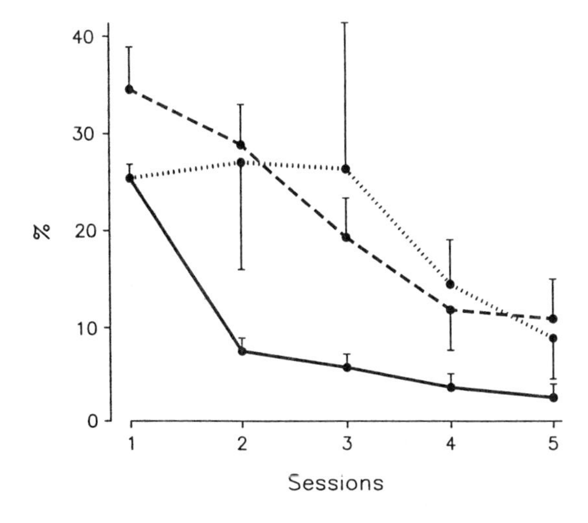

Fig. 7.2 Percentage of time spent by paradise fish (*Macropodus opercularis*) with approach towards the goldfish (*Carassius auratus*) in each of the five sessions (mean + SD). (—) = sham-operated control fish; (- - -) = olfactory nerves destroyed prior to experiment (olf1); or (. . . .) after the first session (olf2). (Redrawn after Miklósi & Csányi, 1989.)

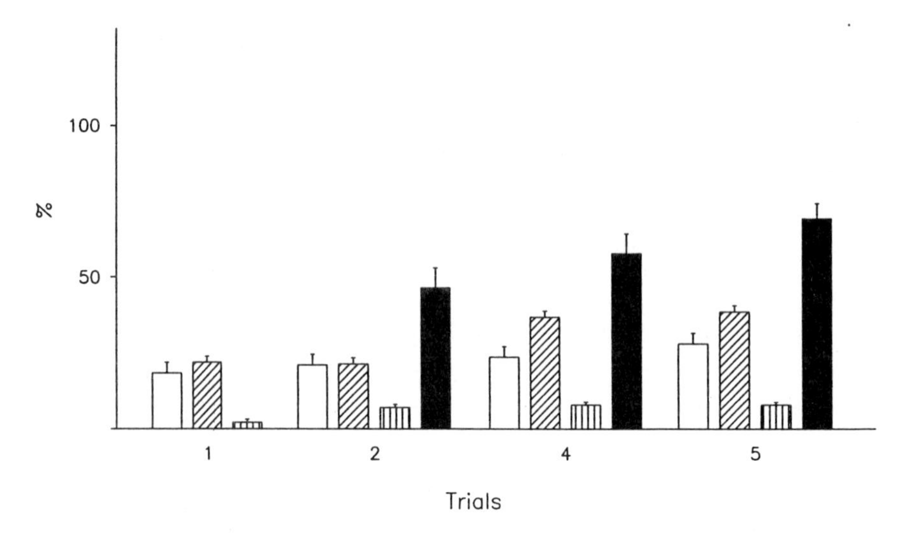

Fig. 7.3 Percentage of time in each of the five trials spent in escape behaviour of paradise fish (*Macropodus opercularis*) in the presence of a satiated pike (*Esox lucius*). Bars represent group mean (± SE). Empty □ = group of fish kept in total isolation from conspecifics or other fish (control); Isolated (▥) = isolated as the empty group but found a satiated pike in each session; Naive = (▨) not isolated, found a satiated pike in each session; Chased (■) = survivors after experiencing the presence of a hungry pike at the first and third sessions. (Redrawn after Csányi, 1985. Published by Brill, Leiden, The Netherlands).

Licht (1989) found that guppy (*Poecilia reticulata*) could discriminate between hungry and satiated predators. He argued that a flexible response to the differential threat can help to lower the costs of anti-predator behaviour. Magnhagen and Forsgren (1991) showed that sand goby (*Pomatoschistus minutus*) did not discriminate between hungry and satiated cod, even though hungry cod ate more gobies. Despite the limited number of experiments on discrimination of level of hunger in the predator, it seems likely that fish are able to learn from the predator's behaviour its level of hunger. Personal *in situ* observations by one of the authors of this book on the behaviour of pelagic or reef-species schools in the presence of barracuda (*Sphyraena barracuda*) indicate that on many occasions the preyfish is not scared by the predator even when it is only a very short distance away.

Behavioural genetic studies on the paradise fish indicate that many elements of defensive behaviour, or whole sequences, can be determined by genetic factors (Csányi & Gervai, 1986; Csányi & Gerlai, 1988; Gerlai & Csányi, 1990). Nevertheless, in many cases this genetically programmed defence behaviour pattern can be modified by learning. Conditioning experiments on the same species show that the duration and frequency of the passive feeding responses were influenced by learning (Csányi, 1993). In another experiment, three groups of fish of the same species were observed in a large shuttle aquarium where a hungry or a satiated pike was resident in the largest part. Members of the control group had not encountered pike before; members of the second and third group had experienced a hungry or a satiated pike respectively. Only fish of the second group quickly learned to avoid the largest part of the aquarium. In the same kind of aquarium, the authors observed the reaction of the same species in front of a large carp. Even though this carp never attacked the paradise fish, at the beginning the avoidance reaction was similar to that towards a hungry pike. But this avoidance decreased rapidly and ceased completely after the fifth or sixth trial, with the paradise fish ignoring the presence of the big carp. This indicates the role of experience and learning in the relationship with other species.

The mechanism of learning still remains largely unknown but it is traditionally accepted that it is gradual (Krechevsky, 1932). In addition, Zhujkov (1995) studied defensive learning and observed large differences in learning capacities among individuals, classified as fast- or slow-learning individuals. He hypothesised the existence of two simultaneously developing independent learning processes: formation of the cause–effect relationship between conditioned and unconditioned stimuli, and formation of the motor avoidance response to the unconditioned stimulus. Nevertheless, recent findings on individual learning in fish suggest an 'all or nothing' type at the level of a particular individual, while variability in a group of fish comes from individual differences in the number of repetitions necessary to reach conditioning (Csányi, 1993). On the same line some authors working on commercial species indicate the possibility that fish that experience enormous stress from contact with fishing gear might undergo a 'one-trial learning' process (Beukema, 1970; Fernö & Huse, 1983; Pyanov, 1993; section 7.2.6).

Decision-making in fish is often a trade-off between foraging profitability and predation risk (Fraser & Huntingford, 1986; Hart, 1986; Huntingford *et al.*, 1988). Even though few specific examples on fish are available at the moment, it is likely that various learning mechanisms play an important role at this level (Lima & Dill, 1990). Huntingford and Wright (1992) show how inherited population differences in predator reaction combine with learning processes and trade-off. Three-spined sticklebacks from a predator-free site learned more slowly than fish from a predator-abundant site the avoidance of patches where simulated attacks of a model avian predator were performed, despite the fact that those patches were previously favoured for feeding.

7.2.2 *Interactions at the group level*

Shoaling is a fundamental behavioural pattern for many pelagic fishes (Chapter 4), but there is little direct evidence for the involvement of learning in shoaling. Recently Gallego *et al.* (1995) tried to assess whether the presence of schooling conspecifics encourages the premature development of schooling in herring (*Clupea harengus*) larvae. Laboratory-reared larvae (length 29–31 mm) were exposed to wild-caught schooling juveniles (length 55 mm) and their behaviour was compared to non-exposed larvae, before and after feeding them. Larvae seemed to opt for the increasing protection of the school of juveniles even when competition for food was disadvantageous for them. Experiments on predator inspection visits in European minnows (*Phoxinus phoxinus*) strongly suggest that the active or passive transfer of information between the inspecting fish and the rest of the school is favoured by social facilitation (Pitcher *et al.*, 1986b; Magurran & Highman, 1988).

Intraspecific and interspecific experiments on fathead minnows (*Pimephales pre-*

melas) and brook sticklebacks (*Culaea inconstans*) demonstrated that individuals that live in groups may have the opportunity to learn to recognise unfamiliar predators by observing the fright response of experienced individuals in the group (Mathias *et al.*, 1996). In this experiment the predator stimulus used for conditioning the prey was only pike (*Esox lucius*) odour. Naive prey gave fright responses to this chemical stimulus when paired with pike-experienced prey of conspecifics or heterospecifics but not when paired with pike-naive prey.

An interesting example of interaction at the group level is given by the fright reaction to an alarm substance released by injured fish belonging to the Ostariophysi and discovered first by von Frisch (1941) in the zebra danio (*Brachydanio rerio*). This alarm substance acts to initiate learning of a co-occurring novel and behaviourally neutral chemosensory stimulus. The new learned chemosensory stimulus can be a base of a cultural tradition (Suboski & Templeton, 1989). It is noteworthy that the fright reaction differs considerably from species to species of the same group. Moreover, it is generally accepted to be mediated by olfaction, but involvement of taste or vision is not entirely ruled out. For instance, when two aquaria with a group of fish in each are placed close together, a fright reaction induced by an alarm substance in one tank would trigger a fright reaction in the other aquarium without water communication. The number of ostariophysans but also non-ostariophysans reported to possess alarm pheromone systems continues to expand and includes some marine species, but at present none of the main pelagic species (reviews: Smith, 1992; Hara, 1993).

7.2.3 *Interactions among the prey*

Social facilitation is well documented in fish (see section 7.3), but few experiments give evidence of stable conditioning. Patten (1977) observed that naive fry of coho salmon (*Oncorhynchus kisutch*) fell prey at a significantly higher rate than did naive fry exposed to the same predators but in the company of conspecific survivors of earlier exposure to predators. Soria *et al.* (1993) found that conditioned tropical sardines (*Opisthonema oglinum*) were able to initiate a global response of a school to a neutral stimulus (series of three sound pulses), even when this school contained 50% of naive fishes. In this experiment the punishment was the hoisting of a horizontal rigid net from the bottom to the surface of the tank (Fig. 7.4).

7.2.4 *Key stimuli in learning*

To investigate whether there were key stimuli connected to predators, Csányi (1986) repeated the shuttle aquarium conditioning mentioned in section 7.2.1 in a different way. The predator was a live goldfish and there were also dummy predators which looked more and more like a fish (Fig. 7.5), associated or not with mild electric shocks (punishment). Neither the dummies, the presence of a live goldfish nor the electric shock alone caused significant changes in latency. Nor did latency increase in response to a combination of punishment and a dummy head with a 'single eye' or without eyes. However, the latency significantly increased if punishment was

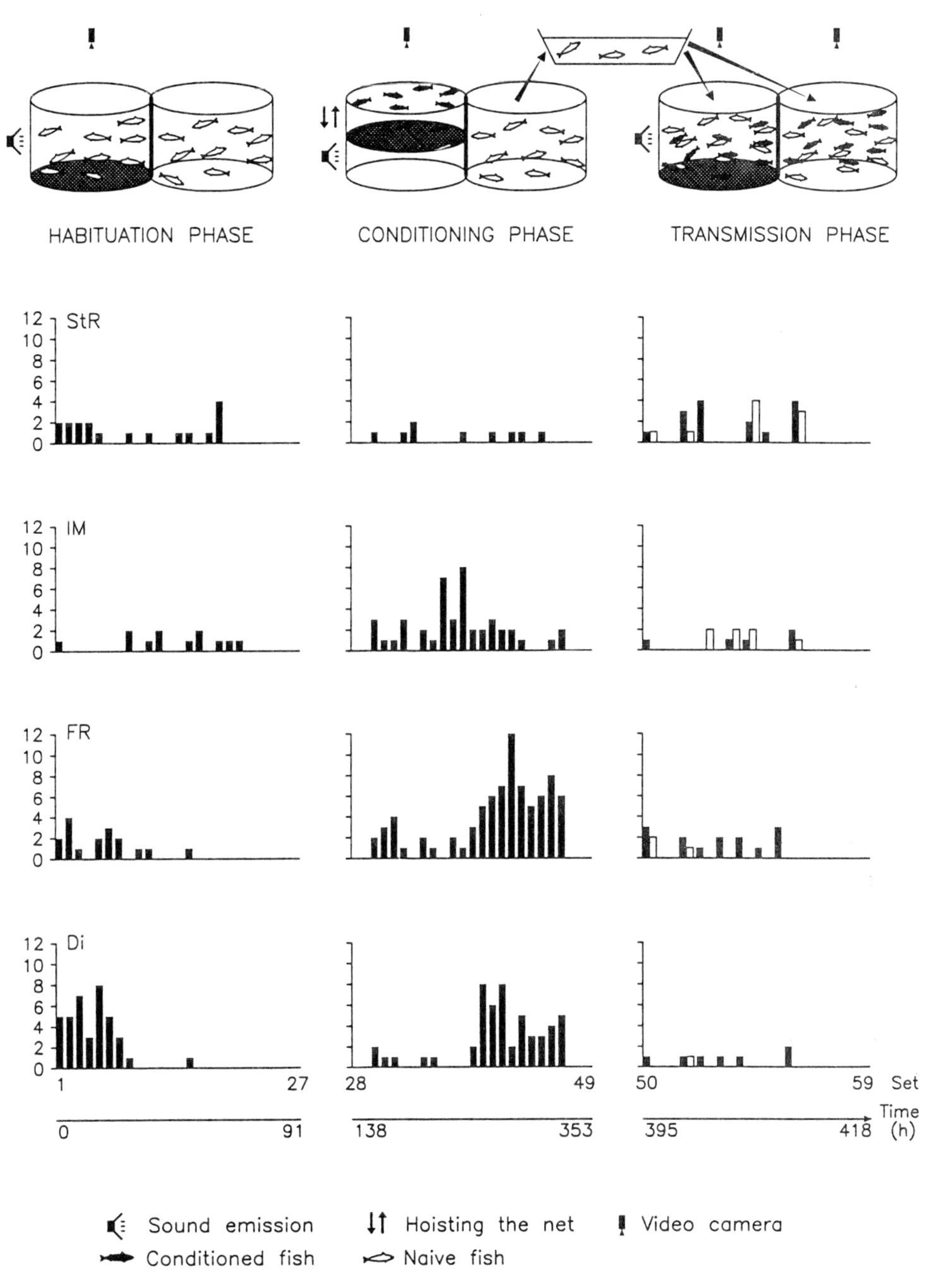

Fig. 7.4 Three phases of the experiment (upper) and results (lower) on thread herring (*Opisthonema oglinum*) expressed as frequency histograms of reactions to the conditioned stimuli in the tank with a net (black bars) and in the other tank (white bars). StR = startle response; IM = imperfect mill; FR = flight reaction; DI = dislocation. Set refers to sets of three sound pulses. (Redrawn after Soria *et al.*, 1993.)

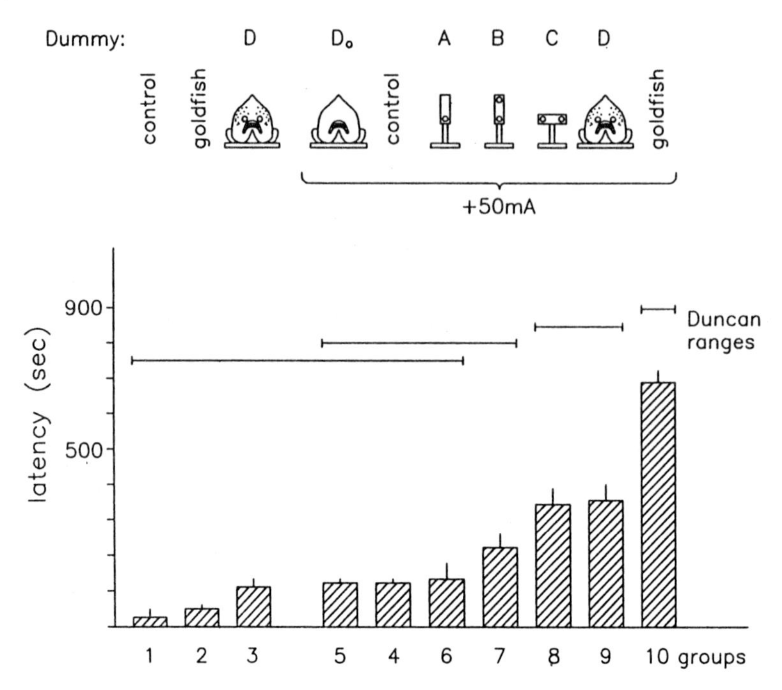

Fig. 7.5 Increase in latency in groups of paradise fish (*Macropodus opercularis*) treated by various dummies. Bars represent means + SE: Duncan ranges are also shown. (Redrawn after Csányi, 1986.)

combined with more realistic dummies. The largest increases were obtained with the live goldfish, and secondarily with a dummy having two lateral eye-like spots arranged horizontally (vertically arranged spots were not so efficient). Other similar experiments with a live pike or catfish and dummies confirm that eyes or lateral eye-like spots on a body have key-stimulus character and are used in learning by the paradise fish (Altbäcker & Csányi, 1990). Nevertheless, Gerlai (1993) observed differences in reactions of paradise fish towards a sympatric predator (*Channa micropeltes*) and an allopatric predator or harmless species and concluded that eyes alone may not differentiate the harmful species from the innocuous. He suggested that additional visual stimuli and olfactory stimuli are also necessary. Altbäcker and Csányi (1990) indicated that motion is very likely to be an additional factor in the key stimulus, or maybe a key stimulus itself.

Part of the previous experimental findings are confirmed by Fréon *et al.*'s (1993a) observations during *in situ* experiments on a pelagic species. A realistic model predator (bonito, *Auxis thazard*) was towed by a dinghy and passed over the same school of *Harengula clupeoa*. A real attack by predators (genus *Euthynnus*, species not identified) was also observed on the same species. The school was observed simultaneously with a vertical sounder, an underwater video camera and aerial photography. One conclusion was that the intensity of the school reaction depends on the intensity, number and duration of stimuli. This is a generalisation of the rule of heterogeneous stimuli summation initially developed on a single individual by Tinbergen (1950) and Lorentz (1974): the intensity of the response of an individual (in our case a group of schooling individuals) fluctuates according to the intensity, number and duration of stimuli.

7.2.5 *Hypothesis on interactive learning*

On the basis of observations and experiments on the paradise fish, Csányi (1987, 1988, 1989) proposed his interactive learning hypothesis. He stated that the main part of the mechanism of predator avoidance is the organised system of the memory traces in the animal brain acquired during daily experiences. This system is connected to certain behaviour mechanisms which are mainly genetically determined. Exploratory behaviour is activated by a new environment or stimulus and enables the animal to gain experience and knowledge about new objects and to store them during the learning process. As far as predators are concerned, eyes on a body, motion and presumably smell are the key stimuli (see section 7.2.1) which promote recognition, recording and learning. If the exploration is not associated with an unpleasant stimulus or pain, habituation occurs. Otherwise, avoidance reactions are activated (but see introduction of the book and sections 3.3, 3.4 and 7.2.1 for a less simplified response, when conflicting behaviours lead to a trade-off). In summary, memory traces representing the predator form links with escape reactions, while those representing harmless fishes link to neutral behaviour units.

We find at least two references related to memory traces in the fish brain acquired during learning – one by Matzel *et al.* (1992), who studied biophysical and behavioural correlates of memory storage, degradation, and reactivation, and another by McElliott *et al.* (1995) on the effect of temperature on the normal and adapted vestibulo-ocular reflex in the goldfish.

7.2.6 *From natural predators to fishing gear*

A large majority (if not all) of fish species seem able to learn, and from a comparative study of learning ability on 14 species, Coble *et al.* (1985) and later Zhujkov and Trunov (1994) indicated that this ability is not related to phylogenetic position and could depend on the level of motor activity of the species. We have seen that fish are able to adapt their behaviour towards predators through learning. In the first analysis, there is no reason to consider that the learning mechanisms towards active fishing gear are different from those towards predators. Nevertheless, the fisherman is a recent predator in the history of the species and can be seen by the fish as a polymorphic predator because he is able to change or improve his fishing tactics by constant technological progress. Therefore the genetic basis of learning in the case of fishing gear, if any, will be difficult to demonstrate (Allendorf *et al.*, 1986) and cannot be compared to the genetic basis of learning about predators. To a certain extent the assumption of optimality of behavioural ethology is violated, especially when the fishing gear is designed to make fish behave maladaptively, as underlined by Fernö (1993). In front of a trawl wire, fish react maladaptively and are herded into the path of the trawl (Wardle, 1986). Similarly, fish are stimulated to bait by the movements of neighbours hooked by a longline (Fernö *et al.*, 1986; Løkkeborg *et al.*, 1989).

An additional reason that makes it difficult to associate learning in fish predation to learning in fishing might be the likely importance of olfaction in predator identification. It is not obvious that fish are able to identify fishing gears by their odour. Although it seems possible for passive gears, especially nets or traps that can be imprinted with fish odour, it does not seem likely that fish have time to detect and react to the odour of fast-moving gear or vessels. It is more likely that auditory stimuli associated with fishing gear or fishing operations can be identified (sounds emitted by the fishing vessel or by the fishing gear). We have seen in Chapter 5 that it is now proved that fish are able to locate the direction of a sound emission, that they are able to perform acoustic distance discrimination at short distance, that sudden changes in revolution or pitch while circling a school might scare the fish and finally that playback of vessel noise results in fish avoidance reaction.

Despite these preliminary remarks, we have spotted in the literature some examples of fish learning to avoid or escape an active fishing gear. Unfortunately most of them concern demersal species. Nevertheless, Soria *et al.* (1993) were able to condition thread herring (*Opisthonema oglinum*) by associating a series of sound pulses with a stress related to a net. Individual reaction and school stability measurements indicated the efficiency of conditioning after 3 to 12 trials (Fig. 7.4). Similarly, Pyanov and Zhujkov (1993) demonstrated the facility of reflex conditioning in aquaria. Engås *et al.* (1991) found that cod captured and marked by an acoustic tag onboard a small trawler always avoided the same vessel during subsequent fishing trials in shallow water the following days. From an angling experiment, Beukema (1970) indicated that carp (*Cyprinus carpio*) are able to substantially decrease their catchability through 'one-trial learning'. Zaferman and Serebrov (1989) reported a decrease in consecutive bottom trawls on marine demersal species.

Pyanov (1993) made repeated sets of experimental trawling on bream (*Abramis brama*) in a reservoir where commercial trawling had never been practised before. He found a significant difference in catch rate between the first and the second repeated hauls in the same area, but no difference between subsequent ones (up to four). Moreover, he showed that the decrease in catchability was greater in old fish than in young ones (Fig. 7.6). The one-trial learning hypothesis was reinforced by the results of simultaneous tagging experiments with ultrasonic tags. All tagged fish displayed avoidance reactions during trawling at about 50 m from the vessel while other naive fish were caught. Pyanov (1993) also performed aquarium experiments on small freshwater species, which reinforced his hypothesis. Small dip nets were used to perform catch trials on groups of tetra (*Hemigrammus caudovittatus*), and also rosy barb (*Barbus conchonius*). After one to three trials, the behaviour of the fish was changed and they displayed school defensive behaviour when the operator and the gear appeared: escapement or increased avoidance distance according to the species. Finally, from commercial fishery data analysis, Le Gall (1975) reported that the catchability of albacore (*Thunnus alalunga*) caught by trolling was significantly higher on the recruited age class (age 2) than in older fish, which can be related to learning.

In a series of passive avoidance conditionings performed on paradise fish, Csányi

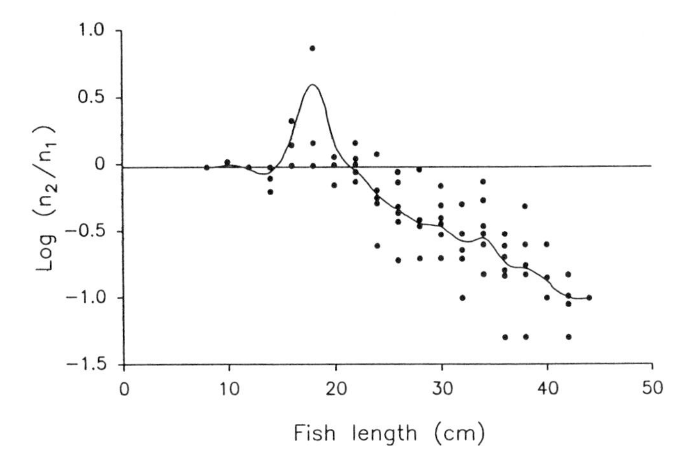

Fig. 7.6 Change in the value of the ratio of the second catch to the first one in different size groups of bream (*Abramis brama*). (Redrawn after Pyanov, 1993.)

and Altbäcker (1990) found that the most easily learned tasks were the ones which had several different solutions, like avoiding by staying on the left side of the aquarium (Fig. 7.7, session B), which can be compared to avoiding the vicinity of a passive gear. When the task was reversed, i.e. staying in the left side was punished by mild electric shocks, fish reacted in the opposite way (Fig. 7.7, session C). Interestingly, this reversibility was not so clear in conditioning to decrease the frequency of escape behaviour; fish had more difficulty learning to decrease this frequency (Fig. 7.8, session B) than to increase it (Fig. 7.8, session C). This last result can be related to the facility of fish to learn avoidance of active gears by active escape.

Soria *et al.* (1993) showed that under certain conditions, experienced thread herring can transmit avoidance behaviour to a naive school. In the above-mentioned experiment, when sound-conditioned fish were mixed with naive fish, the mixed school also displayed a conditioned reaction. This indicated that the learning capabilities of fish are effective and make them able to avoid fishing gear after having escaped. Aversive learning experiments on several freshwater species indicate that fish can retain their learned experience for months, especially striped bass (*Morone saxatilis*) and largemouth bass (*Micropterus salmoides*), although performance deteriorated with time (Coble *et al.*, 1985).

Fish are also able to learn how to avoid or escape from passive gear. Fernö and Huse (1983) observed cod (*Gadus morhua*) learning from baited hooks. Similarly Johannessen *et al.* (1993) observed that during the period of an experiment on the baiting behaviour of cod and haddock, the hooking frequency gradually changed through learning. Zhujkov and Pyanov (1993a,b) demonstrated that tetra (*Hemigrammus caudovittatus*) are able to learn how to escape an experimental trap net. As in the case of the experiment on active nets previously mentioned (Soria *et al.*, 1993), Zhujkov and Pyanov's (1993b) experimental results suggest social facilitation in the tetra school. They noted that the exploratory behaviour depends on the composition of the school: 'In the case of great differences between the school members, the

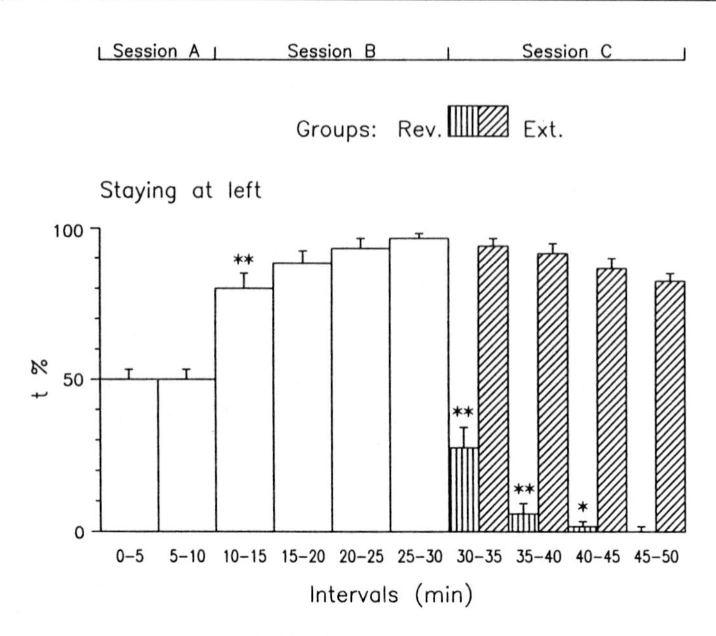

Fig. 7.7 Time percentages spent at the left side of the experimental tank during the consecutive sessions of avoidance conditioning. Group means ± SE are shown. Asterisks indicate significant differences at $p < 0.05$ level. Session A: control (before punishment). Session B: contingent punishment. Session C: reversal conditioning for group Rev (punishment abandoned for group Ext). (Redrawn after Csányi & Altbäcker, 1990.)

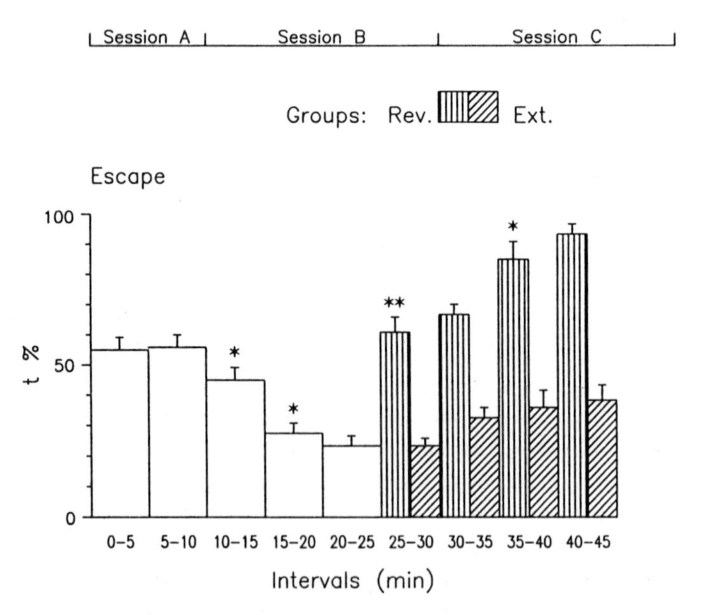

Fig. 7.8 Time percentages spent by paradise fish (*Macropodus opercularis*) in escape reaction (ESC) during the consecutive sessions of avoidance conditioning. See Fig. 7.7 for key. (Redrawn after Csányi & Altbäcker, 1990.)

behaviour of the school is dictated by the reactions of individuals rapidly perceiving new information, rather than an arithmetic mean for the school'.

7.3 Other learning processes

We will see now a limited number of other learning processes that might influence spatial dynamics and stock assessment of pelagic fish. From a literature survey, Dodson (1988) suggested that learning is an important process in orientation and migration of many vertebrates, including fish. He argued that experience and learning may influence the direction of movement and how the goal can be recognised in the following processes: imprinting, early experience and spatial learning, including social transmission of migratory routes and directions. Røttingen (1992) contended that this applied to herring (*Clupea harengus*) in the Norwegian Sea where new generations were guided to feeding grounds by older experienced herring. A similar social facilitation was suggested by Corten (1993) for the choice of spawning grounds, but also overwintering grounds and feeding areas of the North Sea herring. Corten argued that young fish will follow older ones which have already learned how to migrate to selected spawning grounds, and repeat this migration pattern during the following years. The author did not mention how these former preferential grounds were initially selected and how convergence of the population to a limited number of areas is possible.

It is difficult to test the previous hypotheses, but various *in situ* experiments or observations support them. Observations on French grunts (*Haemulon flavolineatus*) clearly indicate the existence of social transmission. This species is known for long-term fidelity to foraging areas (Helfman *et al.*, 1982) but individuals transplanted to new schooling sites and allowed to follow residents at the new sites used the new migration routes and returned to the new site (Helfman & Schultz, 1984). Similar *in situ* observations of searching strategy involving spatial memory and expectation have been observed in zooplankton feeders by Noda *et al.* (1994). The authors noticed that stout-body chromis (*Chromis chrysurus*) usually stayed for more than one hour in given areas, irrespective of the presence or absence of prey, and searched there with a tortuous pattern with reduced speed. In contrast, movements between regions tended to be rapid with almost no feeding. Moreover, laboratory experiments on goldfish (*Carassius auratus*) indicate that fish can learn how to use local landmarks for choosing favourable foraging areas (Warburton, 1990).

In the case of spawning grounds, Corten's hypothesis could appear in competition with the natal homing hypothesis for pelagic fish mentioned in section 3.3.1. If new-born pelagic fish are imprinted by temporally and spatially variable water-mass characteristics, as suggested by Cury (1994), it is unlikely that they will be able to spawn in their native area when adults. Following Cury's hypothesis of the general adaptive value of obstinate reproductive strategy, we do believe that homing is more generalised than expected in pelagic fish. But, as earlier suggested by Mathisen (1989) for the Peruvian anchoveta, we think that permanent chemical stimuli present in the spawning grounds provide imprinting cues for many pelagic species, as in the well known cases of salmonids or eels. These stimuli could be related to the local marine ecosystem and/or to continental inputs (rivers, estuaries, lagoons). Nevertheless, it is now accepted that natal imprinting is not the only way salmon recall the river they left as juveniles. Coho salmon, *Oncorhynchus kisutch*, released from two hatcheries and transported around much of their migratory route, returned primarily to their release site. This suggests that they needed to learn sequences of odours during their seaward migration in order to home in a complex river system (Quinn *et al.*, 1989). Such a sequence of odours is less likely to occur in the sea due to the three dimensional dispersion, except in coastal areas. In their review of learning in fish behaviour, Kieffer and Colgan (1992) suggested that homing behaviour of fish may be partly the result of the development of specific parts of the brain and partly of changes in behaviour with experience, which in a sense conciliates the imprinting and the social facilitation approaches.

Whatever is the dominant ethological process explaining fidelity to spawning grounds, this fidelity is likely to generate intraspecific diversity by favouring the emergence of sympatric subpopulations and therefore substocks. Even when different sub-populations spawn at the same time on the same spawning ground, there is not necessarily crossed fertilisation between them probably due to small spatial and temporal scale differences in their distribution. On the basis of microsatellite DNA, Ruzzante *et al.* (1996a) demonstrated that within a 20 km diameter sampled during three weeks, the aggregation of larval cod was formed by more than one genetically distinct larval group and it was even possible to identify a cohort of larvae resulting from a single spawning event. Genetic heterogeneity in a pelagic population was mentioned early on by Hedgcock *et al.* (1989). Cury and Anneville (1997) reviewed many examples of pelagic stocks of the same species having distinct spawning grounds due to social facilitation or natal homing, but that are found mixed during some period of their life: North Sea herring, Northwest Atlantic herring, northern cod and north Atlantic bluefin tuna, among others. The observations are based either on fishing data or on genetic studies. We will see in Chapter 8 how overfishing may decrease intraspecific diversity and lead to a long-term decline in productivity of an assemblage of stocks.

Interaction at the group level, mentioned in section 7.2.2, is not limited to predation, and many works suggest the importance of interactive learning in the foraging process (see the review of Kieffer and Colgan, 1992). For instance, Ryer and Olla (1992) and Baird *et al.* (1991b) noted that juvenile walleye pollock (*Theragra chalcogramma*) are more successful at exploiting spatially variable ephemeral food patches when foraging in groups than alone. They attribute this result to local enhancement (information from conspecifics) and to social facilitation of feeding motivation (fish initiated feeding sooner when grouped than when alone). Social facilitation might be seen as an evolutionary behaviour which favours rapid patch exploitation. Some foraging models incorporating learning (among other processes) have been proposed (e.g. Clark & Mangel, 1986; Hart, 1986).

Finally, in some species recognition of individuals has been demonstrated. It applies either to the ability of parents to learn to recognise their fry (Noakes & Barlow, 1973; Hay, 1978) or to the ability of any fish to recognise a particular conspecific (review in Kieffer & Colgan, 1992). Such examples are limited and never apply to pelagic fish. Nevertheless, the intriguing examples of school fidelity mentioned (Chapter 3) in the case of tunas regularly associated with FADs raise the question of possible individual recognition in pelagic fish.

7.4 Conclusion

We have seen in this chapter that different forms of learning play an important role in the behaviour of fish. We have focused our attention on learning about predators, which is an associative form of learning favoured by inherited predisposition. The main key stimuli in learning are visual stimuli, but they seem often associated with olfactory stimuli. There is growing evidence that fish are able to learn about fishing, either from repeated associative learning or by one-trial learning in situations of extreme stress. Despite the fact that the genetic basis of learning in the case of fishing gear is uncertain due to the short history of human exploitation (but see section 4.11), we hypothesise that inherited predispositions for learning about predators can favour learning about active fishing gears. As far as passive gears are concerned, especially traps and hooked lines, avoidance learning seems to play a major role. In active gears, sound seems to be an additional key stimulus. In both types of gear, social facilitation seems important either to decrease the gear efficiency or, less frequently, to favour the catch process (biting on hooks for instance).

Social facilitation seems also to play an active role in fish migration, habitat selection and foraging. The role of imprinting in natal homing behaviour is well documented for some long-range migratory species, but its possible application to pelagic fish is an interesting hypothesis. It is difficult to separate the effects of learning from indirect sampling bias or error effects when analysing commercial fishing data, especially because fishermen are also learning about fish behaviour. Nevertheless, both migration route learning and natal homing might change our view of the stock identity in the near future if more evidence is obtained.

In Chapter 8 we will show how the five major traits of pelagic fish distribution and behaviour (habitat selection, schooling, avoidance, attraction or association and learning) might affect fisheries and stock assessment by current population dynamics models. We will look more closely at the basic assumptions and necessary input data for these models, and the effects of behaviour on catchability.

Chapter 8
Effects of Behaviour on Fisheries and Stock Assessment using Population Dynamic Models

8.1 Introduction

In this chapter we first present briefly the main stock assessment models, their input data and basic equations. Models are subdivided into surplus-production models, age-structured models (and the related yield per recruit model) and stock-recruitment relationships (readers familiar with these concepts are recommended to skip section 8.2). For easy comprehension, the structure of the rest of this large chapter reflects the five previous chapters: we will review successively the influence of habitat selection, social behaviour (aggregation), avoidance, attraction, associative behaviour in relation to floating objects and learning on fisheries and stock assessment by population dynamic models.

8.2 Stock assessment by population dynamic models

8.2.1 *Stock assessment by surplus-production models*

Surplus-production models are also commonly called production models or catch-effort models. Theoretically, the surplus production represents the amount of catch that can be taken while maintaining the biomass at constant size in an equilibrium situation. Nevertheless, this situation of equilibrium is practically never reached owing to the rapid variation of fishing effort. Such fluctuations usually result from a 'natural' high growth rate of fishing effort owing to overcapitalisation (Sharp, 1978a; Hilborn & Walters, 1992; Mackinson *et al.*, 1997) or conversely from management actions that freeze the fishing effort or stop it abruptly by a fishing ban. Surplus-production models have been adapted to this non-equilibrium or transition situation; Hilborn and Walters (1992) proposed to rename them 'biomass dynamic models', which also has the advantage of distinguishing them more clearly from the age-structured models. The inputs of these models are limited to time series of catches and fishing effort. One of the most used (and criticised) outputs is the maximum surplus or maximum sustainable yield (MSY), followed by the corresponding 'maximum effort' also improperly termed 'optimum effort' (E_{MSY}).

Surplus-production models were the major assessment tool for many pelagic fisheries from the 1950s to the 1970s, despite the early development of the age-structured models. Nevertheless, structured models were then preferred – at least when the required data were available – because they use a more detailed demographic approach and were therefore believed to be more reliable (but we shall see that fishery management based only on these models does not guarantee stock viability). However, surplus-production models benefited from a second youth after some improvements in their formulation (Fletcher, 1978; Rivard & Bledsoe, 1978; Roff & Fairbairn, 1980; Uhler, 1980; Walter, 1986) leading to a rehabilitation of their performance in terms of parameter estimates (Schnute, 1977, 1989; Hilborn, 1979; Uhler, 1980; Butterworth & Andrew, 1984; Lleonart *et al.*, 1985; Ludwig & Walters, 1985, 1989; Babayan & Kizner, 1988) and to the incorporation of additional variables, as detailed in this chapter.

One reason for the success of surplus-production models is the limited input data they require, but this is also their weakness (Butterworth, 1980; Laloë, 1995). The user just needs a time series of total catches C_i and fishing effort E_i over a time period i, $i = 1 \rightarrow n$. Usually the period used in fisheries is one year (which does not necessarily start on 1 January), in order to limit the seasonal effects. Nonetheless, it is possible to use a shorter period on deseasonalised data sets, or to incorporate the seasonal effect in the model (e.g. Laloë & Samba, 1991). The data series must be as long as possible, especially if many year classes are exploited, and the range of variation of the fishing effort must be wide enough to observe a large range of stock abundance (paradoxically, the ideal situation is to cover a period encompassing both underexploitation and overexploitation). The fishing effort input is assumed to be standardised from nominal fishing such that $qE_t = F_t$, where F_t is the instantaneous coefficient of fishing mortality and q is the catchability coefficient, including availability to the fishery and vulnerability to the fishing gear. The q coefficient is defined as the fraction of the stock caught, on average, by a unit of effort. Conventional units of fishing effort for most pelagic fisheries (purse seiners, midwater trawlers or pole and line boats) are based on the total time at sea, the time spent on the fishing grounds or, better still, the searching time (i.e. the time on the fishing grounds minus the time devoted to the operations of capture themselves). If q is constant, the fishing effort will be proportional to the fishing mortality, and then the annual catch per unit of effort ($CPUE_t = C_t/E_t$ for year t) is an appropriate abundance index ($B_t = CPUE_t/q$, where B is the biomass).

Five main basic assumptions are required to use the production models under non-equilibrium conditions (which is the usual situation):

(1) The model must be applied to a closed, single-unit stock
(2) The age groups being fished have remained, and will continue to remain, the same
(3) Time lags in processes associated with population change have negligible effects on the production rate
(4) Deviations from the stable age structure at any population level have negligible effects on the production rate

(5) The catchability remains constant over the studied period and therefore the catch per unit of effort is proportional to the abundance.

Additional discussion of these assumptions can be found in Schaefer and Beverton (1963), Pella and Tomlinson (1969) and Fox (1974). We will see in further detail how, owing to fish behaviour, assumptions (1), (2) and (5) may fail and therefore introduce bias in the abundance estimates.

Failures of the last assumption might also result from other factors (not directly related to fish behaviour) such as change in the fishing power, which have long been recognised (Gulland, 1955, 1964; Robson, 1966; Rothschild, 1978). Nowadays, a common technique to limit these biases it the estimation of an annual abundance applying a general linear model (gLM), an old method (Draper & Smith, 1966) which is now usable on large data sets thanks to modern computing facilities (e.g. SAS Institute, 1989). Because this procedure is also useful for limiting biases caused by fish behaviour, let us now see its principle. The idea is to express the $CPUE_{t,i}$ of any vessel category i during any year t, as a function of a given fishing unit and year, for instance $CPUE_{1,1}$ (i.e. CPUE of the first fishing unit during the first year) arbitrarily chosen as the reference or standard unit. If a multiplicative model is retained:

$$CPUE_{t,i} = CPUE_{1,1} a_t b_i \varepsilon_{t,i} \tag{8.1a}$$

where a_t is a factor relating the mean CPUE in year t relative to year l, b_i the efficiency of fishing unit i relative to the standard category, and $\varepsilon_{t,i}$ the residuals. If the computational resources allow the whole original data set to be processed, it is recommended to work on individual observation – instead of means as in equation (8.1a) – having the subscript j and to fit the model:

$$CPUE_{t,i,j} = CPUE_{1,1} a_t b_i \varepsilon_{t,i,j} \tag{8.1b}$$

A reasonable estimation of annual CPUE series can be obtained from these models – often termed 'least square mean' – and the corresponding standardised effort is simply obtained by dividing the total catch by this CPUE. A logarithmic transformation allows a linearisation of the model:

$$\log(CUE_{t,i}) = \log(CPUE_{1,1}) + \log(a_t) + \log(b_i) + \varepsilon'_{t,i} \tag{8.2}$$

The consequences of this logarithmic transformation, which limits the influence of the highest values, are discussed in Robson (1966), Laurec and Le Gall (1975) and Laurec and Fonteneau (1978) who preferred it to an additive model without transformation. They also discuss the importance of the basic assumptions underlying this model (independence, normal distribution of the residuals) and the different fitting procedures (least square, maximum likelihood, stochastic model). The generalised linear models (GLM) allow the treatment of a loss of normal error distributions (McCullagh & Nelder, 1989). They are now commonly available owing to recent computer facilities and they will certainly be more and more often used in fishery work (e.g. Gaertner *et al.*, 1966). In certain cases, GLM have been shown to avoid bias in the estimation assuming normal or lognormal distribution (Gauthiez, 1997). In this book we will limit our presentation to general linear models (gLM) – which are particular cases of GLM – to ease the presentation.

We do not develop here the derivations from the basic model (8.2), but we will see in the following sections how other attributes concerning fish behaviour can be incorporated in the linear model.

The different surplus-production models are all derived from the equation for logistic population growth:

$$\frac{dB_t}{dt} = k\ B_t(1 - (B_t/B_\infty)). \tag{8.3}$$

After subtracting the catches ($q\ E_t\ B_t$) this becomes:

$$\frac{dB_t}{dt} = k\ B_t\ (1 - (B_t/B_\infty)) - q\ E_t\ B_t \tag{8.4}$$

where B_t is the biomass at time t, B_∞ the environmentally limited maximum biomass or carrying capacity and k the intrinsic growth rate of the population (these last two parameters are denoted K and r by terrestrial ecologists). Another expression of equation (8.4) is obtained when h is defined as the slope of the relative rate of stock increase ($h = k/B_\infty$):

$$\frac{dB_t}{dt} = \underbrace{h\ B_t\ (B_\infty - B_t)}_{\text{production}} - \underbrace{q\ E_t\ B_t}_{\text{catch}} \tag{8.5}$$

In the Schaefer model (Graham, 1935; Schaefer, 1954), the equilibrium yield $CPUE_e$ derived from (8.5) under equilibrium conditions ($dB_t/dt = 0$) is a linear function of effort:

$$CPUE_e = q\ B_\infty - q^2E/h \tag{8.6}$$

and the equilibrium yield is strictly a parabola (Fig. 8.1, curve $m = 2$, where m is an additional parameter incorporated in the generalised model of Pella and Tomlinson, 1969):

$$Y_e = q\ B_\infty E - q^2E^2/h \tag{8.7}$$

while in other related models (Table 8.1; Fig. 8.1, where $m < 2$) the decline of yield in the right part of the graph is more flexible (generalised model, Pella & Tomlinson, 1969; exponential model, Garrod, 1969; Fox, 1970). Fitting these models can be performed either by integration of the differential equation 8.5 or by fitting equations 8.6 or 8.7 assuming equilibrium conditions (which is usually unacceptable) or using the past-averaging approach (Gulland, 1969; Fox, 1970). For the purposes of this book it is enough to remember that these models all use, directly or indirectly, the CPUE as an index of abundance (except if F is estimated directly by other methods and replaces E in the equation, but then a proper estimation of q is required to reconnect the model to the fishing effort).

Table 8.1 Main surplus production models.

Author	Model	Fitted equation	MSY	E_{MSY}
Graham (1935) Schaefer (1954)	Linear	$CPUE_e = kq/h - q^2E/h$ or: $= CPUE_\infty - b\,E$	$k^2/4h$ or: $a^2/4b$	$k/2q$ or: $a/2b$
Garrod (1969) Fox (1970)	Exponential	$CPUE_e = q\,e^{-(k/h - qE/h)}$ or: $= CPUE_\infty\, e^{-bE}$	$(h/e)\,e^{(k/h)}$ or: $CPUE_\infty/be$	h/q or: $CPUE_\infty/e$
Pella & Tomlinson (1969)	Generalised	$Y_e = qE((k/h) - qE/h))^{(1/(m-1))}$	$k((1-m)/m)(k/mh)^{(1/(m-1))}$	$(k/q)/((1-m)/m)$
Fox (1974)	Density-dependent $q = \alpha B^\beta$	$CPUE_e = (-c'd'/a'b')^{(1/(b'-d'))}$ where: $a' = h/(\alpha)^{((\beta+2)/(\beta+1))}$ $b' = (m-1-\beta)/(\beta+1)$ $c' = -k/(\alpha)^{((\beta+2)/(\beta+1))}$ $d' = -\beta/(\beta+1)$	$\dfrac{c'(1-m)(c'/(a'm))^{(1/(m-1))}}{m}$	$\dfrac{c'(1-m)}{m}\left(\dfrac{c'}{a'm}\right)^{(\beta/(1-m))}$
Fox (1974)	Two mixed and exploited sub-stocks	Equation (8.24) in the text	Depends on the mixing rate and the effort ratio between the two sub-stocks	Depends on the mixing rate and the effort ratio between the two sub-stocks
Clark & Mangel (1979)	Association and aggregation	Model A: $(b\chi\alpha' BK/\beta')\,E$ Model B: $(b\chi Q^*K')\,E$	Not given by the authors, but finite value	Not given by the authors, but finite value
Fréon (1988)	CLIMPROD*	$CPUE_e = q(V)\,B_\infty(V) - q(V)^2\,E/h$	$[B_\infty(V)]^2\,h/4$	$B_\infty(V)\,h/2q(V)$.
Laloe (1988)	Inaccessible biomass**	$\dfrac{dB_t}{dt} = h(\alpha)\,B_t\,(B_\infty - B_t) - q$ $E(B_t - \alpha B_\infty)$. $h(\alpha) = h(1-\alpha)$ or $h(\alpha) = h$***	$\alpha < 0.5$: $-h(a)\,B_\infty^2/4$ $\alpha = 0.5$ and $E \to \infty$: $-h(\alpha)\,B_\infty^2/4$ $\alpha > 0.5$ and $E \to \infty$: $-h(\alpha)\,B_\infty$ $(\alpha(\alpha-1))$	$-h\,B_\infty/(2q(1-2\alpha))$
Die *et al.* (1990)	Variable area	Equations (8.25) and (8.26) in the text	Depends on the mixing rate T No mixing: $k\alpha B_\infty/4$** High mixing ($T = K$): $kB_\infty/4$	Depends on the mixing rate T No mixing: $k/2$ High mixing ($T = k$): $k/2\alpha$

* See Table 8.4 for additional models
** α is the fraction of inaccessible biomass related to B_∞
*** according to the model retained (see text).

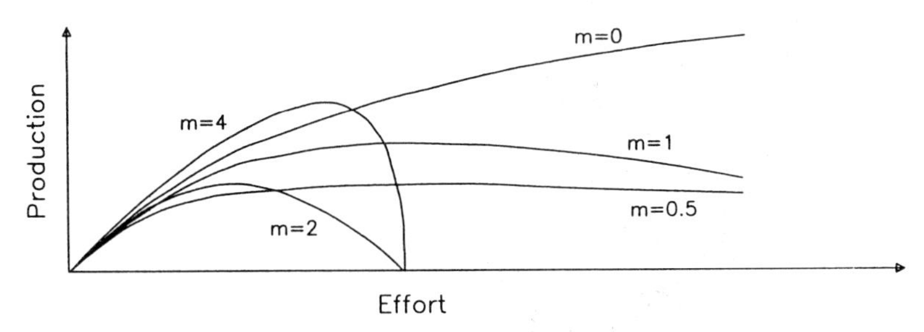

Fig. 8.1 Equilibrium catch-effort curves from the generalised model according to the value of the parameter m (adapted from Pella & Tomlinson, 1969).

To overcome the limitations of the conventional production models, the auxiliary information is nowadays commonly used. It can be either internal information on the exploited stock or external variables of the environment. Examples of internal information are direct estimates of the biomass, catchability estimates or mortality estimates (e.g. Csirke & Caddy, 1983; Caddy & Defeo, 1996). In the delay-difference models (Deriso, 1980; Schnute, 1985, 1987; Fournier & Doonan, 1987), considered as a bridge between the surplus-production models and the age-structured models, the internal information consists of individual growth parameters, natural survival rate and annual recruitment. External variables, often related to fish behaviour, have been included in global production models to take into account the influence of the environment in different ways: on the catchability or the abundance (Fréon, 1986, 1988, 1991), on the accessibility of the biomass (Laloë, 1989), on the rate of exchange between sub-stocks (Fox, 1974) or on the changes in the area covered by the fishing fleet (Die *et al.*, 1990). Among the numerous developments from the original surplus-production model, we will present in subsequent sections only those that can be related to fish behaviour, and particularly those related to change in catchability (failure of assumption 5).

The reader must have in mind that all these models oversimplify reality and that, despite any improvements to them, they cannot pretend to offer by themselves an accurate diagnostic on the fishery. Instead of debating which is the correct approach, our feeling is that fishery biologists and managers must use as many of the available assessment tools as possible and compare the different 'benchmarks' provided by these tools. This comparison, and the analysis of the reasons for observed differences, can be fruitful in the search for the best management advice (see Caddy & Mahon, 1995, for discussion).

8.2.2 *Stock assessment by age-structured models and yield per recruit*

The historical period of stock assessment by structured models is briefly presented in Chapter 1. Megrey (1989) presents a detailed review and comparison of available models (see also Laurec, 1993; Gascuel, 1994). No substantial development has taken place since then, except the improvement of the separability approach, and we present here only the general principles governing age structured models. In this section we will use Megrey's naming conventions, which regroup under the generic name of age-structured stock assessment (ASA) all the methods used to estimate fishing mortality rates and absolute population abundance, given catch-at-age data and possibly some data independent of the commercial fishery. It includes, among others, the well-known VPA (virtual population analysis) and cohort analysis. The term 'cohort' designates all the fish in the population that share the same birth year (or season when there are several reproductive seasons within a year).

ASA methods estimate the number of fish alive in each cohort for each past period – usually a year – simply by assuming that from one period to the next the decrease in abundance results from catches and natural mortality:

$$N_{t+1} = N_t - C_t - D_t \tag{8.8a}$$

where N_t and N_{t+1} are the number of fish alive at the beginning of two consecutive periods, C_t is the catch during the period and D_t is the number dying from natural mortality. Negative exponential functions are used to compute mortality according to elapsed time intervals:

$$N_t = N_i e^{-(M+F)(t-t_i)} \tag{8.8b}$$

where M and F are respectively the natural mortality and the fishing mortality. The main equations of ASA models must be solved interactively or by approximations. Therefore most ASA models implicitly make the assumption that fishing mortality is constant throughout the year (or at least at the middle of the year in the case of pulse fisheries). Nevertheless, Tomlinson (1970) incorporated the effect of seasonal catches to circumvent the standard assumption that fishing mortality is constant throughout the year (see also Mertz & Myers, 1996).

The method requires annual data of catches by age, an estimate of the natural mortality, and a starting point of abundance of the cohort for a given year for initiating the iterative process of estimation. The former methods used the recruitment in the fishery as a starting point (forward methods), but the backward solution proposed by Gulland (1965) is more common owing to its feature of converging during the iterative process. It requires a starting guess of the fishing mortality rate for the oldest age of the cohort ('terminal F').

A major inconvenience of ASA models is that final parameter estimates depend critically on the choice of terminal F in the backward approach and several solutions can provide equally good fits (Pope, 1977; Shepherd & Nicholson, 1986). Former methods were analysing one cohort at a time (or artificially combined cohorts) and estimates of different cohorts cohabiting in the population were not related to each other. The introduction of the separability assumption during the 1970s, in which the fishing mortality is represented as the product of an age-specific and a year-specific coefficient, improved the quality of the estimates substantially (despite the risk of over-parameterisation). It allowed for taking into account variation in availability, vulnerability, exploitation pattern, or selectivity, but did not overcome the problem of estimating F in the most recent years – that is, the most important ones for management implementation – with acceptable precision (Doubleday, 1976; Pope, 1977; Pope & Shepherd, 1982). Usually the terminal F of partially exploited cohorts are assumed equal to the F value obtained from fully exploited cohorts, but this method does not reflect rapid changes in the exploitation level (e.g. Brêthes, 1998). More recently, 'calibrated' or 'tuned' models have been proposed to use fishery-independent data to solve this problem. First, fishing effort (or CPUE) was used in different methods reviewed and compared by Pope and Shepherd (1985). Some of these methods assume a constant catchability (Saville, Hoydal-Jones, modified gamma, partial exploited biomass, Laurec-Shepherd and Armstrong-Cook, but only for the last three years of exploitation), others a monotonic variation with time (linear variation in rho method, exponential variation in log-rho and hybrid methods) or relate it to the population by a power function (gamma method). We will see in this chapter that these assumptions are often violated.

Since the 1980s, other methods of tuning virtual population analysis (VPA) have made use of biomass estimation by different kinds of survey (acoustic surveys on recruits or adults, aerial surveys, egg surveys or experimental surveys by fishing with a gear) or sometimes by recreational fisheries. Acoustic surveys are also influenced by fish behaviour as described in Chapter 9, as are other surveys (Gunderson, 1993; Chapter 10). To estimate the most recent years is therefore still problematic as shown in a review of the different tuning methods used in herring stock assessments presented by Stephenson (1991). Some generalised models incorporate into the separability assumption the possibility of integrating fishing effort and fishery-independent data such as fecundity by age, ageing error or catch estimate error (Fournier & Archibald, 1982; Deriso *et al.*, 1985). Quinn *et al.* (1990) reviewed techniques for estimating the abundance of migratory populations and proposed a new age-structured model using migration rate among regions.

Another major difficulty when using ASA models is to estimate properly the natural mortality. The effect of natural mortality is the same as the effect of fishing mortality and the two factors cannot be distinguished because they appear at the same level in the sets of equations (e.g. equation 8.8). As a consequence, an overestimation of natural mortality will lead to an underestimation of fishing mortality and vice versa.

Another approach derived from the ASA is the length-based VPA first proposed by Jones (1974, 1981). It makes use of the growth equation to follow the cohorts directly from the length distribution. Pauly *et al.* (1984) and Fournier *et al.* (1990) developed respectively the ELEFAN and MULTIFAN software for estimating both the growth equation and the mortality vector from the length data. Despite some weakness underlined by Hilborn and Walters (1992), these methods are useful for tropical areas where age composition is often not available. The new FiSAT software (Gayanilo *et al.*, 1996) integrates the different routines using length- or age-structured data.

The yield per recruit models allow investigation of the consequences of change in size of first capture according to the natural and fishing mortalities (Thomson & Bell, 1934) and are commonly presented as a complement to ASA models. In pelagic species, it is now recognised that growth overfishing is seldom the key problem, mainly because these species usually have fast growth and a high natural mortality. Therefore the critical size is usually smaller than the minimum size of interest to fishermen for commercial or technical reasons, such as minimum size required for canned species, entanglement of small fish in the purse seine, or unavailability of youngest stages because of their different habitat selection. Nevertheless, exceptions might exist, as for instance the large beach seines used in Ghana which for some years overwhelmingly caught fish smaller than 12 cm (Anon. 1976). Growth overfishing can also occur in pelagic species having a medium life span and/or a medium growth rate (e.g. menhaden, horse mackerel, tuna and other scombroids). Note also that in overexploited resources, the mean age of individuals tends to decrease owing to the dynamics of the population, but owing also to changes in the fishing strategy so as to maximise profit (Pauly, 1995). Therefore a survey of the changes in the size of first capture and the influence of these changes on the yield per recruit model is recommended.

Considering that natural mortality is usually roughly estimated, different hypotheses must be tested (e.g. Le Guen, 1971; Fréon, 1994a; Caddy & Mahon, 1995, among others). The results are usually very sensitive to these hypotheses on natural mortality (M) and therefore the knowledge of this parameter is still a bottleneck in stock assessment. Moreover, because of our ignorance of the variability of the natural mortality, M is considered constant in all the exploited year-classes, which is certainly not true.

8.2.3 *Stock-recruitment relationship*

Although studying the stock-recruitment relationship is not an assessment method, this relationship is now recognised as one of the central problems in the population dynamics of most exploited species and can be related to fish behaviour in many instances (see next sections). This density-dependent relationship is implicitly included in the 'black box' of the surplus-production models, while age-structured models ignore it because they take into account only the post-recruitment dynamics. Some auto-regenerative models integrate the whole life cycle by combining the stock-recruitment relationship and the yield per recruit (e.g. Laurec, 1977; Laurec *et al.*, 1980).

The first formulation of the stock-recruitment relationship was given by Ricker (1954), who proposed a flexible function:

$$R = aS\,e^{-bS} \tag{8.9}$$

where R is the recruitment in numbers, S the spawning stock expressed as biomass, and a and b are coefficients. The different curves resulting from the values of a and b are dome shaped. There is an increasing recruitment according to S at low levels of the stock and then, after a maximum, a declining recruitment at higher stock sizes. Among the most common biological processes behind this model are cannibalism of the first stages by adults and density-dependent predation, both related to fish behaviour.

In contrast, Beverton and Holt (1957) proposed an alternative function where the recruitment increases toward an asymptote as the spawning stock increases:

$$R = S/(aS + b) \tag{8.10}$$

where a and b are coefficients. In this model it is assumed that the density-dependent mortality is limited to a critical period in the early life history, mainly owing to competition for food. Deriso (1980) generalised this function by adding a third parameter which gives more flexibility:

$$R = aS(1 - bcS)^{1/c} \tag{8.11}$$

A detailed discussion on these functions can be found in Cushing (1988) and Hilborn and Walters (1992).

8.3 Habitat selection and its influence on catchability and population parameters

Because population dynamics models generally use large databases regrouping on commercial fisheries over long time intervals (from months to years), they are not usually biased by very short-term (less than a couple of hours) variability in habitat selection. Usually, a high number of observations reduce the variance of the estimates. In sections 8.3.2 to 8.3.4 we will deal with the influence of changes in habitat selection on catchability at different time scales (year, season, and day; moon cycle effect is considered only in other sections despite its possible effect on vertical habitat selection). We shall see that this scale classification is somewhat artificial because the different time scales present some interactions. Then we will focus on the spatial variability of the catchability and lastly present two sections more related to ASA models: the influence of habitat selection on growth parameter estimates and on mortality estimates. But before moving to these topics, let us develop one of the key questions in pelagic fisheries which is related not only to fish behaviour: the biomass-dependence of local density and catchability in relation to exploitation and environmental changes. Because this point is related both to habitat selection and to social behaviour, it will be a link between the following sections.

8.3.1 *Yearly changes in abundance, density, habitat and catchability in relation to exploitation and the environment*

The main debate at the end of the nineteenth century was to decide whether fishing activity did or did not have a major influence on fish abundance (Chapter 1). A century later, after much investigation aimed at demonstrating the importance of the effects of fisheries, and after the development of sophisticated direct or indirect methods for measuring their impact and managing the fisheries accordingly, the debate has resumed for both pelagic (e.g. Beverton, 1990) and demersal species (e.g. Hutchings, 1996). This kind of happening is not so scarce in the history of science. Of course the question is now addressed in a different way. There is no longer any doubt that fishing activities influence stock abundance, but the huge variations in stock abundance, from outburst of several hundred thousand tonnes to collapse, are more and more related to other factors that probably interact with the fishery. Among them, changes in environmental conditions and the modification of stock distribution are likely to play a major role. From 1983, at least six international symposia have been devoted to this topic (Csirke & Sharp, 1984; Wyatt & Larrañeta, 1988; Kawasaki *et al.*, 1990; Payne *et al.*, 1992; Pillar *et al.*, 1997; Durand *et al.*, 1998) and from the end of the 1980s the annual ICES meeting has had special sessions related to this problem. It is not the purpose of this book to review in detail the literature on this subject, but we will have a 'behaviouristic look' at it, because a large part of the discussion is related to fish behaviour in terms of habitat selection and social behaviour.

One of the key questions concerning stock assessment of highly variable resources is how the large variations in abundance are expressed in terms of density, or conversely whether the area of distribution is related to abundance. Swain and Sinclair (1994) found that the area containing 90% of the cod from the southern stock of the Gulf of St Lawrence was density dependent, in contrast to the core area (50% of the stock). They modelled fish distribution and proposed a figure showing two extreme cases of variation in fish spatial spread according to the biomass (constant density or density proportional to biomass) and a threshold of density used to define the exploited stock area (A). Gauthiez (1997), using a similar model, proposed a figure where in addition an intermediate case is simulated (Fig. 8.2). In Fig. 8.2a, when the biomass increases the density increases non-uniformly in a manner consistent with density-dependent habitat selection (but the maximum density in the core remains constant) and the stock area increases substantially. In contrast, in Fig. 8.2c, the stock density increases uniformly over all areas when the biomass increases (as in the IFD model) and there is no change in the total area occupied by the stock. Nevertheless the exploited stock area A increased slightly. In Fig. 8.2b, the increase in biomass is associated with an increase both in stock area and in mean density.

Petitgas (1994) had already proposed hand drawing of density-dependence rather similar to Fig. 8.2, with in addition a particular case of very high density located in a small core area. He also proposed the use of a conventional geostatistical 'selectivity

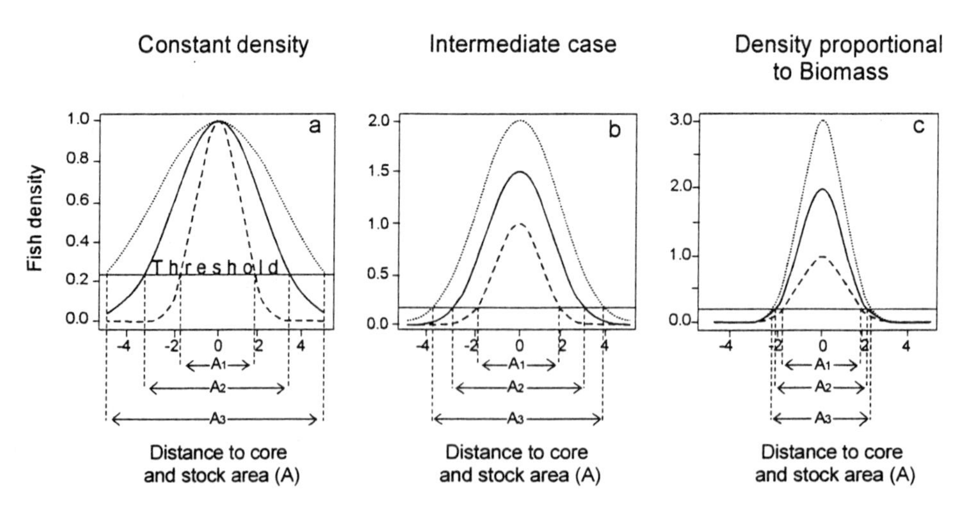

Fig. 8.2 Model of spatial variation in fish density according to the stock biomass (core is the centre of the distribution area) (adapted from Swain & Sinclair, 1994, and Gauthiez, 1997). (....) Biomass high; (—) biomass medium; (- - -) biomass low.

curve' Q(T), better named 'aggregation curve' in Petitgas (1997), to identify the spatial behaviour of the stock. Let us see the simplest case of a regular sampling grid where z(x) denotes the fish density at point x, T(z) the area where the random function Z takes values greater than z and Q(z) the fish quantity that is standing on the surface T(z). The graph Q(T) gives the cumulative integrated fish quantity Q(z) as a function of the occupied area T(z) for densities greater than z:

$$Q(z_p) = \sum_{i=p}^{n} \frac{n_i}{n} z_i;\; T(z_p) = \sum_{i=p}^{n} \frac{n_i}{n} \qquad (8.12)$$

$$z_1 < z_2 < \ldots < z_n$$

Because Q(z) varies between zero and m (the mean of Z), to compare the different Q(z) corresponding to different levels of abundance, Petitgas (1994, 1997) proposed to use relative values P(z) defined as the ratio Q(z)/m. This type of graph is symmetrical to the graphs showing the contribution to the mean of the class values of density proposed by Fréon *et al.* (1993c). It is also related to the Lorenz curves which relate a percentage of biomass to the corresponding occupied area (e.g. Myers & Cadigan, 1995).

Most indirect methods of stock assessment are based on density estimation because they use the CPUE as an index of abundance (e.g. surplus-production models) or as a tuning variable for ASA models. From a theoretical point of view, we can distinguish at least three main mechanisms of change in abundance in pelagic fish which will match better with one or the other of the extreme cases of change in density proposed by Swain and Sinclair.

In the first case (Fig. 8.3a), changes in biomass are mainly due to variations in the fishing mortality F, which can be related either to modifications in the nominal fishing effort E (scenario a_1), or to changes in the catchability coefficient q in relation to changes in the environmental conditions (scenario a_2) without major change in the carrying capacity. But this contrast between a variation in F due to the modification of E or q is an oversimplification of a complex reality because in the first scenario we have:

$$E\nearrow \rightarrow F\nearrow \rightarrow B_t\searrow \qquad (8.13a)$$

where ↗ and ↘ represent increasing and decreasing values respectively, but the variability of q is also likely to occur because the decline of the biomass B_t might increase the catchability (MacCall, 1976, Ulltang, 1976, 1980; Csirke, 1988). This will create a loop in the process, sometimes responsible for the stock collapse:

$$\begin{array}{l} E\nearrow \rightarrow F\nearrow \rightarrow B_t\searrow \rightarrow q\nearrow \\ \uparrow \leftarrow \leftarrow \leftarrow \leftarrow \leftarrow \downarrow \end{array} \qquad (8.13b)$$

The social behaviour of fish is thought to be responsible for this increase in catchability because the population maintains a constant density, even when depleted, and therefore the stock area shrinks. This kind of shrinkage in area is also called 'range collapse' (section 3.4.1).

By contrast, in the second scenario (a_2) the increase in catchability can be attributed to different mechanisms related to environmental modifications (V), such as changes in accessibility (e.g. modification of the vertical distribution owing to changes of the thermocline depth) or in vulnerability (e.g. modification of the avoidance reaction related to changes in turbidity):

$$V\nearrow \rightarrow q\nearrow \rightarrow F\nearrow \rightarrow B_t\searrow \quad \text{(or reversal changes)} \tag{8.14a}$$

where V↗ means an increase in favourable environmental conditions. In this scenario, there is less reason to suspect an acceleration in the process leading to an increase in F than in the previous scenario, or to suspect a shrinkage in the stock area, even though it can occur in the case of prolongation of unfavourable climatic conditions. But an increase of the nominal fishing effort E due to higher catch rates can also create a loop in the process:

$$\begin{array}{l} V\nearrow \rightarrow q\nearrow \rightarrow F\nearrow \rightarrow B_t\searrow \rightarrow q\nearrow \quad \text{(or reversal changes)} \\ \quad\quad \downarrow \rightarrow E\nearrow \rightarrow \uparrow \leftarrow \leftarrow \leftarrow \leftarrow \downarrow \end{array} \tag{8.14b}$$

It is not easy to relate practical examples to these different cases, but we think that the collapse of the northern sardine stock of the eastern Pacific, the North Sea and north-east Atlantic herring stocks, and the northern cod off Newfoundland and Labrador, correspond mainly to scenario a_1. The Peruvian anchoveta stock could correspond mainly to scenario a_2. Let us now recall briefly the history of these stocks.

Northern sardine (*Sardinops sagax caerulea*) collapsed in 1961 and 1962, mainly due to overfishing of this naturally unstable population (review in Troadec *et al.*, 1980). Similarly, the North Sea and north-east Atlantic herring stocks suffered overexploitation and collapsed in 1970 because management actions came too late and were not sufficiently restrictive (review in Saville and Bailey, 1980; Burd, 1985; Jakobsson, 1985).

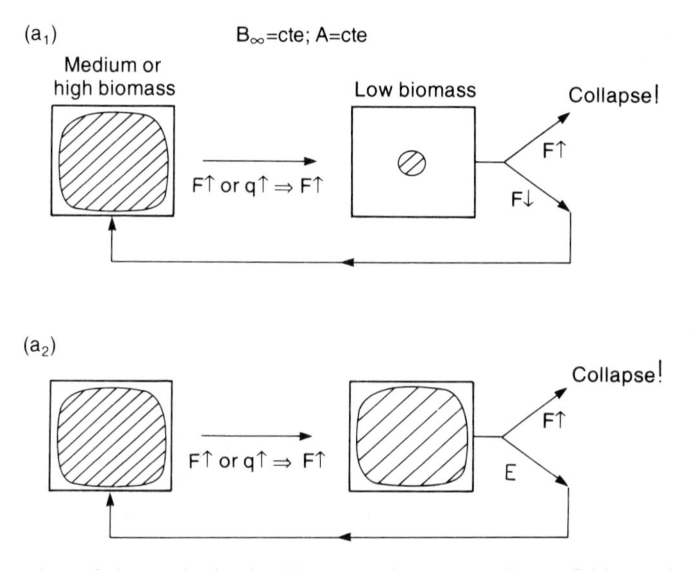

Fig. 8.3 Illustrations of changes in density, biomass and area according to fishing and environmental parameters (rectangles delimit the normal habitat).

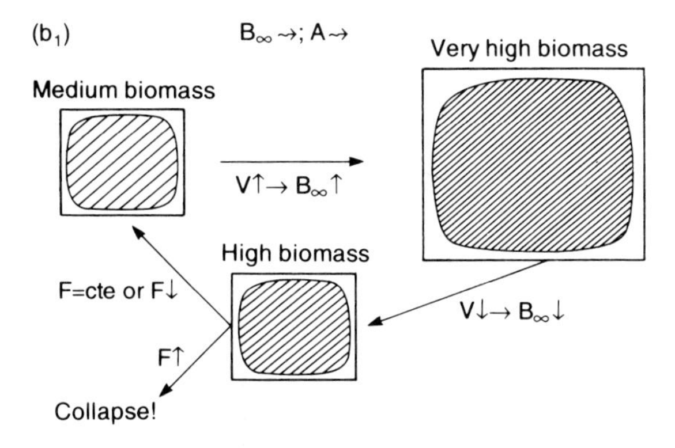

(b1)
B∞ →; A →
Very high biomass
Medium biomass
V↑→ B∞↑
High biomass
F=cte or F↓
V↓→ B∞↓
F↑
Collapse!

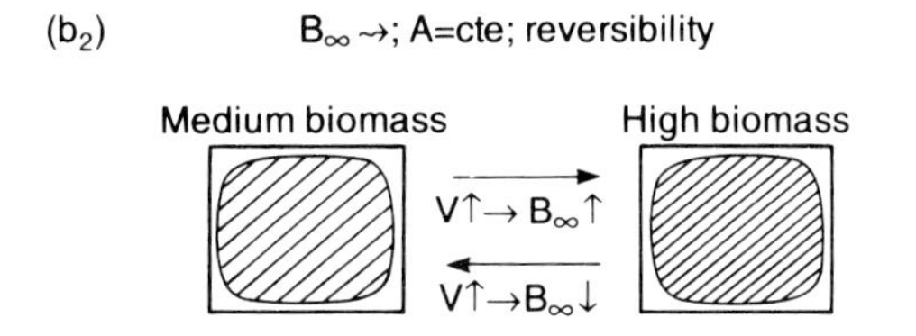

(b2)
B∞ →; A=cte; reversibility
Medium biomass
High biomass
V↑→ B∞↑
V↑→B∞↓

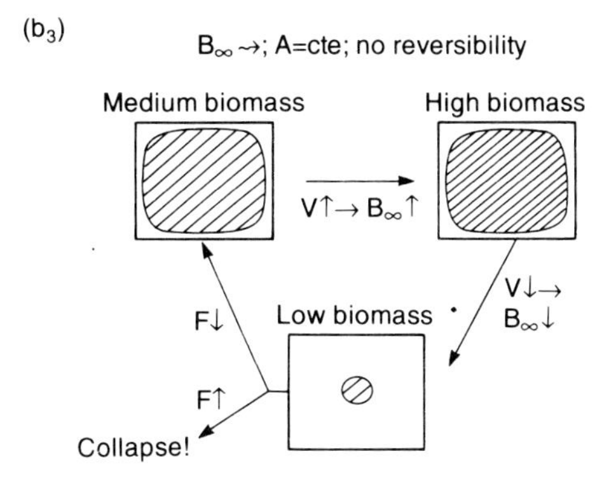

(b3)
B∞ →; A=cte; no reversibility
Medium biomass
High biomass
V↑→ B∞↑
V↓→
B∞↓
F↓
Low biomass
F↑
Collapse!

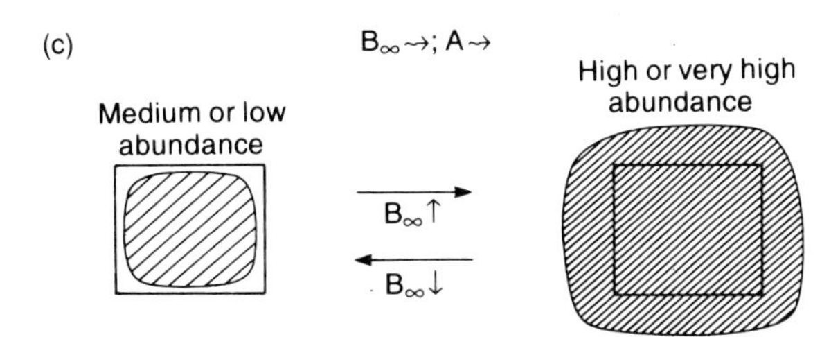

(c)
B∞ →; A →
High or very high
abundance
Medium or low
abundance
B∞↑
B∞↓

Fig. 8.3 *Contd.*

The analysis of catch per tow of northern cod (*Gadus morhua*) off eastern Canada – which collapsed in the early 1990s – indicates that, as expected, the proportion of low-density tows (< 100 kg/tow) increased while the proportion of high-density tows declined during surveys performed between 1981 and 1992. But surprisingly, the proportion of high-density tows (> 500 kg/tow) remained constant in a small, constant number of dense aggregations (Hutchings, 1996). From trawl surveys, Atkinson *et al.* (1997) provided two figures showing the range collapse of this stock: there is a shrinkage of the area where 90% of the stock is located, which also appears in contour ellipses (Warren *et al.*, 1992) of density (Fig. 8.4a and b). Despite a strict management by quota, mainly based on ASA models and secondarily on trawl surveys (Baird *et al.*, 1991a; Lear & Parsons, 1993), it is now accepted by many authors that this shrinkage in area arises mainly from overfishing (Myers & Cadigan, 1995; Myers *et al.*, 1996; Atkinson *et al.*, 1997). Nevertheless, the role of environmental factors on a possible change in habitat selection on this assemblage of substocks is still open to debate (e.g. DeYoung & Rose, 1993; Hutchings & Myers, 1994; Kulka *et al.*, 1995).

The Peruvian anchoveta stock collapse in 1972 appears to have resulted both from a low recruitment in 1971 (partly due to overfishing) and from changes in availability owing to the strong El Niño event of 1972. Nevertheless, a noticeable effect of El Niño on recruitment was observed for some years (Pauly & Tsukayama, 1987; Pauly *et al.*, 1989).

The second case (Fig. 8.3b) corresponds to a variation of the biomass first initiated by environmental changes which modify its carrying capacity:

$$V\searrow \rightarrow B_{\infty}\searrow \rightarrow B_t\searrow \quad \text{(or reversal changes)} \qquad (8.15)$$

In general, this is associated with corresponding changes both in the area (A) of the habitat and in the density, especially in upwelling areas (scenario b_1) which can extend according to the strength of the wind stress. Nevertheless, the possibility of a change only in density could be contemplated, with reversibility (scenario b_2) or not (scenario b_3). Usually the history preceding the collapse of such stocks – especially scenario b_1 – starts with a period of increase in biomass owing to favourable environmental conditions, which usually allows large benefits and consequently investment in new boats and therefore a subsequent increase in the fishing mortality (Fig. 8.5, adapted from Csirke & Sharp, 1984). Then, when the environment becomes unfavourable, the area of the habitat shrinks (as in a basin model, section 3.4.1). As a result, the local density increases suddenly, which in turn causes an increase of catchability. Therefore the fishing mortality is high and the stock collapses:

$$V\nearrow \rightarrow B_{\infty}\nearrow \rightarrow B_t\nearrow \rightarrow E\nearrow\nearrow\nearrow \rightarrow F\nearrow\nearrow\nearrow \rightarrow B_t\searrow \leftarrow V\searrow \qquad (8.16)$$
$$\uparrow \leftarrow q\nearrow \leftarrow \downarrow$$

In scenarios b_1 and b_3, fish behaviour depends indirectly on the environment, unlike scenario b_2 which is not so much dependent on fish behaviour and so is beyond the scope of this book. There are some situations where environmental conditions influence directly both the carrying capacity and the catchability, this is a mixture of cases a and b.

The difficulty of investigating such situations comes from the interacting

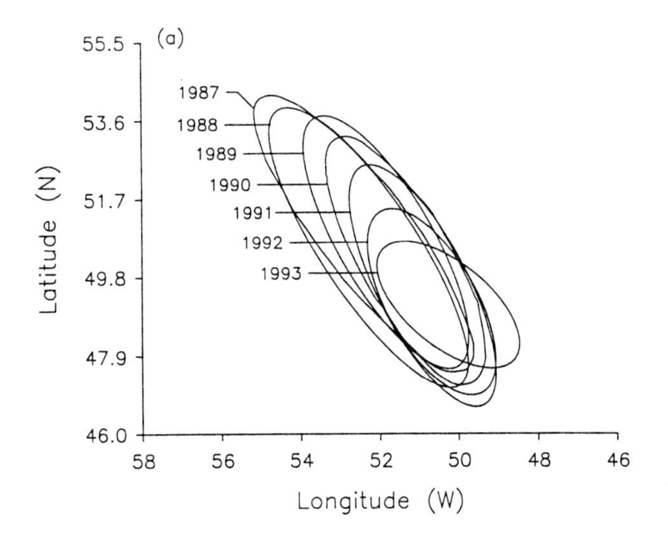

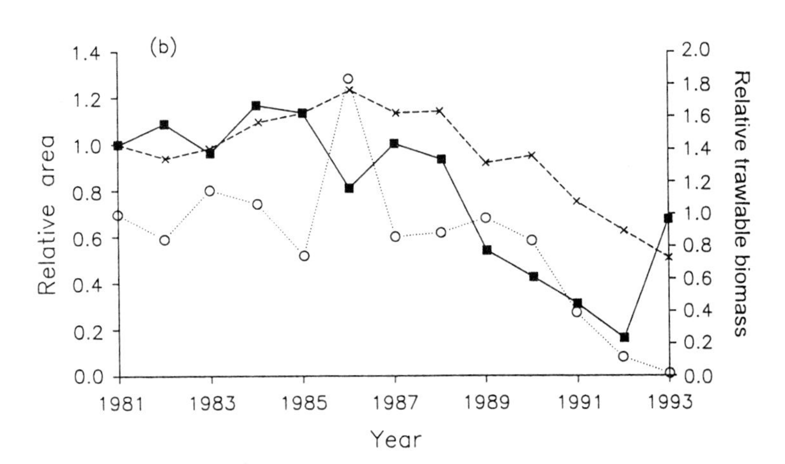

Fig. 8.4 Bivariate ellipses determined from the distribution of cod during the annual autumn surveys of divisions 2J3KJL during 1987–1993: (a) location of the ellipse; (b) relative area of distribution as determined by the ellipses compared to area with 90% of trawlable biomass and trawlable biomass from autumn surveys (redrawn from Atkinson *et al.*, 1997). For part (b): ■ = area with 90% of biomass; x = ellipse area; ○ = RV biomass estimate.

influence on the stock of both fishing effort and environment. Some scientists argue that the fishery plays a minor role compared with the environment (Cushing, 1982, 1995; Lluch-Belda *et al.*, 1989, 1992; Kawasaki, 1992) because the main stocks of sardine and anchovy showed similar worldwide fluctuations attributed to global climatic changes. Moreover, records of past history of some stocks from anaerobic sediments indicate a large variation of biomass without exploitation (Soutar & Isaacs, 1974; Shackleton, 1987, among others), despite some criticisms addressed to this method (Shackleton, 1988). Others claim that fishing mortality

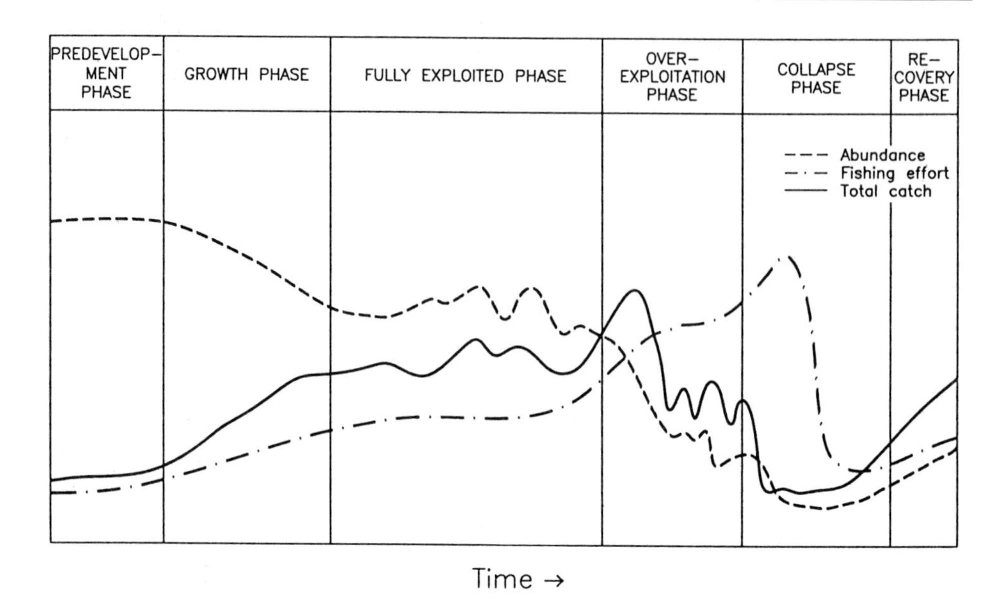

Fig. 8.5 Generalised history of a developed fishery that tends to fluctuate due to environmental conditions (redrawn from Csirke & Sharp (eds), 1984).

dramatically increases the probability of collapse when combined with unfavourable environmental factors (Fréon, 1988; Beverton, 1990; Hutchings & Myers, 1994; Pauly, 1995). Our purpose is not to enter into this debate, but to stress how in such a situation, habitat selection and social behaviour may interfere with the abundance estimations.

A typical example of change in biomass and area (scenario b_1) owing to environmental conditions is the sardine stock of the southern Canary current. *Sardina pilchardus* is usually considered as temperate and was found in abundance north of the 26th parallel along the Moroccan coast. From 1966 to 1977 this limit shifted progressively to the south, reaching the south of Mauritania (17°N) in 1977. An industrial fishery started to develop on new fishing grounds, providing annual catches above 500 000 t (Boëly & Fréon, 1979; Domanevski & Barkova, 1979). Few fish were found far in the south in the Bay of Dakar, Senegal (Fréon & Stéquert, 1982; Fréon, 1986). Then the stock abundance and extension decreased during the 1980s. But at the beginning of the 1990s, the stock displayed a second outburst and spreading, and once more catches were reported in the Senegalese fishery (Binet *et al.*, 1997; Demarcq, 1997). This extension of the area of distribution was attributed to an increase in the wind intensity, providing an intensification of the upwelling and a linkage between Mauritania and Senegal by a tongue of cold water which provided a continuum of habitat between the two countries and the southern transport of the sardine larvae. In addition, the upwelling intensification increased the phytoplankton production which in turn favoured the sardine stock.

The pelagic populations off Senegal are dominated by *Sardinella aurita* and *S. maderensis* and seem representative of scenario b_2. From the results of routine acoustic surveys performed on the southern coast of Senegal (Levenez *et al.*, 1985), Petitgas (1994) showed that the distribution of the fish was not density dependent when the local abundance varied from high to medium (a single value of very low abundance provided a different result, but as it is not reflected in the coastal fishery it remains questionable).

The third case (Fig. 8.3c) is an extension of habitat selection when a species starts to colonise new biotopes vertically and/or horizontally, resulting in an outburst of biomass. Two examples can illustrate this situation: the trigger fish in West Africa and the Japanese sardine.

The demersal trigger fish (*Balistes carolinensis*) was until 1971 considered rare along the West African coast. Then this species started to colonise the pelagic ecosystem during the two first years of its life and was easily detected during acoustic surveys, while larger individuals colonised the demersal ecosystem and were abundant in experimental and commercial trawl catches (Caverivière, 1982). This increase in total abundance expanded along the coast from the core area of Ghana-Côte d'Ivoire mainly to the north (Table 8.2). The higher density was found in Guinea around 1975 and the species invaded Cap-Vert (Senegal) in 1978. During the 1979 summer, incidental catches were reported as far as the south of Mauritania. In 8 years the trigger fish biomass increased from a few thousand to 700 000 t. According to Caverivière *et al.* (1980, 1981), this species largely dominated the catches during three bottom trawl surveys performed in January from 1978 to 1980 in Côte d'Ivoire: 85% to 97% of the catches in the 21–60 m depth ground were trigger fish. The average experimental yield obtained during the 1978 survey performed with a medium bottom trawl (opening 15 × 3 m) was 364 kg ha^{-1} on the sandy-muddy grounds, and reached 1350 kg ha^{-1} in some substrata. The abundance of the species then decreased progressively and it is no longer a dominant species in the ecosystems.

Table 8.2 Estimated biomass of trigger fish (*Balistes carolinensis*) from acoustic surveys performed along the west African coast at the end of the 1980s (adapted from Caverivière *et al.*, 1980).

Country	Year of invasion	Fish biomass (t) end of the 1980s	Trigger fish biomass (t)	Estimated surface of the continental shelf (n.mi^2)
Senegal and Gambia	1979	1 200 000	80 000	35 000
Guinea Bissau	1977	150 000	30 000	52 500
Guinea } Sierra Leone }	1975	850 000	450 000	39 500 26 900
Liberia	*	*	*	17 400
Côte d'Ivoire	1974	71 000	20 000	11 600
Ghana	1972	310 000	50 000	21 700
Togo	1972	*	*	1 700
Burkina Faso	*	*	*	2 600
Nigeria (West of Dod river)	1976	75 000	10 000	37 000
Total observed	—	2 656 000	640 000	245 900
Total extrapolated	—	2 900 000	700 000	—

* No acoustic survey

The spawning grounds of the Japanese sardine (*Sardinops melanostictus*) were located on the continental shelf of the Pacific coast of Japan at the end of the 1970s and expanded progressively into the oceanic water around the Kuroshio current during a period of population increase in the first half of the 1980s (Figs 8.6 and 8.7; Watanabe *et al.*, 1996).

In these examples, the increase in density remained long after the extension of the distribution and therefore they cannot be modelled by MacCall's basin approach (section 3.4.1). The reasons for these sudden changes in abundance, density and behaviour remain speculative, because in most cases no environmental modification was observed, or this modification was hardly related to the species' biology (e.g. the influence of salinity on trigger fish abundance). Nevertheless, in the reported examples, knowledge of the environmental changes was generally limited to a few parameters (mainly temperature and salinity) and the biology of the species was not

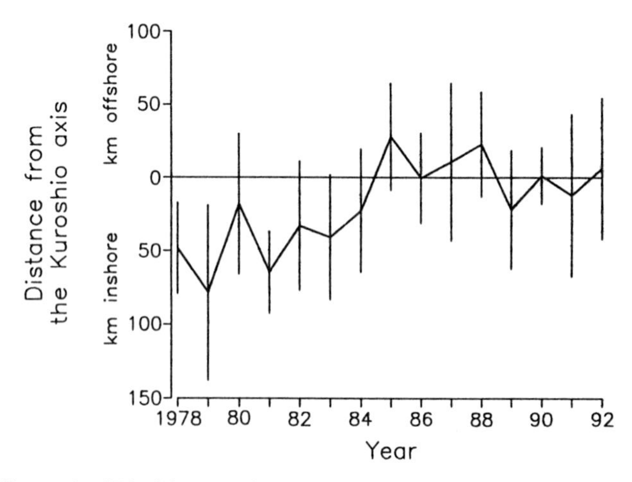

Fig. 8.6 Mean distance (± SD) of the spawning ground centre from the Kuroshio current axis in the main spawning month from 1978 to 1992 (redrawn after Watanabe *et al.*, 1996).

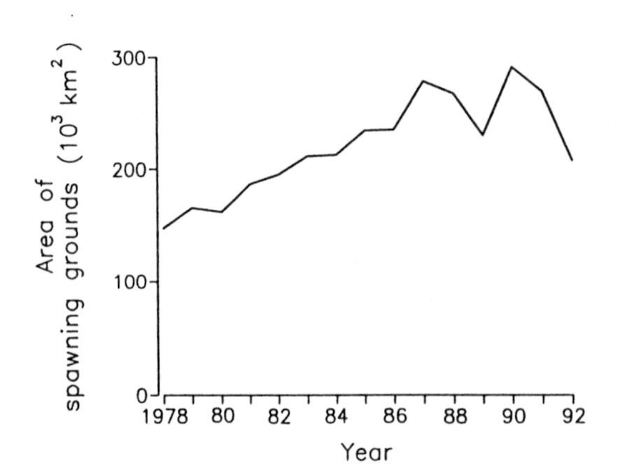

Fig. 8.7 Changes in area of annual spawning grounds of sardine along the Kuroshio current from 1978 to 1992 (redrawn after Watanabe *et al.*, 1996).

completely investigated. Therefore it is still possible to relate this third case to the second one.

8.3.2 *Some attempts at modelling the yearly variability of catchability in relation to habitat selection*

Correction of abundance indices

In the case of unusual extension of habitat, the first step is to compute relevant indices of abundance before modelling (except when the model does not intend to describe the stock abundance but only the CPUE as in the next section). If the changes in stock abundance occur without a trend in the spatial change in the stock distribution ('proportional density', Fig. 8.2c), the mean CPUE will be proportional to abundance, provided that the fishing pattern remains the same and that no other biases exist (but note the effect of a density threshold of the fishery in this figure). In contrast, in the case of change in stock distribution ('constant density' or intermediate case, Fig. 8.2a and b), even if the fishery area follows the extension and retraction of the fluctuating stock, the mean CPUE will not give a reliable index of abundance. In the case of strict constant density and no density threshold, the mean CPUE is expected to be constant. Under these conditions of density dependence, the stock collapse can occur without warning given by the CPUE analysis. But in practice the fishing effort is usually not randomly distributed and the CPUE results from the combination of the stock distribution and the fishing strategy. This is illustrated by simulations performed by Gauthiez (1997) who used the two extreme cases of density distribution of Fig. 8.2 ('proportional density' and 'constant density') and four models of fishing effort distribution:

(1) Model A: uniform distribution of effort over a threshold of density of the stock distribution
(2) Model B: distribution of effort proportional to abundance
(3) Model C: uniform distribution of effort over a constant surface area located around the core of the stock (= core area)
(4) Model D: uniform distribution outside the core area.

As expected, when fish density is proportional to abundance, the relationship between the mean CPUE and abundance is linear, or displays a small departure from linearity (Fig. 8.8a). This departure from linearity is observed only for the model A of uniform effort distribution, which is not a common situation (usually fishermen are able to allocate most of their effort to the more productive areas). In contrast, the relationship between CPUE and abundance is never linear when the fish density model is constant, owing to a decrease in the catchability (models, A, B and C; Fig. 8.8b). The only relatively favourable case is model D, which is an unusual situation simulating an unavailable area of high density (for technical reasons or owing to implementation of a refuge or sanctuary).

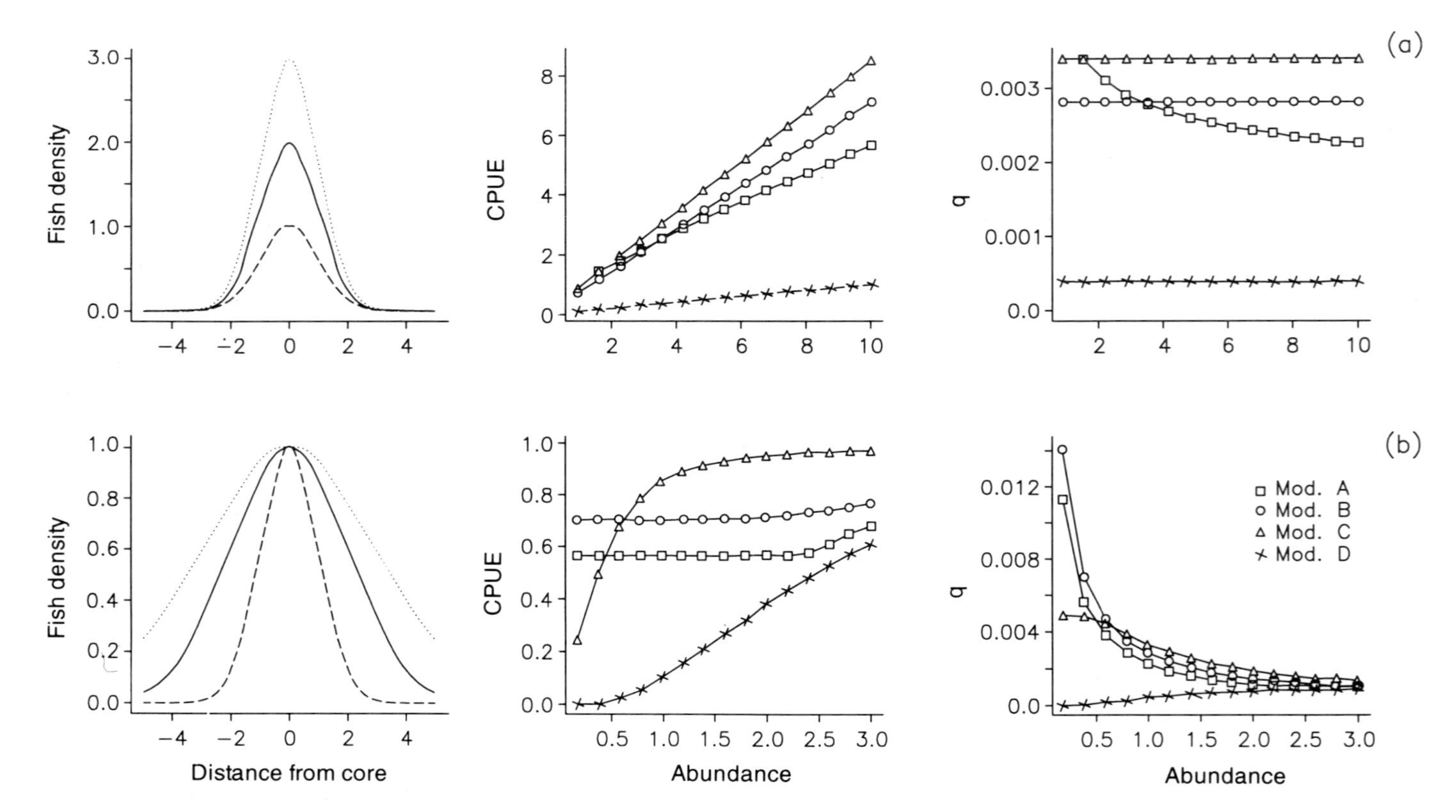

Fig. 8.8 From left to right: fish density distribution according to the distance from the core area of the stock, CPUE versus abundance; and catchability (q) versus abundance according to different models of fishing effort distribution (A,B,C and D): (a) fish density proportional to abundance; (b) constant density (redrawn from Gauthiez, 1997).

Instead of using the mean CPUE as an index of abundance, it is recommended to collect geo-referenced data and to compute a weighted average CPUE where the weighting factor is the stratum $A_{i,t}$ of each fishing sector i during time t:

$$\overline{CPUE_t} = \sum_i \left(\frac{A_i}{A} CPUE_{i,t}\right) \tag{8.17}$$

The difficulty is then to get enough observations in the different strata. This is partly obtained by enlarging the size of the strata, but there then arises the problem of intra-stratum variability of the catchability, especially for highly aggregated pelagic species (see section 8.4.1). But the main problem is how to deal with the lack of data in strata where no fishing operations are performed. To obtain a consistent time series of the abundance index when the density is not proportional to abundance, it is necessary to revisit historical data and to make some assumption on the fishing zone not visited during low biomass periods. Assuming that the abundance index was equal to zero in these fishing zones is usually not realistic because fishermen allocate their effort to a fishing zone only when the guess of yield is above an empirical threshold. Other information apart from commercial data is usually necessary to make proper assumptions (e.g. acoustic surveys, trawl surveys).

Modification of surplus production models

To take into account the influence of environmental factors on stock production, Fréon (1986, 1988) proposed a modification of the conventional surplus-production models. An additional environmental variable (V) has been inserted into the conventional models so as to overcome the failure to meet assumption (5) (section 8.2.1) regarding the constant catchability, or to take account of the changes in the carrying capacity on the biomass. Three kinds of models have been developed according to the mechanism of environmental influence (Table 8.3):

(1) Influence only on the stock abundance
(2) Influence only on catchability coefficient
(3) Influence on both abundance and catchability.

Influence on stock abundance will not be detailed in this book, insofar as fish behaviour is not directly affected by the variability of the carrying capacity of the environment. In the case of an increase of the carrying capacity owing to a change in habitat selection, it could be tempting to use this kind of model. Nevertheless, we think that it is not relevant because such changes occur suddenly, probably when some environmental factor exceeds a threshold, but probably without immediate reversibility when the environmental conditions return to their usual values. We will develop the case of environmental influence on catchability in section 8.4.1 dealing with the role of aggregation. But let us now see the interesting case of influence on both abundance and catchability, which can be related to an increase of the biomass associated with an increase in catchability and vice versa.

Let $B_\infty(V)$ be the function representing the fluctuations of B_∞ owing to environ-

Table 8.3 Surplus production models available in the CLIMPROD software classified according to the environment (V) and fishing effort (E) effect. CPUE is the catch per unit of effort, q the catchability coefficient, B_∞ the carrying capacity, a, b, c and d are parameters. (Adapted from Fréon *et al.*, 1991)

Type of model	Equation	Function CPUE = f(E)	Function CPUE = f(V)
CPUE = f(E)	CPUE = a + bE	linear	—
	CPUE = a exp(bE)	exponential	—
	$CPUE = (a + bE)^{(1/c - 1)}$	generalised	—
CPUE = f(V)	CPUE = a + bV	—	linear
	$CPUE = a\ V^b$	—	exponential
	$CPUE = a + bV^c$	—	exponential
	$CPUE = aV + bV^2$	—	quadratic
CPUE = f(E, V); q = f(V)	CPUE = aV + bE	linear	linear
	CPUE = a + bV + cE	linear	linear
	$CPUE = aV^b + cE$	linear	exponential
	$CPUE = aV + bV^2 + cE$	linear	quadratic
	CPUE = (a + bV) exp(cE)	exponential	linear
	CPUE = aV exp(bE)	exponential	linear
	CPUE = a exp(bE) + cV + d	exponential	linear
	$CPUE = aV^b \exp(cE)$	exponential	exponential
	$CPUE = aV^b \exp(cV^d\ E)$*	exponential	exponential
	$CPUE = aV + bV^2 \exp(cE)$	exponential	quadratic
	$CPUE = ((aV^b) + cE)^{(1/(d - 1))}$	generalised	exponential
	$CPUE = ((a + bV^2)^{(d - 1)} + cE)^{(1/(d - 1))}$	generalised	quadratic
CPUE = f(E, V); B_∞ = f(V)	$CPUE = aV + bV^2\ E$	linear	linear
	$CPUE = a + bV - (c + dV)^2\ E$	linear	linear
	$CPUE = aV^b + cV^{(2b)}\ E$	linear	expontial
	CPUE = aV exp(bVE)	exponential	linear
	CPUE = (a + bV) exp(– c(a + bV)E)	exponential	linear
	$CPUE = aV^b \exp(cEV^b)$	exponential	exponential
	$CPUE = aV(b - cV) - dV^2\ (b - cV^2)E$	linear	quadratic
	CPUE = aV(1 + bV) exp(cV(1 + bV)E)	exponential	quadratic
CPUE = f(E, V); B_∞ = f(V) and q = f(V)	$CPUE = aV^{(b + c)} + dV^{(2b)}E$	linear	exponential exponential
	$CPUE = aV^{(1 + b)} + cV^{(2 + b)} + DV^{(2b)}E$	linear	quadratic exponential
	$CPUE = aV^b \exp(cV^dE)$**	exponential	exponential
	$CPUE = (aV^{(1+b)} + cV^{(2+b)}) \exp(dV^b\ E)$	exponential	quadratic exponential

* without constraint
** with sign constraint on b and d

mental factors, and q(V) the function representing the fluctuations of q; then equations (8.6) and (8.7) become respectively (Fig. 8.9):

$$CPUE_e = q(V)\ B_\infty(V) - q(V)^2\ E/h \tag{8.18}$$

$$Y_e = q(V)\ B_\infty(V)\ E - q\ (V)^2\ E^2/h \tag{8.19}$$

Therefore MSY and E_{MSY} values will be functions of V:

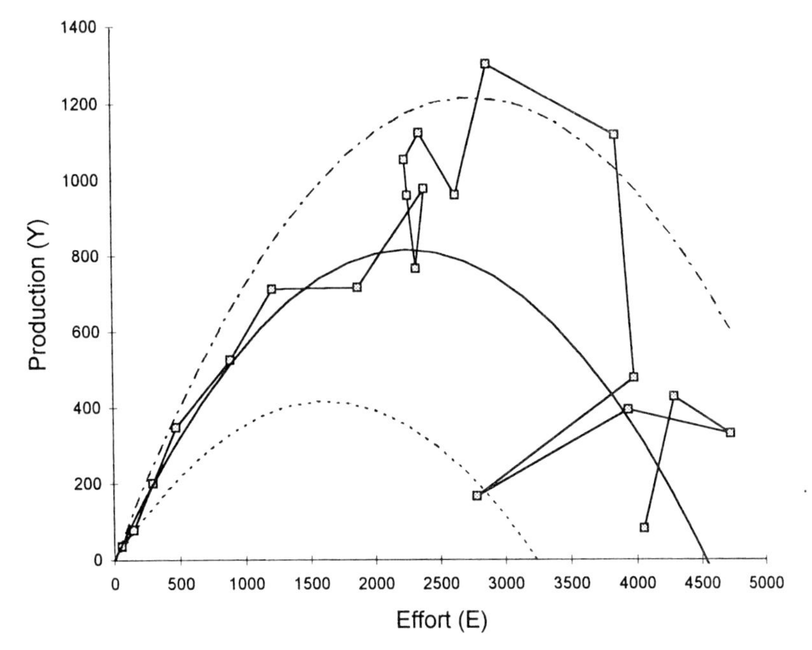

Fig. 8.9 Equilibrium catch-effort curve from a surplus production model incorporating an environmental variable (sea surface temperature) which influence both the production and the catchability (CLIMPROD software, Fréon *et al.* (1991); Fréon & Yáñez (1995)). ▩ = Production (Y); (– · –) fitted function for V_{min} = 16.5; (—) fitted function for V_{middle} = 17.6; (- - -) fitted function for V_{maxi} = 18.8.

$$MSY = [B_\infty(V)]^2 h/4 \tag{8.20}$$

$$E_{MSY} = B_\infty(V)\ h/2q(V) \tag{8.21}$$

Note that in such models, the CPUE does not pretend to represent an index of abundance because it is accepted that q is not constant (nevertheless the fishing effort must be standardised in order to take into account the variation of q owing to non-environmental factors, such as technological changes).

Another possibility in order to take into account the effect of V could be to incorporate it as an effect in the gLM aimed at obtaining an unbiased estimate of CPUE and effort and then to fit a conventional surplus-production model. The first inconvenience of such an approach is a confusion between the effect of V on catchability and on abundance (MSY will no longer depend on V). Second, it is better to estimate E and V effects in a single fit than in a two-step approach. Third, the standardised effort taking into account V would then be difficult to use for management implementation because it would no longer be translatable into a nominal effort. We then prefer to use equations derived from (8.18) to (8.21) which are an attempt to model the biological process.

The functions q(V) and $B_\infty(V)$ are usually not precisely known and have to be chosen according to the data. Moreover, this choice must be combined with the choice of the conventional production model: linear (as above), exponential or generalised. Table 8.3 presents the set of equations according to these different cases.

The expert-system CLIMPROD (Fréon *et al.*, 1991, 1993d) helps users to select the model corresponding to their case according to different ecological criteria and not only to the best fit (see Saila, 1996, for comments on this approach) and uses the past-averaging approach to deal with non-equilibrium situations (for both E and V).

In some instances, a substantial part of the biomass lives in a habitat inaccessible to the fishery (deep grounds, remote areas or zones subject to strong currents, shallow water, etc.) but the proportion of the biomass unavailable to fishermen may vary from year to year, depending on environmental changes which can affect the habitat selection. Laloë (1988) proposed two global production models taking account of this 'inaccessible quantity of biomass'. He applied them to the stock of *Sardinella aurita* of the Côte d'Ivoire and Ghana, mainly exploited by short-range artisanal fisheries. In these fisheries, the distance of the stock from the coast depends on the extension of the river plumes in the area, and therefore the accessibility of the stock is supposed to depend on the river flows (Binet, 1982). The apparent collapse of the stock in 1973 was followed by a fast recovery in 1976 with three year classes immediately available, and therefore the time series of CPUE are difficult to interpret as an accurate abundance index. Even though Laloë (1988) had to define arbitrary thresholds of the environmental variable to get a good fit, his approach is conceptually interesting and the derived equation can be applied when the inaccessible quantity of biomass depends on the surface of the exploited area (Laloë, 1989). The formulations of the models are derived from equation (8.3). When only the second term of this equation is modified it becomes:

$$\frac{dB_t}{dt} = hB_t\ (B_\infty - B_t) - qE(B_t - \alpha B_\infty) \qquad (8.22)$$

where αB_∞ is the inaccessible quantity of biomass. When both the first and second terms of equation (8.3) are modified, supposing that h is a function of α such as $h(\alpha) = h(1-\alpha)$, it becomes:

$$\frac{dB_t}{dt} = h(1 - \alpha)\ B_t\ (B_\infty - B_t) - q\ E(B_t - \alpha B_\infty) \qquad (8.23)$$

The first formulation provides constant MSY regardless of α (Fig. 8.10a), while in the second, MSY varies inversely to α (Fig. 8.10b; Table 8.1). Laloë (1988) recognise that it is not always easy to decide which of these two formulations is the more realistic. Note that if the inaccessible quantity of biomass is defined as a proportion of the actual biomass B_t, equation (8.22) becomes: $dB_t/dt = hB_t(B_\infty - B_t) - qE(\alpha B_t)$. Therefore, if α is a function of V, the resulting models are similar to those presented in section 8.4.1 where q is a function of V. Another approach to the situation of a confined fishery, proposed by Mullen (1994), is based on diffusive transport rather than on a fraction of inaccessible biomass.

The rather common situation of a heterogeneous spatial distribution at the scale of the stock results in different subunits which can be exploited by different fleets (Chapters 3 and 5). In this case, assumption (1) of the surplus-production models is not met. Fox (1974) modified the exponential surplus-production model for the

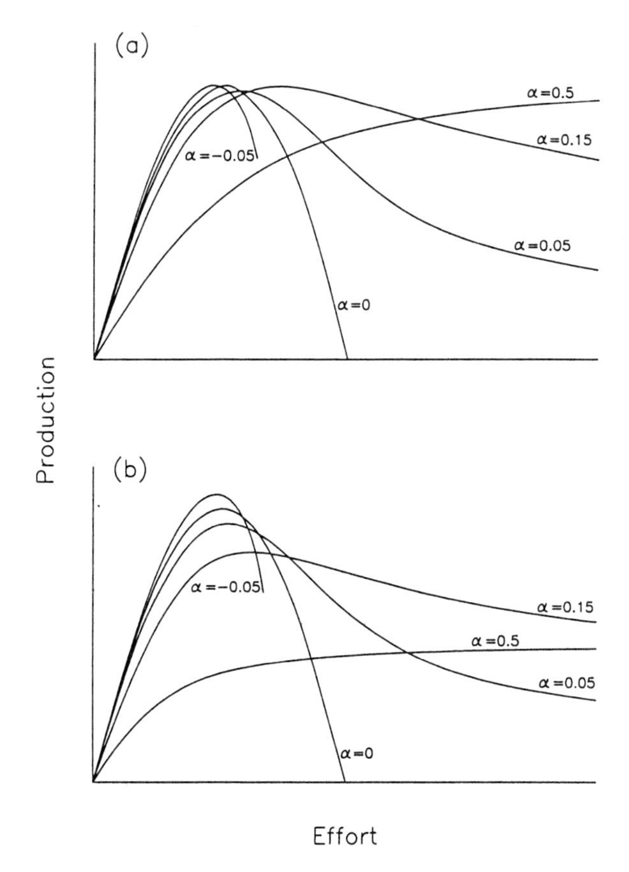

Fig. 8.10 Equilibrium catch-effort curves (arbitrary units) from surplus production models with inaccessible quantity of biomass according to the value of parameter α: (a) model corresponding to equation 8.22; (b) model corresponding to equation 8.23 (redrawn from Laloë 1988).

special situation where there are two substocks having a certain rate of exchange and different levels of exploitation. For substock 1:

$$\frac{dB_1}{dt} = (k_1\ B_{1t} + T_2\ B_{2t})\ [(B_{1\infty} - B_{1t})/B_{1\infty}]$$
$$-T_1\ B_{1t}\ [(B_{2\infty} - B_{2t})/B_{2\infty}] - F_{1t}\ B_{1t} \qquad (8.24)$$

where B_t is the biomass at time, t, k the intrinsic rate of population increase, and T the rates of biomass transfer between stocks 1 and 2 (F and B_∞ have their usual meaning). The equation for substock 2 is obtained by replacing subscript 1 by 2 and vice versa in the previous equation (see Table 8.1 for other equations of this model).

Die *et al.* (1990) adapted this model to the situation where a change in the area of coverage of the fishing fleet results in a change in the fraction of the stock that is available (similar to Laloë (1989)), but conversely a change in the stock area with no change in the fishery can benefit from this adaptation. Die *et al.* (1990) proposed two different models according to the rate of mixing between the exploited and non-exploited fractions of the stock (high-mixing and no-mixing models). Assuming in equation (8.24) that the rates of population growth are identical ($k = k1 = k2$) and that the biomass transfer takes place exclusively from the unexploited to the exploited segment ($T_1 = 0$, $T_2 = T$), it becomes:

$$\frac{dB_1}{dt} = (k\ B_{1t} + TB_{2t})\ [(\alpha_t\ B_\infty - B_{1t})/\alpha_t\ B_\infty] - F_{1t}\ B_{1t} \qquad (8.25)$$

and

$$\frac{dB_2}{dt} = k\ B_{2t}\ [(1 - \alpha_t)\ B_\infty - B_{2t}]/[(1 - \alpha_t)B_\infty]$$
$$-T\,B_{2t}\ [(\alpha_t\ B_\infty - B_{1t})/\alpha_t\ B_\infty] \qquad (8.26)$$

where α_t denotes the fraction of the virgin stock biomass that is potentially affected by fishing at any given time. Die *et al.* (1990) applied these models to the eastern Pacific yellowfin tuna fishery, which has experienced large changes in area coverage over the past two decades. From their results it seems that the no-mixing model is more appropriate, but the high-mixing model could probably find application in other pelagic species.

The use of ASA models does not necessarily solve the problem of the difficulty in detecting catchability trends as stressed by Stokes and Pope (1987), who studied the results of a generalised linear model for the multiplicative, integrated analysis of catch-at-age and commercial effort data. They used a simulated data set comprising three fleets, five years and four ages with a known increasing trend in catchability. Only an increasing trend in catchability equal to or greater than 13% per year might be detectable given a typical level of noise in the data (20%). Their conclusion is that these trends must be estimated externally. Along the same lines, Daan (1991) presented a simulation model that allows an evaluation of possible bias in the results of standard virtual population analysis (VPA) in combination with separable VPA and tuning methods when applied to a heterogeneous unit stock. The results indicate that, if a unit stock defined for assessment purposes is composed in reality of two or more substocks that are exploited differentially, the estimated exploitation curve is dome shaped, even when the exploitation pattern imposed upon each substock is flat topped. This bias may cause significant underestimates of fishing effort owing to an associated change in apparent catchability.

8.3.3 *Seasonal variability of catchability in relation to habitat selection*

Many fish species choose to forage in different areas according to the season, and reproduction often occurs away from foraging areas (Chapter 3). Therefore, if the fishery does not cover the whole area of the stock distribution, or if fishing effort is not equally allocated to the different subareas, then a seasonal component will appear in the CPUE.

A typical example of the effect of migration along the coast is the Senegalese sardine fishery, where adults of *Sardinella aurita* and other species mentioned in Fig. 8.11a are caught only during the cold season whey they migrate south to the nursery

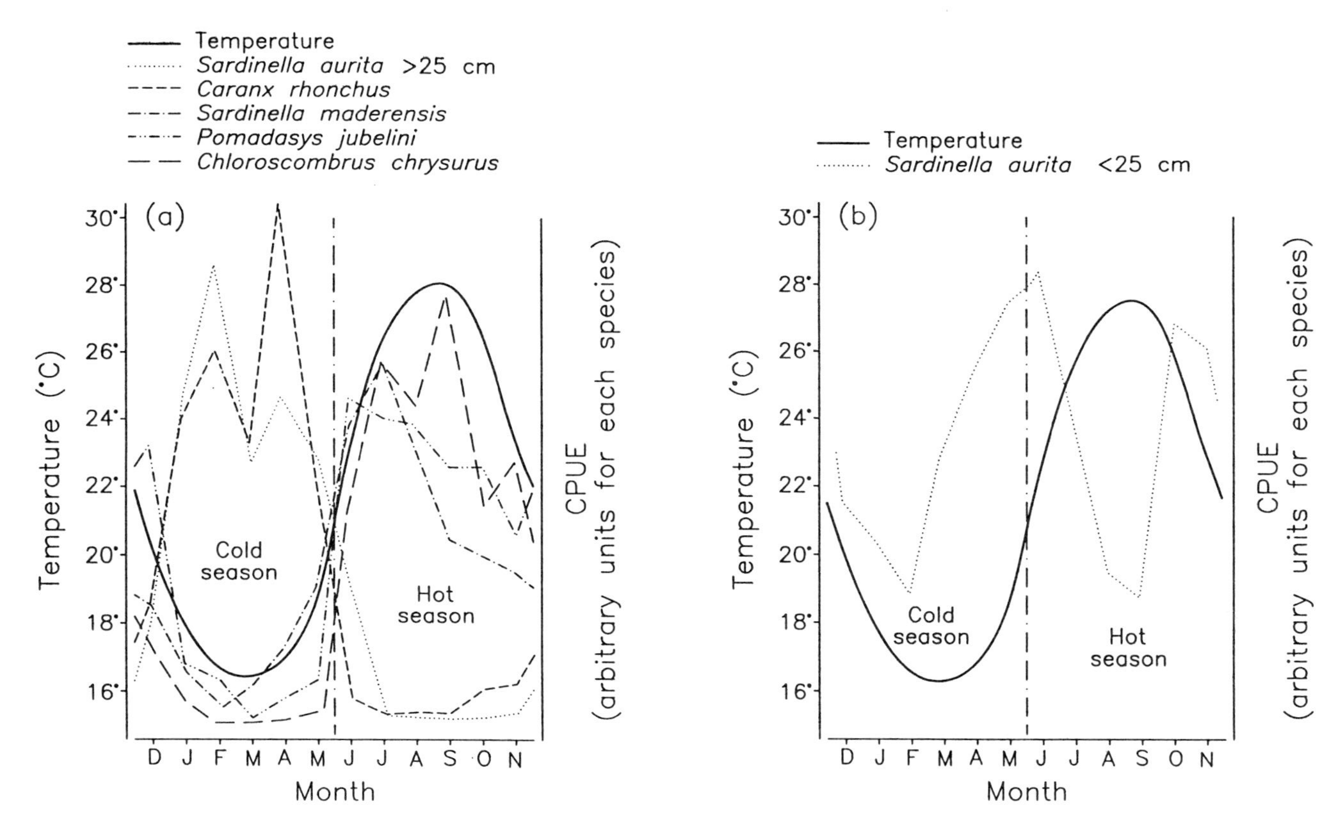

Fig. 8.11 Seasonal variability of the catch per unit of effort (CPUE) and the sea surface temperature in the Senegalese purse seine fishery (redrawn from Fréon *et al.*, 1978).

grounds of the 'Petite Côte'. In contrast, species of the Guinean ecosystem are caught mainly during the warm season (Champagnat & Domain, 1979), while the young of *Sardinella aurita* display a different pattern owing to local recruitment and emigration (Fig. 8.11b). A similar example in a temperate area is provided by Raid (1994), who described the seasonal variation of length distribution of herring caught in the Gulf of Finland. A typical example of inshore–offshore migration is given by the seasonal changes in availability of the jack mackerel (*Trachurus symmetricus murphyi*) for the coastal fisheries of Chile. This species is caught by the coastal purse seiner fleets mainly during the intense feeding season along the coast. Then it migrates from the coastal water to oceanic waters for spawning and there it is available only to the transoceanic trawler fleets (Serra, 1991).

Such situations raise the problem of a proper estimate of an annual index of abundance. Even if mean annual values of catch and effort are used in surplus models, the CPUE will give an unbiased indication of abundance variation under two assumptions:

(1) The seasonal timing of migration must remain exactly the same from year to year (in particular, the time spent by the stock in the area accessible to the fleet must remain constant)
(2) The seasonal pattern of exploitation must also remain constant, particularly the yearly proportion of fishing effort allocated each season to the different fishing sectors and to the different target species in multispecific fisheries.

Let us see first the case where assumption (1) is met but not assumption (2). This problem of seasonal variation in exploitation pattern was addressed very early by Gulland (1955), who recommended that an index of concentration of the fleet (in space or time) be computed. If this index is too high, it is suitable to subdivide the fishing season so as to compute an average estimate of abundance.

Nowadays, an efficient and objective procedure to deal with these problems of abundance estimation when the availability varies seasonally is to use 'deseasonalised' CPUE. Laurec and Le Gall (1975) applied different methods to the data series of the Atlantic Japanese tuna long-line fishery on albacore (*Thunnus alalunga*) in order to estimate the three main components of the abundance index: yearly trend, seasonal component and impact of fishing effort. The multiplicative model was found to be more realistic than the additive one, especially when the interannual variability of CPUE is high, which results in an overestimation of the seasonal component. They also compared two ways to model trends, continuous or stepwise, and found the latter more appropriate because it provided annual indices of abundance directly. They also found that during the fitting procedure by least squares, the variance of the estimator could be reduced by weighting the squared residual according to the fishing effort. Nevertheless, Laurec and Fonteneau (1978) noted that this method could introduce a bias when the fishing effort is not independent of CPUE, which is the usual case (increasing effort during 'good years'). Modern statistical software allows for such choice of weighting factor in the gLM models. The general equation of such a model with a season effect is:

$$\log(CPUE_{tij}) = \log(CPUE_{111}) + \log(a_t) + \log(b_i) + \log(c_j) + \varepsilon_{tij} \qquad (8.27a)$$

where everything is as in equation (8.2), except that now the subscript j refers to season. An example of application of such a model is given by Chang and Hsu (1994) or Mejuto (1994) on tuna fisheries.

But as noted by Zijlstra and Boerema (1964), when the fishing effort is applied only during the core of the season, and when the duration of the fishing season varies according to the abundance, this solution leads to bias. If one uses only the data of the actual fishing season for computing the mean annual CPUE, it results in an overestimate of the lowest abundance levels. In contrast, if the average at the lower level of abundance is computed over the same period as at the high level, assuming the density in the periods of no fishing to have been zero, it results in an underestimation of the abundance during the year of lower level (Fig. 8.12). This density-dependent variation in timing of the fishing season is similar to the density-dependent variation in space of the stock area (Fig. 8.2). The application of the gLM procedure in this case will incompletely solve the problem, especially if there are too many missing values of CPUE per stratum in the data set as a result of changes in the fishing pattern.

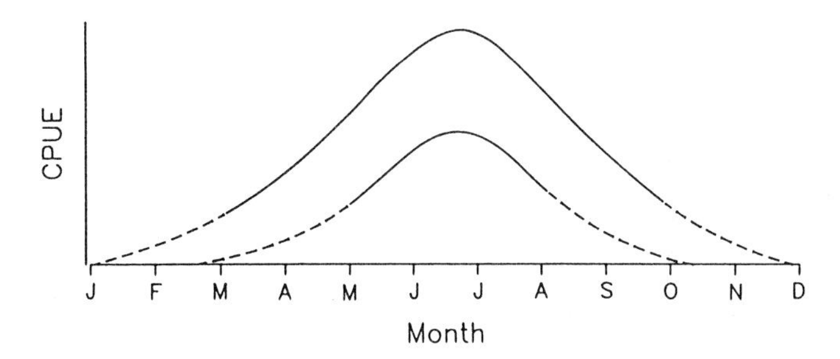

Fig. 8.12 Hypothetical graphs of catch per unit of effort (CPUE) at two levels of abundance. Full lines indicate period of fishing, while in the period of no fishing activity the expected CPUE is indicated by broken lines (redrawn from Zijlstra and Boerema, 1964). (—) Observed; (- - -) expected.

Let us see now the case where the relative abundance from one season of the year to another (e.g. summer and spring) will vary according to the year, in contradiction with assumption (1). This is observed if the pattern of migration changes from year to year, which means that an interaction between the year effect and the season effect will occur. This interaction can be assessed by the incorporation of the term log (d_{tj}) in equation (8.27a):

$$\log(CPUE_{tij}) = \log(CPUE_{111}) + \log(a_t) + \log(b_i) + \log(c_j) + \log(d_{tj}) + \varepsilon_{tij} \qquad (8.27b)$$

If this interaction is significant, it will make impossible the evaluation of an unbiased estimate of the annual CPUE by the model. Then an attempt to obtain an unbiased index can rely on the computation of the CPUE only during the core of the fishing season, that is when CPUE is at a maximum. But once more we face the problem of the definition of the duration of the fishing season. A suitable solution would be to identify the key factor responsible for the habitat selection, or the cue for migration. Then the CPUE could be computed only during this period, assuming zero yield when the absence of fishing activity can be related to nothing other than the absence of fish. But our present knowledge of fish ecology seldom allows the use of such a method. Another attempt could be the use of an empirical threshold of CPUE to define the fishing season each year. Besides the difficulty of choosing an optimal threshold, this approach raises the problem of deciding whether the threshold must be fixed for all years (then it must be low) or may vary from year to year so as to take account of the interannual change in abundance (but then the final abundance estimate will be sensitive to this choice). Moreover, when a high instability of the CPUE is observed within the fishing season, it must be decided whether a decrease of the CPUE beyond the threshold corresponds to a decrease in abundance of the stock or to its unavailability. In the previous example from the Senegalese fishery, adult *S. aurita* migrate southward during the beginning of the fishing season and northward at the end, and during some years they are absent from the fishing area between these two periods.

An additional problem arises when fish of different length groups and age groups arrive in the area at different times, as in the North Sea herring, gulf menhaden or Japanese sardine fisheries. Then if the increasing fishing pressure is reducing the older part of the stock, and so reducing the duration of the fishing season, the abundance of the oldest year classes will be gradually underestimated whereas the abundance of the younger fish will tend to be overestimated. As a result the mortality rate will be overestimated by the ASA models (Zijlstra & Boerema, 1964).

8.3.4 *Circadian variation of catchability related to habitat selection*

Circadian variability is not a problem for indirect methods if the fishing operations are decided randomly by day or by night, or in a constant proportion from year to year. But if an interannual change occurs in the fishing pattern, this might introduce a serious bias in abundance index estimates resulting from simple averaging of the CPUEs, as exemplified in the following sections.

Midwater trawl fisheries

Diel variability in trawl efficiency has been mostly studied in bottom trawl fisheries (Wardle, 1983; Engås *et al.*, 1988; Glass & Wardle, 1989; Walsh, 1991). Nevertheless, circadian variability in midwater trawl catchability was noted many years ago in the North Sea herring fishery, and was related to the size of the fish and their maturation stage (Lucas, 1936; Richardson, 1960; Woodhead, 1964). The efficiency of midwater trawls is not as sensitive to the vertical distribution of fish compared with that of purse seines, which are unable to catch fish below a critical depth related to the height of the net. Nevertheless, the diel vertical migration of pelagic fish might affect the CPUE of midwater trawls when the range of vertical migration of the species is very large. This is the case for herring, which can descend below 300 m during the day east of Iceland and come closer to the surface during the night (Mohr, 1969; Misund *et al.*, 1997). Nowadays the high technology of acoustic detection and trawl positioning limits such differences in availability with depth, but circadian variations in catchability resulting mainly from habitat selection are likely.

Joseph and Somvanshi (1989) found that horse mackerel (*Megalaspsis cordyla*) caught by midwater trawl along the north-west coast of India displayed circadian variations in catchability, associated with migration and shoaling habit. The horse mackerel (*Trachurus symmetricus murphyi*) is caught in the southern Pacific by the large-scale fishery of midwater trawlers operated until recently by the former USSR and now by some countries of eastern Europe in the south-east and south-west Pacific. Ivanova and Khmel'-nitskaya (1991) analysed the catch statistics of midwater trawls on the 1979–1989 data series. They reported large diel variation of school size and school depth according to the season. In summer the vertical range is 2.5 times wider in the daytime than at night. In winter the fish remain in large schools at lower depths and diel migrations are less pronounced. The catchability varies between 0.27 and 0.37 at night and between 0.11 and 0.26 during daytime. This is a typical situation where the circadian effect and its interaction with the seasonal effect must be taken into account during the estimation of abundance indices based on CPUE, especially if some interannual change in the day/night ratio of fishing effort occurs.

Purse seine fisheries

In the multispecific Senegalese purse seine fisheries, some species are caught only or mostly in daytime (*Sardinella maderensis*, *Pomadasys jubelini*, *Chloroscombrus chrysurus*), others mostly at night-time (*Trachurus trachurus*, *T. trecae*, *Scomber japonicus*). The case of the main species (*Sardinella aurita*) is more complex: the young fish are caught mainly during the daytime while the adult migratory fish are caught mostly at night (Fig. 8.13). As the allocation of the fishing effort between day and night may change from year to year by a factor of three, the CPUE gives different abundance indices according to the computation (Fig. 8.14).

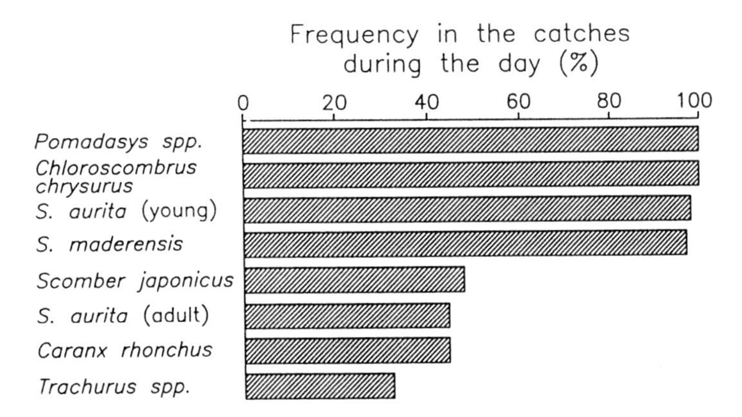

Fig. 8.13 Day-night ratio of the catches of the different pelagic species exploited by the Senegalese purse seine fishery expressed in percentage of catches during the day.

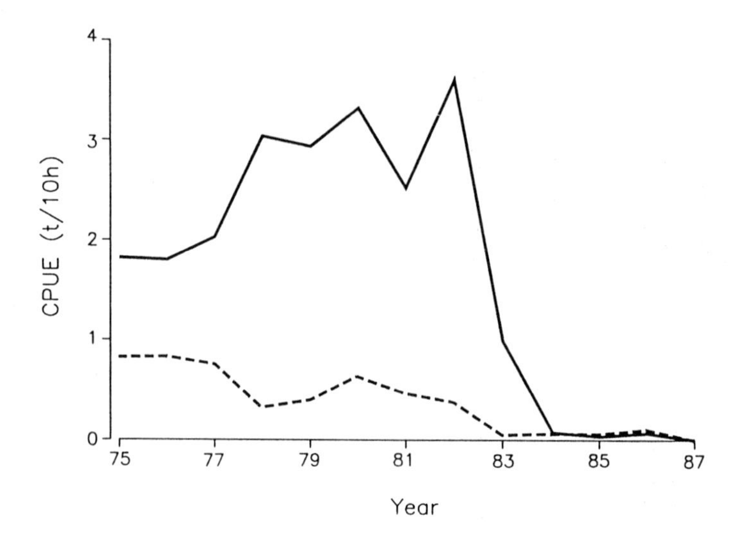

Fig. 8.14 Discrepancy between abundance indices of *Caranx rhonchus* in the Senegalese purse seine fishery. (—) CPUE during the night; (- - -) CPUE during the day.

The mean depth of the captured schools in the Namibian fishery (Thomas & Schülein, 1988) also exhibits circadian cycles which vary according to the species. The cycle is not significant for pilchards and very clear for anchovies, which are caught at about 6 m at night (depth of the top of the shoal) and about 13 m at midday.

In some fisheries, the circadian variation in depth of the schools is the limiting factor for setting the purse seine, as in the Chilean fishery for horse mackerel (*Trachurus symmetricus murphyi*) of the Talcahuano area during wintertime. Hancock *et al.* (1995) reported that in winter, successful sets were made mainly during the night at an average depth of 37 m when some of the schools reached the upper layer, whereas during the day, most of them were located below 50 m, that is out of the range of the purse seine efficiency. The factors governing the variability in the depth of the schools during the day are not documented for this species, but any slight interannual variability of these factors will certainly play a great role in the variation of the abundance indices derived from the fishery.

In the above example, it is relatively easy to compute an unbiased abundance index (see recommendations). But let us see a last example where the bias in the annual abundance cannot be corrected. Dommasnes *et al.* (1979) observed an unusual marked diel vertical migration of the mature capelin in the Barents Sea during winter 1974. During daytime the fish were located in unusually deep water, and therefore were caught mostly by the vessels with the largest purse seines. In this particular year, the CPUE was underestimating the real abundance.

Long-line fisheries

The long-line fishery targeting bigeye tuna is performed by day with deep long-lines. Fedoseev and Chur (1980) indicated that maximum catches of bigeye tuna were observed at 14 00–15 00 h and 17 00–18 00 h and corresponded to the mean values of the stomach filling index. In contrast the swordfish are caught during the night by surface long-lines. Further examples of circadian variation in yield and catch composition in long-line fisheries are provided in section 6.2.

Gill net fisheries

Azuma (1991) investigated by gill net the circadian variations in the salmon catch in the Bering Sea. Chum salmon (*Oncorhynchus keta*) exhibited a significant difference in circadian catches (with catches at night 1.8 times greater than daytime catches), while sockeye salmon (*O. nerka*) and pink salmon (*O. gorbuscha*) did not. From 1989 to 1993, Blaber *et al.* (1995) used gill net and beach seine to sample the fish communities of the shallow inshore waters of a tropical bay in the Gulf of Carpentaria, Australia. They found that water turbidity, tidal range, wind and day versus night were the only abiotic factors that correlated with the relative abundance. According to Cui *et al.* (1991), vision is a major factor affecting catch rate by gill nets, which certainly explains the effect of turbidity and sunlight on these results.

Recommendations

These few examples of different gears indicate that the catchability coefficient can vary greatly according to the hour of the fishing operation in relation to diurnal habitat selection. To overcome the above-mentioned problem of a yearly change in the exploitation pattern, a day–night factor e_k must be included in the gLM procedure (section 8.2.1):

$$\log(CPUE_{tijk}) = \log(CPUE_{1111}) + \log(a_t) + \log(b_i) + \log(c_j) + \log(d_{tj}) + \log(e_k) + \varepsilon_{tijk} \quad (8.28a)$$

where everything is as in equation 8.27b, except that now the subscript k refers to the period of the day (for instance 1 is day and 2 is night) and e is a parameter relating the abundance during the day to abundance during the night. If an interaction f_{ik} between the day/night effect and the fishing unit classes is suspected (for instance if some units might be better able to fish during the night than others but have a different global efficiency), the model can be written:

$$\log(CPUE_{tijk}) = \log(CPUE_{1111}) + \log(a_t) + \log(b_i) + \log(c_j) + \log(d_j) + \log(e_k) + \log(f_{ik}) + \varepsilon_{tijk} \quad (8.28b)$$

Many other interactions can be incorporated in the model if they are suspected to occur. But two important points must be underlined concerning such incorporations. First, over-parameterisation of the model must be avoided. Because of the large amount of data usually available (currently > 100 000 observations), nearly all effects and their interactions tend to be significant at a low level of probability (F test on the parameters), even if their contribution to the model is minor. This is also because the basic assumption of independence of data is usually not met owing to temporal and spatial correlations. Therefore additional criteria of variable selection must be used to build the model, as the proportion of variance explained by each factor and the gain (or loss) of the incorporation (withdrawal) of a factor using stepwise methods. Second, in this respect a particularly crucial decision is whether or not to incorporate an interaction including the year effect, because only this type of interaction will prevent the estimation of an unbiased CPUE time series. As stressed by Lebreton *et al.* (1991), instead of intending to get the ideal model explaining the highest percentage of variance, it is preferable to allow some secondary and hypothetical effects in the residuals and to focus on the main effects.

One example of such an interaction is found in the multispecific midwater trawl fishery of Peru. From the analysis of 25 000 hauls from 1983 to 1987, Icochea *et al.* (1989) concluded that the daily variation in CPUE depended on the El Niño events and on the occurrence of the southward coastal undercurrent. During an El Niño event, fishing depths are usually over 100 m and the CPUE is high and relatively stable all day (i.e. over 24 hours). When a marked southward undercurrent occurs, fishing grounds are close to the coast and trawl sets are close to the surface at night-time and deeper at daytime, with a higher CPUE than during the night. When the undercurrent is weak, fishing grounds become oceanic, catches more superficial and CPUE shows little daily variation, despite an intensification of the fishing effort at night.

8.3.5 *Spatial variability of catchability in relation to habitat selection*

Despite the difficulty of separation between temporal and spatial components of the variability in habitat selection – which also depends on the scale of measurement (Horne & Schneider, 1995) – let us now see the situations where the main effect is strictly spatial and not density-dependent.

Examples of spatial variability are easier to find in demersal fisheries than in pelagic ones; Rocha *et al.* (1991) proposed the use of principal component analysis to define the fisheries zones for each of six species caught by English trawlers in the North Sea: cod, plaice, sole, haddock, saithe and whiting. As far as coastal pelagic species are concerned, most of the abiotic factors that structure habitat selection are related to the water mass (temperature, salinity, dissolved oxygen, water transparency, light intensity, current, etc.; Chapter 3) and therefore habitat is expected to vary more temporally than spatially. In the case of the eastern Atlantic yellowfin tuna fishery, Fonteneau (1978) took advantage of the habitat selection of this species according to the surface temperature to propose an annual index of abundance. Instead of computing the average CPUE in those 1° latitude–longitude squares where the proportion of yellowfin tuna was > 50% in the total catches, as before, he selected only the areas where the surface temperature was over 23°C during a 15 day period. In addition, he fixed a threshold of fishing effort for these spatial and temporal strata (12 or 24 hours). The resulting figures provided a higher decreasing trend in the abundance of the stock compared with the conventional averaging methods. The advantage of this method compared with a threshold of the proportion of the species in the catches is that the resulting abundance index of a given species does not depend on the abundance of the other species.

In pelagic ecosystems, some abiotic factors are rather permanent on a large scale, especially in the vertical plane (e.g. oxycline in tuna's habitat), and other abiotic factors are fixed features in the environment (bottom depth, nature of the sea bed). As a consequence, some permanent productive fishing grounds can be observed at medium and large scale for pelagic species. For coastal pelagic species, these fishing grounds are usually defined (or structured) by a combination of bottom depth interval and sectors of the coast which correspond to permanent or season areas of enrichment by upwelling or river discharge (e.g. Fig. 8.15 from Barría, 1990, or Yáñez *et al.*, 1995). For tunas, the more productive fishing grounds are defined in the horizontal plane by offshore upwelling, or the proximity of coastal upwelling exporting part of their production offshore or by large gyres (Fig. 2.2). In Fig. 8.16, combining the catches of the whole tuna fleet in the eastern Atlantic from 1978 to 1981, it appears that skipjack tuna concentrate in three main areas of upwelling (with maybe an additional effect of Congo river discharge) while the yellowfin tuna areas are more numerous and dispersed. In the vertical plane, the catch rate and the species composition and mean body size are known to vary largely according to the location of the gear relative to the thermocline or oxycline depth (Chapter 3).

Gaertner and Medina-Gaertner (1992) reported that the purse seine tuna fishery in the Caribbean Sea is limited to the south-east margin, because in the rest of the basin the deeper depth of the thermocline and of the oxycline does not favour the concentration of yellowfin tuna schools in the upper layer where they would be accessible to small and medium purse seiners. The abundance index in the rest of the basin is therefore largely underestimated because the vertical extension of the habitat is wider. Since this fishery started to develop during the 1990s, skippers have progressively learnt the location of the best availability of the resource and the fishing effort has become increasingly concentrated in the south-east margin of the basin.

The previous examples strongly suggest that spatial variability in the catches is also an important factor to take into account in the computation of abundance indices in pelagic fisheries and in modelling. Collecting spatial-referenced data for these fisheries is therefore recommended as is incorporating both spatial and temporal variables in the gLM models aimed at producing mean annual CPUE series. A typical model without interactions can be written:

$$\log(CPUE_{tijkl}) = \log(CPUE_{11111}) + \log(a_t) + \log(b_i) + \log(c_j) + \log(d_k) + \log(e_l) + \varepsilon_{tijkl} \quad (8.29)$$

where the incorporated effects are year (a_t), vessel type (B_i), season (c_j), day–night (d_k) and geographic sector (e_l). Many interactions can be incorporated according to the situation, but the most common one is an interaction between the season and the fishing sector (f_{jl}):

$$\log(CPUE_{tijkl}) = \log(CPUE_{11111}) + \log(a_t) + \log(b_i) + \log(c_j) + \log(d_k) + \log(e_l) + \log(f_{jl}) + \varepsilon_{tijkl} \quad (8.30)$$

Let us see now a last illustrative example of the effect on stock assessment of the habitat selection in the vertical and horizontal planes combined with a trend in the fishery. That is the surplus production modelling of the Atlantic yellowfin tuna

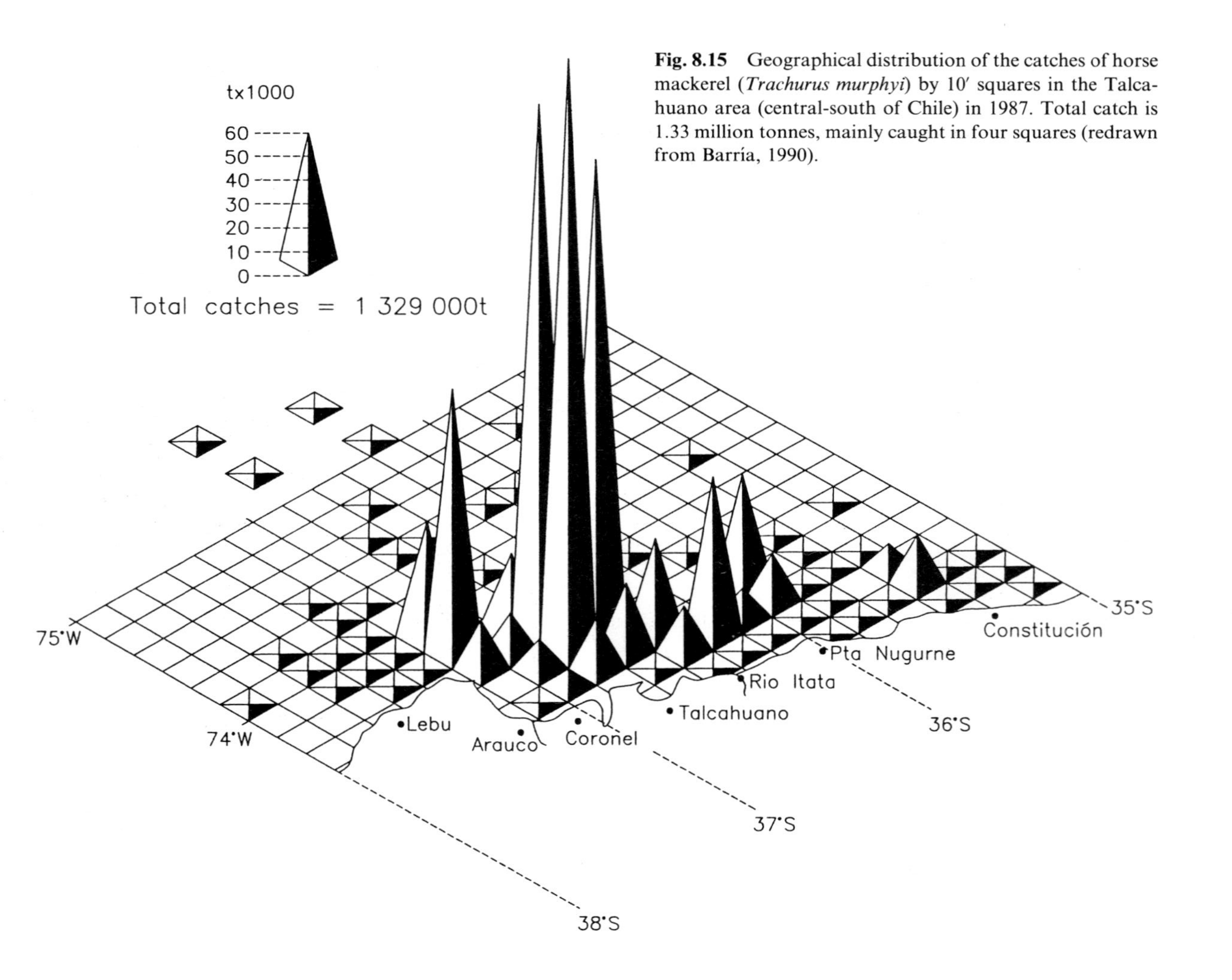

Fig. 8.15 Geographical distribution of the catches of horse mackerel (*Trachurus murphyi*) by 10′ squares in the Talcahuano area (central-south of Chile) in 1987. Total catch is 1.33 million tonnes, mainly caught in four squares (redrawn from Barría, 1990).

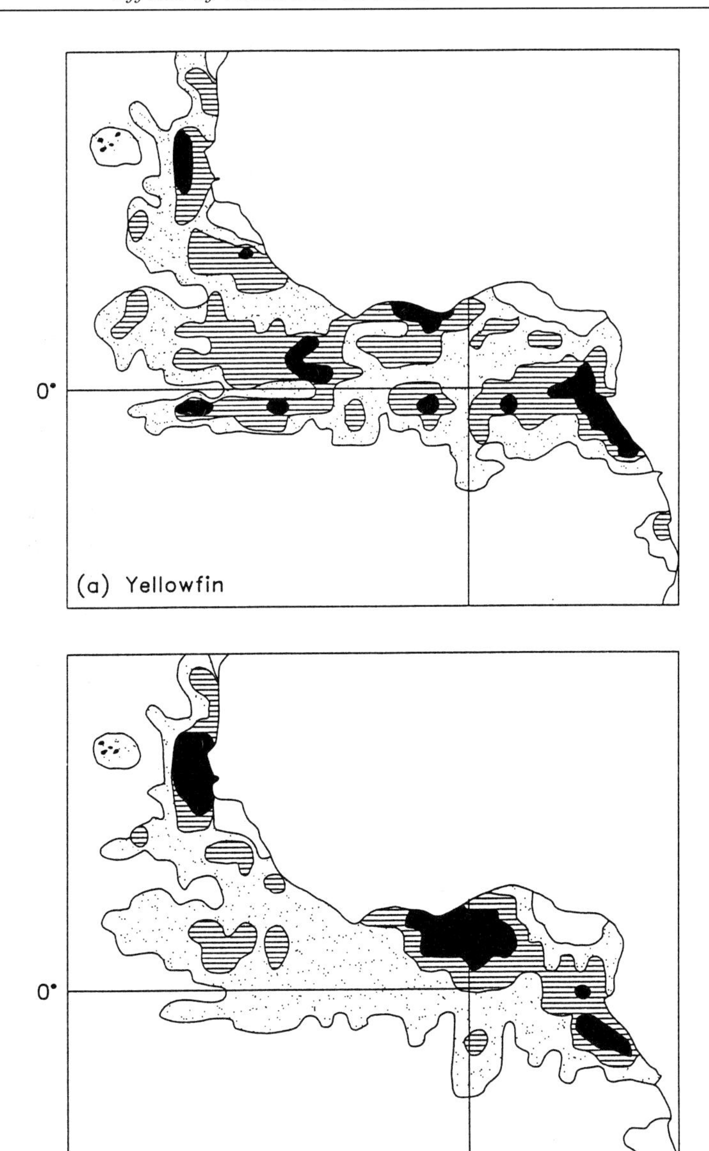

Fig. 8.16 Geographical distribution of the mean annual catches of tuna by 1° squares along the African coasts during the period 1978–1981, all surface fleets together: (a) yellowfin tuna (*Thunnus albacares*); (b) skipjack tuna (*Katsuwonus pelamis*) (redrawn from Fonteneau, 1986a). ■ > 750 t yr^{-1}; ▤ 200–700 t yr^{-1}; ⛶ < 200 t yr^{-1}.

fishery summarised by Fonteneau (1988) (Fig. 8.17). The industrial tuna fishery started at the end of the 1950s with mainly surface long-line and pole-and-line vessels, but from 1968 the purse seiners' catches were predominant (Fonteneau & Marcille, 1988). In 1972 the MSY was estimated at 70 000 tons by the generalised production

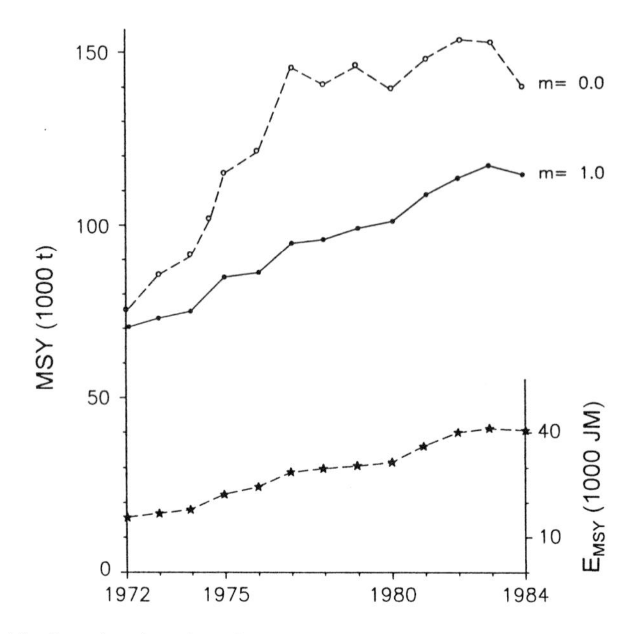

Fig. 8.17 Trend in the estimation of maximum sustainable yield (MSY) and the optimal effort (Fopt = E_{MSY}) of the yellowfin tuna (*Thunnus albacares*) in the eastern Atlantic according to the last year of available data (data series started in 1962) (redrawn from Fonteneau, 1988).

model. In 1983 it reached 120 000 or 150 000 tons according to the value of the coefficient m of the retained model. This increase in MSY is obviously due to the permanent increase of the fishing area of the purse seine and longline vessels and by a progressive access to deeper layers and therefore to older fish as a result of the use of deep longlines and deeper purse seines. This is reflected in the variability of the catchability coefficient according to the age and the period (Fig. 8.18). A similar situation is observed in the eastern Pacific tuna industry. From the 1934–1955 data series on yellowfin tuna catches and effort, Schaefer (1967) estimated at around 100 000 t the MSY value, while the same analysis performed on a series ending in 1994 provided a value of 320 000 t (Anon., 1995). Despite the violation of assumptions (2), (5) and probably (3) of surplus-production models (section 8.2.1), this situation justified the use of the inaccessible biomass model, but here, instead of having α varying according to the environment, it varied according to the horizontal range explored by the fisheries (Laloë, 1989). Similarly the changes in the vertical range could be taken into account by this model.

Despite the limitations in the use of CPUE as an index of abundance, this approach is still valid when fishery data are properly collected and processed in order to take into account the various sources of variability mentioned in this section and the previous one. In the Humboldt current area for instance, two different abundance indices were computed independently by two different institutes: the abundance estimated by VPA and the standardised CPUE (taking into account the boat characteristics and the spatial and temporal factors). The coefficient of correlation between those two series (1961–1977) was highly significant (0.93) for the stock of

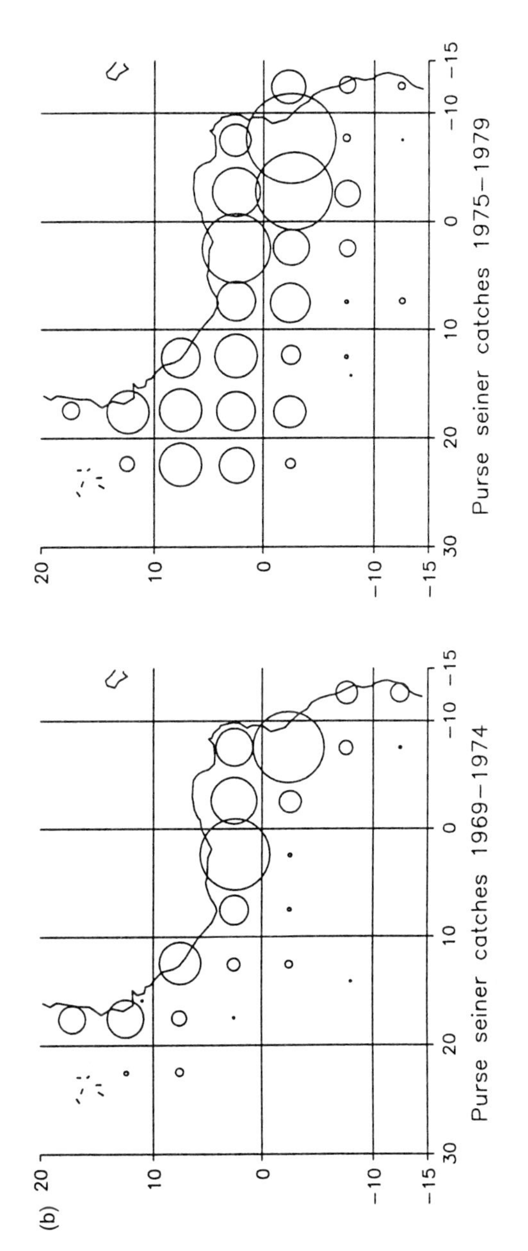

Fig. 8.18 Variation of the catchability coefficient (q) of yellowfin tuna (*Thunnus albacares*) in the eastern Atlantic (a) according to the age of the fish and to the period of exploitation (1969–1974 versus 1975–1979) and (b) corresponding horizontal extension of the purse seine fishery estimated by the catch distribution (redrawn from Fonteneau *et al.*, 1997).

anchoveta (*Engraulis ringens*) considered as shared by Peru and Chile. A similar comparison was performed for the Chilean stock of Spanish sardine (*Sardinops sagax*) for the period 1974–1992 and the value of the coefficient was 0.95 (Yáñez, 1998).

8.3.6 *Influence of habitat selection on growth estimate*

An interesting example of a depth-related trend in growth pattern is given by Kerstan (1995) for the Agulhas Bank horse mackerel (*Trachurus trachurus capensis*). He clearly observed in one of the two years of the study (1992) a change in the mean lengths-at-age from Bhattacharya's modal separation method (which agreed well with otolith ageing). The 2-group increased in mean length from 23.5 cm at about 80 m to 26 cm at depths greater than 140 m, while the 3-group increased from 28.2 cm to 31.6 cm at the same depths. This result cannot be interpreted as a change in growth related to temperature, because it would have a sign opposite to the expected one. More likely this is attributable to a change in habitat selection with body length at the same age. This interpretation can be related to the influence of the ratio body-surface/body-length on the rate of heat exchange with the water and/or by the ratio gill surface/body-weight, both resulting in a selection of deeper habitats for larger fish (section 3.3.1).

Whatever the biological processes of this change in the mean lengths-at-age, which in some cases is also dependent on longitude (e.g. Arruda, 1984; Kerstan, 1995), it can be a potential source of bias in growth parameter estimates. Therefore the sampling strategy for age-length keys must take into account these spatial effects.

8.3.7 *Influence of habitat selection on mortality estimate by ASA models*

Conventional age-structured models estimate the mortality of cohorts from the relative abundance of the different year classes in the catches. If the catchability within the recruited cohorts changes with age, the basic assumption of those conventional ASA models is violated. There are many examples of large variation in catchability by age owing to change in vulnerability related to gear selectivity, but these variations can be easily measured by size selectivity studies and taken into account in the second-generation ASA models which use the separability assumption (section 8.2.2).

A more difficult case concerns variation in catchability by age related to fish behaviour because these variations cannot be modelled in the absence of knowledge of the source of variability. Swain *et al.* (1994) found an inverse relationship between the catchability coefficient and the abundance for the younger ages (3 to 5) of Atlantic cod and a positive relationship for the older ages. Among various hypotheses explaining these results, the authors mentioned age-dependent variation in habitat selection and the possible decrease in competition for density-dependent resources during migration, spawning and overwintering. There are many other examples of similar difficulty related to different migratory pattern according to age (sedentary young fish versus migrant adults). In midwater trawl fisheries, underwater video recording indicated that at high catch rates, mesh blockage occurred for several metres ahead of the catch bulge (codend) during the later part of the tow (Suuronen & Millar, 1992). Because most escapes occur at the front of the catch bulge, the size selectivity can change according to the level of saturation, especially with diamond-mesh codends. In addition, Suuronen and Millar (1992) observed that during some of the higher-catch hauls, the codend was undulating, causing turbulence and hindering the process of active escape.

The relative shortness of this section on the influence of habitat selection on ASA models could give the false impression that the age-structured approach is less sensitive to the effect of fish behaviour. In fact the problems of catchability mentioned throughout this book apply also to age-structured models, not only because some of them are tuned by CPUE data, but also because all of them require a proper estimate of the catchability coefficient when fishing mortality is converted into fishing effort for management implementation. Nevertheless, ASA models do not require an absolute estimate of the catchability coefficient, when the stock management advice required is based on quota.

8.4 Influence of aggregation on fisheries and population dynamic models

We shall first develop the influence of aggregation on catchability, then the influence of mixed species schools on abundance indices, and finally the role of aggregation on growth parameter estimates and age-length keys.

8.4.1 *Influence of aggregation on catchability*

The linear relationship between CPUE and abundance (assumption 5 of the surplus production models) may be altered by the existence of a non-random distribution of fish in space and time, once the fishermen have adapted to this distribution (Gulland, 1964; Paloheimo & Dickie, 1964; Hilborn & Walters, 1987). Pelagic fish populations are usually spatially structured in schools and in clusters of schools (Chapter 4). It is important to consider the variation in school size when abundance is changing because most indices of abundance used in pelagic fisheries are derived from gears aimed at catching schools. Therefore we shall see first the small-scale effect owing to aggregation in schools (saturation of the fishing unit) and in clusters (effect of patchiness, co-operation and competition). Then the effect of circadian, weekly and seasonal change in patchiness will be reviewed, followed by the influence of lunar and annual changes in aggregation patterns.

Saturation of the fishing unit

An additional problem in estimating annual abundance index with CPUE is the saturation of the fishing unit (gear or holding capacity of the boat). For most pelagic gears, the aggregation of pelagic fish in schools or dense layers is responsible for a reduction of the relative efficiency at high levels of abundance. If the mean school size spotted by the fishermen is significantly larger than the capacity of the gear (or the boat capacity), this will lead to a non-linear relationship between abundance and CPUE (CPUE will decline slower than abundance). Such saturation is likely to occur in small-scale coastal pelagic fisheries such as the Senegalese fisheries which used purse seines of 250–300 m length and transport canoes able to carry between 5 and 16 t in the former fishery, and up to 24 t from the end of the 1970s (Fréon & Weber, 1983). Saturation is documented in some coastal industrial fisheries on pelagic fishes such as the jack mackerel (*Trachurus declivis*) purse seine fishery of Tasmania (Williams & Pullen, 1993) in which schools are so big that fishermen have to set the net on only a fraction of the school to avoid net outburst.

In surface tuna fisheries, saturation probably occurred in the past with small and medium purse seiners, but is now limited with the common use of large tuna ships. Nonetheless, large catches are still avoided because the duration of fish loading increases the level of stress in tuna, and therefore the histamine level in the flesh. In gill net fisheries, especially those using surrounding gill nets, the level of saturation is often reached. At a low level of abundance, especially if the mean size of the schools decreases, the saturation effect might be lower or may disappear, which would lead to an underestimation of the long-term decrease of biomass.

Saturation effects are also expected in longline fisheries: even when the distance between hooks is short, all the baits in certain locations along the line might be eaten (with a variable hooking success). As far as we know, this effect has not been studied for pelagic fisheries. An *in situ* experimental (and statistical) comparative study between the CPUE obtained by angling and visual density estimation (D) performed from a submersible is available for the quillback rockfish (*Sebastes maliger*) (Richards & Schnute, 1986). The authors provided first a comprehensive figure of all the possible relationships between CPUE and D and used two methods for fitting the models: an ordinary least squares method (OLS) assuming error only in CPUE, and a more realistic 'errors-in-variable' (EV) method assuming errors both in CPUE and in D. Despite the fact that this study was not specifically aimed at studying saturation, at high density the asymptotic value of CPUE can be interpreted as a saturation effect in some figures.

It follows from these considerations that saturation potentially exists in most of the fisheries, even though it is not a fatality. It is often easy to identify, but more difficult to quantity, especially because it often concerns the onset of a fishery (high abundance, smaller gear and vessel) i.e. historical data.

Influence of patchiness

Different theoretical cases of aggregation according to the density can be contemplated (Fig. 8.19): changes in abundance can be related to changes only in the number of schools, or to changes in the size of the schools (Fréon, 1991), or to both (Petitgas & Lévénez, 1996). From simulations and from the analysis of small-scale (in time and space) data on tropical tuna fisheries, Laurec and Le Guen (1978) drew several conclusions on the influence of the spatial distribution and structure of tuna

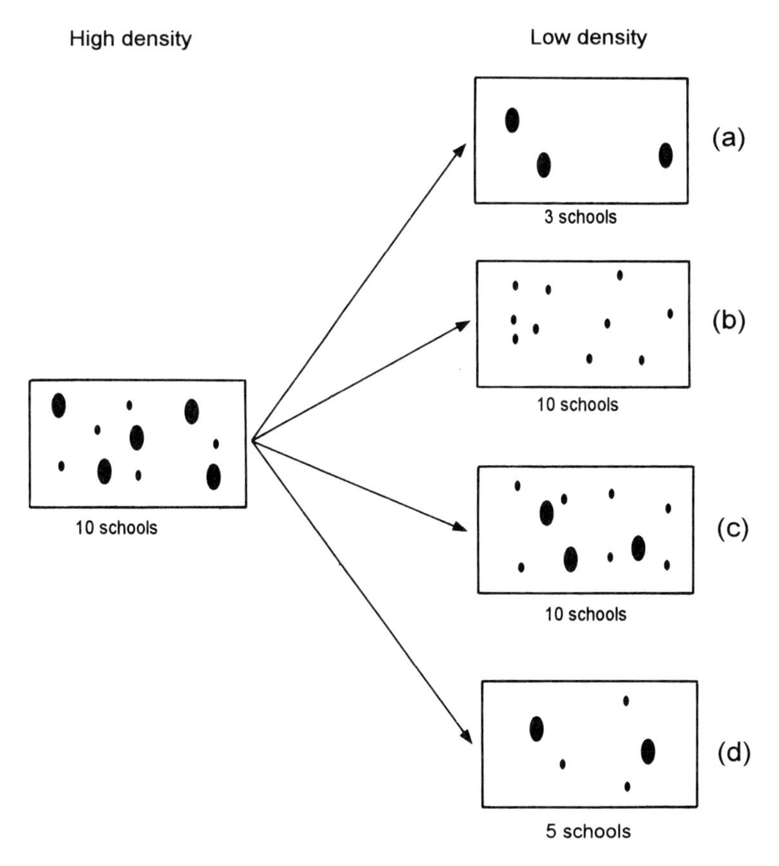

Fig. 8.19 Diagram of possible theoretical changes in school number and school size according to the density: (a) fewer schools, all large (unrealistic); (b) same number of schools, all small; (c) same number of schools, fewer large ones; (d) fewer large and small schools. This figure can also be used for representing clusters.

schools, and these conclusions vary according to the fishermen's strategy. One of these conclusions is that CPUE will be proportional to abundance (assumption (5) of surplus production models) only if the school distribution is random and if searching time is used as the unit of effort, that is days at sea minus the seining time (Greenblatt, 1976) and night time. The searching activity is stopped during fishing operations and at night. (Note that at the time of this study, fishing on floating objects was not frequent, or at least not reported.) Because fish schools are usually distributed in clusters, another important conclusion is that the validity of assumption (5) depends on the kind of response, at small scale, of the changes in stock abundance in conjunction with the ability of fishermen to detect clusters. (Fig. 8.19 can be used also for representing the variability of clusters.)

CPUE will or will not give an acceptable measure of abundance depending on fish behaviour and fishermen's behaviour on a small scale. Those behaviours are far from well known, neither are their interactions. Fish behaviour at the scale of the school concentration is not well documented, because studies on school clusters are just starting (Chapter 4). However, some alternative hypotheses have been expressed (Fig. 8.19). Fishermen's behaviour at sea (i.e. fishing tactics and strategy) is a rather new and promising field of study (Allen & McGlade, 1986, 1987; Sampson, 1991; Ferraris, 1995; Gaertner *et al.*, 1996). Fishermen's skill in recognising the presence and the limits of a concentration is expected to vary according to its size or its density, and depends largely on the vessel's equipment. The increasing use of long-range acoustic equipment (multibeam sonar; high-resolution radars to detect bird flocks) and long-range visual observation (powerful binoculars for tracking surface schools, birds or logs in tuna fisheries) reduces the probability of small or low-density concentrations escaping to the fishermen's skylines. Some drifting FADs are equipped with an echosounder connected by radio to the tuna purse seiner. In addition, some fisheries employ spotter planes or helicopters and receive thermal satellite data onboard (Petit *et al.*, 1995).

Finally, Laurec and Le Guen (1978) recommend studying, among other things:

- The geographic distribution of the fishing fleet (which gives an idea of the number of exploited concentrations)
- The length of the sequences of days with no catch (which could be related to searching for concentration)
- The duration of the sequences without significant displacements (which could be related to the duration of the exploitation of a concentration)
- The catch per set (which is proportional to the school weight when saturation does not occur).

The main conclusions of their work plus personal thoughts are summarised in Table 8.4.

Mangel and Beder (1985) proposed a general theory for the estimation of stock size from search data within a single fishing season. Their search model is an extension of the Poisson process to include depletion and take into account a parameter measuring search effectiveness during a random search (vessel speed, detection width, etc.). The limitations of the model are discussed, particularly the violation of the assumption on the randomness of the search, the unexpected effect of clumped schools in clusters, learning of fishermen and the fact that recruitment is not included. The authors suggest improvements and developments of their models to overcome these limitations, but as far as we know no further application has been proposed.

The use of the mean catch per set as an abundance index in purse seine fisheries was proposed by Fréon (1986, 1991). This implies the two following assumptions:

(1) The catch per set is proportional to the school size
(2) The mean school size reflects the stock abundance.

No major saturation effect is required to satisfy the first assumption and fishermen are supposed to sample schools randomly or to use the same threshold for selecting them. The second assumption requires a proportionality between the school size and the total biomass of the stock. These assumptions seem to be met in the Senegalese *Sardinella* fisheries except that schools are not selected randomly but according to a threshold of minimum estimated size. This threshold varies in the medium and long

Table 8.4 Possible effects of the decrease in stock abundance according to the spatial scale of the structures from the authors' interpretation of Laurec and Le Guen (1978) and personal considerations (CPUE1 = catch per time at sea; CPUE2 = catch per searching time; CPUE3 = catch per set; ↑ increase; ↓ decrease proportional to abundance; ↘ decrease slower than abundance; ↙ decrease faster than abundance; ↔ stable).

Scale	Effect on the fish	Effect on the fishery	Effect on the CPUE
Stock	↓ area	Usually ↓ trip duration[a]	↑ CPUE1; ↔ CPUE2; ↔ CPUE3
	and/or ↓ density	↑ searching time for fishing grounds	↓ CPUE1; ↓ CPUE2; ↔ CPUE3
Cluster	↓ number of clusters	↑ searching time for clusters	↘ CPUE1; ↘ CPUE2; ↔ CPUE3
	and/or ↓ cluster area	↑ searching time for clusters	↘ CPUE1; ↙ or CPUE2[b]; ↔ CPUE3
School	↓ number of schools/ cluster	↑ searching time for schools in cluster	↘ CPUE1; ↓ CPUE2; ↔ CPUE3
		↓ duration of exploitation of the cluster	↘ CPUE1; ↙ CPUE2; ↔ CPUE3
	and/or ↓ school weight	↑ time for fishing operations	↘ CPUE1; ↓ CPUE2[c]; ↓ CPUE3[c]

[a] except if the location of the core of the overexploited stock is far from the fishing harbour(s)
[b] ↘ if the cluster and its limits are directly detected by the fishers, or ↙ in the opposite case
[c] except if there is some saturation effect

term according to the global abundance and in the short term according to the remaining storage capacity on board. To limit these effects, the author retained only the sets above a fixed threshold (2 tonnes) and the trips with a single set or with several sets if the total catch did not exceed 80% of the total loading capacity of the boat.

Wada and Matsumiya (1990) proposed an abundance index P′ in purse seine fisheries which in addition to the mean school size uses the mean school density integrated over the fishing ground area. The school size is here also estimated by the mean catch per set and the school density is assumed inversely proportional to the searching time T:

$$P'_t = \sum_{i=1}^{D_t} (Y_{ti}/X_{ti})/T_t \qquad (8.31)$$

where D_t is the total number of fishing sectors for the period t, Y_{ti} and X_{ti} are respectively the catch and the number of sets at period t and sector i. The application of this abundance index to the sardine fishery of the south-east of Hokkaido from 1976 to 1984 indicated that the interannual variation in abundance was mainly reflected by changes in the school size. Moreover, Wada and Matsumiya (1990) found that mean school size within ten-day periods was negatively correlated to the fishing ground area (and secondarily to the school density) when the abundance was constant. The authors claim that this last result supports the validity of their approach. We think that the validity of P'_t as an index of abundance depends on other implicit assumptions. First, the fishing area is implicitly assumed to be proportional to the stock distribution which is a strong assumption. Second, as in the Senegalese fishery, it is assumed that there is no saturation of the gear and no density-dependent selection of school size by the fishermen, in the estimation of school size. Third, the assumption that the searching time is inversely proportional to school density relies on the random distribution of the school and fishing effort inside the fishing sectors.

Gauthiez (1997) simulated different combinations of small-scale aggregation and fishermen's ability to detect the resource. The distribution of density of the stock was a gamma or a negative binomial density function (providing similar results). In addition, Gauthiez used Taylor's relationship between the variance V and the mean E of the local abundance ($V = aE^b$) to constrain the data variance. In this relationship, he simulated three typical values of b which are related to some of the hypotheses we propose in Fig. 8.19.

- $b = 1$, linear relationship; obtained if the school biomass distribution within clusters does not vary according to the cluster biomass (Fig. 8.19d)
- $b = 2$, low curvature of the relationship, constant coefficient of variation; obtained if the mean school biomass within clusters is proportional to the cluster biomass (Fig. 8.19c)
- $b = 3$, high curvature of the relationship; obtained if the mean school biomass increases faster than the cluster biomass (that is few big schools of clusters having the highest biomass; Fig. 8.19b)

The capability of the fisherman to locate the aggregation was simulated first by performing n simulated random fishing operations in the strata and second by considering that the fisherman actually performed the best of these n operations. Of course this does not correspond to the real situation (except in shrimp fisheries using trial nets) but provides an excellent non-deterministic simulation. If n equals 1, the fisherman is considered to fish randomly. If n is medium (typical value 10), a medium knowledge of the strata is simulated, while high values of n (typically 50) are used to simulate excellent knowledge of the strata because most of the simulated catches will be located in rich concentrations. Figure 8.20 presents the relationships between the mean random catch (that is the real local abundance) and the mean non-random catch ($n > 1$):

- When $b < 2$, the curves are more and more convex when n increases, which reflects a decreasing local catchability when the local abundance increases, especially for low abundance levels. For medium and high abundance, the relationship is close to a straight line having a positive ordinate for zero abundance.
- When $b = 2$, the relationships are linear, that is the simulated non-random CPUE provides an index of abundance without bias.
- When $b > 2$, which is a realistic case in pelagic fisheries, the curves are more and more convex when n increases, which reflects an increasing local catchability when the local abundance increases.

These results underline the key role of the spatial structure of abundance versus the fishermen's strategy and we suggest that b be estimated by experimental fishing or acoustic survey to quantify the bias resulting from the use of CPUE as an index of abundance. In the case of an acoustic survey, the sensitivity of the estimate of b to the length of the elementary sampling distance unit (ESDU) must be studied.

Patchiness in fish distribution is also commonly studied by the geostatistical

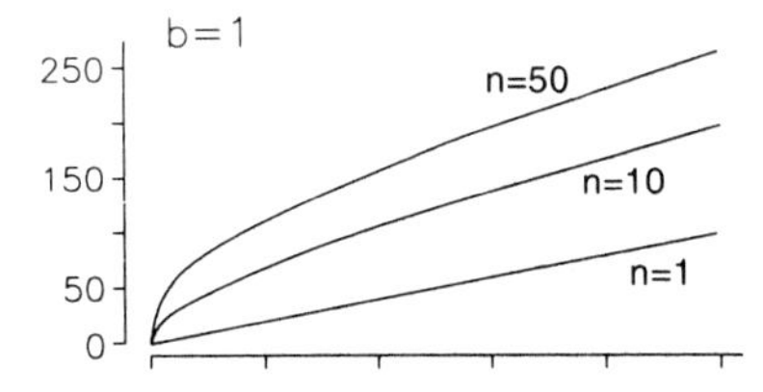

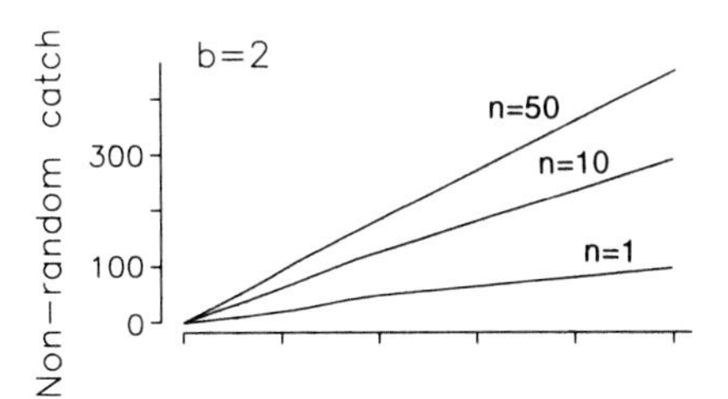

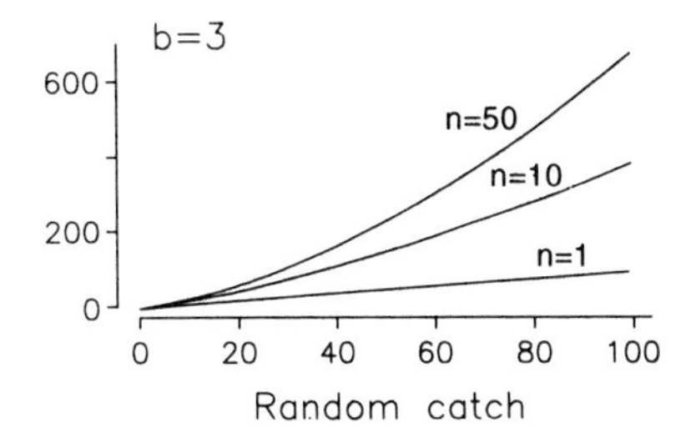

Fig. 8.20 Random mean catch versus non-random mean catch according to the number of simulated 'trials' (n) virtually made by the fisherman and to the exponent (b) of the Taylor law (redrawn after Gauthiez, 1997).

approach. Spatial analysis (by variograms or autocorrelograms) is not adapted to the analysis of large databases of crude yields originated from commercial fishery, without pre-processing. Vignaux (1996) proposed an interesting approach for analysing the small-scale structure in these types of data. It simply consists of studying the spatial and temporal autocorrelation of the residuals of a linear model aimed at removing the seasonal, circadian, latitudinal and technological effects. Autocorrelation in the residuals from the model extended out to 11 n.mi. and persisted only during 14 days. This result suggests that the patches are not related to permanent features in the environment or the bathymetry, even in the case of the deep bottom species, *Macruronus novaezelandiae*, studied.

Finally, unexpected variations of catchability by age can also be related to changes in patchiness. The Chilean jack mackerel (*Trachurus symmetricus murphyi*) provides an example of an unknown reason for change in catchability. It seems that larger fish (fork length > 40 cm) form looser or smaller schools than small fish, possibly in relation to a different diet (Serra, 1991).

Cooperation and competition among fishing units

The tactic of cooperative fishing alters the relationship between the size or density of a concentration and its probability of detection, because once a boat detects a concentration, other boats join it and may exploit the concentration for several days. As a result of all these considerations, the decrease in cluster area is likely to be underestimated by the CPUE.

A detailed study of such an effect of cooperation between boats is given by Hancock *et al.* (1995), who observed the fishermen's tactics in the fishery for the horse mackerel (*Trachurus symmetricus murphyi*) of the Talcahuano area in Chile. They reported that the fleet was mainly operating in one or two fishing groups each including 3–14 boats with their nets set and several other boats searching or cruising among them. Within groups, the overall mean nearest neighbour distance (NND) $\pm$ SD between boats with set nets was 0.75 ± 0.39 n.mi. The core area occupied by a fishing group was defined by the position of the boats with set nets. Additionally, two marginal areas surrounding the core area were defined, each having a width of the mean NND plus one SD (Fig. 8.21). Within a fishing group, 90% of the time was spent searching for schools, while outside the group, 94% of the time was spent cruising into port or directly to fishing areas. The CPUE, expressed as catch per searching time, was 417 t h^{-1} in the core compared with 40–50 t h^{-1} in the margins and 14 t h^{-1} outside the fishing group. These results mean that it is easier for skippers to find fish by first finding other fishing boats than it is to make an independent search for new aggregations. In the Javanese purse seine fishery, there is first a cooperative strategy to locate the fish concentration during the day and then a competition to exploit it during the night, using illuminated FADs (Potier *et al.*, 1997). In such situations, CPUE is a measure of abundance only within the school concentration.

A particular form of cooperation appeared in some tuna fisheries in the 1980s: that between purse seiners and pole-and-line boats. While the pole-and-line boat maintains the tuna school densely aggregated at the surface with baiting (but also through a 'boat-log' effect, see section 6.3.1), the purse seiner can set the net around it with a lower risk of avoidance. In the southern Caribbean Sea, for instance, Gaertner and Medina-Gaertner (1991) reported that the number of unsuccessful sets was only 12% with such cooperation, against 41% without cooperation, but usually the catch per set was lower with cooperation. Analysis of the fishermen's tactics indicates that most prefer the less risky tactic of cooperation to the more productive tactic of fishing alone (except for the large purse seiners because they use deeper and fast-sinking nets). Therefore more than half of the catch of purse seiners is made in association with pole-and-line boats. The association is particularly fruitful for school sets (schools not associated with floating objects or mammals) and when the thermocline or the oxycline are too deep to restrict the distribution of the fish in the upper layers (Gaertner *et al.*, 1996). Before starting fishing operations, the skippers of the two fishing boats agree by radio on how to share the catch. Usually 25% to 33% of the catch value is given to the pole-and-line vessel.

An additional reason for the failure of assumption (5) in relation to aggregation is competition between boats, which may occur only if schools are clustered. Competition has an effect on the relationship between CPUE and abundance opposite to the effect of cooperation. The effect of competition between fishing units can be observed by detailed data analysis using small spatial and temporal resolution. The geographic resolution must be comparable to the cluster size and the temporal resolution com-

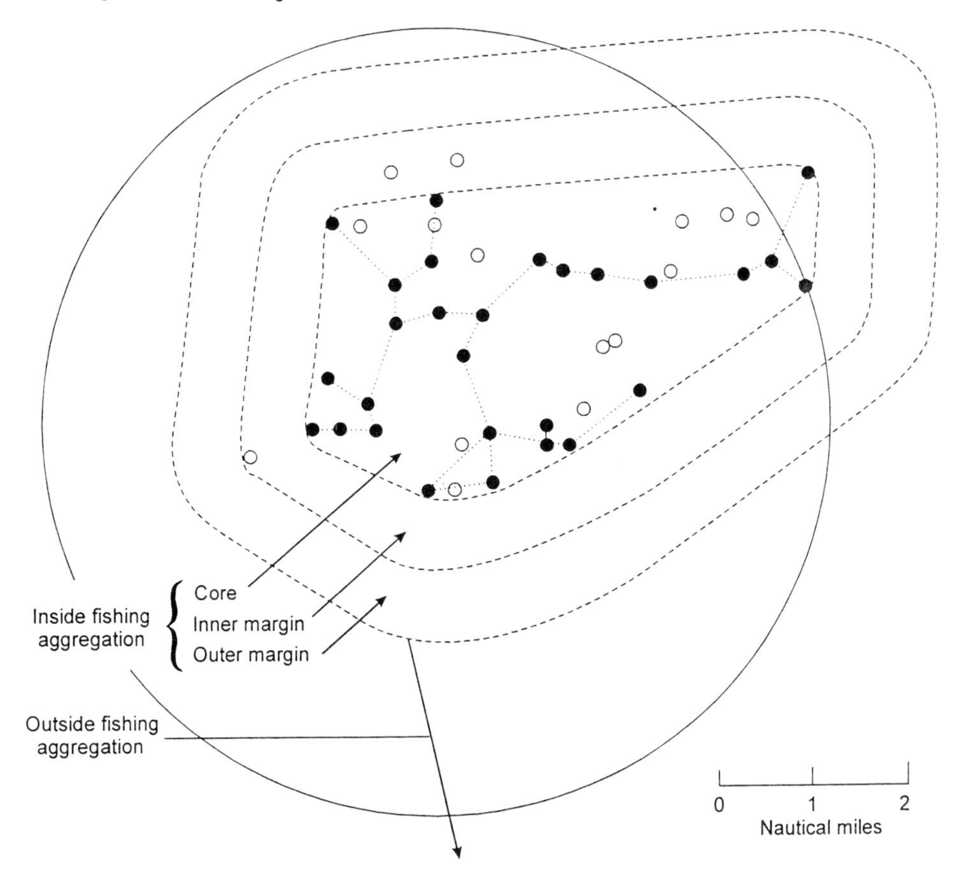

Fig. 8.21 Distribution of purse seiners exploiting a cluster of schools in the horse mackerel (*Trachurus murphyi*) fishery of Talcahuano (see Fig. 8.15). Example of radar recording of boats with set nets (filled circles) or not (open circles). Finely dotted lines show the nearest neighbour distance (NND) of boats with set nets. On this figure NND = 0.57 ± 0.27 n.mi and this value was used to delimit the width of the inner and outer margins (heavily dashed lines). (Modified after Fig. 4 from Hancock *et al.* (1995) by addition of non-fishing ships (data kindly provided by the authors).)

parable to the duration of the exploitation of a cluster, that is usually one or two weeks. Fonteneau (1986a) analysed the CPUE of the tuna purse seine fishery in the eastern Atlantic using fortnight strata of 1° squares in 1980 and 1981. The results clearly showed a decrease of the CPUE in the strata of higher effort, especially in concentrations where skipjack tuna were more abundant. A similar result was obtained in yellowfin tuna: the CPUE in nine clusters exploited continuously during periods varying from 12 to 33 days displayed an increase during the first week and then a decline owing to local depletion (Fonteneau, 1985). An estimate of the number of fish per cluster clearly shows the absence of local recruitment and indicates that nearly all the fish have been caught by the time the fleet leaves the cluster (Fig. 8.22).

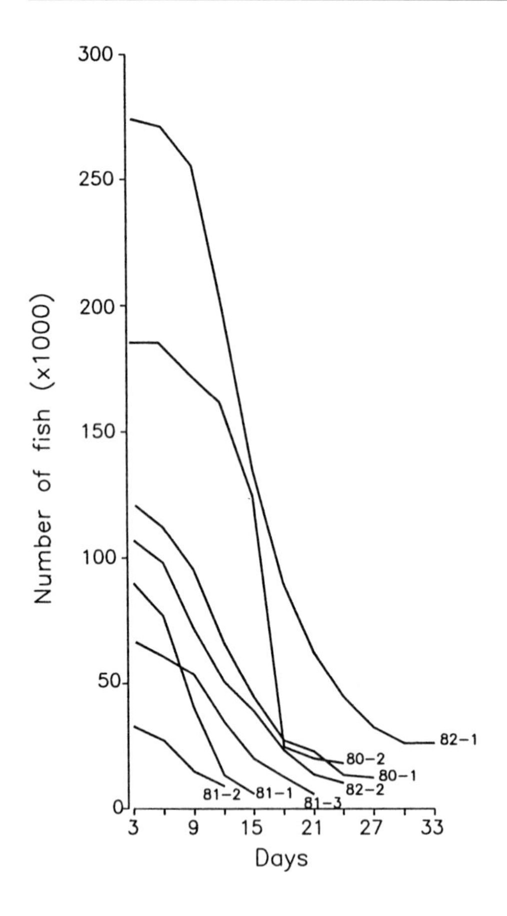

Fig. 8.22 Estimation of variation in the number of fish within nine clusters of yellowfin tuna exploited by purse seiners between 12 and 33 days (redrawn from Fonteneau, 1985).

Because competition increases when the cluster – or the whole stock – abundance decreases, the trend in the CPUE will exaggerate the real decline of the stock. It is attractive to hope that this overestimation of the decrease of abundance will compensate for the numerous reasons for underestimation previously mentioned. But fishermen are as clever at limiting the competition effect as they are at taking advantage of cooperation.

Circadian, weekly and seasonal changes in patchiness

The patchiness of pelagic fish distributions is known to change according to circadian and seasonal rhythmicity, but we will also mention unusual examples of weekly and yearly variation.

In the Namibian purse seine fishery, Thomas and Schülein (1988) found circadian cycles in the catch frequency and catch per set of the different species. Pilchard (*Sardinops ocellatus*) and, to a certain extent, adult anchovies (*Engraulis capensis*) tend to form larger and fewer shoals during the day than at night. Recruited anchovy and horse mackerel (*Trachurus trachurus capensis*) are mainly caught during the day but show no meaningful day/night difference in catch per set, even though the daily cycle of the anchovy exhibits a consistent pattern with two maxima during dusk and dawn. Nevertheless, this pattern is uncommon for *Engraulidae* observed in other areas, where they usually aggregate by day (with a higher packing density observed by acoustic methods) and disperse by night, as for instance *E. anchoita* off Argentina (Matsumiya & Hayase, 1982), or *E. mordax* off California (Mais, 1977).

In the Senegalese purse seine fishery, the mean catch per set varies according to the time of day, with higher values early in the morning (Fig. 8.23). This means that biggest schools occur early in the morning, at least if we assume that the catch per set is proportional to school size. This result is unexpected because pelagic fishes are known to disperse at night and one might expect the reconstruction of large schools to take a long time. In fact it seems that large schools do not disperse completely in layers during the night but remain in large and loose shoals which are able to reform rapidly just before sunrise (Fréon *et al.*, 1996). If the circadian or seasonal fishing pattern is changing from year to year for external reasons (market, changes in target species, etc.), the consequences of circadian variability of q on the mean CPUE will be the same as those mentioned in the sections on habitat selection (8.3.3, 8.3.4), i.e. a bias in abundance estimates.

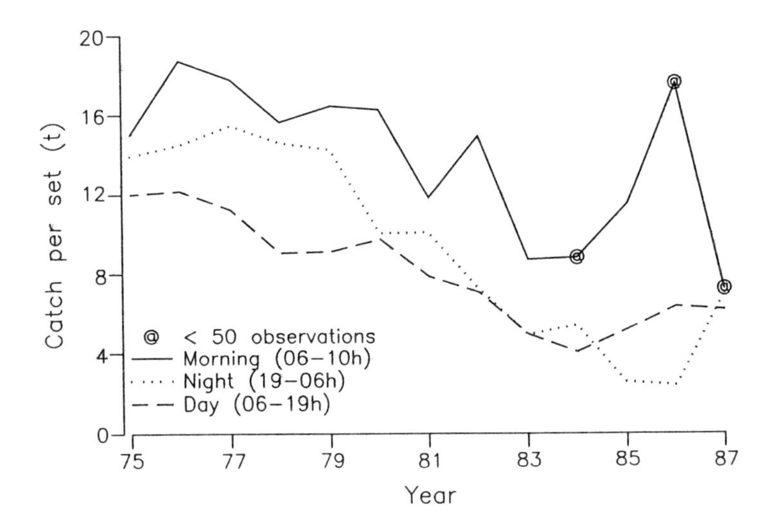

Fig. 8.23 Mean catch per set according to the period of the day in the Senegalese purse seine fishery from 1975 to 1987. Note that the 'day' period overlaps the 'morning' period because, in addition to two identified shorter periods (10–16 h and 16–19 h), it the includes the whole day period (06–19 h). This overlap increases the meaning of the difference between the two series (redrawn from Fréon *et al.*, 1994).

In the Louisiana area of the menhaden fishery, purse seiners' catch per unit of effort changes according to the day of the week: it is higher on Monday than during other days of the week, owing both to an increase of the catch per set and to a higher number of sets per day (Kemmerer, 1980). The underlying mechanism is probably a change in fish behaviour as a result of fishing activity. Fish tracking (by boats and by planes) and fishing operations might increase the school splitting and make more difficult the natural coalescence of small schools into larger ones, except on Sundays, when there is no fishing activity. This situation is not general (in many fisheries, fishermen work all week) and was not observed in the adjacent area of the fishery (Mississippi). But when this kind of phenomenon is relevant, and in the case of marked weekly change in the CPUE, the day-effect must be included in the computation of the mean abundance index through the gLM procedure, especially if a long-term change in the weekly allocation of the fishing effort is observed which will bias the comparison between years.

Large predators (fish or mammals) are also able to split large schools into several small schools, less suitable for commercial fisheries. In contrast to the fishermen's activity, the time and space of a predator's major activity is difficult to predict and therefore to take into account, except on a circadian base. Predation on schooling fish is more likely to occur during twilight periods when stalking predators have an advantage over shoaling prey under low light levels (Pitcher & Turner, 1986). This is not necessarily the case for dolphins, which have been observed preying during the night (Olson & Galván-Magaña, 1995; Scott *et al.*, 1995).

Seasonal variation of patchiness might influence catchability. In the Senegalese fishery, the mean catch per set of *Sardinella maderensis* displays a clear seasonal pattern, from 3 t set^{-1} in winter to 6 t set^{-1} in summer, despite similar body length composition (Fréon, 1991). Once more such a season effect will be easily removed by a linear model.

Influence of the lunar phase on fish aggregation

Fish aggregation may also change according to the phase of the moon, probably in relation to the light intensity (sections 3.2 and 3.3.5). These changes are generally associated with vertical migration and therefore habitat selection, but are reviewed briefly here for ease of presentation.

The lunar and tidal influence on the gill net fishery of the Bay of Bengal was studied by Pati (1981), who concluded that the lunar effect was not obvious but that the difference between spring tide and neap tide CPUE over 6 years was significant. In contrast, Di Natale and Mangano (1986) found a significant correlation between moon phases and CPUE (41% of the variance explained) of swordfish (*Xiphias gladius*) in the Italian driftnet fishery, the better yield being observed during the new moon phase.

In the Namibian purse seine fishery, the catch per set of anchovies and pilchard is significantly greater during full moon than during the other phases, but the pattern is not clear for horse mackerel (*Trachurus trachurus capensis*), even though this species is more often caught during full moon (Thomas & Schülein, 1988). A higher aggregation owing to a higher light level is probably the reason for this variability. In contrast, in the Senegalese purse seine fishery, the total duration of a trip is around 10 hours, and in winter the fishermen choose the time of departure according to the phase of the moon so as to avoid fishing during a period of moonlight (Fig. 8.24): during the phase prior to full moon, fishing operations take place mainly after midnight, and after full moon they are performed at the beginning of the night (Fréon *et al.*, 1994). From interviews with fishermen one might believe that the only reason

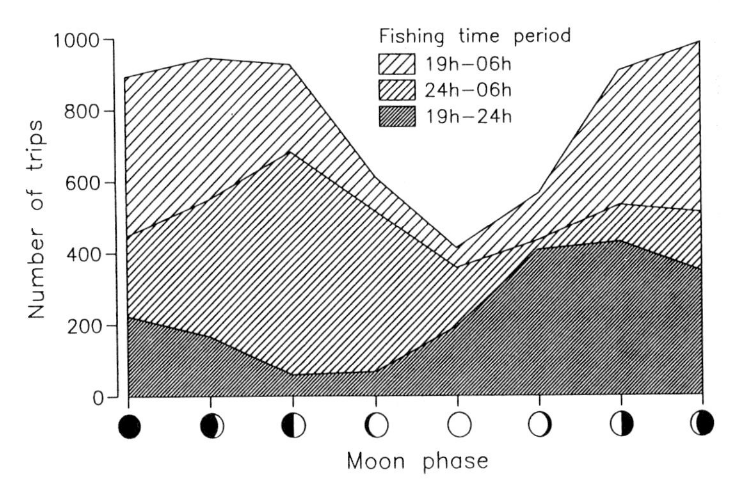

Fig. 8.24 Number of trips according to the moon phase and the fishing time period of day in the Senegalese purse seine fishery from 1975 to 1987 (redrawn from Fréon *et al.*, 1994).

for this fishing pattern is an easier detection of fish schools on dark nights, facilitated by the bioluminescence of plankton disturbed by the fish (no lateral sonar on board). Nevertheless, it is likely that fish are closer to the surface in total darkness, which is favourable both to the detection of the bioluminescence and to the availability to small purse seines.

In summary, the influence of the lunar phase is not always observed, and when it is significant, the phase associated with the highest CPUE depends on the species and the gear. In such examples of lunar phase effect, if there is any interannual change in the pattern of the fishery with respect to the lunar cycle (e.g. according to seasonal closure of the fishery), it could be suitable to take this cycle into account in the gLM procedure.

Annual changes in aggregation

In the above subsections, the conjunction of stable fish behaviour at intra-annual scales and unstable fishermen behaviour at the inter-annual scale was responsible for a bias. In contrast, section 8.3.1 focused on a density-dependent area of the habitat and some modified surplus-production models were proposed in section 8.3.2 to take it into account. At a smaller scale, the aggregative behaviour of the fish might change from year to year and this change can also be density-dependent or related to environmental changes that are not always identified.

Long-term changes in stock abundance of pelagic species are often associated with simultaneous changes in catchability, inversely proportional to the area occupied by the stock (section 8.3.1; Ulltang, 1976, 1980; Shelton & Armstrong, 1983; Csirke, 1988), which could be a social aggregation effect. In contrast, change in catchability may occur without change in abundance when long-term changes in the environment influence the aggregation pattern (Mangel & Beder, 1985). For pelagic stocks exploited by purse seiners, the usual CPUE is computed using the time spent searching for schools. Evidently in such situations this CPUE will not reflect the range shrinkage of a stock, nor will it show variation that is related not to abundance fluctuations but only to change in the catchability.

MacCall (in Fox, 1974, and MacCall, 1976) suggests a non-linear relationship between the catchability coefficient q and the biomass in the California Pacific sardine (*Sardinops sagax caerulea*) fishery from 1932 to 1950. He proposed the following equation, even though he had to remove three years of observations to make it significant:

$$q = \alpha\, B^{\beta} \tag{8.32}$$

where α and β are parameters. He also attempted to take into account the changes in area distribution by a geographical stratification of the data, but used only two large divisions (San Pedro and Monterey), which perhaps were too few for the large geographical extent of this fishery. A wider audience for MacCall's findings was given by Fox (1974), who introduced function (8.32) in the global production model (8.3) to describe the depensatory regime in some fisheries:

$$dB/dt = h\, B_t\, (B_{\infty} - B_t) - \alpha B^{\beta}\, E_t B_t \tag{8.33}$$

Table 8.1 presents the equations at equilibrium derived from this model, which was applied to the California sardine fishery for the period 1932–1955. The model fits the data reasonably well and a negative value of β allowed the collapse of the stock to be explained. It was also applied to other stocks (e.g. Fig. 8.25). Other authors used equation (8.33) to describe changes in catchability related to biomass in different models (e.g. Winters & Wheeler, 1985; Crecco & Overholtz, 1990).

At the same time, Clark (1974) also proposed a structural depensatory production model based on the modification of the stock-recruitment relationship which led to the same kind of figure when the yield was plotted against the fishing effort. In both cases, once the population is reduced past the critical point, it falls to zero even when the fishing effort is reduced. Nevertheless, two different underlying mechanisms are evoked in Fox's or Clark's model. In the first case, we have seen that it was a change in catchability with stock size, while in the second, Clark (1974) supposes that the school size is reduced when the abundance decreases, which leads to an increase in relative mortality by predation.

Let us review now the situation where interannual changes in aggregation are not linked to stock abundance. In the Bay of Biscay, anchovies (*Engraulis encrasicolus*) are usually aggregated in clusters of small schools and their distance from the sea bed is at least 15 m. Massé *et al.* (1995) reported that in 1992 this pattern was completely different and unlike anything previously observed by fisheries scientists or fishermen: the anchovies formed very dense layers, often in contact with the sea bed. This change, supposedly related to an unusually homogeneous vertical temperature profile, modified the catchability of the species. Another example is given by Mais (1977), who reported that northern anchovies (*Engraulis mordax*) of the California current system are mostly located 10 to 50 m from the surface during daylight hours and are too small and evasive to fish profitably. The fish are available to the purse seine

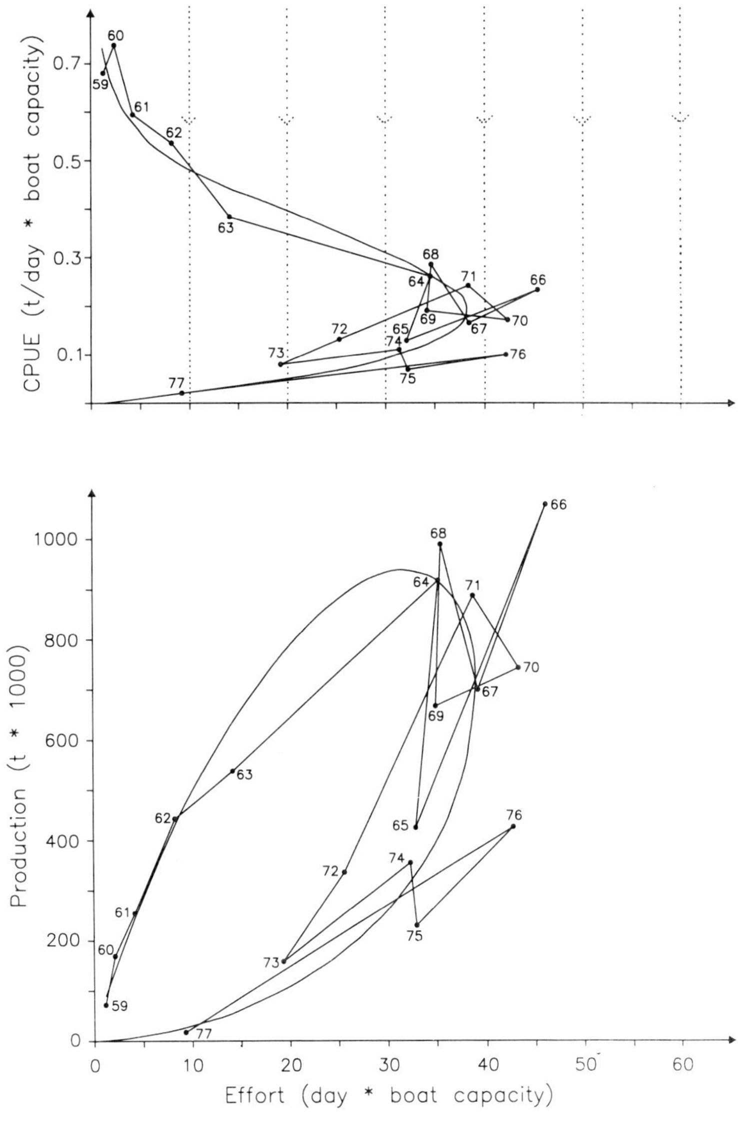

Fig. 8.25 Depensatory surplus-production model applied to the Chilean fishery of anchovy (*Engraulis ringens*) in the northern area, from 1959 to 1977 (redrawn from Drago & Yáñez, 1985).

fishery mainly when schooling densely at the surface, between midnight and dawn, for a period that lasts from 20 minutes to 7 hours, but which fails to occur some years or persists only for a short period of several weeks. A highly favourable behaviour occurs some years in late spring, when large dense surface schools form during daylight hours over the deep-water basins. A marginally favourable period for school capture of this species occurs in the late winter when schools are moving offshore to spawn. Mais (1977) concludes that the success of the fishery in California is entirely unrelated to overall abundance but depends instead on the erratic and unpredictable occurrence of favourable schooling behaviour. We present now a modelling approach of such interannual changes when the environmental factor responsible for such changes is known.

Surplus-production models can be modified to take into account the influence of environmental factors on both interannual changes in the catchability and carrying capacity (equations 8.18 to 8.21), and the case for an influence on just the catchability coefficient (scenario a_1) is a simplification of these equations when B_∞ is constant. In the linear production model it becomes (Fréon, 1986, 1988):

$$CPUE_e = q(V)\, B_e = q(V)\, B_\infty - q(V)^2\, E/h \quad (8.34)$$

$$Y_e = E\, CPUE_e = q(V)\, B_\infty E - q(V)^2\, E^2/h \quad (8.35)$$

The MSY is constant in this model, unlike E_{MSY} which is a function of V:

$$MSY = B_\infty h/4 \quad (8.36)$$

$$E_{MSY} = B_\infty h/2q(V) \quad (8.37)$$

Table 8.3 gives the set of final equations according to the choice of the q(V) function and to the basic production model (linear, exponential or generalised).

It is not always easy to distinguish changes in catchability from changes in abundance in annual data series, and often the same data set can be fitted correctly to different classes of surplus-production model from a statistical point of view. Therefore it is recommended that the data series is analysed over a smaller time scale (month or trimester) and as far as possible ancillary data is used to choose the most appropriate model from a biological and behavioural point of view. This model, and only this one, might survive changes in the fishery and might allow reasonable short-term prediction useful for management.

8.4.2 *Influence of mixed-species schools on abundance indices*

The major upwelling systems of the world are occupied by species communities of three to four main small pelagic species (Parrish *et al.*, 1983) that have variable degrees of interaction. We have seen that one of the consequences of social behaviour (section 4.9) was the formation of mixed-species schools. In those instances, it is quite difficult to obtain reliable abundance indices for each species from the CPUE, especially for the non-dominant species which are not targeted by the fishery (but when there is not such a mixed-species school, the abundance index of some non-dominant species can be less biased than the index of the dominant species).

In such circumstances it is common practice to use a global index of abundance for two or more species, in spite of a violation of assumption (1) of the surplus-production models (single stock). To a certain extent, this violation can be justified when the species forage largely at the same trophic level, which is generally the case in species schooling together (Medina-Gaertner, 1985). It must also be said that the initial applications of the surplus-production approach were to aggregate stocks of different species (Graham, 1935).

Another approach is to compute separate CPUEs per species and per small spatial-temporal strata, using a threshold of relative abundance. Fonteneau (1986a) performed such a pilot study on eight multi-specific concentrations of tuna exploited in the eastern Atlantic in 1981, based on 1° squares and five-day periods. An empirical threshold of 70% of the dominant species fairly discriminated effort targeted on yellowfin tuna from effort targeted on skipjack tuna, in contrast to the threshold of 50% tested in a previous study (Fonteneau, 1978). This 70% threshold was used to compute a series of abundance indices for the whole eastern Atlantic from 1971 to 1981 and gave a more realistic (decreasing) trend than a conventional standardised and deseasonalised CPUE (Fonteneau, 1986a).

A more objective way to substantiate abundance indices is to compare them, as did for instance Wysokinski (1987) for the horse mackerel (*Trachurus trachurus capensis*), which is often mixed with other species in midwater trawls off Namibia. He compared two CPUEs of two different fleets in the area with biomass estimates from VPA. For the first CPUE he used a unit of effort defined as the days fished when horse mackerel constituted at least 30% of the daily catch of a given trawler. The second CPUE was based on hours fished regardless of the species. Only the first CPUE displayed a positive linear relationship with the biomass (Fig. 8.26). Nevertheless, the range of variation of the biomass was not very large (from 1.5 to 2.5 million tonnes) and the rejection of days without catch during the computation of CPUE might overestimate the very low levels of abundance. Therefore one must be very cautious when using CPUE based on a threshold and it is recommended that the sensitivity of the

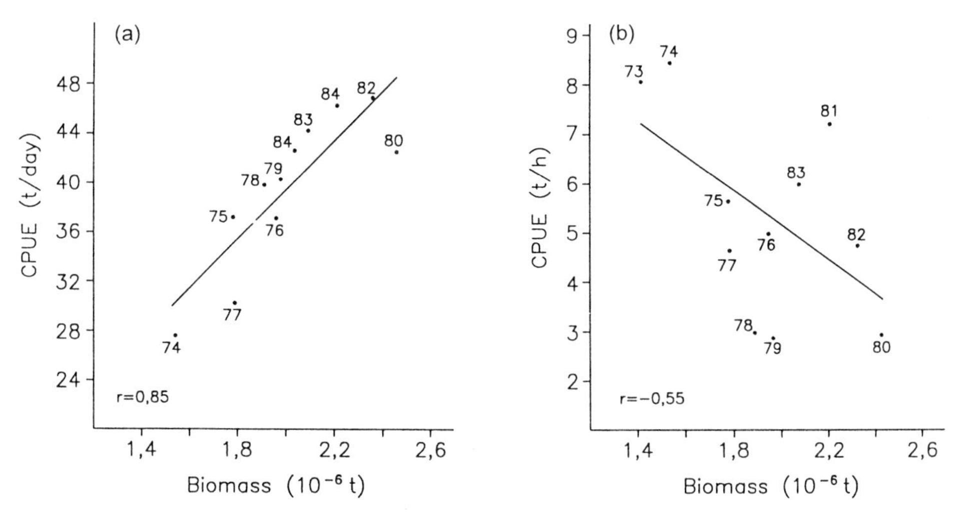

Fig. 8.26 CPUE versus biomass estimates by VPA on age groups 2–6 of horse mackerel (*Trachurus trachurus capensis*) in the Namibian Fisheries: (a) fishing effort expressed as days fished when horse mackerel catches are >30% for standard B-29 Polish trawlers; (b) fishing effort expressed as hours fished by USSR fleet (redrawn from Wysokinski, 1987).

threshold is tested and trend analysis is performed on other variables (number of days without catch, catch per set, etc.).

8.4.3 *Influence of aggregation on growth estimates and age-length key*

All the structured models require the estimation of growth parameters and/or the age structure of the population. This second goal is achieved either by a direct ageing of samples and an extrapolation to the whole population, or by a two-stage sampling using the body length as an intermediate variable (Kimura, 1977). In this latter case, an age-length key is used to describe the statistical distribution of length at a given age. It is supposed to be more efficient because it allows sampling to be concentrated on the less numerous, older year classes. The difficulty with all these techniques arises when the samples used for ageing originate from individual schools, as in most pelagic fisheries (purse seine, midwater trawl, pole-and-line).

From fishery data on menhaden (*Brevoortia tyrannus*), sardinella (*S. aurita*) and from simulations, Fréon (1984) studied the variability in fish schools and cohorts. The length-frequency distribution within a fish school cannot be considered as representative of the length-frequency distribution of a cohort as a whole. As young fish grow, the length variability increases more rapidly within cohorts than within schools with a similar mean length, especially before recruitment and if the period of spawning lasts several months. On the other hand, for older fish the length variability decreases within a given cohort as it gets older, but continues to increase in schools as the mean length of fish in the schools increases (Fig. 8.27). The permanent increase in length variability according to the mean length within schools arises from a constant coefficient of variation of the size, probably in relation to the functional structure of the school (Chapter 4). As a result, schools of young fish represent only part of the length distribution of a single cohort before recruitment. In contrast, schools of older fish are made up of individuals belonging to different age classes, which school

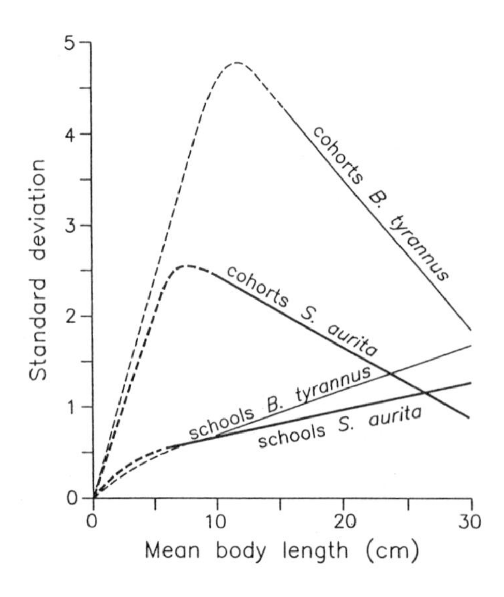

Fig. 8.27 Diagram of the relationship between the mean length and the standard deviation of the length in schools and cohorts of gilt sardine (*Sardinella aurita*) in Senegal and menhaden (*Brevooria tyrannus*) in North Carolina, US (redrawn from Fréon, 1984). (—) Fitted observations; (- - - -) hypotheses.

together according to length similarity (Fig. 8.28). This is because of the higher overlap in size between cohorts of older fish, which is mainly a consequence of asymptotic growth. This difference in cohort number in the schools according to mean length can lead to serious sampling errors in age or length data from pelagic stocks. Consequently it can invalidate the resulting estimates based on these data. An example of the extreme difficulty of length data analysis is provided by the Senegalese pelagic fishery on *S. aurita*, where the combination of schooling behaviour by size, the long duration of reproduction and the permanent immigration/emigration in the fishing area keep the modal length frequency unchanged for 6 months, despite extremely fast growth (19 cm during the first year). In similar cases, the use of any length-based method is meaningless (Fréon, 1994b).

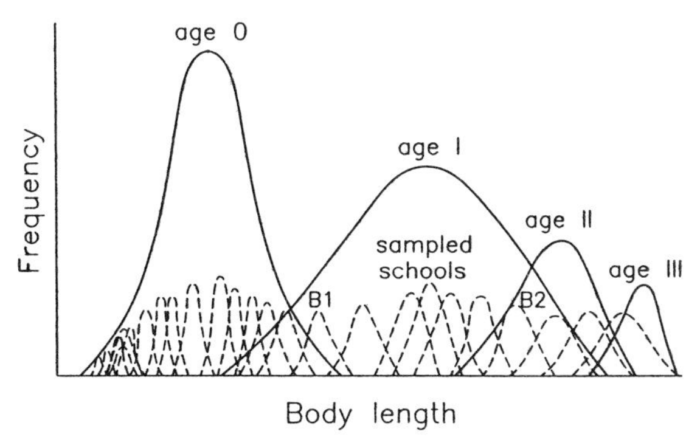

Fig. 8.28 Theoretical diagram of frequency distribution of body length in schools and cohorts of pelagic fish (four year classes) (redrawn from Fréon, 1985).

If time series of length distribution data are used directly to estimate the growth (Petersen's method or its derivations) or in other length-based stock assessment methods (VPA on length etc.), there is a risk that a modal length class corresponds to nothing else than an oversampled school. This risk is greater if this type of method is applied to non-grouped samples, as is a common practice for studying growth of younger fish. To overcome this problem, it is recommended that the number of fish measured in each school is limited (around 30 individuals) and the number of sampled schools increased. The use of the age-length key gives rise to another kind of problem, especially if few catches per stratum are analysed so as to save on the costly work of ageing. From Fig. 8.28 it appears that samples from a given school made of two year classes will overestimate the mean size of the youngest fish and underestimate the mean size of the oldest (e.g. schools B1 and B2): the smallest fish of the school are the biggest of the younger year class, and conversely the largest fish are the smallest of the oldest year class. This situation can be compared to the bias pointed out by Ricker (1975) when a size-selective gear is used: here the schooling behaviour plays the same role as the selective gear. Different trials of estimation of the optimal two-stage sampling strategy have been performed, taking into account the relative costs of ageing and measuring. The results of these simulations show that in the case of short-lived pelagic fish of Senegal, the number of fish to age in order to get an acceptable precision is over 10 000, and that the advantage of the two-stage strategy over direct ageing is not obvious (Fréon, 1985).

When the parameters of the growth equation are directly estimated from individual fish by reading their hard parts, different results can be obtained according to the sampling strategy. Simple random sampling of the catches will give an accurate estimate of growth for those year classes that are fully recruited and still abundant, but a poor estimate of the younger and oldest fish. This is especially true if the growth presents a seasonality or departure from the classical von Bertalanffy growth equation, which requires an even precision in the range of age-length data (various equations of seasonal growth are available in Pauly, 1997a). A better description of the growth will be obtained by systematic sampling of the range of sizes available in the schools, even though this will overestimate the dispersion of the size around the mean age.

Among the various density-dependent regulating factors that explain the natural regulation of fish populations, variability in growth resulting from competition for food is usually mentioned (Beverton & Holt, 1957; War, 1980). But Bakun (1989) proposed an additional explanation of density-dependent growth for schooling species: the effects of potential linkage of school size with population size. If mean school size increases with stock abundance, the average supply of food per individual within the school will decrease. This condition will be met if the ratio of the volume of the school to its cross-sectional area swept when it moves is proportional to the school size. Bakun (1989) mentioned that this situation is likely to occur, even if large schools are strictly limited in the vertical dimension by external constraints, because both horizontal dimensions probably tend to increase or decrease together, 'excluding the case where the length of the school axis parallel to its motion remains constant while school width, normal to the direction of motion, increases'.

Bakun (1989) underlined the fact that little information on school geometry was then available (except in the special case of 2D giant bluefin tuna schools described by Partridge *et al.*, 1983). Nowadays this information is given by multi-beam sonars, while aerial observations provide knowledge on the orientation of moving schools. In the Catalan Sea (Mediterranean basin), for instance, following the three-dimensional measurement technique used by Soria *et al.* (1996), we have estimated the ratio between school length (maximum detected horizontal extension) and school width (minimum detected horizontal extension) and found a median value of 1.7. The mean ratio between length and height (maximum vertical extension) was 2.8. The dominant species in this area was *Sardina pilchardus*. From aerial surveys to the south-east of Hokkaido island, Hara (1985, 1987) indicated that schools of *Sardinops melanostica* are generally elongated in a direction perpendicular to the direction of movement. These observed schooling behaviours compensate for the food limitation in large schools and probably limit the effect of density-dependent growth related to school size. In addition, for night feeders, the dispersion of fish during the night limits the effect of competition for food inside the schools. Nevertheless, these observations do not completely invalidate Bakun's hypothesis, especially because he also suggests that other substances, such as dissolved oxygen, would need to be replenished where the concentration may have been depleted, e.g. by respiration, in the rear part of the school (McFarland, 1967).

In conclusion, the previous examples on the various influences of aggregation clearly indicate the need for behavioural studies to choose between different ethological hypotheses which can interfere with the biological parameter estimations, and therefore invalidate the results of population dynamics estimates.

8.5 Influence of avoidance on abundance indices

The ability of fish to avoid gear may change diurnally or seasonally without any change in habitat. The circadian variation of catchability of midwater trawls is well documented; the main factor is recognised as the variability of avoidance according to light level (Nunnallee, 1991; see also Glass & Wardle, 1989, for bottom trawl). Suuronen *et al.* (1997) performed visual and acoustic observations on 1134 commercial hauls of the Finnish fleet (mainly pair trawlers) fishing on herring off the south-west coast of Finland. The authors indicate that avoidance reactions to the midwater trawl occurred in 34% of the 493 hauls performed with fish observed near the trawl mouth, and they characterise some of these avoidance reactions as 'strong', resulting in escapement of most of the school. The proportion of strong avoidance reactions was significantly higher during daytime (16%) than at night (2%). Herring usually reacted less than 5 m in front of the gear by swimming rapidly downwards and returned to their earlier swimming depth as soon as the trawl had passed.

Seasonal variation of avoidance reaction is often related to physiological changes. The blue whiting (*Micromesistius poutassou*) is a small fish of the cod family; its spawning habits in the North Atlantic make it vulnerable to heavy fishing by midwater trawls towed at or below depths of 300–400 m. The fishing season is short and the ships need to be of large capacity and high powered to compete with the rest of the fleet (Anon., 1979; Monstad, 1990). Herring hibernating in Norwegian fjords form huge schools of several hundred tonnes with a high packing density and have a lower avoidance than during other seasons. Because the whole stock is usually concentrated in one or two fjords, strict regulations are applied to avoid overexploitation and recently day fishing has been prohibited for large purse seines so as to limit the number of net bursts from overloading (Slotte & Johannessen, 1996).

The Atlantic herring in the Gulf of Maine region is more vulnerable to fixed gears (stop seine and weir) during the dark phases of the moon (Anthony & Fogarty, 1985), probably owing to poorer vision and lower avoidance during these periods. The same explanation certainly applies to the avoidance of gill nets according to turbidity (Cui *et al.*, 1991) and probably bioluminescence, both factors known to vary seasonally.

The consequences of this periodic variation on abundance estimates are identical to those presented in the section dealing with changes in habitat selection. They can be overcome by an appropriate use of the general linear model in most cases.

8.6 Influence of attraction and associative behaviour on population modelling

8.6.1 *Influence of attraction*

Because the major factors influencing catch rate when fishing with light are the moonlight, the power of the lamps and their location relative to the sea surface, it is recommended that they are incorporated in a general linear model aimed at estimating an abundance index. Unfortunately the necessary technical information is not always available and there is currently a trend in the technology to increase the fishing power (Potier & Sadhotomo, 1995), which can result in an overestimation of abundance indices derived from CPUE. An additional problem is the size selectivity of attraction by light. As a result, only the smaller fish of the oldest available cohorts are recruited, which can lead to growth-overfishing. Finally, difficulties appear when establishing growth parameters and an age-length key from the landed fish of fisheries, similar to those mentioned for the influence of aggregation.

In pole-and-line fisheries, the bait species and the quantity available play a role in the catch rate and the species composition. This role is not quantified and not incorporated in abundance estimates, despite the likely effect of seasonal and inter-annual changes of the bait species.

In longline fisheries, attraction varies according to the type of bait, its soak time, its movement and the period of the day (see section 6.2), among other factors not related to attraction. In the best cases, only the period of fishing is known and can be included in a gLM on CPUE, but because the gear remains submerged for a long time (12 to 24 hours) this effect is not easy to take into account. Until recently, the species used for baiting was fortunately the same one (*Cololabis saira*), at least in the Japanese, Taiwanese and Korean fisheries (Fonteneau & Marcille, 1988). But new species of bait are now being used and this change is not documented.

There is often a competition for the same bait ('gear saturation') because chemical stimuli from the bait are able to attract the fish at long distance and/or because the fish are aggregated. The efficiency of the long-line also depends on its orientation with respect to the current but this information is seldom available.

Boggs *et al.* (1997) provided an interesting application of gLM and GAM models to the Hawaii-based North Pacific longline fishery. The variance in CPUE of swordfish was mainly explained by latitude and the number of light sticks per hook (among other parameters related to temperature, temperature gradient, lunar index, etc.). These results confirm the necessity to take into account the gain in catchability arising from the recent use of light sticks.

8.6.2 *Influence of association*

The conventional indices of abundance used in pelagic fisheries were defined mainly from the experience gained on tuna fisheries. When fishing on artificial floating objects was not a common practice, the catch per unit of searching time was considered to be the most appropriate index of abundance for these fisheries (Greenblatt,

1976). Nowadays, the proportion of catches under artificial floating objects is predominant in many fisheries, and the yield depends more on the time necessary to find a floating object than on that to find a school. Of course this searching time for an object is not related to the fish abundance, but the catch under an object, combined with the density of the floating objects, might reflect the local abundance. Clark and Mangel (1979) proposed two classes of surplus-production model for the tuna surface fishery in the eastern tropical Pacific Ocean (as well as derived models in appendices). In Model A, tuna associate with a given 'attractor' (such as porpoise schools or floating objects) at a rate $\alpha'B$ proportional to the background population N, and dissociate at rate $\beta'Q$, proportional to the current school size Q:

$$dQt/dt = \alpha' B_t - \beta Q_t \qquad (8.38)$$

In this model, for a fixed B, the resulting equilibrium school size Q^* is given by:

$$Q^* = \alpha' B/\beta' \qquad (8.39)$$

In model B, the maximum school size is a constant, Q'^*, which is independent of the background population. Therefore equation (8.38) is replaced by:

$$\frac{dQ_t}{dt} = \alpha' B_t \ (1 - (Q_t/Q'^*)) \qquad (8.40)$$

These models are combined with the conventional Schaefer model (8.3):

$$\frac{dB_t}{dt} = k\,B_t \ (1 - B_t/B_\infty) - \theta \qquad (8.41)$$

where θ is the net rate of transfer to the surface population obtained from equations (8.38) or (8.39) according to the number of attractors K, the surface tuna population S and the corresponding surface carrying capacity:

$$\theta = \begin{cases} \alpha' B_t\,K - \beta' S & \text{(Model A)} \\ \alpha' B_t\,K\ (1 - S/S_\infty) & \text{(Model B)} \end{cases} \qquad (8.42)$$

(see Table 8.1 for other equations of these models). Clark and Mangel (1979) denote $\alpha'K$ as the 'intrinsic aggregation rate' or 'intrinsic schooling rate'. If this intrinsic schooling rate is less than the intrinsic growth rate of the population ($\alpha'K < k$) the equilibrium yield-effort curves for models A and B are similar, indicating that the yield approaches a positive asymptotic value as effort approaches infinity, which means that the stock can never be overexploited (Fig. 8.29). On the contrary, when $\alpha'K > k$, model A indicates that the yield approaches zero for finite effort level and the shape of the model does not differ very much from conventional surplus-production models (Fig. 8.30a). The worst situation occurs with model B: when schooling rate is greater than the intrinsic growth rate, the yield undergoes a catastrophic transition when effort exceeds a critical level Ec (Fig. 8.30b).

After Clark and Mangel's (1979) pioneer modelling, additional developments were proposed by Samples and Sproul (1985) and Hilborn and Medley (1989), aimed at determining the optimum number of FADs and of vessels in a fishery. These models require field measurements of recruitment and loss of fish associated with FADs,

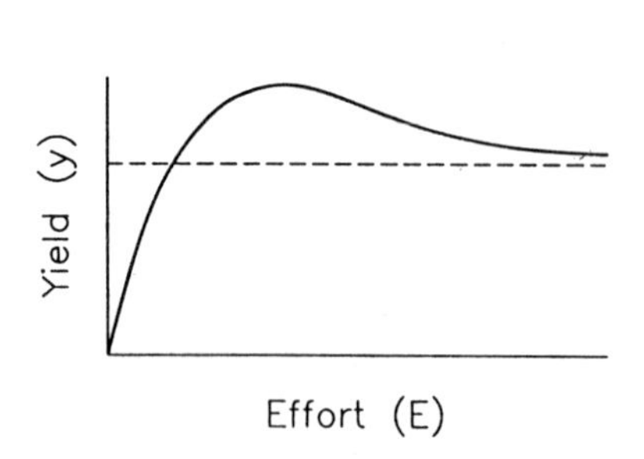

Fig. 8.29 Equilibrium catch-effort curves for equations 8.38 and 8.40, when the 'intrinsic aggregation rate' is lower than the 'intrinsic schooling rate' ($\alpha'K < r$). Models A (school size proportional to tuna abundance) and B (school size independent from tuna abundance) provide similar figures (redrawn from Clark & Mangel, 1979).

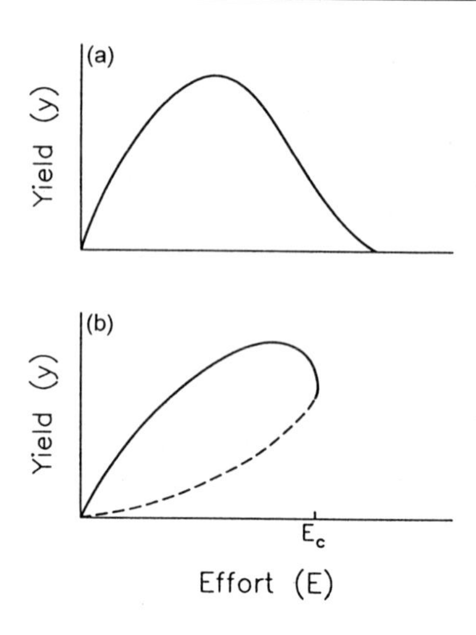

Fig. 8.30 Equilibrium catch-effort curves for equation 8.40, when the 'intrinsic aggregation rate' is greater than the 'intrinsic schooling rate' ($\alpha'K < r$). Note that in this case (a) models A and (b) B provide different shapes (redrawn from Clark & Mangel, 1979).

which are at present roughly estimated. Nonetheless, Hilborn and Medley's model was tentatively applied to the skipjack tuna fishery of the western Pacific. Results show that if the recruitment to FADs is proportional to the biomass not associated with FADs, then increasing the number of FADs beyond some limit will actually decrease the total catch. The model accounts for the fishing costs, including the cost per FAD, but does not take into account catches of schools not related to the FADs ('non-associated schools'), nor does it consider possible recruitment overfishing by the incorporation of surplus production, as in Clark and Mangel's model. Despite their limitations, these models represent a substantial advance in modelling surface tuna fisheries, but it is surprising how few applications they have generated so far. In our opinion, the main reason is lack of knowledge on fish behaviour. Nevertheless, we have seen in Chapter 3 that the increasingly widespread use of ultrasonic tagging brings increasing amounts of information, especially on attraction distance and on fidelity to anchored FADs. If the same scientific effort were to be applied to the association of tuna with logs or drifting FADs, one could expect those models to receive the attention they deserve.

The peak activity of setting on floating objects in the western Pacific occurs before sunrise, and secondarily at dusk (section 6.3). It could be interesting to enter in a gLM model a factor relative to the period of the day as in equation (8.28) and a factor relative to the type of set (non-associated school or log school) as practised in the Inter-American Tropical Tuna Commission (IATTC) workshop in 1992 (Hall, 1992b). Nevertheless, this precaution is insufficient because the time spend searching for floating objects is usually unknown, does not depend on the stock abundance, and varies with the type of object (natural, artificial moored or drifting, set and owned by the boat or not, etc.). In addition, fishermen attach a radio beacon to natural or artificial objects or use supply vessels drifting at sea or anchored on seamounts to attract and maintain schools (section 6.3.1). Finally, we have seen that the attraction effect might change according to the type of object, even though this point is not sufficiently documented. All these factors represent a strong limitation to the use of a single and simple gLM model for the whole fishing activity.

To overcome the difficulties linked to estimating abundance from fishing statistics, Fonteneau (1992) recommends the use of two separate gLM analyses: a conventional one for non-associated schools, aimed at estimating the stock abundance from the catch per searching time per stratum; and a second one for schools associated with floating objects. This second gLM analysis would take into account the catch per set – instead of the catch per searching time – related to different factors such as the local fishing effort, the density of floating objects, their characteristics and any environmental variables influencing the efficiency with which the floating object will attract fish (e.g. water transparency). The results of this last gLM could better represent the local biomass by sector.

Nevertheless, Ariz *et al.* (1992), Fonteneau (1992) and Hallier (1995) recognise that one additional problem of the tuna fisheries using floating objects is related to the fact that the mean age of yellowfin and bigeye tuna associated with floating objects is lower than the rest of the exploited population (unassociated schools or schools associated with dolphins). Therefore assumption (2) of the surplus-production models (the age groups being fished have remained and will continue to remain the same) is not met if one wants to combine old data series with recent ones including a major proportion of catches under floating objects. This problem is especially identified for some of the yellowfin and bigeye tuna stocks which are already heavily exploited and will hardly support an increasing fishing pressure on the first year class owing to increasing use of fishing on logs. Multi-gear yield-per-recruit analyses were conducted simultaneously on the three major species exploited by the eastern Atlantic fisheries (yellowfin, skipjack and bigeye tuna) by Ariz *et al.* (1992). Despite uncertainty concerning the level of exploitation, the results suggest that the log fishery provides an efficient way to increase the yield, especially for the skipjack tuna – which is the dominant species caught under logs – because in the far offshore area the vulnerability of this species is dramatically increased by the use of drifting FADs.

8.7 Influence of learning on stock assessment

We have seen in Chapter 7 that fish are able to learn, especially in relation to predation. There is growing evidence that fish are also able to learn about fishing gears and tactics, probably in a way related to learning about predation. Predator recognition can be related to gear recognition, and predator avoidance to gear avoidance or escapement, despite some differences (see section 7.2.6 on maladaptive fish behaviour). In some cases, one-trial learning is likely to occur when the stress applied by fishing operations is huge. Finally, we mentioned the possibility of social facilitation in gear avoidance when experienced fish are mixed with naive fish. Let us see now how learning might influence stock estimation by surplus-production models and by structural models.

8.7.1 *Influence of learning on surplus-production models*

To be effective at the stock level, fish learning related to fishing must be associated with a substantial rate of gear contact, and with a high rate of survival to such contact.

The importance of the escapement of small fish through the mesh of midwater trawls or beach-seines is well known and depends on the mesh size relative to the body height of the fish (e.g. Glass *et al.*, 1993; Lamberth *et al.*, 1995). But entire schools can escape a midwater trawl by movement in the vertical plane, usually by swimming downward when the gear is approaching (Suuronen *et al.*, 1997). Fewer observations are available for purse-seining. In Senegal, a limited number of *in situ* visual observations of *Sardinella aurita* during purse seining operations indicate that in many sets a substantial part of the school is able to escape the net before the purse seine closure is complete (Soria, 1994). In this fishery, as in most purse seine fisheries, the purse seine is set only in case of visual or acoustic detection of a school the biomass of which is estimated to exceed a certain threshold (here around 1 t). Nevertheless, the average proportion of unsuccessful sets was around 20% from 1969 to 1987 (Fréon *et al.*, 1994). In addition, some unsuccessful sets result from unexpected behaviour of the net owing to current or to the rupture of some elements, allowing the fish to escape. The average occurrence of such incidents in Senegal was 3% of the total number of sets over the same period. Tuna schools are probably also able to learn how to escape from a purse seine before the complete closure of the net. A yellowfin tuna school of about 100 tonnes escaped successively from ten purse seines within one hour (A. Fonteneau, pers. comm.).

The survival of fish avoiding or escaping a pelagic gear is far from negligible. Treschev *et al.* (1975) and Efanov (1981) studied the survival of Baltic herring (*Clupea harengus*) escaping from a midwater trawl codend of 24–32 mm mesh sizes. They concluded that about 90% of the fish survived after escaping. In contrast, the study of Suuronen *et al.* (1993, 1996a), performed on the same species, indicated that the survival was 10–40%. In another study, Suuronen *et al.* (1996b) found an even lower rate of survival. Mortality did not depend on the mesh size or on the mounting of a rigid sorting grid (except a slight decrease for larger fish) but was higher (96–100%) for small fish (12 cm) than for large ones (12–17 cm; 77–100%) 14 days after their transfer to large holding cages. Mortality was mainly caused by loss of scales and skin infection. It is interesting to note the survival of the control fish in Suuronen *et al.*'s (1996a) experiment because to some extent it can be related to survival in purse seines. These fish were caught either by handline fitted with barbless hooks or by a small purse seine made of small-mesh knotless nylon netting and transferred in the holding cages. The cumulative mortality of these fish reached only 9% in spring and 55% in autumn. Similar results of high survival (93%) were obtained by Misund and Beltestad (1995) when transferring fish from a purse seine to a large net pen (> 1000 m^3) and storing them for 5 days. Very high survival rates were found during bottom trawling by Soldal *et al.* (1993) on gadoids: around 90% for saithe (*Pollachius virens*) and haddock (*Melanogrammus aeglefinus*), while no mortality was observed on cod (*Gadus morhua*). Misund and Beltestad (1995) studied the survival of herring after two different simulated net burst experiments by first purse seining a school, then transferring two groups of fish to two net pens: one for the experiment, the other as a control. The first net pen was pulled up by hydraulic power block until the net was torn by the weight and force of the herring. In this extreme situation of successive stresses and compression, survival after a few days in two small net pens of different size (30 and 100 m^3 respectively) was very low (5% and 0%) and was mainly due to massive scale loss (75% of the skin surface on the side of the fish). Nevertheless, in this experiment it is also interesting to note what occurred to the two control groups because these fish also suffered stress and were damaged by the initial purse seining operation and during transfer to the control pen net, in a way equal to or higher than fish escaping from the purse seine before its final closure. The survival rates were 2% and 88% respectively for the 30 m^3 and 100 m^3 net pens and the surface of scale loss was 40% and 25%.

Despite their variability, these results suggest that in a heavily exploited pelagic fishery, a substantial number of experienced fish will survive contact with the gear. Nevertheless, this experience is usually collective and many individuals from the same school are 'trained' at the same time. If we hypothesise that these fish have learnt how to avoid the fishing gear or to escape from it and that they are able to facilitate the avoidance of escapement of naive fish by social transmission, the remaining question is: what is the rate of exchange between experienced and naive schools? The answer depends on the turnover rate in schools and is related to the recurrent question of school fidelity.

Studies in school fidelity are scarce and give contrasting results. In tunas, fidelity is demonstrated only in the special case of anchored FADs (section 6.3.1) and in minnows schools of few individuals observed in aquaria (Seghers, 1981). Fidelity was also observed on sedentary species living on coral reefs (McFarland & Hillis, 1982) or in bays of small lakes (Helfman, 1984). Moreover, several aquarium experiments suggest kinship in small shoals of freshwater or coral species (e.g. Shapiro, 1983) and it has now been demonstrated by several studies that chemical cues are used to discriminate natural shoalmates from unfamiliar conspecifics (e.g. the study of Brown and Smith (1994) on fathead minnows *Pimephales promelas*). Such kinship seems unlikely to exist in pelagic fish, which spawn in groups and in the water column, and to a lesser extent in bottom spawners, because the ontogeny of schooling behaviour occurs after the dispersal of the larvae. Nevertheless, the question of kin recognition and school fidelity in pelagic fish is still open for three reasons. First, natal homing might occur, at least for some of these species (section 3.3.1). Second, recent observations of Japanese sardine (*Sardinops melanostictus*) mating behaviour performed in tanks and *in situ* indicates that, even in large schools, this species mates in single pair units at night (Shiraishi *et al.*, 1996). Third, intriguing deficiencies of heterozygotes have been obtained in *Sardinella maderensis* in West Africa by Chikhi (1995), similar to those obtained on the freshwater shiner (*Notropis cornutus*) by Ferguson and Noakes (1981). Chikhi suggested several possible explanations for these deficiencies of heterozygotes, one of them being kinship within schools.

Despite these uncertainties, it seems likely that the turnover rate in pelagic schools is large. Most *in situ* observations performed on coastal pelagic fish suggest that there is a high rate of mixture between schools as a result of daily dispersal at night (even though dispersal is probably partial for large schools; Fréon *et al.*, 1996) and to permanent school splitting and merging under predation (Fréon *et al.*, 1993b; Pitcher *et al.*, 1996). In addition, the high variability in growth performance suggests that fish born at the same time are so heterogeneous in body length when recruited in the fishery that they will not be able to remain in the same school (Fréon, 1984). Finally, notwithstanding the lack of quantitative data, we can conclude that the turnover in pelagic schools is high enough to significantly propagate gear-avoidance learning in an exploited stock.

The consequences of learning may be a long-term decrease in efficiency of the fishing gears, and resulting biases in the use of CPUE as indices of abundance. For instance, the conflicting results between the trends in long-line and purse seine tuna fisheries of the eastern Pacific reported at the beginning of the 1970s (Hayasi, 1974) could be the result of learning. Similar observations were reported for the North Atlantic fishery on albacore (*Thunnus alalunga*) by Shiohama (1978). The conventional linear surplus-production model can be modified to reflect a decrease of the catchability coefficient q according to the fishing pressure during the previous years. We have simply used a linear function and simply apply the averaging method recommended by Fox (1975) in the case of conventional models of n exploited year-classes to estimate $\bar{E}$:

$$q_i = a - b \left[\sum_{j=0}^{j=n-1} (n-j)\, E_{i-j} \Big/ \sum_{j=0}^{j=n-1} (n-j) \right] = a - b\bar{E} \tag{8.43}$$

where a and b are coefficients. Then, if we replace q by this function in equation (8.6) and replace the initial value of E by the same past-averaged fishing effort $\bar{E}$ (Fox, 1975), we get:

$$CPUE_i = (a - b\,\bar{E})\, B_\infty + (a - b\,\bar{E})^2 \bar{E}/h \tag{8.44}$$

$$Y_i = ((a - b\,\bar{E})\, B_\infty + (a - b\,\bar{E})^2 \bar{E}/h)\, E_i \tag{8.45}$$

The corresponding theoretical curve of production is very flexible and it can display a bimodal shape owing to the power-four function of E. Nevertheless, this bimodal shape is obtained only when the variation of q is greater than two orders of magnitude, which is not realistic. When a more realistic variation of q (less than one order of magnitude) is set (Fig. 8.31), a stock collapse is obtained by a slow decrease in production when optimum effort is exceeded (then the second mode of the production curve, which appears for E values much greater than the collapse value, is no longer taken into account). Similar figures are obtained by an age-structured model of simulation, which also provides figures where the production is asymptotic for

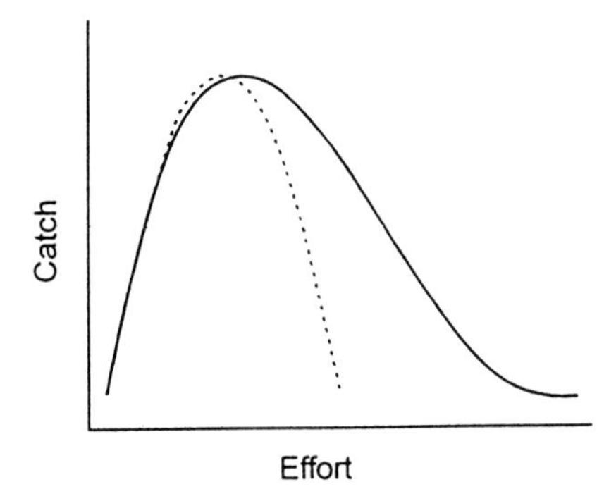

Fig. 8.31 Equilibrium catch-effort curves when the catchability is a linear function of the fishing effort (equation 8.43, solid line) due to learning, compared to a conventional Schaefer model (dotted line). (—) q = f (effort); (- - -) q = constant.

increasing effort. Fig. 8.32 is obtained assuming 'one-trial learning' for 5% of the fish escaping from the gear during fishing operations and a given stock-recruitment relationship following the Beverton and Holt equation (8.10). This situation theoretically prevents the stock from collapse.

The right-hand part of the curve in Figs 8.31 and 8.32 contrasts with the sharp decrease displayed by the conventional linear model. It can be similar to that of the exponential model or of the generalised model for c values close to one. When these two models were proposed in the early 1970s, they received great attention despite the fact that their basic equations were largely empirical and based on observations of

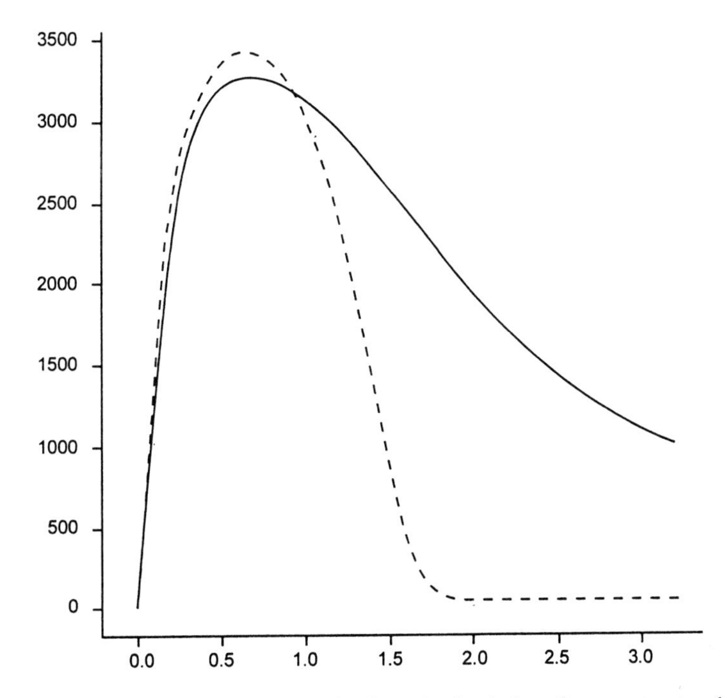

Fig. 8.32 Equilibrium catch-effort curves resulting from the simulation of an age structured model when different proportions of fish learn how to escape the fishing gear ('one-trial learning'). From Fréon, Laloë, Gerlotto and Soria (in preparation). (- - -) = No learning; (—) learning and a 6% level of escapees.

case studies. Learning could be one of the major justifications for the superiority of these models over the linear model, even though in this case equation (8.44) would be more appropriate. Nevertheless, we would definitely not recommend the general use of such models when the catch-effort curve does not suggest overfishing, because many other factors might give an opposite bias to fish learning. Two such factors are fishermen's learning and constant improvement in fishing technology, both of which can compensate for fish learning and usually result in an overestimation of the abundance during overfishing periods, which in turn leads to catastrophic management mistakes. In case of doubt, the precautionary approach must be used (Garcia, 1994).

8.7.2 *Influence of learning on intraspecific diversity and stock identification*

Social facilitation during spawning migration or natal homing – which may occur in pelagic fish more frequently than expected – might favour intraspecific diversity (Chapter 7 and section 8.7.4). Overfishing such composite stocks might result in a loss of genetic diversity if the different substocks suffer different fishing pressures and/or display different resilience levels in response to exploitation (Sharp, 1978a; Hilborn, 1985). Cury and Anneville (1997) reviewed many composite pelagic stocks and noticed that once a spawning site has been abandoned or fished out, it takes a long period of time to be reoccupied by following generations. One of the most convincing examples is the bluefin tuna (*Thunnus thynnus thynnus*). This species is strongly suspected of performing natal homing in two separate nurseries (south central Mediterranean Sea and Gulf of Mexico) and the spatial distribution of the catches has displayed dramatic changes, with some traditional fisheries disappearing. Cury and Anneville (1997) suggest that the elimination of substocks by overfishing is responsible for long-term decreasing productivity (Fig. 8.33), as observed by Beverton (1990). This last author reviewed nine pelagic stocks experiencing collapse and recovery and noticed that only one had fully regained its original size. We present now in more detail different scenarios of overfishing followed by a return to a suboptimal effort in the example of an assemblage of seven substocks having different MSY and E_{MSY}.

In the first scenario, the exploitation rate of each substock at the time of maximum overfishing (1.8 fold total E_{MSY}) is proportional to its E_{MSY} (Fig. 8.34a). In this ideal situation, the total production curve is reversible: once each substock is overexploited up to 180% of its own E_{MSY}, the total production is again able to reach its initial maximum when the total effort is decreased. Such a situation is likely to occur if the fish from the different substocks have a density proportional to the abundance of their substocks and are equally available to the fishery (randomly mixed on the same fishing grounds or separated on equally accessible fishing grounds). This scenario can also occur, but with more difficulty, when the density of each substock is not proportional to its abundance, if the substocks are exploited mainly when they do not overlap (e.g. during reproduction) but are still equally available to the fishery. In this situation, the fishery should first exploit the substock with highest productivity, but then, long before reaching the total collapse of this substock (2.0 E_{MSY}), fishermen ought to shift part of the effort toward less productive but less exploited substocks so

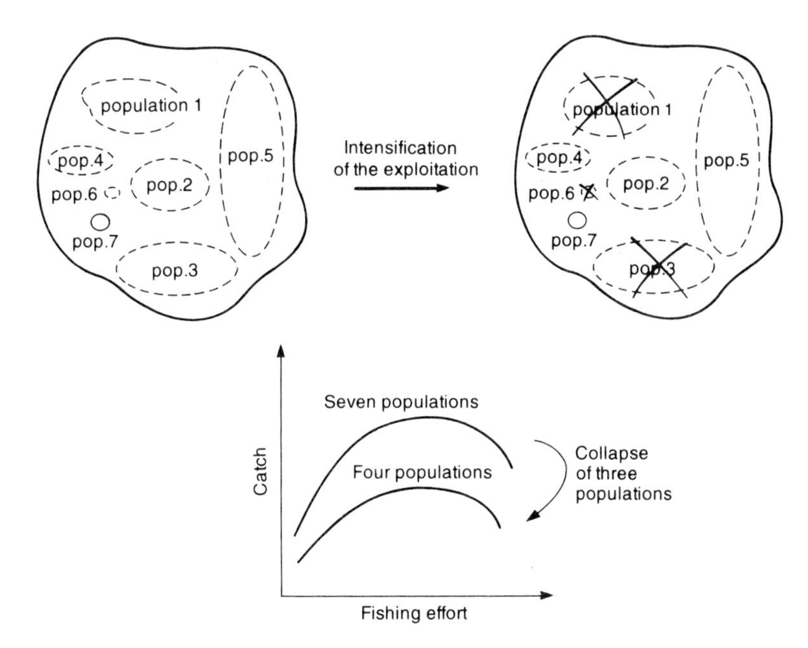

Fig. 8.33 The erosion of the intraspecific diversity may affect the productivity of the fisheries as the disappearance of populations under an intense exploitation (from seven to four populations) can have a detrimental effect on long-term marine fish catch. The associated decreasing productivity is illustrated by the catch-effort diagram where two catch levels are possible for a given fishing effort according to the two stocks' composition (for the higher level the stock is composed of seven populations and for the lower level of four remaining populations). (Redrawn from Cury & Anneville, 1997).

as to maximise their profits. At the end, these permanent small shifts of fishing effort from one substock to another will lead to roughly equal levels of exploitation among the different subunits. Mathisen (1989) proposed such a scenario for the anchoveta stock off Peru.

In the second scenario, the fishing effort on each substock is proportional to their catchability and strictly proportional to individual E_{MSY}. Therefore all substocks will collapse, starting from the substock having the lowest E_{MSY}, except the one with the highest E_{MSY} which will reach 1.8 E_{MSY} (Fig. 8.34b). This situation might theoretically occur if individuals from different substocks are not mixed when they are exploited. They can be separated either at the level of the schools of different mean biomass (bigger schools of a substock providing higher catchability), or clusters or fishing areas having different fish density but equally available to the fishery.

In the third scenario, which is the most realistic, the exploitation rate is randomly distributed among the substocks. Then the shape of the total catch-effort curve will depend on the allocation of fishing effort on the different substocks and Fig. 8.34c is only one of the possible realisations. When coming back to the suboptimum exploitation period after overexploitation, the exploitation rate of a given substock will possibly increase (e.g. substock 2) despite a general decrease of effort. In this situation also the total production is not reversible, but the loss of production when effort is then decreased to its optimum value is less than in the previous scenario.

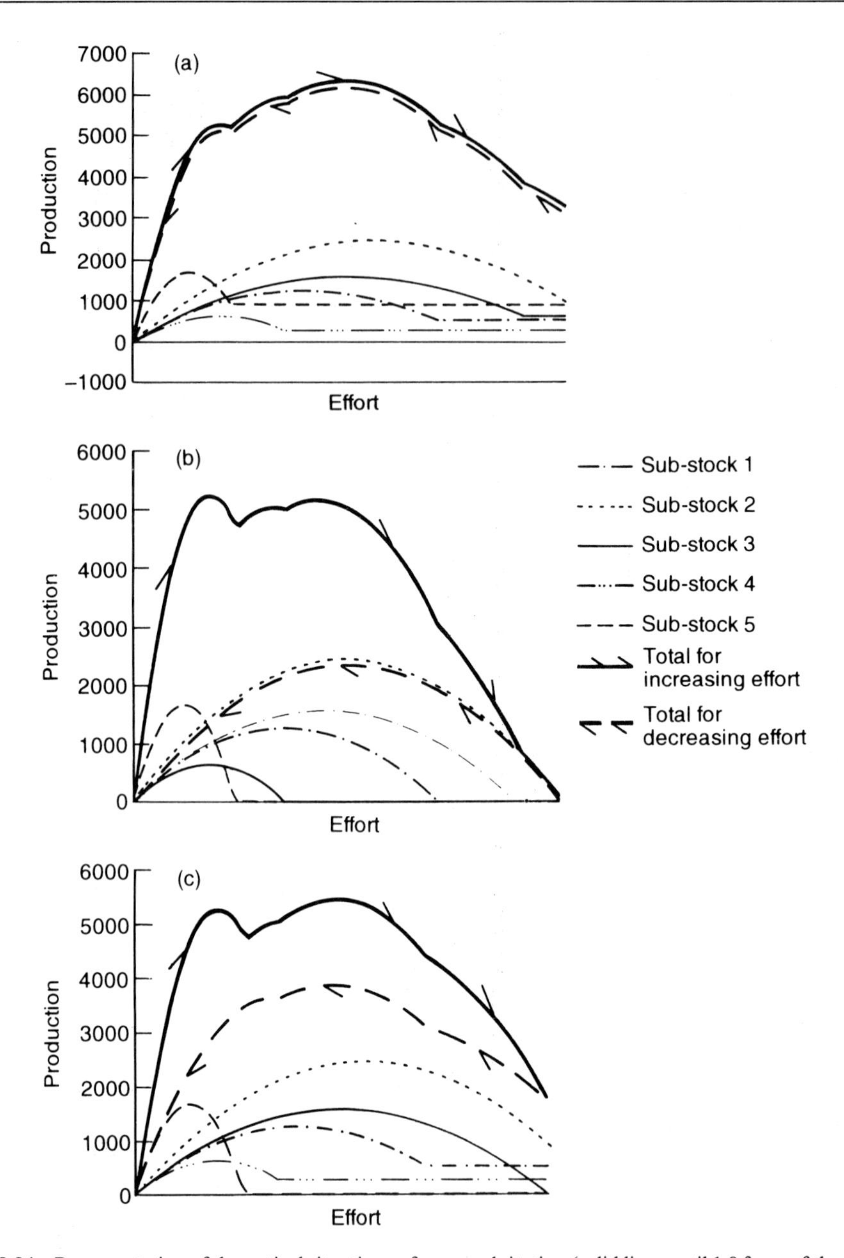

Fig. 8.34 Representation of theoretical situations of overexploitation (solid lines until 1.8 f_{MSY} of the total stock) followed by a reduction of the fishing effort (broken lines) of a stock made of five substocks: (a) exploitation rate of each substock at the moment of maximum overfishing (1.8 E_{MSY}) is proportional to its E_{MSY}; (b) fishing effort on each substock is proportional to their catchability and therefore to individual E_{MSY} if E_{MSY} is strictly proportional to q; (c) the exploitation rate is randomly distributed among the substocks.

The problem of assessment and management of such substock assemblages is similar to the mixed-species fisheries, especially when these species are competitors rather than predator and prey. Conventional surplus-production models cannot be used because assumption (1) on stock unity is violated. An optimal total effort can be computed according to different constraints (optimal yield, biodiversity preservation, etc.) as proposed by Murawski and Finn (1986) for the trawl fisheries exploiting demersal fish stocks on Georges Bank, US (see also Polovina, 1989). But of course such methods require an identification of the commercial catches from the different substocks, which is usually not possible routinely. At present the first step is to identify stock-by-stock the existence of such an assemblage of genetically independent substocks mixed in a single exploited stock, to validate the generalisation of this conceptual framework. This can be done by tagging spawners or by studies of the genetic diversity on microsatellite DNA of larvae and adults so as to assess the importance of this problem. Such a study was performed by Ruzzante *et al.* (1996b) for the identification of inshore and offshore Newfoundland overwintering substocks of cod, separated by a few hundred kilometres in winter but intermingled inshore during the summer feeding period. It is also possible to study indirectly the overfishing of substocks by episodic surveys aimed to quantify loss of genetic diversity, even by methods less expensive than microsatellite DNA. For instance, a significant decrease in the heterozygosity of loci was observed in the orange roughy (*Hoplostethus atlanticus*) stock off New Zealand in a six year period corresponding to a 70% decrease of the virgin biomass (Smith *et al.*, 1991). Nevertheless, the authors did not retain the hypothesis of elimination of substocks, but instead interpreted their results as the consequence of the removal of the oldest – and therefore most heterozygous – individuals by the fishery. This interpretation is probably right for such long-lived species, but the same method is more likely to identify possible elimination of substocks in short-lived species such as small pelagic fish. The second step is then to study the possible colonisation of spawning sites either by a few survivors from the original substock, or by strays from neighbouring substocks (particularly if there is an increase in abundance of one of these substocks owing to changes in environmental conditions or decrease of fishing effort). Such information is necessary to infer the resilience of a given pelagic stock to overexploitation, and particularly its ability to recover from collapse.

8.7.3 *Influence of learning on structural models*

If the oldest fish are more experienced with fishing gear because of higher cumulative numbers of contacts with the gear, their fishing mortality may be lower than expected. This will affect structured models using the backward approach because they require an *a priori* estimate of the fishing mortality of the oldest age groups ('terminal F'). The corresponding error in this estimate will propagate in the vector of mortality, especially when the exploitation rate is low (Pope, 1977; Mesnil, 1980; Bradford & Peterman, 1989). From a sensitivity analysis of VPA and yield per recruit, Pelletier (1990) indicates that terminal F plays a major role in both cases, especially for recent-years estimates in the case of VPA.

We would like to give here a more optimistic note on the potential role of learning which is not completely reflected by surplus-production models. If we assume (or better still, have evidence) that older fish have a very low catchability coefficient, learning should prevent the stock from collapse owing to the high fecundity of these specimens out of reach of the fishermen. Unfortunately it is likely that fishermen's skills and advancing technology compensate for (and possibly interact with) fish learning. This is obvious when a fishery starts to exploit a stock (or use a new gear). In the Senegalese pelagic fishery for instance the CPUE increased steadily from 7 to 32 t $10\,h^{-1}$ from 1962 to 1967 despite a decrease of the upwelling index and an increase of the fishing effort (from one to three purse seiners) which both were supposed to decrease the abundance (Fréon *et al.*, 1994). Therefore, despite the likely effect of fish learning on stock viability, the precautionary approach must remain the rule.

8.7.4 *The strength of paradigms*

Some of the shapes of the catch-effort relationship obtained from surplus-production models incorporating learning (Figs 8.31 and 8.32) or an assemblage of substocks (Fig. 8.34) might look unfamiliar to fishery biologists. These models might appear in contradiction with the paradigm implicit in the conventional linear production models under equilibrium conditions: a collapse must occur if the fishing effort reaches a value close to twice the effort level necessary to obtain the maximum production (MSY). Despite the present disfavour of the conventional surplus-production models (mainly owing to failures related to the violation of basic assumptions), they have been successfully applied in the assessment of fish stocks and we think that the above-mentioned paradigm has some biological justification. Nevertheless, we think that the generalisation of this paradigm is not likely. We have mentioned previous attempts to give more flexibility to the catch-effort relationship, but only the exponential model (Garrod, 1969; Fox, 1970) received significant attention. This model provides a slower decrease of the catch than the linear model when E_{msy} is exceeded, but the improvement of the quality of the fit does not always result from the different shape of the catch curve. The improved fit obtained with the exponential model often comes from a reduction of the variance of the highest CPUEs owing to the logarithmic transformation of the data. The exponential model is not so different from the linear model, contrary to the high flexibility of the generalised model of Pella and Tomlinson (1969). The generalised model has been widely used and was promoted by the computational facilities provided by Fox (1975), who proposed the PRODFIT software.

Nevertheless, a survey of the literature indicates that papers presenting results from the generalised model either mention a use with the option of constraining the parameter m to fix values equal to 2 (equivalent to a linear model) or to 1 (equivalent to the exponential models), or a result where the software itself fitted m to values close to 1 or 2 (example of exception: Guerrero & Yáñez, 1986). There are two interpretations of this observation. The first one is that it demonstrates that most stocks obey a linear or exponential model and that learning always has a secondary effect, or is counterbalanced by faster learning of fishermen and technological improvement.

The second one is that it only demonstrates the strength of a paradigm, largely because alternative values of m, especially outside the interval 1–2, were not supported by theoretical considerations. As a consequence, we suppose that scientists finding such results would be reluctant to publish them, if there are likely to be difficulties having them published owing to referees' caution. In addition, we have mentioned that the situation where catchability could be a function of effort was likely to occur in passive-gear fisheries, i.e. not in the major commercial fisheries which are the most studied.

We suggest that learning could be one explanation for departure from the conventional linear models. We have seen that in the catch curve obtained from equation 8.43 (Fig. 8.31), the stock collapse could occur with effort values much greater than $2E_{msy}$. We have already mentioned other possible explanations for such a departure from the conventional linear model, and underlined that the weak point of the surplus-production models is that several causes might give similar shapes to the catch-effort function. This results from the oversimplification of the surplus-production approach and/or from undetected violation of the basic assumptions (inaccessible part of the biomass, exchange with an underexploited substock, or collapse of unidentified substocks). Despite these restrictions, we encourage scientists who find such indications of effort-related learning to report them. The comparative approach is a way of validating a new concept.

8.8 Conclusion

We have shown in this chapter how different the causes of biases or errors in population dynamics related to fish behaviour can be. Some of these problems can be easily resolved by a proper use of conventional tools of population dynamics. This is particularly the case for the periodicity in the data (circadian, weekly, moon cycle, seasonal) and for some of the spatial effects. If these factors do not interact with the annual effect, a proper abundance index can be obtained from a general linear model. The three most difficult problems are the interannual change in catchability, the effects of aggregation and the identification of a unit stock.

Interannual change in catchability can be due to different reasons such as change in stock abundance, environmental change, the combination of these two changes, etc. As a result, conventional models are not able to distinguish a change in catchability from a change in recruitment, which can be easily shown by simulation (e.g. Fonteneau, 1986b).

The effects of aggregation are numerous and vary according to the scale (school, cluster, population). Fishermen only exploit a selected part of the structure, which is the easiest available with the highest yields. Few gears are designed to catch dispersed fish, and fisherman resort to more efficient alternatives. This enables them to catch large schools located in the right tail of the positively skewed frequency distribution of school size occurring in the largest clusters. In addition, they tend to explore more intensively areas located nearest the harbour. Then we need to study not only fish behaviour, but also how fishermen's behaviour is adapted to it. The tuna fisheries suffer from a specific problem related to the association of tuna with floating objects.

Finally, it is difficult to verify if the assumption of a single unit of stock – common to all approaches – is respected, and if not to relax it. In many instances, a single stock is erroneously modelled as several independent stocks or, more frequently, several stocks are confounded in a single one. As a result of this second mistake the over-exploitation of one of several single stocks cannot be detected by any model. This problem can be solved by a better knowledge of the species distribution and its migratory pattern in relation to the fishing effort distribution, and by high resolution genetic studies.

In most cases, appropriate solutions to correct those biases, errors or violations of assumptions do exist, even if they are not perfect, but few authors use them. This is partly due to the dominance of some conventional methods and to the strength of paradigm. But a likely additional reason could be that these methods require a detailed understanding of the behavioural mechanisms, which is not always available. There are good reasons to hope that the progress now being made in the understanding of fish behaviour and spatial structure (especially at medium and large scale) will favour the use of such approaches, the creation of new ones or the transfer of existing approaches to fishery biology (e.g. GIS (geographic information system); Meaden & Do Chi, 1996). For that purpose, modern techniques of observation, such as acoustic tracking, multi-beam sonar, archive and 'pop-up satellite' tags, both storing data for several months (Block *et al.*, 1997), electronic microchip hook timers, remote sensing, and amplified mitochondrial DNA analysis are of major interest. At the same time, more information on fishermen's strategy and tactics must be collected, as fine time and space scales. Then, but only then, additional simulations (including the IBM approach) can also be very useful in better understanding fish and fishermen's behaviour from new input provided by *in situ* data. I turn these simulations will provide new insights and stimulate additional field research.

Conventional acoustic studies are also of interest for a better understanding of fish behaviour and for direct estimation of biomass independent from the indirect methods of assessment reviewed in the present chapter (except when ASA models are tuned by acoustic data). Acoustic surveys suffer also from biases and error owing to fish behaviour, but – as we shall see in the next chapter – the most important problems in acoustics come from behavioural traits different from those mentioned in this chapter.

Chapter 9
Effects of Behaviour on Stock Assessment using Acoustic Surveys

9.1 Introduction

The hydroacoustic method for estimating abundance of fish was mainly developed during the last three decades of the twentieth century. The basis for the method was laid by the invention of the echo integration system (Dragesund & Olsen, 1965), and initial studies of the reflecting properties of the fish (Midtun & Hoff, 1962; MacCartney & Stubbs, 1971; Nakken & Olsen, 1977). During the 1980s there was a rapid development of computerised echo integration systems (Bodholdt *et al.*, 1989; Dawson & Brooks, 1989), accurate calibration procedures (Foote *et al.*, 1987), and sophisticated post-processing systems (Foote *et al.*, 1991).

Despite its rather recent development, the hydroacoustic method is widely used for providing fishery independent estimates of fish abundance. Many of the most important pelagic stocks of blue whiting, capelin, herring, sardinella and pilchard are surveyed by this method. However, due to the many uncertainties connected with the method, especially related to fish behaviour, the abundance estimates are mostly used as relative indices to tune structured assessment models by which the stock assessments are made (see section 8.2.2). For a short-lived species such as capelin, which dies after spawning, the acoustic abundance estimates are considered absolute, and the stocks assessed accordingly (Vilhjálmsson, 1994).

In this chapter we will consider effects of fish behaviour on acoustic abundance estimates, starting with a short description of the basic principles of the hydroacoustic assessment method. A complete introduction to the method is given by MacLennan and Simmonds (1992).

9.2 The hydroacoustic assessment method

The sound intensity (I_0) of a pulse emitted from an underwater transducer decreases due to geometrical spreading and absorption as it propagates through the sea. The sound intensity (I) returning to the transducer from a fish density of N individuals per unit volume at a certain range (R) in the sea, is expressed by the sonar equation (Forbes & Nakken, 1972; Burcynski, 1982):

$$I = I_0 \cdot (c\tau/2) \cdot \int b^2(\theta,\varphi)d\Omega \cdot e^{-2\beta R} \cdot R^{-2} \cdot (\sigma/4\pi) \cdot N \qquad (9.1)$$

where c = speed of sound underwater ($\sim$ 1500 m/s), τ = duration of sound pulse ($\sim$ 0.001 s), b = beam directivity function, θ = angle relative to the acoustic axis, φ = angle relative to a reference plane through acoustic axis, Ω = solid angle, β = attenuation coefficient, and σ = back scattering cross section of the objects (m^2).

The transmission loss due to geometrical spreading is two way, which means that the total geometrical spread equals R^4. A compensation equal to R^4 is therefore used during target strength measurements of single fish or of solid spheres during calibration of the echo integration system. When a sound pulse emitted from the transducer propagates through the sea, the area of the pulse increases with range at an amount equal to R^2. The larger the area of the pulse, the greater the number of fish in a layer with constant density that may be insonified. To correct for this range dependency in the number of fish that may be insonified, the compensation for transmission loss is reduced from a factor proportional to R^4 to a factor proportional to R^2. To compensate for the transmission loss due to both geometrical spread and absorption during echo integration surveys for fish abundance estimation, the returning intensity is magnified by a time-varied gain function equal to $R^2 \cdot e^{2\beta R}$.

In an echo integrator the voltage (V) of the returning signals is squared and summed over defined range intervals to give an output (M) according to the equation:

$$\begin{aligned} M &= \int (R^2 \cdot e^{2\beta R} \cdot V^2)\,dR \\ &= \int (R^2 \cdot e^{2\beta R} \cdot K \cdot G \cdot I)\,dR \end{aligned} \tag{9.2}$$

where K = receiving voltage response, and G = gain function.

If the echo integration unit is calibrated by the use of a metal sphere with known backscattering properties (Foote *et al.*, 1987), the echo integrator output can be converted to fish density per unit area (ρ_A) by the equation (Dalen & Nakken, 1983):

$$\begin{aligned} \rho_A &= (C_I/\sigma) \cdot M \\ &= s_A/\sigma \end{aligned} \tag{9.3}$$

where C_I = calibration constant of the equipment, M = echo integration output (mm/nautical mile), and s_A = area back scattering coefficient (m^2/(nautical mile)^2).

During surveys, the echo integrator is set to produce outputs for predetermined depth channels over a certain distance sailed (e.g. 0.1 nautical mile, 1 nautical mile or 5 nautical miles). The integrated output has to be partitioned to actual species according to fishing samples or other means of identifying the recordings (Dalen & Nakken, 1983). The number of fish of a given species in an area surveyed can then be found by multiplying the area density of the species by the distribution area.

9.3 Effects of habitat selection

Conventional acoustic assessment requires that fish be distributed from a few metres below the surface to a small distance off bottom (Mitson, 1983). An echo sounder transducer is normally mounted in the bottom of the hull at a depth of 5–6 m on large survey vessels. For technical reasons the echo integration process is set to start 4–5 m from the transducer. Therefore, when surveying with large vessels there is an upper dead zone near the surface of about 10 m (Aglen, 1994). Similarly, there is a bottom dead zone because the resolution of the acoustic beam near the bottom is dependent on pulse length, beam width and bottom depth (Ona & Mitson, 1996). Species living permanently close to the surface or on the bottom must therefore be assessed by other methods. Fish schooling near the surface may be recorded by horizontal guided sonar (Hewitt *et al.*, 1976; Misund, 1993a), and abundance indices of bottom fish are obtained by trawl surveys (Gunderson, 1993).

Several economically important species like cod and pollack live semipelagically, and to obtain the best possible coverage in such cases a combination of acoustic and trawl surveys is applied (Wespestad & Megrey, 1990; Godø, 1994). A substantial difficulty for this approach is that the degree to which such species are distributed demersally or pelagically may vary locally, seasonally and from year to year (Godø & Wespestad, 1993). Diurnal variation in availability of semi-pelagic species to the sampling gear may indirectly influence acoustic abundance indices (Engås & Soldal, 1992). On certain occasions, semi-pelagic fish may also choose to be either completely demersal or pelagic (Godø & Wespestad, 1993). Similarly, species usually living pelagically, like herring, may on certain occasions distribute themselves so close to the surface or bottom that they become inaccessible to conventional acoustic assessment (Jakobson, 1983; Misund & Aglen, 1992).

The air-filled swimbladder accounts for about 90–95% of the backscattering cross section of fish (Foote, 1980a). The swimbladder can be open (physostomous) or closed (physoclist). The physoclist species like the gadoids have a rete mirabile which is an organ for secretion/resorption of gas from the blood to the closed swimbladder. However, the buoyancy-regulating process is slow, and physostomous fishes are not believed to be neutrally buoyant though they often perform substantial diel vertical migrations (Blaxter & Tyler, 1978). Such migrations may affect the swimbladder volume and thereby also the backscattering cross section of the fish. Nevertheless, Harden Jones and Scholes (1981) estimated that the effect of varying swimbladder volume and thereby backscattering cross section through vertical migrations of physoclist fishes was probably less than that of tilt-angle-induced variations in the back-scattering cross section.

Physostomous species like the clupeoids lack a gas-producing organ. Most physostomous species have a posterior canal from the swimbladder to the oesophagus, and an anterior canal from the swimbladder to the anus (Blaxter & Batty, 1990). It is therefore supposed that the swimbladder is filled by swallowing air at the surface, even to such an extent that an above atmospheric pressure is built up in the swimbladder (the big gulp hypothesis, Thorne & Thomas, 1990). However, opening of physoclist swimbladders of fish caught at shallow depths often indicates a substantial gas pressure in the swimbladder that is difficult to explain just by swallowing. Nevertheless, the physostomous fishes are often characterised by substantial diel vertical migrations which undoubtedly may affect the backscattering cross section. Ona (1990) found that the swimbladder volume of herring followed a depth relation according to Boyle's law, and Halldorsson and Reynisson (1983) and Olsen and Ahlquist (1989) found a certain reduction in the backscattering cross section of herring with increasing depth. Acoustic and photographic observations indicate that this can be the case for Norwegian spring spawning herring when hibernating in the deep fjords in northern Norway (Huse & Ona, 1996). Herring distributed in deep water (300–400 m) seem to adapt a rise and glide strategy to compensate for their negative buoyancy. Such swimming results in bimodal tilt angle distribution which in addition to reduced swimbladder volume when at great depth should result in a reduced backscattering cross section of herring (Huse & Ona, 1996).

9.4 Effects of social behaviour

As described in Chapter 4, fish can distribute and behave individually, assemble socially in shoals (for instance during feeding and spawning), or swim in synchronised and polarised schools (Pitcher, 1983). These behaviour patterns determine to a large extent the volume occupied by the individuals in a stock (Pitcher, 1980), and consequently have significant influence on sampling.

9.4.1 *Distribution function of densities*

Some fish species live mostly individually, others aggregate in shoals, while the highest densities are formed by species that organise themselves into schools. The aggregative behaviour induces a large dispersion and skewness in the distribution function of the data collected, both for direct and for indirect sampling methods. This is especially evident when the fish assessed is schooling (Aglen, 1994). The confidence limits of the estimates when using classical statistics may therefore be very large, but are usually supposed to conform to the central limit theorem, and therefore the two first moments (mean and variance) of the distribution function are still finite.

However, there are examples where the distribution function of the data sampled does not have these properties, but instead an infinite variance (Levy, 1925). In such a case, an increase of the sampling rate has a limited effect on the precision of the estimator. The distribution function of data from 18 acoustic surveys in tropical areas resembles that of a Pareto distribution with an infinite variance (Fréon *et al.*, 1993c). Even though curve fitting using the likelihood approach was not perfect for the highest values of biomass, the results suggest that the arithmetic mean of the samples may not give a good estimate of the population.

9.4.2 *Target strength*

To convert the echo integral over a specific depth interval and distance sailed it is necessary to know the backscattering cross section (σ) of the fish. In classic literature of underwater acoustics this quantity is usually referred to as the target strength (TS) of the fish, and is defined (Urick, 1983) as $TS = 10 \log_{10} (\sigma/4\pi)$. The target strength of fish can be measured experimentally on tethered, anaesthetised individuals (Nakken & Olsen, 1977), on live fish in cages (Edwards & Armstrong, 1983) or *in situ* by the dual beam or the split beam technique (MacLennan & Simmonds, 1992). The various methods for target strength measurement of fish and the results obtained are reviewed by Foote (1987).

Social aggregation affects the orientation of fish. The tilt angle distribution is supposed to become narrower and the mean more horizontal when fish are schooling, compared to when they are shoaling. The result is probably a higher average target strength for fish schooling than for fish shoaling (Foote, 1980b; 1981). This indicates a diurnal variation in target strength which has been observed for caged cod, herring and mackerel (Edwards & Armstrong, 1983; MacLennan *et al.*, 1989, 1990; MacLennan, 1990). The variation is especially large (about 5 dB) for Atlantic mackerel because this swimbladder-less species tilts upwards at low swimming speed when shoaling at night to keep an upward hydrodynamic lift (He & Wardle, 1986).

According to these findings, different target strength values should be applied for fish that school during the day and shoal at night during acoustic surveys. This is not yet practised because the validity of applying cage measurements to wild fish is questioned (MacLennan *et al.*, 1990). However, there are several indications of diurnal changes in social aggregation patterns of fish that affect their orientation and thereby target strength. Tilt angle measurements of herring in a large net cage (Beltestad, 1974) and when free swimming (Buerkle, 1983) showed a change from slightly negative tilt during daytime to slightly positive tilt during night-time. For walleye pollock (*Theragra chalcogramma*), the target strength has been observed to be 3 dB lower by night than by day (Traynor & Williamson, 1983).

Due to the difficulties in measuring target strength in dense aggregations *in situ*, the necessary information to differentiate target strength between the different social aggregation patterns is at present lacking (Foote, 1987). One possibility to overcome the problem is to measure the tilt angle of fish when passed over by a survey vessel, both when schooling and when shoaling, by using a special tag (Mitson & Holliday, 1990). The average target strength can be calculated according to the swimbladder morphometry of the species and the tilt angle distribution observed (Foote, 1985). This may also be done directly from the tilt angle distribution if the scattering properties of the species are known (Foote, 1980b).

An independent estimate of the target strength of fish in shoals or schools may be obtained by the comparison method. The principle is first to transect a shoal or a school with an echo integration system to obtain an estimate of the area backscattering cross section (s_A), and then capture a known part of the shoal or the whole school by purse seine to obtain an independent estimate of fish density. The backscattering cross section (σ) can then be found by rearranging equation (9.3). The method has been applied to obtain estimates of target strength of herring both when shoaling at night (Hagstrøm & Røttingen, 1982; Hamre & Dommasnes, 1994) and when schooling during the day (Misund & Beltestad, 1996). At present, the variation in the data collected by this method is too large to reveal any systematic difference in target strength when shoaling or schooling.

9.4.3 *Acoustic shadowing*

The conventional echo integration method has been experimentally proven to be linear (Foote, 1983), which means that the acoustically measured density and the real density are strictly proportional. This principle was found to be valid for measurements of herring about 27 cm long in a $0.7\,m^3$ cage at densities up to about 60 fish m^{-3} (Foote, 1983). Free-swimming herring of this size usually organise schools at an average density of about 2 fish m^{-3} (Misund, 1993b). Røttingen (1976) observed a linear proportionality between echo intensity and fish number up to a density of about 100 individuals m^{-3} for saithe about 35 cm long and about 2000 individuals m^{-3} for sprat about 10 cm long held in a net cage of about $2.7\,m^3$. For higher densities, the linearity decreased steadily and even declined for extremely high densities. For the saithe measurements, Foote (1978) was able to deduce the mean extinction cross section of the fish. Thereby the observed relationship between echo intensity and fish density could be reproduced fairly accurately as a function of the backscattering cross section and the extinction cross section (Foote, 1978).

However, in natural fish aggregations that have a high density and a large vertical extent, substantial extinction of the sound energy emitted from an acoustic transducer can occur (Lytle & Maxwell, 1983). This is often observed during surveys by a weakening, and in some cases disappearance, of the bottom echo under large dense schools. Olsen (1986) quantified the effect of shadowing by a fish school on the echo energy of a calibrated sphere. The energy decreased to 15% of its nominal value when the fish school passed between the transducer and the sphere. Similarly, Armstrong *et al.* (1989) demonstrated a linear relationship between the acoustic cross section or density of fish in a cage and the extinction of the echo energy of a reference sphere below the cage. Appenzeller and Legget (1992) suggested that sound extinction when fish occur in dense schools may cause an underestimation of biomass by up to 50%. For acoustic stock assessment the consequence of such extinction is underestimation of the fish density, but still the phenomenon has been neglected in survey work as adequate methods to compensate for the extinction have been lacking.

As an additional theorem to the linearity principle, Foote (1983) suggested an equation involving knowledge of the fish density and extinction cross section to compensate for the extinction. Similarly, Lytle and Maxwell (1983) proposed three approximations to correct estimates of high density, all involving knowledge of the extinction cross section. As this parameter is difficult to quantify at sea, Olsen (1986) suggested that the extinction problem might be overcome by measuring the packing density in a small upper part of the school and applying this as the average density for the whole school. The validity of this correction might be affected by an increase in packing density of subsurface schools when passed over by a research vessel (Gerlotto & Fréon, 1992). By recording the attenuation in the bottom echo as a function of the vertical extent of herring aggregations in a Norwegian fjord (Fig. 9.1), Toresen (1991) was able to fit equations to correct the fish density estimates for extinction. This method can be used to derive the extinction cross section of fish *in situ* (Foote *et al.*, 1992). Foote (1990) has extended his extinction correction solution for application to aggregations which vary in density both horizontally and vertically. This solution can

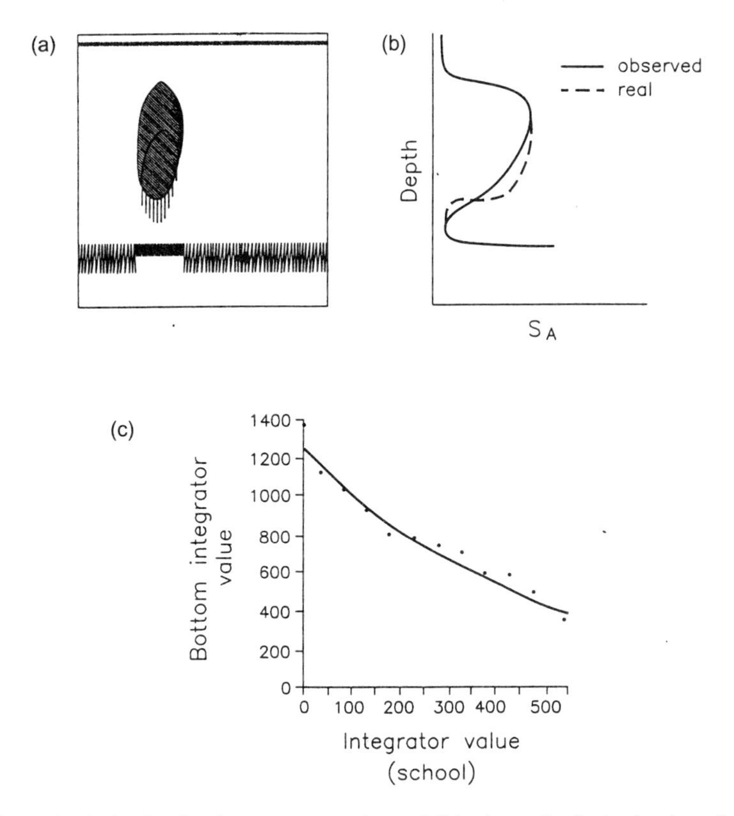

Fig. 9.1 Acoustic shadowing by dense concentrations of fish. Acoustic shadowing by a large herring school will often result in (a) a reduced bottom echo under the school and (b) a reduced area backscattering strength throughout the school. (c) Proportional reduction of the bottom echo integrator value with increasing density of herring in the water column (after Toresen, 1991).

be implemented on modern postprocessing equipment (Knudsen, 1990; Foote *et al.*, 1991), and used to compensate for the extinction during postprocessing of the echo signal data.

The algorithms suggested in the preceding paragraph have been applied to quantify the influence of extinction on acoustic abundance estimates. Gerlotto (1993) estimated the effect of acoustic shadowing during a survey in Venezuela to just 4% for the total biomass of schools by applying the correction factor proposed by Olsen (1990) on the basis of an equation given by Foote (1983). Misund (1993b) found that the packing density of herring schools recorded in fjords in northern Norway increased by a factor of 1.3 in average and maximally by 2.4 when applying the correction derived by Toresen (1991). Røttingen *et al.* (1994) found that abundance estimates of herring overwintering in Ofotfjord-Tysfjord in northern Norway increased by factors of about 1.4 when applying the extinction correction algorithms suggested by Foote *et al.* (1992) and Foote (1994). Clearly, these results demonstrate that extinction may significantly influence acoustic abundance estimates of pelagic schooling fish, and thus should be taken into account when calculating acoustic abundance estimates of such fish stocks.

9.5 Effects of avoidance

Avoidance is a general term for behavioural patterns that fish perform to move away from stimulation that elicits fright or is sensed as unpleasant. During stock assessment, avoidance can be significant if the vessel, the monitoring instruments or the sampling gear generate stimuli to which fish react in a way that causes the measured density to be lower than the actual, or the recorded composition of species and distribution of size groups to be different from the real ones.

9.5.1 *Methods to study vessel avoidance*

The behaviour of a fish in its natural habitat is usually not easy to observe, especially when the fish is at some depth offshore. An additional difficulty when observing vessel avoidance reactions may be the substantial dynamic which may occur during such situations. Avoidance reactions may alter the distribution of fish completely and induce movements that may rapidly bring the fish far from the position of observation. As a result, visually based observation techniques are usually inadequate for studying vessel avoidance. Together with the need for night-time observations, this has encouraged development of non-visual observation methods.

Due to the reflection properties of fish and the great range and volume coverage at very short intervals, the use of underwater acoustics has proved successful for observations of fish behaviour in the field (Mitson, 1983). To study vessel avoidance of herring shoals during night-time, Olsen (1979) developed a technique based on submerged, downwards-or upwards-directed transducers connected to a small observation vessel positioned some distance away. The position of the transducers was marked by a buoy with a light signal. When herring shoals were detected underneath or above the transducers, a large research vessel was ordered to pass in a straight course as close as possible to the marker buoy. The reactions of the fish when the survey vessel approached and passed over could then be deduced from the recordings by the transducer. This method has been applied in sheltered waters in fjords, and in the Barents Sea in good weather conditions (Olsen *et al.*, 1983a).

A more convenient version of the method when dealing with fish that occur in more dynamic distributions is to position the transducer directly under a small vessel. The technique requires only a small transducer that is easily handled overboard. This takes advantage of the mobility of a small observation vessel to search for a favourable position relative to the actual fish distribution. When the small vessel has taken up a favourable position over the fish, the survey vessel can be ordered to take up a straight course so that it passes close to the small vessel. The behaviour of the fish as the survey vessel comes closer and passes, can then be monitored directly from the small vessel. This technique has been applied on distributions of cod and haddock in the Barents Sea (Ona & Chruickshank, 1985; Ona & Godø, 1990). In the Caribbean, the technique has been used to study the avoidance behaviour of pelagic species occurring both in shoals and in distinct schools (Fréon & Gerlotto, 1988a,b; Fréon *et al.*, 1990, Gerlotto & Fréon, 1990, 1992). The latter was possible because the schools were visible from the surface so that the observation vessel with the transducer could be manoeuvred into a position straight on top of schools even if they were moving. By maintaining this position relative to the school, the behaviour of the fish during the passage of the survey vessel could be recorded.

Another method for observing changes in distributions of fish when passed over by a survey vessel is to use a narrow-beamed sector-scanning sonar. Such equipment is best mounted and operated from the survey vessel itself due to the many components involved and the requirements of electric current. Ona and Toresen (1988a) mounted the mobile transducer of such a sonar on various positions on the hull of a research vessel when surveying herring concentrations in a Norwegian fjord. The transducers were directed so as to give a vertical section of the water column. By this method it was possible to record and compare the distributions of fish underneath and to the sides of the vessel. Similar observations are possible with a portable multi-beam sonar of high frequency (455 kHz) which gives the advantage of high resolution and simultaneous observation of a vertical plane from 0 to 100 m from the vessel (Gerlotto *et al.*, 1994). By combining several successive such cross-sections, three-dimensional presentation of fish distribution in relation to the survey vessel can be obtained (Plate 2, opposite page 21).

Fish swimming in schools out of visible range of the surface are best observed by horizontally guided sonar (with the transducer mounted directly on the hull of the research vessel). The behaviour of the schooling fish can thereby be studied from on board the vessel itself. The sonar can be omnidirectional or multibeam, thereby covering a large volume in each ping. This ensures that quick movements of the school are detected. The whole horizontal extent of a school will also be projected so that changes in the external appearance of the school in vessel avoidance situations can be investigated. Some sonars even have an additional beam fan so that the vertical extent and position in the water column of the school can be recorded (Ona, 1994). However, for multibeam sonars that operate with rotational directive transmission, a blind zone is created at close range to the vessel (< 50 m) and the reactions of the school just before the vessel passes over are not observed. Several sonars can display the movements both of the school and the vessel in true motion (Bodholdt & Olsen, 1977). This enables detailed quantification of the avoidance behaviour. The horizontal and vertical swimming speed of the schools can be measured fairly accurately. To describe the avoidance behaviour, parameters that relate the movement of the school to that of the approaching vessel, such as the radial swimming speed, are easily estimated.

A horizontal guided sonar can also be used to measure the Doppler shift in returning echoes from fish schools (Holliday, 1977). This can be useful for avoidance studies because the frequency shift of the returning echo will reveal whether the fish are avoiding the approaching vessel or not. If the fish are avoiding the ship, the frequency shift will be positive, but if the fish is approaching the vessel the frequency shift will be negative. The swimming speed of the fish towards or away from the vessel can be calculated from the frequency shift. The method may also be used on echo-sounders to measure the vertical swimming speed of fish underneath the vessel, as done by Olsen *et al.* (1983a) on capelin concentrations in the Barents Sea.

Studies of the avoidance behaviour of individual fish are more difficult with conventional acoustic equipment. However, they are partly possible with split-beam sounders using measurement of the phase angle to position the returning echoes in the acoustic beam (Brede *et al.*, 1990). This requires that there be some space between individuals within the acoustic beam to obtain the resolution necessary for positioning of single targets. If the distribution is dense, the resolution can be increased by lowering the split-beam transducer closer to the fish. The split-beam sounders enable tracing of the swimming behaviour of individual fish as they move through the beam (Ona, 1994). Nevertheless, to follow the movements of single, identified individuals through the whole avoidance behaviour phase during passage by a research vessel will still be difficult.

Another alternative method for studying the avoidance behaviour of individuals is to use acoustic tags on the fish. Then it is possible to trace the movement of the fish by recording the position of the tag through a system of passive listening buoys or a hull-mounted sonar (Mitson, 1983). There are several techniques to place the tags on the fish. It can be done simply and safely by catching the fish first, and then mounting the tags internally or externally before releasing the fish again. A disadvantage of this technique is that the fish can be exposed to such stress during the capturing and marking process that it will modify its behaviour strongly to avoid coming into a similar situation during subsequent experiments ('one-trial learning', see Chapter 7). Fish may be kept unaffected by the tagging process by putting the tag into food that they ingest in their natural habitat. However, to base an investigation on this method can be a time-consuming process because it is very dependent on appetite. Moreover, this method cannot be applied to plankton feeders.

9.5.2 *Observations of vessel avoidance during surveys*

According to the avoidance behaviour model of Olsen *et al.* (1983a), fish will react to the noise stimuli by increasing speed and swimming radially with a downward component away from an approaching vessel. Such swimming behaviour may result in a horizontal dilution of fish density in front of the vessel, and a reduction in target strength both due to compression of the swimbladder if rapid diving occurs, and because of an increase in the tilt angle (Ona, 1990). The total effect may be a substantial underestimation of fish density as recorded during the conventional acoustic surveys.

Echosounder observations from an independent small vessel showed that polar cod (*Boreogadus saida*) in the Barents Sea avoided at about 150 m in front of an approaching survey vessel (Olsen *et al.*, 1983a). It is possible that this species is very sensitive to vessel sound, because similar investigations on other species indicate little or no avoidance more than 50 m in front of research vessels running at normal survey speed (about 5 m s^{-2} or 10 knots). This is concluded from a large number of recordings on cod and haddock (*Melanogrammus aeglefinus*) in the Barents Sea (Olsen *et al.*, 1983a; Ona & Chruickshank, 1985; Ona & Godø, 1990). Similarly, there have been no indications of changes in the distributions of shoaling herring at some distance in front of the approaching survey vessel during experiments in Norwegian fjords (Olsen, 1979, 1980, 1990; Ona & Toresen, 1988b). The same pattern with no detectable changes in the fish concentrations in front of survey vessels has also been found during studies with echosounders operated from an independent vessel on small clupeoids in the Caribbean Sea (Fréon & Gerlotto, 1988a,b; Fréon *et al.*, 1990; Gerlotto & Fréon, 1990, 1992).

It is possible that recordings with echosounders on fish concentrations do not reveal slight changes in swimming behaviour of the individual fish as a survey vessel approaches. Recordings made by omnidirectional or multibeam sonars indicate that horizontal dilution of the density of schooling fish more than 50 m in front of survey vessels may occur depending on species and geographic area. Goncharov *et al.* (1989) observed that schools of jack mackerel (*Trachurus symmetricus*) increased the swimming speed remarkably linearly within an interval of about 350 to 50 m in front of a large, approaching vessel. This pattern was observed in three different areas (Fig. 9.2). There were clear differences in the reaction distance among the three areas, but the slope of the increase in swimming speed with decreasing distance from vessel to school was much the same. Goncharov *et al.* (1989) argued that differences in the sea temperature caused the variations in reaction distance among the three areas. Another possibility is that the conditions for sound propagation varied so that the fish became aware of the vessel at different distances in the three areas.

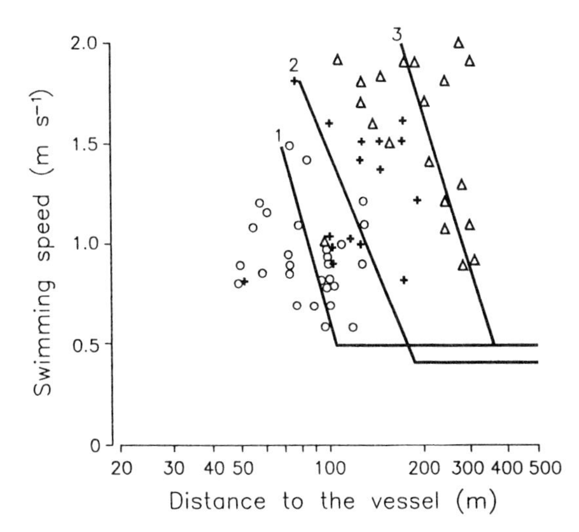

Fig. 9.2 Swimming speed of jack mackerel schools as a function of distance to an approaching vessel in three adjacent areas denoted 1, 2 and 3 (after Goncharov *et al.*, 1989).

By use of an omnidirectional sonar, Diner and Masse (1987) observed that some schools of herring in the English Channel, and sardine and anchovy in the Bay of Biscay, tended to avoid horizontally at about 150 m in front of an approaching survey vessel. However, most of the schools seemed not to avoid the path of the vessel. Misund *et al.* (1996) found that the distance of first reaction of herring schools in the North Sea varied from 50 m to 1000 m (Fig. 9.3). The schools that reacted were distributed within 20° of each side of the bow.

Schools of herring (*Clupea harengus*) and sprat (*Sprattus sprattus*) in the North Sea that reacted to the survey vessel tend to be guided by the approaching vessel (Fig. 9.4). It is suggested this is due to a supposed noise directivity pattern of the vessel. Due to

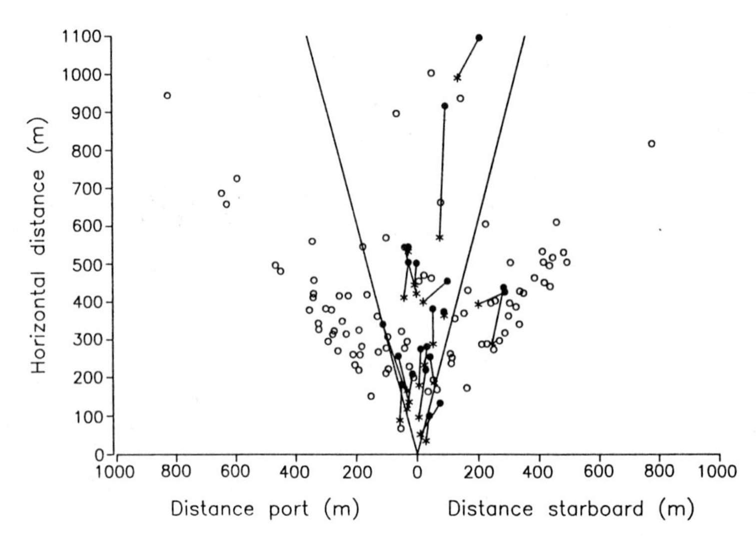

Fig. 9.3 Distance and direction to herring schools in the North Sea at first detection by a Simrad SR240 sonar onboard RV *Johan Hjort*. (○) schools not avoiding the vessel; (●) schools avoiding the vessel; (*) position when school reacts to the vessel (after Misund *et al.*, 1996).

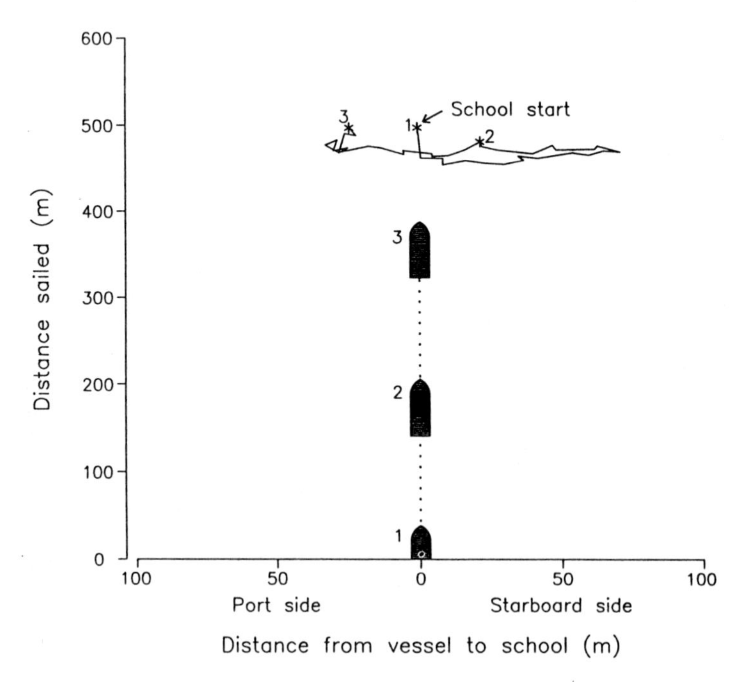

Fig. 9.4 Swimming patterns of herring school presumably reacting to the sound field of RV *Johan Hjort* when approaching at 11 knots (after Misund *et al.*, 1996).

shadowing by the hull, there will be lobes of maximum intensity to each side of the vessel (Fig. 5.2) as suggested by Gerlotto and Fréon (1990) by combining results from Misund (1990) and Urick (1983), and confirmed by Misund *et al.* (1996). Just in front of the vessel there will be a minimum intensity. Fish that sense and react to the increasing sound intensity in front of an approaching vessel will thereby be guided in front of it. However, there seemed to be no relation between swimming speed, avoidance pattern and distance between vessel and school in the interval from about 600 m to 50 m in front of the vessel (Misund & Aglen, 1992). Schools of spawning migrating capelin (*Mallotus villosus*) showed no avoidance in front of a survey vessel (Misund *et al.*, 1993).

Just before, and as the vessel passes over, the avoidance model (Olsen *et al.*, 1983b) predicts sudden escape reactions. This can affect acoustic stock assessment, both through a horizontal dilution of fish density in the path of the acoustic beam and by a lower target strength than predicted, both because the fish is tilted downwards (Nakken & Olsen, 1977) and due to a compression of the swimbladder during fast diving (Ona, 1990). The strength of the reactions depends upon the depth of the fish, the size of the vessel, and the vessel speed.

Echosounder recordings from small, independent vessels have shown that this is the case for shoaling herring in Norwegian Fjords, and cod, polar cod, and capelin in the Barents Sea (Olsen 1979, 1980, 1990; Olsen *et al.*, 1983a). The resulting reductions in density underneath the survey vessels have been quantified to vary from 40% to 90%. In accordance with the avoidance behaviour model, the reactions seemed very dependent on the swimming depth of the fish. This is clearly seen from Table 9.1, in which the observations of herring are incorporated into a depth dependent factor for correcting the acoustic fish density estimate for avoidance behaviour. That the avoidance reaction depends on the vessel size (and engine power) is deduced from comparison of reactions of herring to the 300 GRT RV *Johan Ruud* and to the 900 GRT RV *G.O. Sars* (Olsen *et al.*, 1983a). How the speed of the vessel may influence

Table 9.1 Estimated avoidance of herring to a 300 tonnes research vessel running at 10 knots (after Olsen *et al.*, 1983b). Mean tilt angle (β_D), ratio of acoustic backscattering cross section disturbed/undisturbed ($k_{\sigma D} = \sigma_{\beta D}/\sigma_{\beta 0}$), ratio of fish density disturbed/undisturbed ($k_{\rho D} = \rho_D/\rho_{D0}$), and ratio of relative acoustic backscattering strength disturbed/undisturbed ($k_{MD} = k_{\sigma D} \cdot k_{\rho D}$).

Depth (m)	β_D (degree)	$k_{\sigma D}$	$k_{\rho D}$	k_{MD} * 100 (%)
10	63	0.40	0.16	0.6
20	48	0.30	0.41	12
30	42	0.44	0.67	29
40	40	0.50	0.92	46
50	33	0.61	1.00	61
60	23	0.78	1.00	78
70	15	0.91	1.00	91
80	9	0.98	1.00	98
90	5	0.99	1.00	100
100	2	1.00	1.00	100

the avoidance reactions has been studied by Doppler shift measurements of capelin when approached by a survey vessel (Olsen *et al.*, 1983b). The vertical swimming speed increased more or less proportionally with increasing vessel speed.

That shoaling herring avoid a passing survey vessel has been confirmed by recent experiments on large herring wintering at depths from 40 to 400 m in fjords of northern Norway (Vabø *et al.*, in pre.). Independent echo integration from a stationary transducer submerged from a float and connected to a small observation vessel, revealed depth dependent avoidance to the 2000 GRT vessel RV *Johan Hjort* when passing at survey speed (8–11 knots). For herring distributed between 40 and 100 m depth, the avoidance reactions resulted in a loss of echo energy in the order of 80–90%. The effect of the avoidance reactions decreased with depth below 100 m, and was insignificant below 150 m depth. The results suggested that the avoidance reactions above 150 m could cause an underestimation of the biomass of herring in the area by about 50% for an acoustic survey run day and night by the specific vessel.

Other investigations on cod, haddock and herring in the same areas revealed no or just weak avoidance reactions to fisheries research vessels running at normal survey speeds. By echosounder recordings from a small observation vessel, Ona and Godø (1990) detected no avoidance reactions of cod and haddock that were approached and passed over by a survey vessel. When recording herring concentrations by a vertically scanning sonar mounted at various positions on the hull, Ona and Toresen (1988a) detected no changes in the herring distributions when the sonar transducers were mounted on the bow and just in front of the echosounder transducers. This indicates that the herring performed no avoidance reactions that significantly affected the biomass estimation by conventional acoustics.

Similar studies on clupeoid species in shallow, tropical areas showed no reduction in fish density during passage of small survey vessels either when sailing or motor driven (Fréon *et al.*, 1990; Gerlotto & Fréon, 1990, 1992). Schools of *Sardinella aurita* off Venezuela avoided vertically, but only in the upper 20 m when passed over by a medium sized survey vessel. This resulted in only slight underestimation of the density as the average tilt angle during the dive was estimated to be less than 10° (Gerlotto & Fréon, 1992).

This significance of avoidance studies using an echosounder from a small observation vessel would have been easier to interpret if recordings from the survey vessel during the experiments were also shown, and compared with the recordings from the independent vessel. In future experiments, this should be done by using identical, calibrated echosounders on both vessels, and applying small echo integration intervals horizontally and vertically so that eventual avoidance reactions can be quantified.

Recordings by omnidirectional and multibeam sonars show that both horizontal and vertical avoidance of schools can occur close to the vessel. Schools of anchovy and sardine in the Mediterranean seemed to avoid laterally or compressed vertically when close to an approaching survey vessel, and fewer schools than expected were recorded in the simulated sample volume of a vertically directed echo sounder (Soria *et al.*, 1996). About 15% of the herring schools in the North Sea that occurred in the path of the survey vessel avoided horizontally when closer than 50 m in front of the vessel and were not recorded by the echosounder (Misund & Aglen, 1992). The schools that avoided were smaller and swimming shallower than those that were recorded by the echosounder (Misund *et al.*, 1993). Depending on the area, 35–65% of the jack mackerel schools observed by Goncharov *et al.* (1989) avoided horizontally and dived about 30 m when approached by a large vessel. Diving of herring schools has also been observed when they were passed over by purse seiners (Misund, 1990). However, all schools of spawning migrating capelin that occurred in the path of the survey vessel were recorded by the echosounder (Misund *et al.*, 1993).

For better quantification of the underestimation of biomass due to avoidance of schools, sonars with better resolution and ability to quantify school size are needed. As underestimation due to vessel avoidance reactions may vary temporarily and spatially, such instruments could be used as a basis in methods for continuous *in situ* monitoring of vessel avoidance reactions of schools (Misund *et al.*, 1993; Gerlotto *et al.*, 1994; Soria *et al.*, 1996).

As vessel avoidance is claimed to be the outcome of vessel sound stimuli and natural stimuli acting on the fish (Balchen, 1984), substantial variation may be expected. Several biotic and physical factors may influence the response threshold and response strength to vessel sound stimuli. Neproshin (1979) observed that the distance at which Pacific mackerel (*Pneumatophorus japonicus*) reacted to approaching trawlers varied substantially with time of day. The reaction distance was less in the morning and the evening than in the middle of the day. This was explained by arguing that jack mackerel avoids most when well fed and swimming in schools in the middle of the day, and least when shoaling and feeding in the morning and evening. When the Pacific mackerel occurred in low-temperature water, the reaction distance was less than when the fish were swimming in warmer water.

Avoidance behaviour may also depend on the life history stage of the fish. Herring have been observed to react more strongly to an approaching vessels when on a spawning migration than when hibernating, feeding or on a feeding migration (Misund, 1990). This is indicated from a greater swimming speed horizontally, relative to the approaching vessel, and downwards, by herring that are on a spawning migration compared with herring in other life stages (Table 9.2).

Herring have been observed to react less to approaching vessels, the better the sonar conditions (Misund & Aglen, 1992). This is concluded from a lower horizontal and relative swimming speed by herring schools occurring in areas with good conditions, compared with areas with worse conditions (Table 9.3). The sonar conditions

Table 9.2 Swimming speeds of herring school when approached by a vessel (L, average fish length; V_H, Horizontal speed; V_{RH}, horizontal speed relative to vessel; V_V, vertical speed, negative if downwards; after Misund, 1990).

Life stage	Region	L (cm)	V_H (m s^{-1})	V_{RH} (m s^{-1})	V_V (m s^{-1})
Feeding/migrating	North Sea	27.2	1.26	0.31	0.00
Hibernating	Norwegian coast	35.5	1.49	0.13	0.02
Prespawning	Norwegian coast	37.3	1.94	0.79	–0.27

Table 9.3 Influence of sound propagation conditions on the swimming behaviour of herring when approached by a survey vessel in the North Sea. (V_H = horizontal swimming speed, V_{RH} = horizontal swimming speed relative to the vessel. After Misund and Aglen (1992)).

Sonar conditions	Detection range (m)	V_H (m s^{-1})	V_{RH} (m s^{-1})
Very bad	< 400	1.10	0.88
Bad	(400–700)	1.21	0.88
Good	(700–1000)	1.00	0.73
Very good	> 1000	0.94	0.68

in an area reflect the properties for sound propagation. If there is no thermocline, there will be linear sound propagation (Smith, 1977) and good sonar conditions. If there is a thermocline, the horizontally emitted sound waves will bend upwards or downwards and the sonar conditions will be more difficult. During the investigation on herring in the North Sea, there were some areas with warm surface water but colder water below about 30 m to 40 m deep. This thermocline could cause a drastic downwards bending of the sonar sound waves, and consequently a reduced detection range of schools. The sonar conditions in different areas were therefore categorised according to the detection range of schools, which is proportional to the degree of sharpness of the thermocline and thereby the downwards sound-ray bending. The reason that this categorisation also groups the avoidance behaviour of the herring schools is that the emission of noise from the vessel follows the same physical laws as those of the sonar sound weaves. If the sonar conditions are bad, the noise from the vessel does not proceed very far due to downwards bending of sound waves. The herring will then detect the vessel at a much shorter distance, than if the conditions for sound propagation were better. Probably the herring react more strongly to an approaching vessel during such situations than if they had detected the vessel at a long distance and experienced a gradually increasing noise intensity.

9.5.3 *Avoidance of vessel light*

At night, avoidance can occur as a response to visual stimulation by the light of the vessel. This is well known to fishermen, and purse seiners fishing on dense shoals are very careful in using light during circling, shooting and pursing of the seine. In most cases only the necessary regulation lights are used. The consequences for stock assessment by acoustics because of avoidance reactions due to the light of survey vessels may be the same as for avoidance reactions to vessel noise. Lévénez *et al.* (1990) observed significant vertical avoidance by tropical clupeoids off Venezuela during new moon when running a survey vessel with a 500 W light on the bridge. With regulation lights only, the fish were distributed more or less evenly from surface to bottom at about 50 m. When the light was turned on, the fish descended and concentrated in a narrow depth interval close to the bottom (Fig. 9.5). This distribution pattern could be generated by switching the light on and off. However, the mean echo integral was the same with the lamp on or off. When conducting a similar experiment

(a)

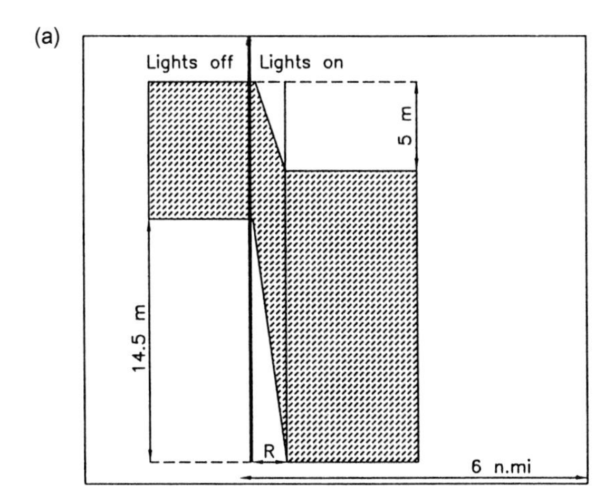

(b)

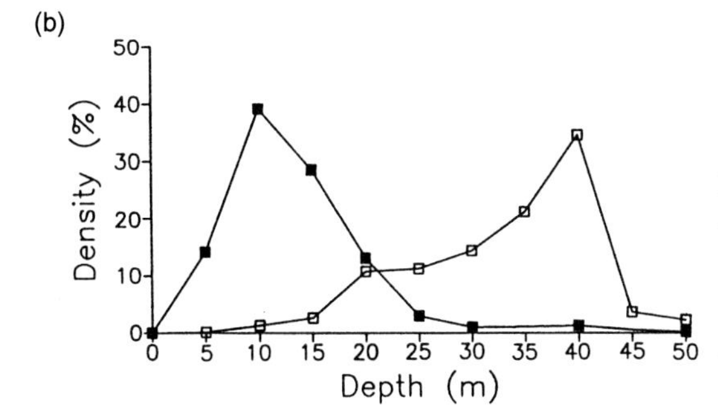

Fig. 9.5 Reaction of pelagic fish off Venezuela, mainly *Sardinella aurita*, to 500 W bridge light aimed forward onboard RV *Capricorne* surveying at 7.5 knots. (a) Change in vertical distribution; (b) relative density distribution with lights off (■) and lights on (□) (after Lévénez *et al.*, 1990).

in the same area, Gerlotto *et al.* (1990) observed no vertical avoidance, but found indications of species-or size-dependent horizontal avoidance when the lamp was on (Fig. 9.6). The last experiment was conducted during cloudy, dark nights, but it was still during a full moon period. This might have had an influence on the fish so that they reacted less to light than during other moon phases. Even if these few observations indicate minor effects on acoustic stock assessment due to avoidance reactions to vessel light, survey vessels should be operated as dark as possible at night-time to minimise the influence of such reactions. It is the experience of Icelandic survey vessels that excess lights from the decks and cabins can cause such strong avoidance reactions of herring that the acoustic measurements may be significant underestimates (Halldorsson & Reynisson, 1983).

9.5.4 *Vessel avoidance during sampling by trawls*

To allocate recordings to species and size groups during acoustic stock estimation, sampling has to be conducted, usually by trawl (Dalen & Nakken, 1983). During this operation, vessel avoidance reactions may make sampling of certain species difficult. Acoustically tagged cod in shallow water (20 m depth) tend to swim in the same direction as the approaching trawler and avoid rapidly sideways when close to the vessel (Engås *et al.*, 1991). Shoaling herring avoided horizontally, and were not

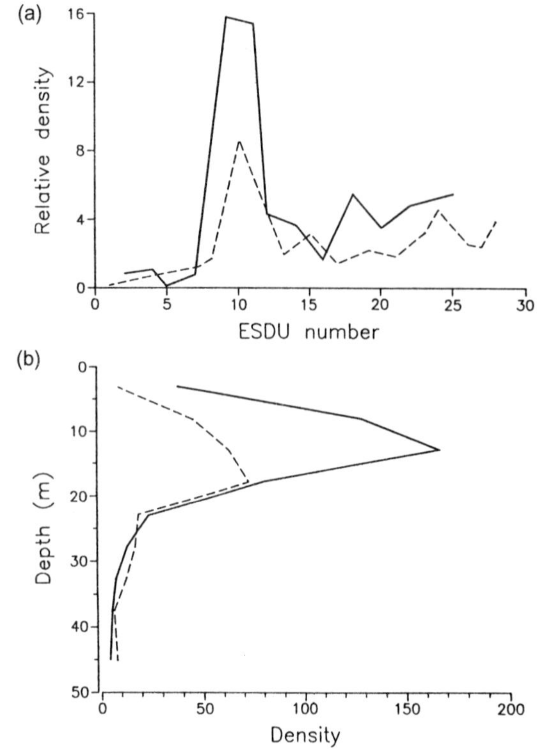

Fig. 9.6 Reaction of pelagic fish, mainly *Sardinella aurita*, in Gulf of Caraico, Venezuela, to a 500 W lamp fixed above the towed transducer body of a 25 m stern trawler running at 5 knots. (a) relative fish density for each elementary sampling distance unit (ESDU); (b) mean relative density versus depth. (- - -) light on; (—) light off (after Gerlotto *et al.*, 1990).

available to a pelagic trawl towed directly in the path of the vessel (Ona & Toresen, 1988b). Schooling herring may perform strong horizontal and vertical avoidance reactions, and precise sonar-guided pelagic trawling is often necessary for successful sampling (Misund & Aglen, 1992; Suuronen *et al.*, 1997). The experience from pelagic trawl sampling in the North Sea is that the herring schools are most difficult to catch when they occur at midwater and avoid both horizontally and vertically. Schools that are recorded at midwater when passed over by the survey vessel often avoid downwards, and can often be found at the sea bottom when passed over by the trawl. However, by taking advantage of this behaviour and lowering the trawl so that the ground rope touches the bottom, the sampling success on such herring schools can be substantially increased.

Vessel avoidance reactions during sampling may also affect acoustic stock estimation indirectly by altering the distribution of species and size groups available to the gear. Similarly, such reactions may cause difficulties in combining the results from swept-area and echo integration estimates of semipelagic fish (Nunnallee, 1991). When exposed to playback of the noise of an approaching trawler, penned cod polarised and dived slowly, but the reactions varied substantially (Engås *et al.*, 1991). Ona and Chruickshank (1985) and Ona and Godø (1990) observed that the avoidance reactions of cod and haddock were stronger when the survey vessel was towing a trawl, and that diving made the fish more available to bottom trawl and less available to pelagic trawl. Similarly, the *in situ* target strength of Cape horse mackerel (*Tra-*

churus capensis) shifted downwards by about 12 dB during trawling, indicating increased tilt angle and downwards avoidance (Barange & Hampton, 1994). For Pacific whiting (*Merluccius productus*), Nunnallee (1991) observed no change in vertical distribution, but a higher echo value probably due to a change in the orientation of the fish when a trawler passed over.

From studies made by the 1043 GRT RV *Eldjarn*, Misund and Aglen (1992) found that the North Sea herring maintained the same horizontal swimming speed when approached by the vessel when either cruising or trawling (Table 9.4). However, the swimming speed relative to the vessel increased during trawling. This indicates an increased and more precise avoidance of the vessel when trawling than when cruising. A trawling vessel tends to produce more cavitation, which generates a high intensity of low-frequency noise (Chapman & Hawkins, 1969). It is therefore quite probable that the more definite avoidance behaviour of the North Sea herring during trawling was a response to a more intense, low frequency noise from the vessel. Nevertheless, possible learning of trawl noise cannot be eliminated, and if this is the case the change in avoidance behaviour may be elicited by the changed noise spectrum generated by the vessel when trawling.

Table 9.4 Swimming speed of North Sea herring during cruising and sampling by pelagic trawl. V_H = Horizontal swimming speed; V_{RH}, horizontal swimming speed relative to the vessel; *, when behind vessel the relative swimming speed is related to the trawl. After Misund and Aglen, 1992.

Vessel mode	School position	V_H (m s^{-1})	V_{RH} (m s^{-1})
Cruising	In front	1.06	0.22
Trawling	In front	1.06	0.46
Trawling	Behind	1.01	0.58*

In the region between the vessel and trawl, the horizontal swimming speed of the North Sea herring schools was about the same as that in front of the vessel. Nevertheless, there was a clear increase in the relative swimming speed (Table 9.4), which for the observations in this region was related to the approaching trawl rather than to the departing vessel. The observations were made in the region just in front of and partly between the trawl doors, and the swimming behaviour may therefore have been the result of visual orientation to the front part of the trawl such as the warps, trawl doors and bridles. The mean relative swimming angle is about 40° to starboard. This also supports the view that the swimming behaviour is influenced by visual orientation, because the trawl tends to approach from the port side. This results from a slight starboard turn of the vessel as part of the tactic to catch herring schools. The vessel is then manoeuvred in a slight curve to the side of the school when passing, to try to reduce the effect of the avoidance behaviour.

9.5.5 *Instrument avoidance*

Anecdotal information from the fishermen using the first generation of fisheries sonars to locate herring schools in the Norwegian Sea indicated that the herring sensed and reacted to the sonar sound. Many fishermen therefore used to turn off the sonar on the main vessel when the master fisherman was in position over the school and recorded it through his echo string (a thin copper string with a small lead weight in the end). It is not unlikely that the herring really detected the sound from some of the first-generation sonars because they were emitting at about 10 kHz. Herring detect sounds at least up to about 4 kHz (Enger, 1967), and the possibility exists that they may sense higher frequencies of a high intensity. Another possibility is that sonars produce loud, low frequency sounds due to the high energy transmitted.

Most fish are unable to detect sound in the ultrasonic region (> 20 kHz), where modern fish-detection instruments operate (Hawkins, 1993). Nevertheless, Bercy and Bordeau (1987b) reported that mullets (*Mugil* sp.) increased swimming activity substantially, and that the cardiac rhythms of turbot (*Psetta maxima*) and wrasse (*Labrus bergylta*) changed when exposed to sound from transducers of 38 or 50 kHz in tanks. Measurements of the output of these transducers showed high-intensity, low-frequency components associated with the carrier frequencies, and which were claimed to elicit the responses. It is still doubtful that the sound from acoustic fish-detection instruments affects fish behaviour, and from numerous net cage experiments reactions to fish-detection instruments have not been reported. Similarly, countless observations at sea indicate that fish do not respond to modern echo-sounders and sonars.

9.5.6 *Sampling gear avoidance*

A basic assumption of fisheries acoustics is that samples by trawl are representative of the populations recorded. This principle thereby enables an unbiased allocation of integrated echo intensity to different species and size groups (Dalen & Nakken, 1983). Accordingly, fish are assumed not to react to the approach of sampling gear in a way that alters the species and size distribution present.

There is increasing evidence that the sampling principle is an oversimplification (Engås, 1994; Godø, 1994). Main and Sangster (1981) observed that cod search towards the bottom when approached by a bottom trawl, haddock rise and may escape over the headline in substantial numbers, while whiting tend to aggregate more in the middle of the trawl opening. Such species-dependent reactions may clearly result in a species composition in the trawl catch that is not representative for the area sampled. The reactions of cod differ among size groups so that the smallest fish escape under the ground gear of ordinary sampling trawls and are grossly underestimated in the catches (Engås & Godø, 1989; Walsh, 1991). Buerkle and Stephenson (1990) obtained substantial differences in the length distributions of herring depending on whether the sampling was conducted by a pelagic trawl or a bottom trawl (Fig. 9.7). As the herring occurred in a dense aggregation extending up from the bottom, this result was probably caused partly by different vertical distributions but also by a fish size-dependent reaction to the gear. Gear avoidance seems mainly to be elicited by visual stimuli (see Chapter 5), and the reactions decrease at night-time (Glass & Wardle, 1989).

(a)

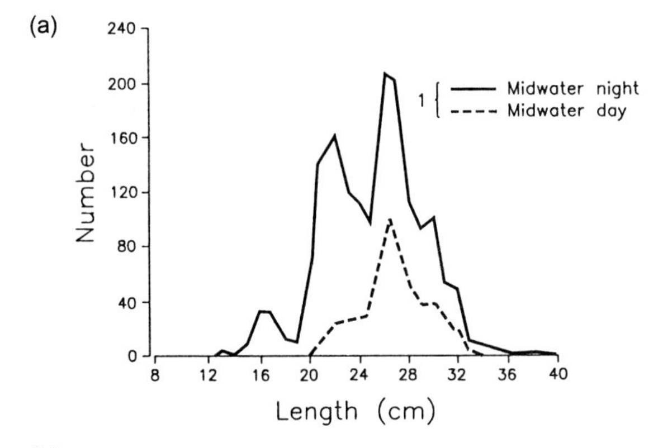

(b)

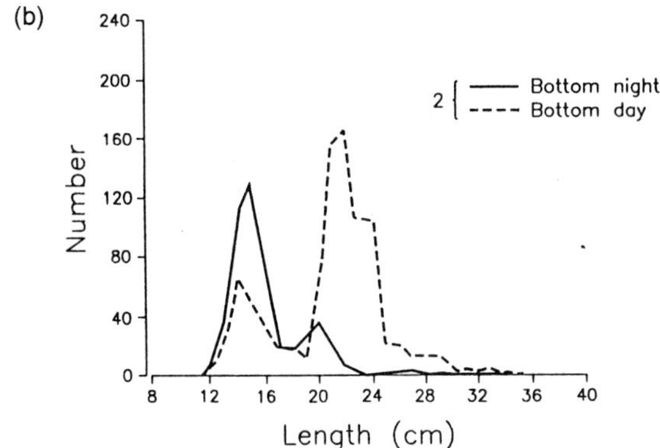

Fig. 9.7 Length distribution of herring sampled by (a) midwater trawl and (b) bottom trawl in Chedabucto Bay, Nova Scotia, Canada (after Buerkle & Stephensson, 1990).

9.6 Replicability of acoustic survey estimates

The very skewed distributions of pelagic stock density measurements by acoustics indicate that there may be substantial variation in population estimates between surveys. In accordance with this, Buerkle and Stephenson (1990) found large variations in biomass estimates from repeated night-time surveys of herring in Chedabucto Bay, Canada. They claimed that the variations were caused by a great dynamic in the aggregations of the population. To some extent this variation may also have been caused by the application of fixed transects so that parts of the stock may have been out of the area, and thus not properly covered from one survey to another.

Aglen (1983) showed that the variability in acoustic survey estimates will decrease with increasing degree of coverage. This is also indicated by Gerlotto and Stéquert (1983), who demonstrated lower variability with shorter distance among transects. These factors have a certain implication for the design of acoustic surveys. If the purpose of the survey is mostly exploratory, a rather large level of variation (50%) can be tolerated (Pope, 1982). If the survey is carried out to get an index for use in time series analysis, a more moderate variation (25–50%) may be allowed. The variation level for a survey designed to yield a population estimate that is to be used to calculate a total allowable catch (TAC) is at most 25% (Pope, 1982).

To reach such a level in precision for acoustic surveys, several criteria must be fulfilled. The surveys must be designed to have a high degree of coverage of the whole fish stock. This requires detailed knowledge of where the fish are at the time of the survey. The species identification and allocation of echo recordings to species must be based on adequate sampling gears that give representative samples (Dalen & Nakken, 1983). The avoidance reactions must be monitored, the plankton contribution limited by proper thresholding, and surveying must be conducted only at an appropriate time of day if necessary. In addition, the survey must be carried out by calibrated, stable instruments, and the target strength of the fish must be known.

It is the experience of many surveys conducted to give absolute population estimates that there is a fairly good precision. Hansson (1993) found that the average coefficient of variation between replicates from one night along a single transect when recording herring in a relatively closed Baltic coastal area was 20%. For replicates of a transect from different nights, the average coefficient of variation increased to 29% due to migration. By use of the delta distribution, MacLennan and Mackenzie (1988) estimated a precision of approximately 25% in acoustic estimates of North Sea herring. Strømme and Sætersdal (1987) observed a variation of less than 20% when repeating surveys of one to seven days duration on pelagic stocks off Senegal and Morocco. Vilhjálmsson *et al.* (1983) showed a high replicability of acoustic estimates of Icelandic capelin, and noticeable deviations were reasonably explained by ice cover or surface schooling. A three-year time series of repeated acoustic surveys of Norwegian spring-spawning herring in fjords of northern Norway show that the abundance estimates may vary by factors of 1.06–2.14 for surveys one to two months apart (Foote & Røttingen, 1995; Foote *et al.*, 1996).

9.7 Conclusion

As described in this chapter, fish behaviour may have a dramatic effect on acoustic abundance estimates. By selecting habitats near the surface or sea bottom, fish may be distributed in the upper or bottom dead zones for conventional echo integration. In the first case the method may be extended by use of horizontal guided sonar, and for fish near the bottom a combination of acoustics and bottom trawling may be applied. Diurnal variation in habitat selection may also influence acoustic abundance estimates by affecting the swimbladder volume and thereby the reflecting properties of the fish.

The target strength of the fish is dramatically influenced by fish orientation. The tilt angle distribution of fish schooling is supposed to be narrower than for fish shoaling, and the target strength may change accordingly. At present, this variation is not taken into account in acoustic surveys. Social aggregation resulting in high densities of fish may cause acoustic shadowing, an effect which is quantified and possible to correct when surveying herring.

Low frequency vessel noise may elicit avoidance reactions to the approaching survey vessel, and cause significant underestimation of fish density. In front of the vessel, herring schools seem to be guided towards a noise minimum in the vessel path caused by shadowing by the vessel hull. Near the vessel, strong sideways and downwards avoidance reactions may cause substantial underestimation of fish density due to both dilution and unfavourable tilt angles. At night, lights from the vessel may also generate avoidance reactions. However, avoidance reactions seem to vary according to factors such as local hydrographic conditions, time of day, maturation stage, between areas, or when conducting trawl sampling, but often fish do not seem to react to the presence of survey vessels at all.

Generally, there is a fairly good precision of acoustic abundance estimates, but variations by a factor of about 2 have been observed for surveys on the same stock conducted only 1–2 months apart.

Chapter 10
Other Methods of Stock Assessment and Fish Behaviour

In this chapter we shall focus on fishing gear surveys (trawl, purse seine, beach seine and long-line), on aerial surveys, on ichthyoplankton surveys and finally on capture–recapture surveys. Underwater visual surveys (scuba, snorkelling or underwater camera) are not adapted to the survey of large pelagic stocks and therefore are not analysed in this book.

10.1 Fishing gear surveys

10.1.1 *Methodology and assumptions*

The principle of this method is similar to any direct method of determining abundance, i.e. based on the extrapolation of density (or density index) observed per unit of area, using a given sampling gear, to the entire survey area (review in Gunderson, 1993). The main assumptions in a trawl survey (note that they are close to those of an acoustic survey and can be used for an aerial survey with slight modification) are:

(1) The survey covers the whole area of the stock distribution
(2) The individual will be sampled a single time and if migration occurs during the survey, the mean speed of migration must be negligible compared with the speed of progression of the survey in the same direction
(3) The target species is completely vulnerable to the fishing gear
(4) The area sampled by the fishing gear is known exactly (absolute abundance estimation) or is constant from one survey to the next (relative abundance estimation)
(5) The sampling sites are selected randomly, or systematically if they are representative and if appropriate statistics are used for computing the variance.

The advantage of fishing gear surveys over indirect methods based on commercial fishery data is that it is relatively easy to satisfy assumptions (1) and (5) owing to the full control of the fishing sites. Nevertheless, in the case of drastic changes in habitat selection, unexpected failure of assumption (1) may occur. To benefit from one of the main advantages of gear surveys, it is recommended that the survey be carried out beyond the bounds of the known stock area, even though this is seldom done owing to financial limitations. Assumption (4) is also easy to satisfy for bottom trawl surveys applied, for instance, to gadoids. But as far as pelagic species are concerned, fishing gears are not commonly used alone for surveying the stocks because it is difficult to evaluate the area sampled by the gear – or by the combination of the gear and detection devices such as sonar – and to get it constant.

Let us see now the influence of fish behaviour according to the different pelagic sampling gears. The case of bottom trawl surveys is fully covered by Gunderson (1993) and is not reviewed in this book. The real new developments since then lie in statistical analysis and survey design using stratification or geostatistics. For instance, Smith and Gavaris (1993) proposed a stratified random design based on historical spatial distribution for estimating the abundance of Atlantic cod. Gear survey can also benefit from the use of geostatistics to compute variance estimation (Conan *et al.*, 1994), even when the survey design is not random (Petitgas, 1993, 1996; Pelletier & Parma, 1994).

10.1.2 *Influence of fish behaviour on fishing gear surveys*

Godø and Valdemarsen (1993) designed a three level pelagic trawl for near surface sampling of juvenile fish. Joseph and Somvanshi (1989) suggested assessment of biomass by 'swept volume' of a midwater trawl, similar to the technique used for bottom trawls. The authors intended to apply this method to the pelagic stocks of horse mackerel (*Megalaspsis cordyla*) along the north-west coast of India. Nevertheless they recognised the limitations of this method owing to unexpected variations in catchability coefficient, associated with the shoaling habit and migration of small pelagic fish, including diurnal variability. Smith and Page (1996) argue that the late inter-annual variability in midwater trawl surveys performed on the Atlantic cod stock of the 4VsW division results mainly from age-and area-specific habitat selection with near-bottom water temperature and salinity. They proposed a model relating the variability in the cold intermediate layer to the estimates of age-specific cod abundance obtained by trawl surveys to derive new survey indices of abundance. These new indices do not have year-effects and seem more consistent than the original time series. All the limitations owing to vessel and gear avoidance during sampling trawls associated with acoustic surveys apply fully to the midwater trawl survey method (Chapter 9) and we think that this method by itself is not adapted to abundance estimation. It must be combined with other direct methods (e.g. acoustic survey or aerial survey) if it is to be efficient.

As far as we know, purse seining is not used in scientific surveys. We have found only one example of its application, on post-larval lingcod (*Ophiodon elongatus*) in the Strait of Georgia, which provided too large a sample variance for a proper monitoring of abundance (Hand & Richards, 1991). The main reason is probably the limited water volume sampled by this gear, which is considered useful only when fish are detected by acoustic equipment or visual observation. Another limitation, compared with the midwater trawl, is the limited vertical range of the water column that it is able to sample. Nevertheless, we think that it could be interesting to investigate the possibility of using this gear in association with acoustic survey for sampling when the biomass is not too deep (a large purse seine can easily sample up to 100 m depth).

The advantages of the purse seine over the midwater trawl vary according to the type of echo sampled. On schools – that is usually during the day – it allows a better separation of individual schools, and therefore can improve the allocation of biomass to different species based on small spatial and temporal scales. The automatic classification of echoes is usually not efficient without such permanent calibration and it is nearly impossible in the case of mixed-species schools. In addition, size-dependent selectivity is limited when sampling with a purse seine. But the greatest advantage expected from a purse seine is likely to occur on shoals or scattering layers of mixed species, in particular in tropical areas where usually at least four species contribute substantially to the total biomass. In such cases, especially on scattering layers, the different species of the assemblage are expected to be mixed randomly. Midwater trawls are species selective in such a situation, in contrast to a purse seine. Of course the yield of a purse seine on a layer will be poor (a few kg) but it can be expected that the samples will be less biased than with other gears.

Beach seine surveys can be used to estimate the abundance of juvenile fish (e.g. Hattala *et al.*, 1988), but the main difficulty is to satisfy assumption (1) on the distribution area. If juveniles are too far from the coastline, they are no longer accessible to this gear. Moreover, this method is time consuming and can be applied only on a limited area.

Longline surveys are sometimes used to assess the abundance of demersal or pelagic fish (e.g. Sigler & Fujioka, 1988; Pelletier & Parma, 1994; Bjordal & Løkkeborg, 1997). To relax the assumption of longline saturation and interspecific competition for hooks, Somerton and Kikkawa (1995) performed an experimental survey of pelagic armourhead (*Pseudopentaceros wheeleri*) using a long-line equipped with hook timers. This method is certainly interesting but is probably too expensive for large-scale surveys.

Finally, we have to remember that all the fishing gear survey methods presented in this section are sensitive to migration (failure of assumption (2) if the stock is migrating actively), aggregation behaviour, avoidance and learning, in a way often difficult to quantify (but see Gauthiez, 1997, on the effect of aggregation on trawl survey).

10.2 Aerial surveys

10.2.1 *Methodology and assumptions*

Aerial survey is a common practice for pelagic species (coastal pelagic fish or tunas) and for mammals. Except for large animals like whales or the giant bluefin tuna (*Thunnus thynnus*) which can be counted individually (e.g. Lutcavage & Kraus, 1995), the unit of observation is the school, associated with a qualitative or semi-quantitative estimate of its biomass. The most commonly used method is to perform a systematic survey along transect lines (e.g. Hara, 1986; Marsac *et al.*, 1988). In Alaska the herring fishery is spread over 5000 km along the coast and therefore aerial surveys have been used to assess and manage the fishery (Belinay, 1994). The Alaska

Department of Fish and Game decides the duration of the fishing season (usually between 10 months and 4 hours!) according to previous aerial census and to maturity stage studies. Purse seine skippers also use planes to locate the schools rapidly, and this is a good example of an overcapitalised fishery.

In addition to direct visual observation, other techniques are used: photography (e.g. Nakashima, 1983), video recording, low-light television (e.g. Cram & Hampton, 1976), or more sophisticated techniques using spectral analysis (Nakashima *et al.*, 1989), microwave from synthetic aperture radar (SAR) (Petit *et al.*, 1992) or laser from LIDAR (light detecting and ranging) (Gauldie *et al.*, 1996). Three types of LIDAR are available and are reviewed and compared by Churnside and Hunter (1996). The radiometric LIDAR has no scanning system and a single element detector. It produces a two-dimensional picture (one axis is depth, the other is the target strength of the school). The imaging LIDAR produces a horizontal image at a fixed depth, without scanning. The volumetric LIDAR uses a scanning system to construct a three-dimensional image. The limitation of this system is that it must be set for a particular depth, and then the user knows only whether the detected school is above or below that depth. The authors conclude that the radiometric LIDAR is probably the most cost-effective device for school fisheries. The typical penetration depth of LIDAR is 20–30 m and depends mainly on water transparency. A large fraction of pelagic schools are found below this depth during the day, and this technique is particularly adapted to shallow waters. Churnside and Hunter (1996) suggest that surveys should be performed during night-time because fish move closer to the surface at night and are able to remain in school due to stellar light or moonlight.

The assumptions of aerial surveys are similar to those presented in the previous section on fishing gear surveys, except that the complete vulnerability to sampling gears (assumption (3)) is now replaced by complete visibility from aircraft. Moreover, in addition to the altitude, the sampled area depends mainly on the size of the school and its illumination (atmospheric condition, sun position with respect to the observer, the sea surface conditions). The quantification of these effects is not easy to achieve (review in Gunderson, 1993; Rivest *et al.*, 1995). Some of them are largely relaxed by the use of LIDAR, but this promising technique is still in an experimental phase as far as stock assessment is concerned. The greatest advantage of aerial survey is that a large area can be surveyed in a short time. Therefore assumption (2), of negligible speed of fish migration compared with the progress of the survey, is usually met. Moreover, the method is adapted to stocks encompassing a large area, such as tunas, which cannot be surveyed by a vessel in a reasonable time.

10.2.2 *Influence of fish behaviour on aerial surveys*

The most important shortcoming of aerial survey is that it is assumed that all the fish are detectable, or at least that the proportion of detectable fish does not vary substantially from one survey to the other (assumption (4) of the previous section). In pelagic fish, this assumption is violated in most cases because only surface or subsurface schools are detectable. Unexpected variations in the fraction of the biomass visible from an aircraft are caused mainly by spatial and temporal changes in vertical habitat selection and in the level of aggregation, as exemplified below.

Kemmerer (1980) indicated that the proportion of surface schools of menhaden may vary by the hour of day. The level of fish aggregation can either make the fish impossible to spot or can modify the area of the school for a given biomass. From a four year experience of aerial surveys in south-eastern Australia (including Tasmania), Williams (1981) concluded that the variability in the number of observed schools was related to moon phase and period of the day. More than double the amount of jack mackerel (*Trachurus declivis*) was sighted during the new moon phase than during the full moon phase and there was a tendency for larger sightings to occur in the early morning and late afternoon. Moreover, in the case of jack mackerel as for bluefin tuna (*Thunnus maccoyii*), the sea surface temperature is a relevant factor for the species distribution. In this study, Williams (1981) underlined these sources of variability but did not include them in the abundance estimates. From an analysis of the purse seine commercial fishery on jack mackerel of Tasmania, Webb (1977) arrived at the conclusion that jack mackerel showed a trend towards an increase in tonnage landed during the new moon and first transitional periods. Williams and Pullen (1993) reached complementary conclusions on the diurnal and seasonal variability in the school size or ratio between surface and subsurface schools encountered by fishermen, but moon phase influence was not significant. From our experience in acoustic surveys, not only the school density is usually lower at night, but often the fish are dispersed in a dense layer, even when the light intensity provided by moonlight or starlight allows visual contact for schooling (occasionally layers have been observed by day). The reasons for this dispersion remain speculative (feeding, physical factors). In such cases the fish cannot be detected by aerial survey, even with LIDAR or SAR technologies.

Large size schools are easier to detect at large distances than smaller ones, and this is one reason for violation of assumption (4), which assumes a known area of detection. Chen (1996) proposed the use of a bivariate detection function with distance and size of the school and considered the use of the kernel smoothing method. He proposed an example of application to the aerial survey of southern bluefin tuna.

The bias from diurnal (and secondarily lunar and seasonal) variation of vertical habitat selection can be avoided by combining aerial and acoustic surveys. From aerial and sonar observation of northern anchovy (*Engraulis mordax*) in north-western Mexico, Squire (1978) indicated that the percentage coverage of the school's area in relation to a circle drawn tangent about the school averaged 42% during the day and 29% during the night. The school area estimate by visual survey was over-estimating the actual school area by 1.72:1. Another way to combine aerial and acoustic surveys for stock management is given by the East Canada stock of capelin (*Mallotus villosus*), which is first surveyed by echosounding before the spawning season, when the fish are located in deep water offshore, and secondly by aerial survey during the spawning season (Nakashima, 1983). The first surveys were done with conventional colour photography, but later Borstad *et al.* (1992) and Nakashima and Borstad (1993, 1996) improved the methodology using a compact airborne spectro-graphic imager (CASI) which permits work when the visibility is not optimal for conventional photography (clouds, high sun angle). Moreover, the CASI system offers the advantage of storing digitally and in real time the data, which can later be incorporated in a geographic information system (GIS). Strangely this promising method does not seem to be used in other quarters. Better validation may be expected, especially from the combination with field experiments to assess the species identification and the probability of school detection according to the depth, size and sea surface condition.

The SAR observations are totally free of the influence of cloud coverage and sunlight effects (SAR can be operated by night). Concurrent independent aerial observations, one with the SAR, the other by a trained spotter pilot, indicated that when tuna schools were at the surface, small variations of water roughness induced by internal waves made them detectable (Petit *et al.*, 1992). The method also permits the detection of vessels and of set purse seines. Nevertheless this method is not currently used for stock assessment, and once more this is probably because only surface schools can be detected.

10.3 Ichthyoplankton surveys

10.3.1 *Objectives, methodology and assumptions*

As far as stock assessment is concerned, ichthyoplankton surveys can be designed according to two primary objectives: the estimation of the recruitment some years in advance and/or the back-calculation of the size of the spawning stock, knowing the fecundity of females (Heath, 1992). The first objective is usually difficult to reach in pelagic fish stocks because the natural mortality in early stages following eggs and larvae is still very high and often unpredictable in species having a high fecundity and spawning pelagic eggs. Presently most of the egg and larval surveys of pelagic species are aimed at the second objective, that is the evaluation of the parental stock biomass, and we will focus mainly on this approach. Egg surveys are often preferred to larval surveys for estimating the parental stock because this last approach requires estimates of growth and mortality rates of the early stages (Hauser & Sissenwine, 1991; Heath, 1993). Larval surveys are usually preferred for predicting recruitment for the same reason.

Ichthyoplankton surveys are designed to cover the whole distribution of eggs and/or larvae of the target population by a sampling grid. This grid is usually systematic (e.g. Priede & Watson, 1993) instead of random, but if the location of the spawning ground can be known in advance or is not variable from year to year, the survey can be prestratified. This is for instance the case for a few pelagic species like herring or capelin which are benthic spawners which deposit their eggs on characteristic spawning sites, and are known to display a natal homing. Then the egg density can be estimated by SCUBA egg deposition surveys (e.g. Schweigert, 1993).

A combination of prestratification and random sampling inside strata is proposed by Hampton *et al.* (1990). The approach by adaptive sampling is generally used for deciding when to stop a transect (usually after two or three adjacent stations without eggs or larvae) but it can be also used to oversample the dense areas, despite the difficulties of avoiding biases in the estimates of the abundance and of its variance. In order to take into account the spatio-temporal variability in egg density, Petitgas (1997) proposed a geostatistical model providing an estimation of the mean density and its variance.

Several sampling gears can be used for collecting eggs and larvae. Larvae are more difficult to sample than eggs because they are able to detect and avoid the gear. Eggs are sampled by plankton nets, horizontally, vertically or obliquely deployed (double-oblique tow). Horizontal or oblique tow are usually performed with a pair of nets deployed in a single frame (bongo). A bongo net is more stable than a single net, and limits larval avoidance of the frame (the two nets of the bongo might have different mesh size). A multiple opening–closing net and environmental sensing system (MOCNESS) was designed in order to better estimate the vertical distribution of the ichthyoplankton and of its local environment (Wiebe *et al.*, 1976; Snyder, 1983). Vertically deployed gears sample a limited volume and can be used only when egg density is not too patchy at small scale. The CalVet (California Vertical Egg Tow) was designed for sampling anchovy and sardine eggs during the long-term CalCOFI survey off California (Smith *et al.*, 1985). In all these gears, a flowmeter provides a measure of the distance travelled by the net in the water column. It must be calibrated properly in order to avoid biases in the ichthyoplankton abundance estimates (Brander & Thomson, 1989).

Because operating research vessels is very expensive, over the past 60 years continuous plankton recorders (RPCs) have been towed behind merchant ships on regular routes in the north-east Atlantic Ocean. In early RPCs the plankton was retained on a fixed filter screen; this has been replaced by a continually moving length of silk mesh. RPCs are aimed at providing information about plankton abundance and distribution, but some authors intended to use them to monitor long term variation in ichthyoplankton abundance (e.g. Dickson *et al.*, 1992).

In order to increase the sampled volume and reduce the time consuming process of egg counting, plankton pumping and optical plankton counter (OPC) has been investigated since the 1980s. Omori and Jo (1989) proposed a vortex plankton pump capable of sampling up to 100 m depth. From a literature survey, Checkley *et al.* (1997) claim that although the depth of spawning varies within and between species, the distribution of eggs achieves a surface maximum in their early stage of development. Therefore they designed a continuous underway fish egg sampler (CUFES) which combines subsurface pumping and OPC. The performance of CUFES has been compared *in situ* with that of MOCNESS and CalVET samplers. The CUFES system provides lower retention of eggs than the other samplers, but the relationship between log-transformed abundance estimates are linear and highly correlated.

There are three main methods for back-calculating the parental stock biomass of pelagic stocks (review in Gunderson, 1993; Hunter & Lo, 1993): the annual egg production method (AEPM), annual larval production method (ALPM), daily fecundity reduction method (DFRM), egg deposition method for benthic spawners (EDM) and finally the daily egg production method (DEPM) which is the most popular for multiple spawning pelagic fish (review in Alheit, 1993). All of them require an estimate of the number of eggs spawned per day ($\hat{E}_t$) during the survey:

$$\hat{E}_t = (\hat{E}_{ij}/d_{ij})\ e^{Z_{ij}t_{ij}} \qquad (10.1)$$

where $\hat{E}_{ij}$ is the number of eggs stage i in the survey area during survey j, d_{ij} the duration in days of egg stage i, Z_{ij} the instantaneous rate of daily mortality for eggs stage i, and t_{ij} the age of the egg in days (review and other equations in Gunderson, 1993).

10.3.2 *Influence of fish behaviour on ichthyoplankton surveys*

As far as fish behaviour is concerned, the main problem in egg surveys is directly related to sampling errors and biases. Matsuura and Hewitt (1995) reviewed the contagion in the dispersion of ichthyoplankton. They showed that the patchy distribution of eggs comes from the schooling behaviour of adults during spawning. Then patchiness decreases due to the dispersive process and increases in the subsequent stages due to the schooling behaviour of juveniles. As a result, the patchiness-at-age curve for several species of schooling fish is U-shaped. Therefore, strictly from a sampling strategy point of view, it seems logical to prefer larval surveys to egg surveys, especially for recruitment estimation. Nevertheless, an additional problem in larval surveys is avoidance of the gear. As a result, the choice between an egg or larval survey will depend on a trade-off between the advantages and inconveniences of both methods, and also on the final objective (spawning biomass or recruitment estimation). An additional behavioural effect is cannibalism by adults on their eggs and larvae, which can lead to an underestimation of abundance, especially when the spawning area is not productive. This is typically the case of the Cape anchovy (*Engraulis capensis*) spawning over the Agulhas Bank (Valdes *et al.*, 1987; Valdes, 1993).

The difficulty in designing an appropriate ichthyoplankton survey often comes from the variability of the location and surface of the spawning area, in relation to environmental effects (Muck, 1989; Boehlert & Mundy, 1994). In the vertical plan, another difficulty is to assess correctly the depth where the ichthyoplankton is mainly located, particularly in the case of larvae which usually display circadian vertical migration (Kendall *et al.*, 1994). An example of the complexity of the environmental influence on the ichthyoplankton distribution is provided by Moser and Smith (1993) who studied the horizontal and vertical distributions of larval fish assemblages of the California current region. They concluded a major influence of the frontal zone between subarctic-transitional and central water masses and observed different patterns of vertical distribution for the same species at the north and south sides of the front.

Larval avoidance is usually more important during the day, due to visual detection of the gear, which is another factor explaining the circadian variability in the catches. Such examples are provided by Roepke (1989) on *Scomber scombrus* larvae and by Boehlert and Mundy (1994) on *Thunnus albacares* and other tunas.

In order to limit the effect of patchiness on ichthyoplankton surveys, Mangel and

Smith (1990) proposed to estimate the spawning area and an index of abundance of the spawning biomass only from the collection of presence/absence data. The models are based on an underlying probabilistic description of the aggregation of eggs or larvae, a search process, and a description of habitat structure. They suggested using this inexpensive method as a continuous monitoring of a stock, alternately with conventional counts of eggs and larvae. They applied this approach to the Pacific sardine (*Sardinops sagax*) in the California current, but as far as we know, no other application has been intended. This method seems especially appropriate to survey the recovery of overexploited stocks.

Combining an ichthyoplankton survey and acoustic survey, which have different advantages and inconvenience, is a good way to overcome their limits. Since 1984, the abundance of spawning Cape anchovy (*Engraulis capensis*) on the South African continental shelf has been assessed by this approach (Hampton *et al.*, 1990). Such combined surveys provide a basis for annual adjustment of a total allowable catch (TAC) aimed at maintaining constant risk of spawning escapement relative to 50% of the estimated unexploited biomass. The management method is adaptive in the sense that initial estimates of risk based on spawner stock size are updated once the strength of the recruiting year class has been determined by a second acoustics survey (Shelton *et al.*, 1993; Hampton, 1996).

10.4 Capture–recapture

10.4.1 *Methodology and assumptions*

The field methodology consists of catching fish, tagging them by an external tag which is easy to detect or an internal steel tag that can be detected magnetically, releasing them immediately, and finally, once they are diluted in the stock, recapturing them (by experimental survey or commercial fishing) following different sampling strategies. According to the number of tagging periods and recapture periods, different estimates can be expected: stock abundance, exploitation rate, survival rate during a time period or recruitment (see Ricker, 1975, or Burnham *et al.*, 1987 for a review). All these techniques are based on the following assumptions:

- the mixing rate between tagged and non-tagged fish is random
- the catchability of tagged fish is identical to the catchability of non-tagged fish
- the natural mortality of tagged fish is identical to the mortality of non-tagged fish (or the additional mortality caused by tagging is correctly estimated)
- there is no tag loss or this loss is correctly estimated
- all recaptured fish are reported.

Other assumptions must be added according to the type of study. For instance, in the case of a single tagging experiment and single period of recapture, the recruitment in between tagging and recapture is supposed to be negligible. Then the rate of exploitation (u) is simply estimated by the ratio between the number of recaptured (R) and the number of tagged fish (T):

$$u = R/T \tag{10.2}$$

The abundance in number N can be estimated by the ratio u/C, where C is the total catch associated with the recapture of R tagged fish. More sophisticated methods and data analysis have been developed around this simple principle, in particular by terrestrial ecologists (e.g. Lebreton *et al.*, 1992).

One of the main problems of the capture–recapture method – not related to fish behaviour – is the difficulty of obtaining a good proportion of tag returns and estimating the proportion of tags not being detected or returned by fisherman. As stressed by Hilborn and Walters (1992), a common mistake is to underestimate the necessary investment of time and money needed to achieve this second phase of the project.

10.4.2 *Influence of fish behaviour on capture–recapture methods*

We reported in Chapter 6 that fish are able to learn by a single experience ('one-trial learning') when they experience a strong stress, such as tagging (Fig. 7.6). Tagging operations certainly represent a huge stress for the fish, especially when performed after a traumatic capture (by hand line or purse seine) and out of the water. Anaesthesia is usually practised because it is supposed to limit the stress owing to the attachment of the tag to the fish's body (or similarly the stress through forced ingestion of an acoustic tag or intra-abdominal injection of a magnetic tag). Unfortunately, anaesthesia itself is an important stress (Carragher & Rees, 1994), especially when carbon dioxide gas (MacKinlay *et al.*, 1994) or MS-222 (Olsen *et al.*, 1995) are used. Some workers on tagging experiments find that anaesthesia is not necessarily helpful when tagging trauma is low (Harris, 1989). In addition, anaesthesia is suspected to increase mortality related to tagging (Hansen, 1988).

Whatever the reduction of stress by anaesthesia, fish manipulation related to the tagging operation, including the anaesthesia itself, is likely to favour 'one-trial learning' and therefore to decrease the probability of recapture. This lower catchability, compared with naive fish, is a potential source of bias in stock abundance estimations performed from tagging experiments. Hampton (1986) observed a relatively low number of recaptures during the first five days after tagging southern bluefin tuna (*Thunnus maccoyii*) and our interpretation is that this result is mainly explained by 'one-trial' learning. The stress could also explain the decrease in the conditioning factor of fish 5–20 days after tagging in this experiment. Nevertheless, other tagging experiments on skipjack tuna caught by pole-and-line do not suggest learning but, surprisingly, an increased catchability of the tagged fish during the day following tagging (Cayré, 1982). This was interpreted as caused either by dominance of some fish and/or by a persistent level of hunger. We think that in apex predators, particularly in non-sedentary ones, the trade-off between feeding and predation risk (fisherman being one of the predators) is more often in favour of feeding than it is in plankton feeders. For a plankton feeder, it is obviously more important to escape a predator than to feed on a plankton patch, since it is relatively easy to come back to this patch or to find another one. In contrast, in apex predators the probability of encountering a prey is low, energy consumption is high due to permanent swimming and these species can suffer hunger and probably die of starvation on many occasions (Olson & Boggs, 1986). As a result they probably take more risk in feeding than plankton feeders.

In addition to learning, other fish behavioural traits can influence the capture–recapture method during the process of recapture (in particular habitat selection and aggregation). Because these aspects are identical to those presented in Chapter 8, they are not reviewed here.

10.5 Conclusion

None of the other methods of stock assessment briefly reviewed in this chapter are free of bias, but the limiting factors are different according to the methods. Fishing gear surveys are sensitive to avoidance reactions, which vary according to the species. They are also sensitive to changes in catchability (mainly owing to changes in habitat selection), aggregative behaviour and probably learning. An investigation is recommended as to whether the use of a purse seine as a sampling gear for acoustic surveys can solve the problem of midwater trawl selectivity (for species and size). The results of aerial surveys strongly depend on the assumption of constant vertical selection of habitat, in addition to proper correction for changes in visibility conditions. The use of LIDAR could partly solve these problems in the near future because this technology allows a larger depth range (from 10 to 40 m according to the water transparency) to be sampled and is less sensitive to environmental conditions. Aerial surveys would certainly benefit from an association with acoustic surveys, but recommending such an association is not always realistic. The two methods are time and fund consuming and difficult to implement owing to their heavy logistical demands. Combining them means more than twice the difficulties and sometimes the final result is not as good as expected, as stressed by Pullen *et al.* (1992). Finally, capture–recapture methods suffer mainly from the influence of learning and aggregation, in addition to the difficulty of assessing tag return. There is no doubt that tagging is a valuable technique for studying fish migration, association with floating objects and other behavioural traits. If tagging is used as a reliable method of stock assessment of pelagic species, there must be a good system for recording the tagged fish in the catches.

Chapter 11
Conclusion

Behavioural ecology is still a young branch of science which had its origins in the middle of the nineteenth century, and even later as far as fish are concerned. Despite the benefits gained recently from technological advances in the study of fish behaviour, this branch of science is still immature compared with other branches such as physics or medicine. In this book we have often mentioned conflicting results, but this does not mean that one represents the truth, the other not. Unexpected (and therefore not measured) factor(s) can explain such differences in the results. Certainly what we consider 'noise' or 'random effects' at the level of the population hides a complex structure of individual behaviour. Three examples can illustrate this apparent complexity.

First, the homing behaviour of tunas under anchored FADs indicates that the apparent randomness of the time of return to a given FAD arises from the juxtaposition of different deterministic behaviours of different schools, if not of individuals, having a precise time interval of arrival (section 6.3.1). Second, there is growing evidence that what is usually considered a single exploited stock is an assemblage of spatially overlapping or mixed populations, each relatively independent from a genetic point of view, which results in a hysteresis effect when some of these populations collapse owing to local overfishing; when fishing effort is then decreased, the catch-effort curve is no longer reversible (sections 3.3.1, 7.3 and 8.7.2). Third, some complex patterns of migration result from length (or age) dependence at the level of seasonal migration – as in Atlantic menhaden (section 3.2) – or at the level of daily migration of specific groups which have their own route – as in gadoids (section 3.3.8) or grunts (section 7.3). Archive and 'pop-up' tags will soon permit understanding of spatial fish behaviour at the individual level, at least for fish large enough to carry such tags. Different simulation approaches, especially the IBM coupled with GIS, are already available to incorporate the resulting findings and will certainly provide in return interesting suggestions for research through an interactive dialogue with field research, as occurs in terrestrial ecology (e.g. Drogoul *et al.*, 1992).

Another weak point of our present knowledge concerns the interactions between species. Despite the fact that ecological literature is rich in equations relative to these interactions at the population level, the underlying mechanisms at the individual fish level are often not understood. How fish compete, interact or associate with others in different ecological niches definitely needs further investigation.

All stock assessment methods may suffer large biases owing to fish behaviour and more particularly to spatial distribution at different scales (Table 11.1). Age

Table 11.1 Summary of the biases in stock assessment methods due to fish behaviour (modified after Fréon *et al.*, 1993a).

Behavioural pattern	Age structured analysis (ASA)	Surplus production models	Acoustic survey	Gear survey	Aerial survey
Habitat selection					
Short term	0	0	+ + or – –	+ or –	+ or –
Circadian	0	0*	+ + or – –	+ + or – –	from – to – – –
Seasonal	0	0*	+ or –	+ + or – –	from – to – – –
Yearly	+ or –	+ + + or – – –	0	+ + or – –	from – to – – –
Aggregation	0	+ + + or – – –	+ + or – –	+ + or – –	+ or –
Avoidance	0	0	– –	– – –	0
Attraction/ association	0	+ +	0	+ +	0
Learning	–	–	0	–	0

+ overestimation
– underestimation
* except if interannual changes in the fishing pattern are not taken into account.

structured models seem to be an exception because the influence of fish behaviour on them is more cryptic, but they can be directly responsive to habitat choice, and to learning in extreme circumstances. In addition, owing to the difficulty in estimating the proper value of the terminal F of partly exploited cohorts, this method is not always appropriate to reflect rapid changes in the exploitation level and can lead to a permanent underestimation of the rate of exploitation (Fig. 11.1, from Brêthes, 1998). Moreover, if age structured models can be useful for stock management, one of their main outputs – the fishing mortality coefficient – must be converted to fishing effort. This requires a proper estimate of the catchability coefficient (but we concede that this is not directly the case for another important output, the TAC). In addition, many ASA models must be tuned by at least one parameter (fishing effort, CPUE, catchability coefficient or recruitment) estimated by other methods. Therefore, if this parameter is overly dependent on fish behaviour, the management advice will be inappropriate. Better knowledge of the key factors governing catchability and habitat selection would in many instances be helpful for limiting such biases. Special attention must be paid to:

(1) The changes in factors governing the selection of habitat at different life history stages
(2) The importance of physical gradients to fish orientation and distribution
(3) Possible long-term changes in habitat selection.

We still ignore which senses are predominant in the orientation of most pelagic species, and therefore we scarcely know which environmental cues they are using to find their suitable habitat at a given time and to move to another one when necessary.

Direct stock assessment methods are sensitive to small-scale variations in natural fish behavioural patterns, and reactions to the vessel or the fishing gear may have a large effect on them. On some occasions, the variability in biomass estimate can also

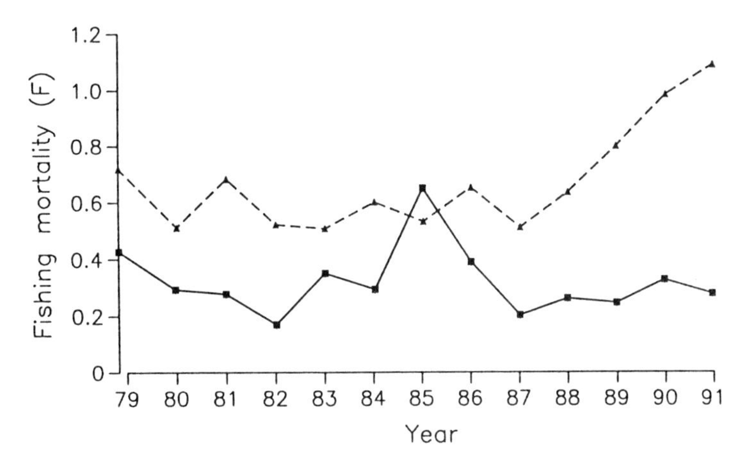

Fig. 11.1 Comparison between mortality estimates of cod in the Gulf of St Lawrence from 1979 to 1991. ■ = F (year) – year by year estimate; ▲ = F (comv.) – a posteriori estimation for fully recruited cohorts by convergence of the VPA. (Redrawn after Brethês, 1988.)

be explained by environmental or intrinsic (species, physiological stage) factors. To understand the causes of the natural behavioural patterns and avoidance reactions, more detailed studies on the interactions within and between these two groups of factors are needed. Experiments in large tanks or mesocosms with a controlled environment are necessary, but to study large-scale effects, *in situ* observations remain indispensable, especially for pelagic fish, which most of the time are schooling or shoaling in great numbers. We argue that such research must be escalated as there is great potential for increasing the accuracy of stock assessment methods through better understanding and knowledge of fish behaviour. Recent technological improvements in acoustics and in visual observation make it possible to reach this goal, especially if fishery biologists benefit from the rigour in experiment design, definition of quantitative indicators and data analysis in use in experimental ethology. Field research is expensive because it requires time at sea. To place observers on board commercial fishing vessels is a more economical way to make observations, but we think that in this field of investigation, research vessels are still indispensable. Note that, paradoxically, most of the increase in catchability comes from increasing knowledge on fish behaviour (from fishermen or scientists) or from sophisticated technological developments aimed at studying this behaviour or at least detecting fish. In turn this increase in catchability, if not properly controlled, leads to an increase in catches and therefore favours stock collapse.

An appreciation of the interaction between the behaviour of fish and that of fishermen is necessary to better understand the links between stock abundance and CPUE, and also between the large-scale geographical distribution of the stock and the usually smallest area covered by the fleet. These spatial aspects are poorly known but it is likely that they depend on the level of aggregation of the fish (school, cluster, layers), on their availability (depth, distance to the harbour, etc.) and on social and economic factors. Here scientific observers on board commercial vessels can be very efficient at gaining some insight into fishermen's tactics and strategy which can be incorporated in a fishery GIS for instance. This is a promising field of investigation even though modelling (both fish population dynamics, fishermen, and sometimes fish commercialisation) requires so many parameters that the corresponding models are mainly useful for simulating scenarios rather than for prediction (Hilborn & Walters, 1987). Adaptive management (Walters, 1986) will probably remain an efficient way of avoiding stock collapse for many years.

Despite the fact that this book is largely devoted to stressing the importance of fish behaviour on stock assessment, we are very concerned by other bottlenecks less related to fish behaviour. It is not realistic to propose a universal prioritisation among the research needs related to stock assessment, because the priority depends on the method and on the stock considered. Therefore the following list of weak points related to stock assessment, mixing behaviour-related studies with others, does not pretend to be ordered:

- Stock-unit identification (in too many cases the first assumption of population dynamics models is violated because they are applied either to substocks or to several isolated stocks)
- Questionable value of the generalisation of natal homing and learning of migration routes in the population structure of pelagic species (micro satellite DNA analysis and modern tagging techniques can be helpful in these investigations)
- Analysis of spatial effects at different scales, which first requires small temporal and spatial scale data collection
- Estimation of the absolute value of natural mortality, its interannual variation and its variability according to age (ASA models)
- Identification of the source of variation of catchability (availability, gear efficiency, cooperation, saturation, influence of learning, effect of the environment, etc.)
- Quantification of the effects of environmental changes on abundance (recruitment, mortality and growth)
- Identification of the key factors of the stock-recruitment relationship
- Dynamics of the interactions within the whole ecosystem (food web, competition between species, etc.)
- Reason(s) for and variability of the association of pelagic fish with floating objects (fixed or drifting) or to spatial structure (seamounts, islands, etc.).

Most of these weak points are directly related to the assumptions of direct or indirect stock assessment methods. All applied methods have reached a high level of sophistication, either at the conceptual level (modelling, mathematics, statistics) or at the technological level (acoustic or visual sensors, signal processing, etc.). As a result, a fishing biologist can be compared to a driver who wishes to cross an unknown desert in a racing car with a GPS (global positioning system), assuming that there are good roads and many petrol stations. A closer comparison to fish biology could be to imagine an entomologist studying the population dynamics of bees with a sophisticated demographic model and with a new brand bee detector, ignoring the existence of bee-hives and queens and assuming that all individuals in a given area belong to the same population, that they are able to reproduce and can be counted in the field when gathering pollen. Nowadays our limits in stock assessment are more on the side of basic knowledge than on methods of representation and modelling.

Fisheries management is the next step following stock assessment and it requires additional inputs not contemplated in this book but that cannot be ignored: social, economic and political aspects, ecosystem management, trade-offs between fisheries management and management of other users of the sea or the coastal area (aquaculture, tourism, navigation, etc.). This makes fishery management a complex task which requires a close cooperation of a broad category of actors as underlined by Caddy and Mahon (1995). The use of reserves for management is an old method, common in terrestrial ecology but curiously limited for marine species. It has enjoyed a renewal of interest (Roberts, 1995; Bohnsack & Ault, 1996; Pauly, 1997b; Roberts, 1997), but as far as pelagic fish are concerned, the limit of the proper use of this method is once more our poor knowledge of population structure (number of stock-units and nurseries, diffusion of fish, migration pattern, natal homing, etc.).

The lack of basic knowledge on fish ecology (in the broad sense of the word) has left the door open for increasing fishing mortality and dramatic overexploitation. At present the large majority of commercially exploited stocks are either overexploited or fully exploited (FAO, 1996), and in most marine ecosystems, exploitation has shifted from the apex predators (now largely overexploited if not extinguished) to lower levels of the food chain, that is mainly small pelagic species (Pauly *et al.*, 1997). These 'forage fish', like any schooling pelagic species, can be overexploited without giving the 'warning' of a sharp decrease in yield, because their catchability often increases when they are depleted. The overexploitation of forage fish can endanger the whole ecosystem and modify it irreversibly, at least in the medium-term. The role of fishery biologists is to stress the priority of resource preservation for future generations. Captain. J.-Y. Cousteau died a few days before we wrote these lines. He was a great populariser of science and inspired many vocations of marine scientists by promoting underwater visual observation, but most of all he was an active defender of our marine heritage. It is also the responsibility of fishery biologists to continue this work and to produce accurate stock assessments. A close cooperation between fishery biologists and ethologists is one (and only one) way of increasing our knowledge of fish behaviour. 'Knowledge is a patrimony of humanity', said Louis Pasteur (1822–1895), and this second patrimony is necessary to preserve the first and concrete one: the life on mother earth.

References

Abrahams, M.V. & Colgan, P. (1985) Risk of predation, hydrodynamic efficiency and their influence on school structure. *Environ. Biol. Fish.*, **13**, 195–202.

Aglen, A. (1983) Random errors of acoustic fish abundance estimates in relation to the survey grid density applied. *FAO Fish. Rep.*, **300**, 293–8.

Aglen, A. (1994) Sources of error in acoustic estimation of fish abundance. In: *Marine Fish Behaviour in Capture and Abundance Estimation* (eds. Fernø, A. & Olsen, S.), pp. 107–33. Fishing News Books, Oxford.

Ahrenholz, D.W., Nelson, W.R. & Epperly, S.P. (1987) Population and fishery characteristics of Atlantic menhaden, *Brevoortia tyrannus. Fish. Bull.* (USA), **85** (3), 569–600.

Akishige, Y., Yoshimura, H., Nishida, H., Kuno, T., Morii, Y. & Aoshima, T. (1996) Flotsam in the equatorial counter current region of the western Pacific with reference to tuna purse seine fishing. *Bulletin of the Faculty of Fisheries*, Nagasaki University, **77**, 97–102.

Aksnes, D.L. & Giske, J.A. (1993) Theoretical model of aquatic visual feeding. *Ecol. Model.*, **67** (2–4), 233–50.

Alderdice, D.F. & Hourston, A.S. (1985) Factors influencing development and survival of Pacific herring (*Clupea haregus pallasi*) eggs and larvae to beginning of exogenous feeding. *Can. J. Fish. Aquat. Sci.*, **42** (Suppl. 1), 56–68.

Alekseev, A.P. (ed) (1968) *Fish Behaviour and Fishing Techniques*. All-Union Conference held in Murmansk, 27 February to 1 March, 1968. Ministry of Fisheries of the USSR.

Alevizon, W.S. (1976) Mixed schooling and its possible significance in a tropical western Atlantic parrotfish and surgeonfish. *Copeia*, (4), 796–8.

Alheit, J. (1993) Use of the daily egg production method for estimating biomass of clupeoid fishes: a review and evaluation. In: *Advances in the Early Life History of Fishes. Part 2. Ichthyoplankton Methods for Estimating Fish Biomass* (eds. Hunter, J.R., Lo, N.C.H. & Fuiman, L.A.), pp. 750–67. *Bull. Mar. Sci.* (US), **53** (2).

Ali, M.A. (Ed.) (1992) *Rhythms in Fishes*. Plenum Press, New York.

Allan, J.R. (1986) The influence of species composition on behaviour in mixed-species cyprinid shoals. *J. Fish Biol.*, **29** (Suppl. A), 97–106.

Allan, J.R. & Pitcher, T.J. (1986) Species segregation during predator evasion in cyprinid fish shoals. *Freshwater Biol.*, **16**, 653–9.

Allen, J.M., Blaxter, J.H.S. & Denton, E.J. (1976) The functional anatomy and development of the swimbladder-inner-ear-lateral line system in herring and sprat. *J. Mar. Biol. Assn.* (UK), **56**, 471–86.

Allen, P.M. & McGlade, J.M. (1986) Dynamics of discovery and exploitation: the case of the Scotian shelf groundfish fisheries. *Can. J. Fish. Aquat. Sci.*, **43**, 1187–200.

Allen, P.M. & McGlade, J.M. (1987) Modelling complex human systems: a fisheries example. *Eur. J. Oper. Res.*, **30**, 147–67.

Allen, R.L. (1985) Dolphins and the purse-seine fishery for yellowfin tuna. In: *Marine Mammals and Fisheries* (eds. Beddington, J.R., Beverton, R.J.H. & Lavigne, D.M.), pp. 236–52. Allen and Unwin, Winchester.

Allendorf, F., Ryman, N. & Utter, F. (1986) Genetics and fishery management: past, present and future. In: *Population Genetics and Fishery Management* (ed. Ryman, N.), pp. 1–22. University of Washington City, Washington.

Altbäcker, V. & Csányi, V. (1990) The role of eye-spots in predator recognition and antipredatory behaviour of the paradise fish (*Macropodus opercularis*). Ethology, **85**, 51–7.

Anderson, J.J. (1980) A stochastic model for the size of fish schools. *Fish. Bull.* (US), **79** (2), 315–23.

Aneer, G., Florell, G., Kautsky, U., Nellbring, S. & Sjøstedt, L. (1983) *In-situ* observations of the Baltic herring (*Clupea harengus membras*) spawning behaviour in the Askø-Landsort area, northern Baltic proper. *Mar. Biol.*, **74**, 105–10.
Anon. (1976) *Rapport du groupe de travail sur la sardinelle* (S. aurita) *des côtes ivoiro-ghanéennes*. Fishery Research Unit Tema, Centre Rech. Océanogr. Abidjan, Orstom Editions, Paris.
Anon. (1979) Small fish – big trawls. *Fishing News International*, **18** (2), 22.
Anon. (1988) *Report of the mackerel working group*. ICES CM/1988, Assess: 12.
Anon. (1992) *Dolphins and the Tuna Industry*. Washington, DC. National Academy Press.
Anon. (1995) *Annual report of the Inter-American Tropical Tuna Commission (1995)*. La Jolla.
Anon. (1996) *Report of the northern pelagic and blue whiting fisheries working group*. ICES CM 1996/Assess: 14.
Anon. (1997) *Annual report of the Inter-American Tropical Tuna Commission*, La Jolla.
Anthony, P.D. (1981) Visual contrast thresholds in the cod *Gadus morhua* L. *J. Fish. Biol.*, **19**, 87–103.
Anthony, V.C. & Fogarty, M.J. (1985) Environmental effects on recruitment, growth and vulnerability of atlantic herring (*Clupea harengus harengus*) in the Gulf of Maine region. *Can. J. Fish. Aquat. Sci.*, **42** (Suppl. 1), 158–73.
Aoki, I. (1982) A simulation study on the schooling mechanism in fish. *Bull. Japan. Soc. Sci. Fish.*, **48** (8), 1081–8.
Appenzeller, A.R. & Legget, W.C. (1992) Bias in hydroacoustic estimates of fish abundance due to acoustic shadowing: evidence from day–night surveys of vertically migrating fish. *Can. J. Fish. Aquat. Sci.*, **49** (10), 2179–89.
Argue, A.W., Williams, M.J. & Hallier, J.-P. (1987) *Fishing performance of some natural and cultured baitfish used by pole-and-line vessels to fish tunas in the central and western Pacific Ocean*. South Pacific Commission, Nouméa. Tech. Rep. 18.
Ariz, X., Delgado, A. & Fonteneau, A. (1992) Logs and tunas in the eastern tropical Atlantic. A review of present knowledge and uncertainties. In: *International Workshop on the Ecology and Fisheries for Tunas Associated with Floating Objects and on Assessment Issues Arising from the Association of Tunas with Floating Objects* (ed. Hall, M.) Inter-American Tropical Tuna Commission Intern. Rep., San Dieto (mimeo).
Armstrong, F., Simmonds, E.J. & MacLennan, D.N. (1989) Sound losses through aggregations of fish. *Proc. I.O.A.*, **11**, 35–43.
Armstrong, M.J. & Thomas, R.M. (1989) Clupeoids. In: *Oceans of Life off Southern Africa* (eds. Payne, A.I.L. & Crawford, R.J.M.), pp. 105–12. Vlaeberg, Cape Town.
Armstrong, M.J., Chapman, P., Dudley, S.F.J., Hampton, I. & Malan, P.E. (1991) Occurrence and population structure of pilchard *Sardinops ocellatus*, round herring *Etrumeus whiteheadi* and anchovy *Engraulis capensis* off the East coast of Southern Africa. *S. Afr. J. Mar. Sci.*, **11**, 227–49.
Arnold, G.P., Greer Walker, M., Emerson, L.S. & Holford, B.H. (1994) Movements of cod (*Gadus morhua* L.) in relation to the tidal streams in the southern North Sea. *ICES J. Mar. Sci.*, **51**, 207–32.
Arruda, L.M. (1984) Sexual maturation and growth of *Trachurus trachurus* (L.) along the Portuguese coast. *Investigación Pesq.*, **48** (3), 419–30.
Atema, J. (1980) Chemical senses, chemical signals, and feeding behaviour in fishes. In: *Fish Behaviour and Its Use in the Capture and Culture of Fishes: Proceedings of the Conference on the Physiological and Behavioral Manipulation of Food Fish as Production and Management Tools* (eds Bardach, J.E., Magnuson, J.J., May, R.C. & Reinhart, J.M.), pp. 57–101. Bellagio, Italy, 3–8 November 1977. ICLARM Conference Proceedings (Manila).
Atema, J., Holland, K. & Ikehara, W. (1980). Olfactory response of yellowfin tuna (*Thunnus albacares*) to prey odours: chemical search image. *J. Chem. Ecol.*, **6**, 457–65.
Atkinson, D.B., Rose, G.A., Murphy, E.F. & Bishop, C.A. (1997) Distribution changes and abundance of northern cod (*Gadus morhua*), 1981–1993. *Can. J. Fish. Aquat. Sci.*, **54** (Suppl. 1), 132–138.
Atz, J.W. (1953) Orientation in schooling fishes. In: *Proc. Conf. Orientation Anim.*, section 2, 103–30. Off. Nav. Res. Washington D.C.
Au, D.W. (1991) Polyspecific nature of tuna schools: shark, dolphin, and seabird associations. *Fish. Bull.* (US), **89** (3), 343–54.
Au, D.W.K. & Pitman, R.L. (1988) Seabird relationships with tropical tuna and dolphins. In: *Seabirds and Other Marine Vertebrates* (ed. Burger, J.), pp. 174–212. Columbia University Press, New York.

Azuma, T. (1991) Diurnal variations in salmon catch by surface gillnets in the Bering Sea during the summer. *Bull. Japan. Soc. Sci. Fish.*, **57** (11), 2045–50.

Babayan, V.K. & Kizner, Z.I. (1988) Dynamic models for TAC assessment: logic, potentialities, development. *Coll. Scient. Pap. Int. Comm. SE Atl. Fish.*, **15** (1), 69–83.

Bahri, T. (1995) *Etude de la structure spatiale de la biomasse des poissons pélagiques côtiers et analyse préliminaire des relations avec l'environnement: le cas de la Mer Catalane.* Diplôme d'Etudes Approfondies, Université Paris VI (France).

Bailey, K.M. (1989) Interaction between the vertical distribution of juvenile walleye pollock *Theragra chalcogramma* in the eastern Bering Sea, and cannibalism. *Marc. Ecol.-Progr. Ser.*, **53**, 205–13.

Baird, J.W., Bishop, C.A. & Murphy, E.F. (1991a) Sudden changes in the perception of stock size and reference catch levels for cod in north-eastern Newfoundland shelves. *NAFO Sci. Coun. Studies*, **16**, 111–19.

Baird, T.A., Ryer, C.H. & Olla, B.L. (1991b) Social enhancement of foraging on an ephemeral food source in juvenile walleye pollock *Theragra chalcogramma. Environ. Biol. Fish.*, **31** (3), 307–11.

Baker, R.R. (1978) *The Evolutionary Ecology of Animal Migration.* Hodder & Stoughton, London.

Bakken, E (1983) Recent history of Atlanto-Scandian herring stocks. *FAO Fish. Rep.*, **291** (2), 521–36.

Bakun, A. (1985) Comparative studies and the recruitment problem: searching for generalisations. *CalCOFI Rep.*, **26**, 30–40.

Bakun, A. (1989) Mechanisms for density-dependent growth in Peruvian anchoveta; alternatives to impact on the regional-scale food supply. In: *The Peruvian Upwelling Ecosystem: Dynamics and Interactions* (eds Pauly, D., Muck, P., Mendo, J. & Tsukayama, I.), pp. 235–43. ICLARM, Manila.

Bakun, A. (1996) *Patterns in the Ocean: Ocean Processes and Marine Population Dynamics.* University of California Sea Grant, San Diego, California, USA, and Centro de Investigaciones Biológicas de Noroeste, La Paz, Baja California Sur, Mexico.

Bakun, A. & Parrish, R.H. (1990) Comparative studies of coastal pelagic fish reproductive habitats: the Brasilian sardine (*Sardinella aurita). J. Cons. Int. Explor. Mer*, **46** (3), 269–83.

Bakun, A. & Parrish, R.H. (1991) Comparative studies of coastal pelagic fish reproductive habitats: the anchovy *Engraulis anchoita* of the south-western Atlantic. *ICES J. Mar. Sci.*, **48** (3), 343–61.

Balchen, J.G. (1984) Recent progress in the control of fish behaviour. *Modelling, Identification and Control*, **52**, 113–21.

Barange, M. (1994) Acoustic identification, classification and structure of biological patchiness on the edge of the Agulhas Bank and its relation to frontal features. *S. Afr. J. Mar. Sci.*, **14**, 333–47.

Barange, M. & Hampton, I. (1994) Influence of trawling on *in situ* estimates of Cape horse mackerel (*Trachurus trachurus capensis*) target strength. *ICES J. Mar. Sci.*, **51**, 121–6.

Baranov, F.I. (1918) On the question of the biological basis of fisheries. *Nauchnye Issledovaniya Ikhtiologicheskii Instituta Izvestiya*, **1**, 81–128.

Baras, E. (1996) Commentaire à l'hypothèse de l'éternel retour de Cury (1994): proposition d'un mécanisme fonctionnel dynamique. *Can. J. Fish. Aquat. Sci.*, **53**, 681–4.

Bard, F.-X., Stretta, J.-M. & Slepoukha, M. (1985) Les epaves artificielles comme auxiliaires de la pêche thônière en océan atlantique. Quel avenir? *La Pêche Maritime*, 655–9.

Bardach, J.A., Magnuson, J.J., May, R.C. & Reinhart, J.M. (eds) (1980) *Fish Behaviour and Its Use in the Capture and Culture of Fishes: Proceedings of the Conference on the Physiological and Behavioural Manipulation of Food Fish as Production and Management Tools*, Bellagio, Italy, 3–8 November 1977. ICLARM Conference Proceedings, Manila.

Barlow, G.W. (1993) Fish behavioural ecology: pros, cons and opportunities. In *Behavioural Ecology of Fishes* (eds Huntingford, F.A. & Torricelli, P.), pp. 7–27. Harwood Academic Publishers, Chur.

Barria, P. (1990) Situación nacional de los principales recursos pelágicos. In: *Perspectivas de la Actividad Pesquera en Chile* (ed. Barbieri, M.A.), pp. 67–71. Escuela de Ciencias del Mar, Universidad católica de Valparaíso.

Barut, N.C. (1992) The payao fisheries in the Philippines and some observation on the behaviour of tunas around payao. In: *International Workshop on the Ecology and Fisheries for Tunas Associated with Floating Objects and on Assessment Issues Arising from the Association of Tunas with Floating Objects* (ed. Hall, M.). Inter-American Tropical Tuna Commission Int. Rep., San Diego.

Batalyants, K.Ya. (1992) On the hypothesis of comfortability stipulation of tuna association with natural and artificial floating objects. *ICCAT Rec. Doc. Sci.*, **40** (2), 447–53.

Batty, R.S., Blaxter, J.H.S. & Libby, D.A. (1986) Herring (*Clupea harengus*) filter-feeding in the dark. *Mar. Biol.*, **91**, 371–5.
Bayliff, W.H. (1984) Migrations of yellowfin and skipjack tuna released in the central portion of the eastern Pacific Ocean, as determined by tagging experiments. *Inter-American Tropical Tuna Commission. Intern. Rep.*, 18.
Bayliff, W.H. (1988) Integrity of schools of skipjack tuna, *Katsuwonus pelamis*, in the eastern Pacific ocean, as determined from tagging data. *Fish. Bull. (US)*, **84** (4), 631–43.
Baylis, J.R. (1981) The evolution of parental care in fishes, with reference to Darwin's rule of male sexual selection. *Environ. Biol. Fish.*, **6**, 223–51.
Belinay de, A. (1994) Hydros contre harengs. *Info-Pilote*, February, 20–22.
Beltestad, A.K. (1974) *Feeding behaviour, vertical migration and schooling among O-group herring* (Clupea harengus L.) *in relation to light intensity*. MSc. Thesis, University of Bergen (in Norwegian, unpublished).
Beltestad, A.K. & Misund, O.A. (1988a) *Behaviour of Norwegian spring spawning herring in relation to underwater light*. ICES FTFB Working Group 1988/04/18–19, Ostende.
Beltestad, A.K. & Misund, O.A. (1988b) Attraction of Norwegian spring-spawning herring to underwater light. In: *Proceedings of World Symposium on fishing gear and fishing vessel design*, pp. 190–2. The Marine Institute, St. John's, Newfoundland.
Beltestad, A.K. & Misund, O.A. (1990) On the danger of incidental fishing mortality in herring purse seining. In: *Proceedings of the International Herring Symposium*, pp. 617–628. Anchorage, Alaska, USA, 23–25 October, 1990.
Ben-Tuvia, A. & Dickson, W. (eds) (1968) Proceedings of the conference on fish behaviour in relation to fishing techniques and tactics, Bergen, Norway, 19–22 October, 1967. *FAO Fish Rep.*, **62**.
Ben-Yami, M. (1976) *Fishing with Light*. FAO of the United Nations. Fishing News Books, Oxford.
Ben-Yami, M. (1995) Tuna fishing – a review: (part 1 – General). *FAO Infofish International*, **5/95**, 69–76.
Bercy, C. & Bordeau, C. (1987a) Effects of underwater noise radiated by tuna fishing boats on fish behaviour. *International Symposium on Fisheries Acoustics*, 22–26 June, 1987, Seattle, Washington, USA, pp. 1–8.
Bercy, C. & Bordeau, A. (1987b) Physiological and ethological reactions of fish to low frequency noise radiated by sounders and sonars. *International Symposium on Fisheries Acoustics*, 22–26 June, 1987, Seattle, Washington, USA, pp. 1–14.
Berg, A.B. van den (1985). Analysis of the phase difference between particle motion components of sound by teleosts. *J. Exp. Biol.*, **119**, 183–97.
Berkeley, S.A. & Waugh, G.T. (1989) Considerations for regional swordfish management. *Proceedings of the Gulf and Caribbean Fisheries Institute, Miami*, **39**, 171–80.
Bernstein, C., Kacelnik, A. & Krebs, J. (1988) Individual decisions and the distribution of predators in a patchy environment. *J. Anim. Ecol.*, **57**, 1007–26.
Bernstein, C., Kacelnik, A. & Krebs, J. (1991) Individual decisions and the distribution of predators in a patchy environment. II. The influence of travel costs and structure of environment. *J. Anim. Ecol.*, **60**, 205–25.
Bertram, B.C.R. (1978) Living in groups: predators and prey. In: *Behavioural Ecology* (eds Krebs, J.R. & Davis, N.B.), pp. 64–96. Blackwell Science, Oxford.
Beukema, J.J. (1970) Angling experiments with carp (*Cyprinus carpio* L.), 2. Decreasing catchability through one-trial learning. *Neth. J. Zool.*, **20** (1), 81–92.
Beverton, R.J.H. (1954) *Notes on the use of theoretical models in the study of the dynamics of exploited fish populations*. U.S. Bureau of Commercial Fisheries, Fisheries Laboratory, Miscellaneous Contribution 2, Beaufort, North Carolina.
Beverton, R.J. (1990) Small marine pelagic fish and the threat of fishing; are they endangered? *J. Fish Biol.*, **37** (Supp.), 5–16.
Beverton, R.J.H. & Holt, S.J. (1957) On the dynamics of exploited fish populations. *Fish. Invest. London*, Ser. II, **19**, 1–533.
Binet, D. (1982) Influence des variations climatiques sur la pêcherie des *Sardinella aurita* ivoiro-ghanéennes: relation sécheresse-surpêche. *Oceanol. Acta*, **5** (4), 443–52.
Binet, D., Samb, B., Taleb Sidi, M., Lévénez, J.L. & Servain, J. (1997) Sardine and other pelagic fisheries changes associated with multi-year trade wind increases in the southern Canary current (26°N–14°N) 1964–1993. In: *Global versus Local Changes in Upwelling Systems* (eds Durand, M.H., Cury, P., Mendelssohn, R., Roy, C., Bakun, A. & Pauly, D.). Orstom Editions, Paris.

Bjordal, Å & Løkkeborg, S. (1997) *Longlining*. Fishing News Books, Oxford.

Bjørndal, T. & Conrad, J.M. (1987) The dynamics of an open access fishery. *Can. J. Econ.*, **20**, 75–85.

Blaber, S.J.M., Brewer, D.T. & Salini, J.P. (1995) Fish communities and the nursery role of the shallow inshore waters of a tropical bay in the Gulf of Carpentaria, Aust. Estuar. Coast. *Shelf Sci.*, **40** (2), 177–93.

Blackburn, M. (1965) Oceanography and the ecology of tunas. *Oceanogr. Mar. Biol.*, **3**, 299–322.

Blaxter, J.H.S. (1985) The herring: a successful species. *Can. J. Fish. Aquat. Sci.*, **42** (Suppl. 1), 21–30.

Blaxter, J.H.S. & Batty, R.S. (1984) The herring swimbladder: loss and gain of gas. *J. Mar. Biol. Assoc.* (UK), **64** (2), 441–59.

Blaxter, J.H.S. & Batty, R.S. (1990) Swimbladder 'behaviour' and target strength. *Rapp. P.-v. Réun. Cons. Int. Explor. Mer*, **189**, 233–44.

Blaxter, J.H.S. & Holliday, F.G.T. (1963) The behaviour and physiology of herring and other clupeids. *Adv. Mar. Biol.*, **1**, 261–393.

Blaxter, J.H.S. & Hunter, J.R. (1982) The biology of clupeoid fishes. *Adv. Mar. Biol.*, **20**, 1–223.

Blaxter, J.H.S. & Tytler, P. (1978) Physiology and function of the swimbladder. *Adv. Comp. Physiol. Biochem.*, **7**, 311–67.

Blaxter, J.H.S., Gray, J.A.B. & Denton, E.J. (1981) Sound and startle responses in herring shoals, *J. Mar. Biol. Assn.* (UK), **61**, 851–69.

Block, B.A., Dewar, H., Williams, T., Prince, E., Farwell, C. & Fudge, D. (1997) Archival and pop-up satellite tagging of Atlantic bluefin tuna, *Thynnus thynnus*. In: *48th Tuna Conference at Lake Arrowhead* (eds Scott, M. & Oson, R.). Inter-American Tropical Tuna Commission Int. Rep., San Diego.

Bodholdt, H. (1982) A multi-beam sonar for fish school observation. *ICES/FAO Symposium on Fisheries Acoustics*, Bergen, Norway, 21–24 June 1982, Doc. no. 55.

Bodholdt, H. & Olsen K. (1977) Computer-generated display of an underwater situation: applications in fish behaviour studies. *Rapp. P.-v. Réun. Cons. Int. Explor. Mer*, **170**, 31–5.

Bodholdt, H., Ness, H. & Solli, H. (1989) A new echo-sounder system. *Proc. Inst. Acoust.*, **11**, 123–30.

Boehlert, G.W. & Genin, A. (1987) A review of the effects of seamount on biological processes. In: *Seamounts, Islands and Atolls* (eds Keating, B.H., Fryer, P., Batiza, R. & Boehlert, G.W.), pp. 319–334. *J. Mar. Res.*, **43**.

Boehlert, G.W. & Mundy, B.C. (1994) Vertical and onshore–offshore distributional patterns of tuna larvae in relation to physical habitat features. *Mar. Ecol. Prog. Ser.*, **107**, 1–2, 1–13.

Boëly, T. (1980a) *Biologie des deux espèces de Sardinelle:* Sardinella aurita *(Valenciennes 1847) et* Sardinella maderensis *(Lowe 1841) des côtes sénégalaises.* Thèse de Doctorat de l'Université de Paris VI (France).

Boëly, T. (1980b) Etude du cycle sexuel de la sardinelle plate *Sardinella maderensis* (Lowe 1841) des côtes sénégalaises. *Cybium*, **8** (3), 77–88.

Boëly, R. & Fréon, P. (1979) Les ressources pélagiques cotières. In: *Les ressources halieutiques de l'Atlantique Centre-Est. lère partie: les ressources du Golfe de Guinée, de l'Angola à la Mauritanie* (eds Troadec, J.-P. & Garcia, S.), pp. 12–78. FAO Tech. Paper 186 (1).

Boëly, T., Chabanne, J. & Fréon, P. (1978) Schémas des cycles migratoires, lieux de concentration et périodes de reproduction des principles espèces de poissons pélagiques côtiers dans la zone sénégalo-mauritanienne. In: *Rapport du groupe de travail ad hoc sur les poissons pélagiques côtiers ouest-africains de la Mauritanie au Libéria (16°N à 5°N)*, pp. 63–70. COPACE/FAO, PACE series 78/10 fr.

Boëly, T., Chabanne, J., Fréon, P. & Stéquert, B. (1982) Cycle sexuel et migrations de *Sardinella aurita* sur le plateau continental ouest africain des îles Bissagos à la Mauritanie. *Rapp. P.-v. Réun. Cons. Int. Explor. Mer.*, **180**, 350–5.

Boggs, C.H. (1992) Depth, capture time, and hooked longevity of longline-caught pelagic fish: timing bites of fish with chips. *Fish. Bull.* (US), **90** (4), 642–58.

Boggs, C.H. (1997). Influence of sea surface temperature fronts and other environmental factors on swordfish and blue shark catches in the Hawaii-based longline fishery. In: *8th Tuna Conference at Lake Arrowhead* (eds Scott, M. & Oson, R.). Inter-American Tropical Tuna Commission, Int. Rep., San Diego.

Boggs, C.H., Bigelow, K.A. & He, X. (1997) Influence of sea surface temperatue fronts and other environmental factors on swordfish and blue shark catch rate in the Hawaii-based longline fishery. In: *48th Tuna Conference at Lake Arrowhead* (eds Scott, M. & Oson, R.). Inter-American Tropical Tuna Commission, Int. Rep., San Diego.

Bohnsack, J.A. & Ault, J.S. (1996) Management strategies to conserve marine biodiversity. *Oceanography*, **9** (1), 73–82.

Borstad, G.A., Hill, D.A. & Kerr, R.C. (1989) *The compact airborne spectrographic imager (CASI): flight and laboratory examples.* IGARRS' 89/12th. Canadian Symposium on Remote Sensing, Vancouver, B.C., 9–14 July (mimeo).

Borstad, G.A., Hill, D.A., Kerr, R.C. & Nakashima, B.S. (1992) Direct digital remote sensing of herring schools. *Int. J. Remote Sens.*, **13** (12), 2191–8.

Boyd, A.J., Taunton-Clark, J. & Oberholster, G.P.J. (1992) Spatial features of the near-surface and mid-water circulation patterns off western and southern South Africa and their role in the life histories of various commercially fished species. In: *Benguela Trophic Functioning* (eds Payne, A.I.L., Brink, K.H., Mann, K.H. & Hilborn, R.), pp. 189–206. *S. Afr. J. Mar. Sci.*, **12**.

Boyer, D., Boyer, H., D'Almeida, G., Cloete, R. & Agnalt, A.-L. (1995) The state of the northern Benguela pilchard stock. In: *Annual Research Meeting* (ed. Ministry of Fisheries and Marine Resources, Namibia), pp. 31–43, Rep. 1.

Bradford, M.J. & Peterman, R.M. (1989) Incorrect parameter values used in virtual population analysis (VPA) generate spurious time trends in reconstructed abundancies. In: *Effect of Ocean Variability on Recruitment and an Evaluation of Parameters Used in Stock Assessment Models* (eds Beamish, R.J. & McFarlane, G.A.), pp. 87–99. *Can. Spec. Pub. Fish. Aquat. Sci.*, **108**.

Brander, K. & Thomson, A. (1989) Diel differences in avoidance of three vertical profile sampling gears by herring larvae. *J. Plankton Res.*, **11**, 775–84.

Brandt, A. von (1984) *Fish Catching Methods of the World.* Fishing News Books, Oxford.

Brede, R., Kristensen, F.H., Solli, H. & Ona, E. (1990) Target tracking with a split-beam echo sounder. *Rapp. P.-v. Reun. Cons. Int. Explor. Mer*, **189**, 254–63.

Breder, C.M., Jr. (1929) Field observations on flying fishes: a suggestion of methods. *Zoologica* (USA), **9** (7), 295–312.

Breder, C.M., Jr. (1954) Equations descriptive of fish schools and other animal aggregations. *Ecology*, **35**, 361–70.

Breder, C.M., Jr. (1959) Studies on social groupings in fishes. *Bull. Am. Mus. Nat. Hist.*, **117**, 393–482.

Breder, C.M., Jr. (1967) On the survival value of fish schools. *Zoologica*, **52**, 25–40.

Breder, C.M., Jr. (1976) Fish schools as operational structures. *Fish Bull.* (US), **74**, 471–502.

Breder, C.M., Jr. & Halpern, F. (1946) Innate and acquired behaviour affecting the aggregation of fishes. *Physiol. Zool.*, **19**, 154–90.

Breitburg, D.L. (1989) Demersal schooling prior to settlement by larvae of the naked goby. *Environ. Biol. Fish.*, **26**, 97–103.

Brêthes, J.-C. (1998) Le poids de la science et de la gestion dans l'effondrement des stocks de poisson de fond: l'example de la morue du sud du Golfe du Saint-Laurent. *Aquat. Living Resour.* (in press).

Brett, J.R. (1979) Vol. VIII: Environmental factors and growth. In: *Fish Physiology* (eds Hoar, W.S., Randall, D.J. & Brett, J.R.), pp. 579–675. Academic Press, New York.

Brill, R.W. (1997) How water temperature limits the vertical movement of pelagic fish. *48th Tuna Conference at Lake Arrowhead.* Inter-American Tropical Tuna Commission, San Diego.

Brill, R.W., Block, B., Boggs, C., Bigelow, K., Freund, E. & Marcinek, D. (1996) Horizontal and vertical movements of adult yellowfin tuna near the Hawaiian islands observed by acoustic telemetry. In: *Sustaining Tuna Fisheries: Issues and Answers* (eds Jackson, A., Rasmussen, R. & Bartoo, N.) Proceedings of the 47th tuna conference, Southwest Fisheries Science Center/National Marine Fisheries Service, NOAA, La Jolla (mimeo).

Brock, R.E. (1985) Preliminary study of the feeding habits of pelagic fish around Hawaiian fish aggregation devices or can fish aggregation devices enhance local fisheries productivity? *Bull. Mar. Sci.*, **37** (1), 40–49.

Brock, V.E. (1954) Some aspects of the biology of aku, (*Katsuwonis pelamis*), in the Hawaiian Islands. *Pac. Sci.*, **8** (1), 94–104.

Brodeur, R.D. (1988) Zoogeography and tophic ecology of the dominant epipelagic fishes in the northern North Pacific. Japan-United States of America Seminar on the Micronection of the Subarctic Pacific, Honolulu, HI (USA), 18 Mar 1985. *Bull. Ocean Res. Inst.* (Univ. Tokyo), **26**, 1–28.

Brossut, R., Dubois, P. & Rigaud, J. (1974) Le grégarisme chez *Blaberus craniifer:* isolement et identification de la phéromone. *J. Insect. Physiol.*, **20**, 529–43.

Browman, H.I., Novales-Flamarique, I. & Hawryshyn, C.W. (1994) Ultraviolet photoreception contributes to prey search behaviour in two species of freshwater zooplanktivorous fishes. *J. Exp. Biol.*, **186**, 187–98.

Brown, G.E. & Smith, R.J.F. (1994) Fathead minnows use chemical cues to discriminate natural shoal-mates from unfamiliar conspecifics. *J. Chem. Ecol.*, **20** (12), 3051–61.

Brown, W.L. & Wilson, E.O. (1956) Character displacement. *Syst. Zool.*, **5**, 49–64.

Buckley, T.W. & Miller, B.S. (1994) Feeding habits of yellowfin tuna associated with fish aggregation devices in American Samoa. *Bull. Mar. Sci.*, **55** (2–3), 445–59.

Buerkle, U. (1968) Relation of pure tone thresholds to background noise level in the Atlantic cod. *J. Fish. Res. Board Can.*, **25** (6), 1155–60.

Buerkle, U. (1969) Auditory masking and the critical band in Atlantic cod (*Gadus morhua*). *J. Fish. Res. Board Can.*, **26**, 1113–19.

Buerkle, U. (1974) Gill-net catches of cod (*Gadus morhua* L.) in relation to trawling noise. *Mar. Behav. Physiol.*, **2**, 277–81.

Buerkle, U. (1977) Detection of trawling noise by the Atlantic cod (*Gadus morhua* L.). *Mar. Behav. Physiol.*, **4**, 233–42.

Buerkle, U. (1983) First look at herring distribution with a bottom referencing underwater towed instrumentation vehicle 'BRUTIV'. *FAO Fish. Rep.*, **300**, 125–30.

Buerkle, U. & Stephenson, R.L. (1990) Herring school dynamics and its impact on acoustic abundance estimation. *Proceedings of the International Herring Symposium*, Anchorage, Alaska, USA, 23–25 October, 1990, pp. 185–208.

Burczynski, J. (1982) Introduction to the use of sonar system for estimating fish biomass. *FAO Fish. Tech. Pap.*, **191** Rev. 1.

Burd, A.C. (1985) Recent changes in the central and southern North Sea herring stocks. *Can. J. Fish. Aquat. Sci.*, **42** (Suppl. 1), 192–206.

Burd, A.C. (1990) The North Sea herring fishery: an abrogation of management, pp. 1–22. In: *Proceedings of the International Herring Symposium*, Anchorage, Alaska, USA, 23–25 October, 1990.

Burger, A.E., Wilson, R.P., Garnier, D. & Wilson, M.P.T. (1993) Diving depths, diet, and underwater foraging of rhinoceros auklets in British Columbia. *Can. J. Zool.*, **71** (12), 2528–40.

Burgess, J.W. & Shaw, E. (1979) Development and ecology of fish schooling. *Oceanus*, **27**, 11–17.

Burnham, K.P., Anderson, D.R., White, G.C., Brownie, C. & Pollock, K.H. (1987) Design and analysis methods for fish survival experiments based on release–capture. *American Fish. Soc. Monograph*, **5**. Bethesda, Maryland.

Bushnell, P.G. & Brill, R.H. (1992) Oxygen transport and cardiovascular response in skipjack tuna (*Katsuwonus pelamis*) and yellowfin tuna (*Thunnus albacares*) exposed to acute hyposia. *J. Comp. Physiol. B.*, **162**, 131–43.

Butterworth, D.S. (1980) The value of catch-statistics-based management techniques for heavily fished pelagic stocks with special reference to the recent decline of the South-West African pilchard stock. *NATO Symposium on Applied Operations Research in Fishing*, Trondheim (Norway), 14 Aug 1979, **2**, pp. 97–136.

Butterworth, D.S. & Andrew, P.A. (1984) Dynamic catch-effort for the hake stocks in ICSEAP divisions 1.3 to 2.2. *Coll. Sci. Pap. ICSEAF/RECL. Doc. Sci. CIPASE*, **11** (1), 29–58.

Buwalda, R.J.A., Schuijf, A. & Hawkins, A.D. (1983) Discrimination by the cod of sounds from opposing directions. *J. Comp. Physiol.*, **150**, 175–84.

Caddy, J.F. & Defeo, O. (1996) Fitting the exponential and logistic surplus yield models with mortality data: some explorations and new perspectives. *Fish Res.*, **25**, 39–62.

Caddy, J.F. & Mahon, R. (1995) Reference points for fishery management. *FAO Tech. Pap.* 347.

Cahn, P. (1972) Sensory factors in the side-to-side spacing and positional orientation of tuna, *Euthynnus affinis*, during schooliong. *Fish. Bull.* (US), **70**, 197–204.

Caraco, T. (1980) Stochastic dynamics of avian foraging flocks. *Am. Nat.*, **115**, 262–75.

Carey, F.G. (1992) Through the thermocline and back again – heat regulation in big fish. *Océanus*, autumn, 79–85.

Carey, F.G. & Olson, R.J. (1982) Sonic tracking experiments with tunas. *Int. Comm. Conserv. Atl. Tuna, Collect. Vol. Scient. Papers*, **17** (2), 458–66.

Carey, F.G. & Robinson, B.H. (1981) Daily patterns in the activities of the swordfish *Xiphias gladius*, observed by acoustic telemetry. *Fish Bull.* (US), **79**, 227–92.

Carlisle, J. Jr., Turner, C. & Ebert, E. (1964) Artificial habitat in the marine environment. Calif. fish and game, *Fish Bull.* (US), **124**, 1–93.

Carlson, F.T. & Reintjes, J.W. (1972) Suitability of internal tags for Atlantic menhaden. *Fish. Bull.* (USA), **70** (2), 514–17.

Carlson, T.J. & Jackson, D.R. (1980) *Empirical evaluation of the feasibility of split beam methods for direct in situ target strength measurement of single fish.* Seattle Applied Physics Laboratory, University of Washington (APL-WW 8006).

Carragher, J.F. & Rees, C.M. (1994) Primary and secondary stress responses in golden perch, *Macquaria ambigua. Comp. Biochem. Physiol.*, **107A** (1), 49–56.

Carscadden, J.E., Frank, K.T. & Miller, D.S. (1989) Capelin (*Mallotus villosus*) spawning on the south-east shoal: influence of physical factors past and present. *Can. J. Fish. Aquat. Sci.*, **46**, 1743–54.

Castonguay, M. & Gilbert, D. (1995) Effects of tidal streams on migrating Atlantic mackerel, *Scomber scombrus* L. *ICES J. Mar. Sci.*, **52**, 941–54.

Castonguay, M., Rose, G.A. & Legget, W.C. (1992) On movements of Atlantic mackerel (*Scomber scombrus*) in the northern Gulf of St Lawrence: association with wind-forced advections of warmed surface waters. *Can. J. Fish. Aquat. Sci.*, **49**, 2232–41.

Caton, A.E. (1991) Review of aspects of southern bluefin tuna: biology, population and fisheries. In: *World Meeting on Stock Assessment of Bluefin Tunas: Strengths and Weaknesses* (eds Deriso, B. & Bayliff, W.), pp. 181–250. Inter-American Tropical Tuna Commission (USA), Special report 7.

Caverivière, A. (1982) Le baliste des côtes africaines (*Balistes carolinensis*): biologie, prolifération et possibilités d'exploitation. Oceanol. *Acta*, **5** (4), 453–9.

Caverivière, A. Gerlotto, F. & Stéquert, B. (1980) *Balistes carolinensis*, nouveau stock africain. *La Pêche Maritime*, August, 3–8.

Caverivière, A., Kulbicki, M., Konan, J. & Gerlotto, F. (1981) Bilan des connaissances actuelles sur *Balistes carolinensis* dans le golfe de Guinée. *Doc. Sci. Cent. Rech. Océanogr. Abidjan*, **12** (1), 1–28.

Cayré, P. (1982) Qu'est-ce qu'un banc de listao (*Kasuwonus pelamis*)? Quelques réflexions à partir d'observations faites lors des campagnes de marquage. *ICCAT coll. vol. Scient. Pap.*, **17** (2), 467–70.

Cayré, P. (1985) *Contribution à l'étude de la biologie et de la dynamique du listao (Katsuwonus pelamis), Linnaeus 1958) de l'Océan Atlantique.* Thèse de Doctorat, Université Paris VI (France).

Cayré, P. (1987) L'oxygène dissous et la répartition des thons (albacore, listao et patudo) dans l'océan Atlantique. *La Pêche Maritime*, **1306**, 92–5.

Cayré, P. (1990) Dispositifs de concentration de poissons et pêche artisanale. In: *Actes de la Conférence Thonière Régionale. Commission De L'Océan Indien* (eds Le Gall, J.-Y., Reviers, X. & Roger, C.), pp. 54–64. Colloques et Séminaires, Orstom Editions.

Cayré, P. (1991) Behaviour of yellowfin tuna (*Thunnus albacares*) and skipjack tuna (*Katsuwonus pelamis*) around fish aggregating devices (FADs) in the Comoros Islands as determined by ultrasonic tagging. *Aquat. Living Resour.*, **4**, 1–12.

Cayré, P. & Chabanne, J. (1986) Marquage acoustique et comportement de thons tropicaux (albacore: *Thunnus albacares*, et listao: *Katsuwonus pelamis*) au voisinage d'un dispositif concentrateur de poissons. *Océanogr. Trop.*, **21**, 167–83.

Cayré, P. & Marsac, F. (1991) *Report and preliminary results of the tagging programme of natural drift logs in the tuna purse seine fishery area of the Western Indian Ocean.* FAO Indo-Pacific Tuna Development and Management Progamme. Collect. Vol. Work. Doc., pp. 1–10.

Cayré, P. & Marsac, F. (1993) Modelling the yellowfin tuna (*Thunnus albacares*) vertical distribution using sonic tagging results and local environmental parameters. *Aquat. Living Resour.*, **6**, 1–14.

Cayré, P., Amon Kothias, J.B., Diouf, T. & Stretta, J.M. (1988) Biologie des thons. In: *Ressources, pêche et biologie des thonidés tropicaux de l'Atlantique centre-est* (eds Fonteneau, A. & Marcille, J.), pp. 157–266. FAO Fish. Tech. Pap. 292.

Champagnat, C. & Domain, F. (1979) Migrations des poissons démersaux le long des côtes Ouest-Africaines de 10 à 24° de latitude nord. *Doc. Sci. Cent. Rech. Océanogr. Dakar-Thyaroye*, **68**, 78–110.

Chang, S.K. & Hsu, C.C. (1994) *An updated assessment of Indian Ocean albacore stock by ASPIC. 5. Expert Consultation on Indian Ocean Tunas, Mahe (Seychelles), 4–8 Oct 1993.* IPTP Work. doc., pp. 102–7.

Chapman, C.J. (1970) Ship noise and hearing in fish. *Scottish Fish Bull.*, **33**, 22–4.

Chapman, C.J. (1973) Field studies of hearing in teleost fish. *Helgoländer wiss. Meeresunters.*, **24**, 371–90.

Chapman, C.J. & Hawkins, A.D. (1969) The importance of fish behaviour in relation to capture by trawls. *FAO Fish. Rep.*, **62** (3), 717–29.

Chapman, C.J. & Hawkins, A.D. (1973). A field study of hearing in the cod, *Gadus morhua*. *J. Comp. Physiol.*, **85**, 147–67.
Chapman, C.J. & Johnstone, A.D. (1974) Some auditory discrimination experiments on marine fish. *J. Exp. Biol.*, **61**, 521–8.
Chapman, P. (1988) On the occurrence of oxygen-depleted water south of Africa and its implications for Agulhas-Atlantic mixing. *S. Afr. J. Mar. Sci.*, **7**, 267–94.
Charles-Dominique, E. (1982) Exposé synoptique des données biologiques sur l'ethmalose (*Etmalosa fimbriata*, S. Bowdich, 1825). *Rev. Hydrobiol. Trop.*, **15** (4), 373–97.
Checkley, D.M., Ortner, P.B., Settle, L.R. & Cummings, S.R. (1997) A continuous, underway fish egg sampler. *Fish. Oceanogr.*, **6** (2), 58–73.
Chen, S.X. (1996) Studying school size effects in line transect sampling using the kernel method. *Biometrics*, **52**, 1283–94.
Chiappa-Carrara, X. & Gallardo-Cabello, M. (1993) Estudio del regimen y habitos alimentarios de la anchoveta *Engraulis mordax* Girard (Pisces: *Engraulidae*), en Baja California, Mexico. *Cienc. Mar.*, **19** (3), 285–305.
Chikhi, L. (1995) *Différenciation génétique chez* Sardinella aurita *et* S. maderensis. *Allozymes et ADN mitochondrial.* Thèse de Doctorat, Université Paris VI (France).
Chivers, D.P. & Smith, R.J.F. (1994a). The role of experience and chemical alarm signalling in predator recognition by fathead minnows, *Pimephales promelas*. *J. Fish Biol.*, **44** (2), 273–85.
Chivers, D.P. & Smith, R.J.F. (1994b) Fathead minnows, *Pimephales promelas*, acquire predator recognition when alarm substance is associated with the sight of unfamiliar fish. *Anim. Behav.*, **48** (3), 597–605.
Chivers, D.P. & Smith, R.J.F. (1995) Free-living fathead minnows rapidly learn to recognise pike as predators. *J. Fish. Biol.*, **46**, 949–54.
Christensen, V. & Pauly, D. (1992) ECOPATH II, a software for balancing steady-state ecosystem models and calculating network characteristics. *Ecol. Model.*, **61**, 169–85.
Churnside, J.H. & Hunter, J.R. (1996) Laser remote sensing of epipelagic fishes. *Int. Soc. Opt. Engenir., CIS Select. Pap.*, 2964, 38–53.
Cillaurren, E. (1994) Daily fluctuations in the presence of *Thunnus albacares* and *Katsuwonus pelamis* around fish aggregating devices anchored in Vanuatu, Oceania. *Bull. Mar. Sci.*, **55** (2–3), 581–91.
Clark, C.W. (1974) Possible effects of schooling on the dynamics of exploited fish population. *J. Cons. Int. Explor. Mer*, **36** (1), 7–14.
Clark, C.W. & Mangel, M. (1979) Aggregation and fishery dynamics: a theoretical study of schooling and the purse seine tuna fisheries. *Fish. Bull.* (US), **77** (2), 317–37.
Clark, C.W. & Mangel, M. (1986) The evolutionary advantages of group foraging. *Theor. Popul. Biol.*, **30**, 45–75.
Clark, E.L., Aronso, R.L. & Gordon, M. (1954) Mating behaviour patterns in two sympatric species of Xiphophorin fishes: their inheritance and significance in sexual isolation. *Bull. Am. Mus. Nat. Hist.*, **103**, 135–226.
Coble, D.W., Farabee, G.B. & Anderson, R.O. (1985) Comparative learning ability of selected fishes. *Can. J. Fish. Aquat. Sci.*, **42**, 791–6.
Collette, B.B. & Nauen, C.E. (1983) FAO species catalogue: Scombrids of the world. An annotated and illustrated catalogue of tunas, mackerels, bonites and related species known to date. *FAO Fish. Synop.* 125.
Conan, G.Y., Maynou, F., Stolyarenko, D. & Mayer, L. (1994) Mapping and assessment of fisheries resources with coastal and depth constraints, the case study of snow crab in the Bay of Island Fjord (Newfoundland). *ICES C.M. 1994/Joint Session on Estimating Abundance from Fishing Surveys and Acoustic Measurements.*
Conand, F., Boely, T. & Petit, D. (1988) Spatial distribution of small pelagic fish in the lagoon of New Caledonia. *Proceedings of the 6th International Coral Reef Symposium, Townsville*, **2**, 65–69.
Constantz, G.D. (1974) Reproductive effort in *Poeciliopsis occidentalis* (Poecilidae) Southwest. *Nature*, **19**, 17–42.
Corten, A. (1993) Learning processes in herring migrations. *ICES CM 1993/H18.*
Corten, A. & van de Kamp, G. (1992) Natural changes in pelagic fish stocks of the North Sea in the 1980s. *ICES Mar. Sci. Symp.*, **195**, 402–17.
Cram, D.L. (1974) Rapid stock assessment of pilchard populations by aircraft-borne remote sensors. In: *Proc. 9th Int. Symp. Remote Sensing Environ.* Ann Arbor, 15–19 April, 1043–1050.

Cram, D.L. & Hampton, I. (1976) A proposed aerial/acoustic strategy for pelagic fish stock assessment. *J. Cons. Int. Explor. Mer*, **37** (1), 91–7.
Crecco, V. & Overholtz, W. (1990) Causes of density-dependent catchability for Georges Bank haddock, *Melanogrammus aeglefinus*. *Can. J. Fish. Aquat. Sci.*, **47**, 385–94.
Csányi, V. (1985) Ethological analysis of predator avoidance by the Paradise fish (*Macropodus opercularis*): I. Recognition and learning of predators. *Behaviour*, **92**, 227–40.
Csányi, V. (1986) Ethological analysis of predator avoidance by the paradise fish (*Macropodus opercularis*): II. Key stimuli in avoidance learning. *Anim. Learn. Behav.*, **14**, 101–9.
Csányi, V. (1987) The replicative evolutionary model of animal and human minds. *World Future: J. Gen. Evol.*, **24**, 174–214.
Csányi, V. (1988) Contribution of the genetical and neural memory to animal intelligence. In: *Intelligence and Evolutionary Biology* (eds Jerison, H. & Jerison, I.), pp. 299–318. Springer-Verlag, Berlin.
Csányi, V. (1989) *Evolutionary Systems and Society: A general Theory*. Duke University Press, Durham.
Csányi, V. (1993) How genetics and learning make a fish an individual: a case study on the paradise fish. In: *Perspectives in Ethology* (eds Bateson, P.P.G., Klopfer, P.H. & Thompson, N.S.), pp. 1–52. Behaviour and Evolution, 10, Plenum Press, New York.
Csányi, V. & Altbäcker, V. (1990) Variable learning performance: the levels of behviour organization. *Acta Biol. Hung.*, **41**, 321–32.
Csányi, V. & Dóka, A. (1993) Learning interactions between prey and predator fish. In: *Behavioural Ecology of Fishes* (eds Huntingford, F.A. & Torricelli, P.), pp. 63–78. Harwood Academic Publishers, Chur.
Csányi, V. & Gerlai, R. (1988) Open-field behaviour and the behaviour-genetic analysis of the Paradise fish (*Macropodus opercularis*). *J. Comp. Psychol.*, **102**, 226–36.
Csányi, V. & Gervai, J. (1986) Behaviour-genetic analysis of the Paradise fish (*Macropodus opercularis*): II. Passive avoidance conditioning of inbred strains. *Behav. Genet.*, **16**, 553–7.
Csányi, V., Csizmadia, G. & Miklósi, Å. (1989) Long-term memory and recognition of another species in the paradise fish. *Anim. Behav.*, **37**, 908–11.
Csirke, J. (1988) Small schoaling pelagic fish stocks. In: *Fish Population Dynamics*, 2nd edn, (ed. Gulland, J.A.), pp. 271–302. John Wiley and Sons, New York.
Csirke, J. & Caddy, J.F. (1983) Production modelling using mortality estimates. *Can. J. Fish. Aquat. Sci.*, **40** (1), 43–51.
Csirke, J. & Sharp, G.D. (ed) (1984) Reports of the experts consultation to examine changes in abundance and species composition of neritic fish resources. *FAO Fish. Rep.*, **291** (1).
Cui, G., Wardle, C.S., Glass, C.W., Johnstone, A.D.F. & Mojsiewicz, W.R. (1991) Light level thresholds for visual reaction of mackerel, *Scomber scombrus* L., to coloured monofilament nylon gillnet materials, *Fish. Res.*, **10**, 255–63.
Cullen, J.M., Shaw, E. & Baldwin, H.A. (1965) Methods for measuring the three-dimensional structure of fish schools. *Anim. Behav.*, **13**, 534–43.
Cury, P. (1994) Obstinate nature: an ecology of individuals. Thoughts on reproductive behaviour and biodiversity. *Can J. Fisheries Aquat. Sci. (Perspectives)*, **51** (7), 1664–73.
Cury, P. & Anneville, O. (1998) Fisheries resources as diminishing assets: marine diversity threatened by anecdotes. In: *Global versus Local Changes in Upwelling Systems* (eds Durand, M.H., Cury, P., Mendelssohn, R., Roy, C., Bakun, A. & Pauly, D.). Orstom Editions, Paris.
Cushing, D.A. (1982) *Climate and Fisheries*. Academic Press, London.
Cushing, D.A. (1988) *The Provident Sea*. Cambridge University Press, Cambridge.
Cushing, D.A. (1995) *Population Production and Regulation in the Sea: a Fisheries Perspective*. Cambridge University Press, Cambridge.
Cushing, D.H. (1952) Echo-surveys for fish. *J. Cons. Int. Explor. Mer*, **18**, 45–60.
Cushing, D.H. (1977) Observations of fish schools with the ARL scanner. *Rapp. P.-v. Réun. Cons. Int. Explor. Mer*, **170**, 15–20.
Cushing, D.H. & Harden Jones, F.R. (1968) Why do fish school? *Nature*, **218**, 918–20.
Daan, N. (1991) Bias in virtual population analysis when the unit stock assessed consists of sub-stocks. *ICES CM 1991/F:30*.
Daget, J. (1954) Les poissons du Niger supérieur. *Mém. IFAN*, 36.
Dagorn, L. (1994) *Le comportement des thons tropicaux modélisé selon les principes de la vie artificielle*. Thèse de l'Ensar. TDM 133, Orstom Editions, Paris.

Dagorn, L., Petit, M. & Stretta, J.-M. (1997) Simulation of large-scale tropical tuna movements in relation with daily remote sensing data: the artificial life approach. *BioSystems*, **44**, 167–80.

Dalen, J. & Nakken, O.M. (1983) On the application of the echo integration method. *ICES CM 1983/B:19.*

Dawson, J.J. & Brooks, T.J. (1989) An innovative acoustic signal processor for fisheries science. *Proc. Inst. Acoust.*, **11**, 131–40.

DeBlois, E.M. & Rose, G.A. (1996) Cross-shoal variability in the feeding habits of migrating Atlantic cod (*Gadus morhua*). *Oecologia*, **108** (1), 1992–96.

Delgado de Molina, A., Santana, J.C., Ariz, J. & Delgado de Molina, R. (1995) Seguimiento de la pesca sobre 'manchas' de túnidos en las Islas Canarias. *ICCAT Collect. Vol. Sci. Pap.*, **45** (3), 215–23.

Demarcq, H. (1998) Spatial and temporal dynamics of the upwelling off Senegal and Mauritania: local change and trend. In: *Global versus Local Changes in Upwelling Systems* (eds Durand, M.H., Cury, P., Mendelssohn, R., Roy, C., Bakun, A. & Pauly, D.). Orstom Editions, Paris.

Depoutot, C. (1987) *Contribution à l'étude des dispositifs de concentration de poissons à partir de l'expérience polynésienne.* Rapp. Orstom Papeete (Tahiti), 33 (mimeo).

Deriso, R.B. (1980) Harvesting strategies and parameter estimation for an age-structured model. *Can. J. Fish. Aquat. Sci.*, **37**, 268–82.

Deriso, R.B., Quinn, T.J. & Neal, P.R. (1985) Catch-age analysis with auxiliary information. *Can. J. Fish. Aquat. Sci.*, **42**, 815–24.

Derzhavin, A.N. (1922) The stellate sturgeon (*Acipenser stellatus Pallas*), a biological sketch. *Byulletan' Bakinskoi Ikhtiologicheskoi Stantsii*, **1**, 1–393.

Devold, F. (1953) Tokter med *G.O. Sars* i. Norskehavet vinteren 1952/53 (Cruises with G.O. Sars in the Norwegian Sea, winter 1952–3). *FiskDir. Småskr.*, **6**, 1–19 (in Norwegian).

Devold, F. (1963) The life history of the Atlanto-Scandian herring. *Rapp. P-v. Réun. Cons. Int. Explor. Mer*, **154**, 98–108.

Devold, F. (1968) The formation and the disappearance of a stock unit of Norwegian herring. *FiskDir. Skr. Ser. Havunders.*, **15**, 1–22.

Devold, F. (1969) The behaviour of the Norwegian tribe of the Atlanto-Scandian herring. *FAO Fish. Rep.*, **62**, 534–9.

Dewsbury, D.A. (1984) *Comparative Psychology in the twentieth century.* Hutchinson Ross, Stroudsburg, PA.

DeYoung, D. & Rose, G.A. (1993) On recruitment and distribution of Atlantic cod (*Gadus morhua*) off Newfoundland. *Can. J. Fish. Aquat. Sci.*, **50**, 2729–41.

Di Natale, A. & Mangano, A. (1986) Moon phases influence on CPUE: a first analysis of swordfish driftnet catch data from the Italian fleet between 1990 and 1991. *ICATT Coll. Vol. Sci. Pap.*, **44** (1), 264–7.

Dickson, R.R., Maelkki, P., Radach, G., Sætre, R. & Sissenwine, M.P. (eds) (1992) Continuous plankton records: the North Sea in the 1980s. *ICES Mar. Sci. Symp.*, **195**, 243–8.

Die, D.J., Restrepo, V.R. & Fox, W.W. Jr. (1990) Equilibrium production models that incorporate fished area. *Amer. Fish. Soc.*, **119**, 445–54.

Diner, N. & Masse, J. (1987) Fish school behaviour during echo survey observed by acoustic device. *International Symposium on Fisheries Acoustics*, 21–24 June, Seattle, USA.

Dingle, H. (1996) Migration. *The Biology of Life on the Move.* Oxford University Press, Oxford.

Dizon, A.E. (1977) Effect of dissolved oxygen concentration and salinity on swimming speed of two species of tunas. *Fish. Bull.* (US) **75**, 649–53.

Dodson, J.J. (1988) The nature and role of learning in the orientation and migratory behaviour of fishes. *Environ. Biol. Fish.*, **23** (3), 161–82.

Domanevski, L.N. & Barkova, N.A. (1979) Particularité de la répartition et état du stock de sardine (*Sardina pilchardus*) au large de l'Afrique nord-occidentale. *COPACE/FAO, PACE Series 78/10*, 86–88.

Dommasnes, A., Midttun, L. & Monstad, T. (1979) Capelin investigations in the Barents Sea during the winter 1978. *FiskenHav.*, **1**, 1–16.

Dommasnes, A., Rey, F. & Rottingen, I. (1994) Reduced oxygen concentrations in herring wintering areas. *ICES J. Mar. Sci.*, **51**, 63–69.

Donguy, J.R., Bour, W., Galenon, P. & Gueredrat, J.A. (1978) Les conditions océanographiques et la pêche de la bonite (*Katsuwonus pelamis*) dans le Pacifique Occidental. *Cah. Orstom Sér. Océanogr.*, **16** (3–4), 309–17.

Doubleday, W.G. (1976) A least squares appraoch to analysing catch at age data. *Int. Comm. Northw. Atlant. Fish. Res. Bull.*, **12**, 69–81.

Døving, K.B. & Knutsen, J.A. (1993) Feeding responses and chemotaxis in marine fish larvae. In: *Fish Nutrition in Practice, Proceedings of the International Symposium on Fish Nutrition and Feeding*, Biarritz (France), 24–27 June 1991 (eds Kanshik, S.J. & Luquet, P.), pp. 579–587. *Inst. Nat. Rech. Agron.*, 61.

Dragesund, O. (1957) Reactions of fish to artificial light with special reference to large herring and spring herring in Norway. *J. Cons. Int. Explor. Mer*, **23** (2), 213–27.

Dragesund, O. & Hognestad, P. (1960) Småsildundersø kelsene og Småsildfisket 1959/60. *Fiskets Gang*, **46**, 703–14.

Dragesund, O. & Olsen, S. (1965) On the possibility of estimating year-class strength by measuring echo-abundance of O-group fish. *FiskDir. Skr. Ser. Havunders*, **13**, 47–75.

Dragesund, O., Hamre, J. & Ulltang, Ø. (1980). Biology and population dynamics of the Norwegian spring spawning herring. *Rapp. P.-v. Réun. Cons. Int. Explor. Mer*, **177**, 43–71.

Drago, G. & Yáñez, E. (1985) Dinámica de los principales stocks pelágicos. (*Engraulis ringens y Sardinops sagax*) explotados en la Zona Norte de Chile (18°20'S–24°00'S) entre 1959 y 1983. In: *Estudios en pesquerias chilenas* (ed. Melo, T.), pp. 97–106. Escuela de Ciencias del Mar, Universidad Católica de Valparaíso.

Draper, N.R. & Smith, H. (1966) *Applied Regression Analysis*. John Wiley and Sons, New York.

Drew, S. (1988) Trolling: a simple, effective and widely used method of line fishing. *Infofish International*, (3), 35–6.

Drickamer, L.C. & Vessey, S.H. (eds) (1992) *Animal Behaviour: Mechanisms Ecology and Evolution*, 3rd edn. Wm. C. Brown Publishers, Dubuque.

Drogoul, A., Corbara, B. & Fresneau, D. (1992) Applying etho-modelling to social organization in ants. In: *Biology and Evolution of Social Insects* (ed. Leuven University Press), pp. 375–383. Leuven University Press, Leuven.

Druce, B.E. & Kingsford, M.J. (1995) An experimental investigation on the fishes associated with drifting objects in coastal waters of temperate Australia. *Bull. Mar. Sci.*, **57** (2), 378–92.

Duffy, D.C. & Wissel, C. (1988) Models of fish school size in relation to environmental productivity. *Ecol. Model.*, **40**, 201–11.

Dufour, P. & Stretta, J.-M. (1973) Fronts thermiques et thermohalins dans la région du Cap Lopez (Golfe de Guinée) en juin-juillet 1972: phytoplancton, zooplancton, micronecton et pêche thonière. *Doc. Sci. Cent. Rech. Océanogr. Abidjan, Orstom*, **4** (3), 99–142.

Dulzetto, F. (1928) Osservazioni sulla vita sessuale di *Gambusia holbrooki*. *Atti R. Acc. Lincei, Rend.*, **8**, 96–101.

Dunning, D.J., Ross, Q.E., Geohegan, P., Reichle, J.J., Menezes, J.K. & Watson, J.K. (1992) Alewives avoid high-frequency sound. *N. Amer. J. Fish. Manag.*, **12**, 407–16.

Dupouy, C. & Demarcq, H. (1987) CZCS as an aid for understanding modalities of the phytoplankton productivity during upwelling off Senegal. *Adv. Space Res.*, **7** (2), 63–71.

Durand, M.H., Cury, P., Mendelssohn, R., Roy, C., Bakun, A. & Pauly, D. (eds) (1998) *Global versus Local Changes in Upwelling Systems*. Orstom Editions, Paris.

Ebersole, J.P. (1980) Food density and territory size: an alternative model and a test on the reef fish *Eupomacentrus leucostrictus*. *Amer. Naturalist*, **115**, 492–505.

Eckmann, R. (1991) A hydroacoustic study of the pelagic spawning behavior of whitefish (*Coregonus lavaretus*) in Lake Constance. *Can. J. Fish. Aquat. Sci.*, **48**, 995–1002.

ECOTAPP project (1995) Etude du comportement des thonidés par l'acoustique et la pêche à la palangre en Polynésie Française. *Rapp. Orstom Papeete* (Tahiti) (mimeo).

Edwards, J.I. & Armstrong, F. (1983) Measurements of the target strength of live herring and mackerel. *FAO Fish. Rep.*, **300**, 69–77.

Efanov, S.F. (1981) Herring of the Gulf of Riga: the problem of escapement and mechanical impact of the trawl. *ICES CM 1981/J:7*.

Ehrenberg, J.E. (1974) Two applications for a dual-beam transducer in hydro-acoustic fish assessment systems. In: *Proc. Conf. Engineering in the Ocean Environment*, 21–23 August, 1974, Halifax, Nova Scotia, Canada, pp. 152–5. IEEE, New York.

Ehrlich, P.R. & Ehrlich, A.H. (1973) Coevolution: heterospecific schooling in Caribbean reef fishes. *Am. Nat.*, **107**, 157–60.

Eibl-Eibesfeldt, I. (ed) (1984) Ethologie. Biologie du comportement. 3rd edn. Naturalia et Biologia, Paris.

Ekloev, P. & Hamrin, S.F. (1989) Predatory efficiency and prey selection: interactions between pike *Esox lucius*, perch *Perca fluviatilis* and rudd *Scardinus erythrophthalmus*. *Oikos*, **56** (2), 149–56.

Enger, P.S. (1967) Hearing in herring. *Comp. Biochem. Physiol.*, **22**, 527–38.

Engås, A.E. (1994) The effects of trawl performance and fish behaviour on the catching efficiency of demersal sampling trawls. In: *Maring Fish Behaviour in Capture and Abundance Estimation* (eds Fernö, A. & Olsen, S.), pp. 45–68. Fishing News Books, Oxford.

Engås, A.E. & Godø, O.R. (1989) Escape of fish under the fishing line of a Norwegian sampling trawl and its influence on survey results. *J. Cons. Int. Explor. Mer*, **45**, 269–76.

Engås, A.E. & Soldal, A.V. (1992) Diurnal variation in bottom trawl catches of cod and haddock and their influence on abundance indices. *ICES J. Mar. Sci.*, **49**, 89–95.

Engås, A. Jacobsen, J.A. & Soldal, A.V. (1988) *Diurnal changes in bottom trawl catches and vertical fish distribution.* ICES FTFB Working Group Meeting, Ostende, Copenhagen (mimeo).

Engås, A.E., Misund, O.A., Soldal, A.V., Horvei, B. & Solstad, A. (1995) Reactions of penned herring and cod to playback of original, frequency-filtered and time-smoothed vessel sound. *Fish. Res.*, **22**, 243–54.

Engås, A.E., Soldal, A.V. & Øvredal, J.T. (1991) Avoidance reactions of ultrasonic tagged cod during bottom trawling in shallow water. *ICES CM 1991/B:41.*

Enquist, M., Leimar, O., Ljungberg, T., Mallner, T. & Segerdahl, N. (1990) A test of the sequential assessment game: fighting in the cichlid fish *Nannocara anomala*. *Anim. Behav.*, **40**, 1–15.

Erickson, G.J. (1979) Some frequencies of underwater noise produced by fishing boats affecting albacore catch. *J. Acoust. Soc. Am.*, **66** (1), 296–99.

Esconomidis, P.S. & Vogiatzis, V.P. (1992) Mass mortality of *Sardinella aurita* Valenciennes (Pisces, Clupeidae) in Thessaloniki Bay (Macedonia, Greece). *J. Fish Biol.*, **41** (1), 147–9.

Fabre, J.H. (1930) *Souvenirs Entomologiques.* Delagrave, Paris.

FAO (1970) Report on a meeting for consultations on underwater noise, Rome, Italy, 17–19 December, 1968. *FAO Fish. Rep.*, **76**.

FAO (1996) *La Situation Mondiale des Pêches et de l'Aquaculture* (Sofia). FAO, Rome.

FAO (1997) *FAO Yearbook, Fishery Statistics, Catches and Landings.* FAO Fisheries Series, No. 46, FAO Statistics Series, No. 128. FAO of the United Nations, Rome.

Farman, R.S. (1985) *The dynamics of fish aggregating under anchored rafts.* South Pacific Commission Int. Rep., 8 (mimeo).

Farr, J.A. (1977) Male rarity or novelti, female choice behaviour, and sexual selection in the guppy, *Poecilia reticulata* Peters (Pisces: Poeciliidae). *Evolution*, **31**, 162–8.

Fay, R.R. (1970) Auditory frequency discrimination in the goldfish (*Carassius auratus*). *J. Comp. Physiol. Psychol.*, **73**, 175–80.

Fay, R.R. & Popper, A.N. (1974) Acoustic stimulation of the ear of the goldfish (*Carassius auratus*). *J. Exp. Biol.*, **61**, 243–60.

Fay, R.R. & Popper, A.N. (1975) Modes of stimulation of the teleost ear. *J. Exp. Biol.*, **62**, 379–87.

Fedoryako, B.I. (1982) Langmuir circulation as a possible mechanism of formation of fish association around floating object. *Oceanology*, **22**, 228–32.

Fedoseev, Y.P. & Chur, V.N. (1980) Relationship between feeding intensity of bigeye tuna (*Thunnus obesus*) of the Gulf of Guinea and their catches in the day time. *ICCAT Coll. Vol. Sci. Pap.*, **9** (2), 287–90.

Ferguson, M.M. & Noakes, D.L.G. (1981) Social grouping and genetic variation in common shiners, *Notropis cornutus* (Pisces, Cyprinidae). *Environ. Biol. Fish.*, **6** (3/4), 357–60.

Fernö, A. (1993) Advances in understanding of basic behaviour: consequences for fish capture studies. *ICES Mar. Sci. Symp*, **196**, 5–11.

Fernö, A. & Huse, I. (1983) The effect of experience on the behaviour of cod (*Gadus morhua* L.) towards a baited hook. *Fish. Res.*, **2**, 19–28.

Fernö, A. & Olsen, S. (eds) (1994) *Marine Fish Behaviour in Capture and Abundance Estimation.* Fishing News Books, Oxford.

Fernö, A., Solemdal, P. & Tilseth, S. (1986) Field studies on the behaviour of whiting (*Gadus merlangus* L.) towards baited hooks. *FiskDir. Skr. Ser. Havunders.*, **18**, 83–95.

Ferraris, J. (1995) Démarche méthodologique pour l'analyse des comportements tactiques et stratégiques des pêcheurs artisans sénégalais. In: *Question sur la Dynamique de l'Exploitation Halieutique* (eds. Laloë, F., Rey, H. & Durand, J.L.), pp. 263–295. Coll. Sémin., Orstom Editions, Paris.

Fiedler, P.C. & Bernard, H.J. (1987) Tuna aggregation and feeding near fronts observed in satellite imagery. *Contin. Shelf Res.*, **7** (8), 871–81.

Fiedler, P.C. & Laurs, R.M. (1990) Variability of the Columbia River plume observed in visible and infrared satellite imagery. *Int. J. Remote Sens.*, **11**, 999–1010.

Finn, D.B. (1964) *Modern Fishing Gear of the World*, vol. 2. Fishing News Books, Oxford.

Fitzgerald, G.J. & Wootton, R.J. (1986) Behavioural ecology of sticklebacks. In: *Behaviour of Teleost Fishes*, 2nd edn. (ed. Pitcher, T.J.), pp. 409–32. Chapman and Hall, London.

Fletcher, R.I. (1978) On the restructuring of the Pella-Tomlinson system. Fish. Bull. (US), **76** (3), 515–21.

Fonteneau, A. (1978) Analyse de l'effort de pêche des thoniers senneurs franco-ivoiro-sénégalais. Cah. Orstom, Sér. Océanogr., **16** (3–4), 285–307.

Fonteneau, A. (1985) Analyse de l'exploitation de quelques concentrations d'albacore par les senneurs durant la période 1980–1983, dans l'Atlantique Est. *ICCAT SCRS/85/79*, 81–98.

Fonteneau, A. (1986a) Eléments relatifs à l'effort de pêche exercé sur le Listao de l'Atlantique (*Katsuwonus pelamis*) et calcul d'indices d'effort spécifiques. In: *Proceedings of the ICCAT conference on the international skipjack year program, Madrid* (eds Symons, P.E.K., Miyake, P.M. & Sakagawa, G.T.E.), pp. 127–139. ICCAT.

Fonteneau, A. (1986b) Le modèle global et la dynamique du Listao. In: *Proceedings of the ICCAT conference on the international skipjack year program, Madrid* (eds Symons, P.E.K., Miyake, P.M. & Sakagawa, G.T.E.), pp. 198–207, ICCAT.

Fonteneau, A. (1988) Modélisation, gestion et aménagement des pêcheries thonières dans l'Atlantique Centre-Est. In: *Ressources, pêche et biologie des thonidés tropicaux de l'Atlantique Centre-Est* (eds Fonteneau, A. & Marcille, J.), pp. 317–356. FAO Doc. Tech. Pêches, 292.

Fonteneau, A. (1991) Monts sous-marins et thons dans l'Atlantique tropical est. *Aquat. Living Resour.*, **4**, 13–25.

Fonteneau, A. (1992) Pêche thonière et objets flottants: situation mondiale et perspectives. In: *International workshop on the ecology and fisheries for tunas associated with floating objects and on assessment issues arising from the association of tunas with floating objects* (ed. Hall, M.) Inter-American Tropical Tuna Commission Int. Rep., San Diego (mimeo).

Fonteneau, A. (1997) Scientific Atlas of Tropical Tuna Fisheries 'world wide'. Orstom Editions, Paris.

Fonteneau, A. & Diouf, T. (1994) An efficient way of bait-fishing for tunas recently developed in Senegal. *Aquat. Living Resour.*, **7**, 139–51.

Fonteneau, A. & Marcille, J. (eds) (1988) Ressources, pêche et biologie des thonidés tropicaux de l'Atlantique Centre-Est. *FAO Doc. Tech. Pêches*, **292**, 1–391.

Fonteneau, A., Gascuel, G. & Pallares, P. (1997) *Vingt-cinq Ans d'Évaluation des Ressources Thonières dans l'Atlantique: Quelques Réflexions Méthodologiques.* ICCAT.

Foote, K.G. (1978) Analysis of empirical observations on the scattering of sound by encaged aggregations of fish. *FiskDir. Skr. Ser. Havunders.*, **16**, 423–56.

Foote, K.G. (1979) Comment on 'Some frequencies of underwater noise produced by fishing boats affecting albacore catch'. *J. Acoust. Soc. Am.*, **66**, 296–9.

Foote, K.G. (1980a) Importance of the swimbladder in acoustic scattering by fish: a comparison of gadoid and mackerel target strengths. *J. Acoust. Soc. Am.*, **67**, 2084–9.

Foote, K.G. (1980b) Effects of fish behaviour on echo energy: the need for measurements of orientation distributions. *J. Cons. Int. Explor. Mer*, **39**, 193–201.

Foote, K.G. (1981) Evidence for the influence of fish behaviour on echo energy. In: *Meeting on hydroacoustical methods for the estimation of marine fish populations*, 25–29 June 1979, vol. 2 (ed. Soumala, J.B.), pp. 201–28. Charles Stark Draper Laboratory Inc., Cambridge, Massachusetts.

Foote, K.G. (1983) Linearity of fisheries acoustics, with addition theorems. *J. Acoust. Soc. Am.*, **73**, 1932–40.

Foote, K.G. (1985) Rather-high-frequency sound scattering by swimbladdered fish. *J. Acoust. Soc. Am.*, **78**, 688–700.

Foote, K.G. (1987) Fish target strengths for use in echo integrator surveys. *J. Acoust. Soc. Am.*, **82**, 981–7.

Foote, K.G. (1990) Correcting acoustic measurements of scatterer density for extinction. *J. Acoust. Soc. Am.*, **88**, 1543–6.

Foote, K.G. (1994) Extinction cross section of herring: new measurements and speculation. *ICES CM 1994/(B+D+G+H):2*, pp. 1–10.

Foote, K.G. & Knudsen, H.P. (1994) Physical measurements with modern echo integrators. *J. Acoust. Soc. Jpn.*, **15**, 393–5.

Foote, K.G. & Røttingen, I. (1995) Acoustic assessment of Norwegian spring spawning herring in the wintering area, December 1994 and January 1995. *ICES CM 1995/H:9*, pp. 1–22.

Foote, K.G., Kristensen, F.H. & Solli, H. (1984) Trial of a new, split-beam echo-sounder. *ICES CM 1984/B:21*.

Foote, K.G., Knudsen, H.P., Vestnes, G., MacLennan, D.N. & Simmonds, E.J. (1987) Calibration of acoustic instruments for fish density estimation: a practical guide. *ICES Coop. Res. Rep.* no. 144.

Foote, K.G., Knudsen, H.P., Korneliussen, R.J., Nordbø, P.E. & Røang, K. (1991) Postprocessing system for echo sounder data. *J. Acoust. Soc. Am.*, **90**, 37–47.

Foote, K.G., Ona, E. & Toresen, R. (1992) Determining the extinction cross section of aggregating fish. *J. Acoust. Soc. Am.*, **91**, 1983–9.

Foote, K.G., Ostrowski, M., Røttingen, I. Engås, A., Hansen, K., Hiis Hauge, K., Skeide, R., Slotte, A. & Torgersen, Ø. (1996) Acoustic abundance estimation of the stock of Norwegian spring spawning herring, winter 1995–1996. *ICES CM 1996/H:33.*

Forbes, S.T. & Nakken, O. (1972) *Manual of Methods for Fisheries Resource Survey and Appraisal. Part 2. The Use of Acoustic Instruments for Fish Detection and Abundance Estimation.* (FAO Man. Fish. Sci. 5). FAO, Rome.

Foster, S.A., Garcia, B. & Town, M.Y. (1988) Cannibalism as the cause of ontogenic niche shift in habitat use by fry of the threespine stickleback. *Oecologica*, **74**, 577–85.

Foster, W.A. & Treherne, H.E. (1981) Evidence for the dilution effect in the selfish herd from fish predation on a marine insect. *Nature*, **293**, 466–7.

Fournier, D. & Archibald, C.P. (1982) A general theory for analysing catch at age data. *Can. J. Fish. Aquat. Sci.*, **39**, 1195–207.

Fournier, D. & Doonan, I.J. (1987) A length-based stock assessment method utilising a generalized delay-difference model. *Can. J. Fish. Aquat. Sci.*, **44**, 422–37.

Fournier, D.A., Sibert, J.R., Majkowski, J. & Hampton, J. (1990) MULTIFAN a likelihood-based method for estimating growth parameters and age composition from multiple length frequency data sets illustrated using data for southern bluefin tuna (*Thunnus maccoyii*). *Can. J. Fish. Aquat. Sci.*, **47** (2), 301–17.

Fox, W.W. Jr. (1970) An exponential surplus-yield model for optimising exploited fish populations. *Trans. Amer. Fish. Soc.*, **99** (1), 80–8.

Fox, W.W. Jr. (1974) An overview of production modelling; Workshop on population dynamics of tuna, ICCAT meeting, Nantes, 2–7 September 1974. *ICATT Rec. Doc. Sci.*, **3**, 142–6.

Fox, W.W. Jr. (1975) Fitting the generalised stock production model by least-squares and equilibrium approximation. *Fish. Bull.* (US), **73** (1), 23–36.

Francis, R.C. (1980) Fisheries science now and in the future: a personal view. *N.Z.J. Mar. Freshwater Res.*, **14** (1), 95–100.

Frank, K.T., Carscadden, J.E. & Simon, J.E. (1996) Recent excursions of capelin (*Mallotus villosus*) to the Scotian Shelf and Flemish Cap during anomalous hydrographic conditions. *Can. J. Fish. Aquat. Sci.*, **53**, 1473–86.

Fraser, D.F. & Huntingford, F.A. (1986) Feeding and avoiding predation hazard: the behavioural response of the prey. *Ethology*, **73**, 56–68.

Fréon, P. (1984) La variabilité des tailles individuells à l'intérieur des cohortes et des bancs de poissons: observations et interprétation. *Oceanol. Acta*, **7**, 457–68.

Fréon, P. (1985) La Variabilité des tailles à l'intérieur des cohortes et des bancs de poissons: 2: Application à la biologie des pêches. *Oceanol. Acta*, **8**, 87–99.

Fréon, P. (1986) *Réponses et adaptations des stocks de clupéides d'Afrique de l'Ouest à la variabilité du milieu et de l'exploitation.* Thèse de Doctorat, Université d'Aix-Marseille II (France).

Fréon, P. (1988) Introduction of environmental variables into global production models. In: *International Symposium on Long Term Changes in Marine Fish Populations Vigo* (eds Wyatt, T. & Larrañeta, M.G.), pp. 481–528. Consejo Superior de Investigaciones Cientificas, Madrid.

Fréon, P. (1991) Seasonal and interannual variations of mean catch per set in the Senegalese sardine fisheries: fish behaviour or fishing strategy? In: *Long-term Variability of Pelagic Fish Populations and their Environment* (eds Kawasaki, T., Tanaka, S., Toba, Y. & Taniguchi, A.), pp. 135–45. Pergamon Press, Oxford.

Fréon, P. (1994a) Rendements par recrue et biomasse féconde du stock de *Sardinella aurita*. In: *Groupe de travail ad hoc sur les sardienelles et autres espèces de petits pélagiques côtiers de la zone Nord du COPACE* (ed. Do Chi, T.), pp. 221–8. COPACE/FAO, PACE series 91/58.

Fréon, P. (1994b) Simulation du cycle de vie de *Sardinella aurita* au Sénégal: comment concilier croissance rapide, stabilité des tailles modales et absence d'un mode. In: *Groupe de travail ad hoc sur les sardinelles et autres espèces de petits pélagiques côtiers de la zone Nord du COPACE* (ed. Do Chi, T.), pp. 181–219. COPACE/FAO, PACE series 91/58.

Fréon, P. & Gerlotto, F. (1988a) *Influence of fish behaviour on fish stock abundance estimations.* ICES FAST Working Group, Ostende, Belgium, 19–22 April 1988, pp. 1–7.
Fréon, P. & Gerlotto, F. (1988b) Methodological approach to study the biases induced by the fish behaviour during hydro-acoustic surveys. *ICES CM 1988/B:52*, pp. 1–16.
Fréon, P. & Stéquert, B. (1982) Note sur la présence de *Sardina pilchardus* (Walb.) au Sénégal: étude de la biométrie et interprétation. *Rapp. P.-v. Réun. Cons. Int. Explor. Mer.*, **180**, 435–9.
Fréon, P. & Weber, J. (1983) Djifére au Sénégal: la pêche artisanale et mutation dans un contexte industriel. *Rev. Trav. Inst. Pêches Mar.*, **47** (3–4), 304.
Fréon, P. & Yáñez, E. (1995) Influencia del medio ambiente en evaluacion de stock: una aproximación con modelos globales de producción. *Investigaciones Marinas.*, **23**, 25–47.
Fréon, P. Gerlotto, F. & Misund, O.A. (1993a) Consequences of fish behaviour for stock assessment. *Int. Coun. Explor. Sea Mar. Sci. Symp.*, **196**, 190–5.
Fréon, P., Gerlotto, F. & Soria, M. (1990) *Evaluation of the influence of vessel noise on fish distribution as observed using alternatively motor and sails aboard a survey vessel.* ICES FAST Working Group, Rostock, 1990.
Fréon, P., Gerlotto, F. & Soria, M. (1992) Changes in school structure according to external stimuli: description and influence on acoustic assessment. *Fish. Res.*, **15**, 45–66.
Fréon, P., Gerlotto, F. & Soria, M. (1993b) Variability of *Harengula* spp. school reactions to boats or predators in shallow water. *ICES Mar. Sci. Symp.*, **196**, 30–35.
Fréon, P., Gerlotto, F. & Soria, M. (1996) Diel variability of school structure with special reference to transition periods. *ICES J. Mar. Sci.*, **53** (2), 459–64.
Fréon, P., Mullon, C. & Pichon, G. (1991) CLIMPROD: a fully interactive expert-system software for choosing and adjusting a global production model which accounts for changes in environmental factors. In: *Long-term Variability of Pelagic Fish Populations and their Environment* (eds Kawasaki, T., Tanaka, S., Toba, Y. & Taniguchi, A.), pp. 347–57. Pergamon Press, Oxford.
Fréon, P., Mullon, C. & Pichon, G. (1993d) CLIMPROD: experimental interactive software for choosing and fitting surplus production models including environmental variables. *FAO Computerised Info. Series, (fish.)*, 5.
Fréon, P., Soria, M. & Gerlotto, F. (1989) Short-term variability of *Sardinella aurita* aggregation and consequences on acoustic survey results. *CIEM Statutory Meetings CM 1989/H:53.*
Fréon, P., Soria, M., Mullon, C. & Gerlotto, F. (1993c) Diurnal variation in fish density estimate during acoustic surveys in relation to spatial distribution and avoidance reaction. *Aquat. Liv. Resour.*, **6**, 221–34.
Fréon, P., Sow, I. & Lévénez, J.J. (1994) Trois décennies de pêche sardinière semi-industrielle au Sénégal. In: *L'évaluation des Ressources Exploitables par la Pêche Artisanale Sénégalaise* (eds Barry-Gérard, M., Diouf, T. & Fonteneau, A.), pp. 265–312. Colloques et Séminaires, Orstom Editions, Paris.
Fréon, P., Stéquert, B. & Boely, R. (1978) La pêche des poissons pélagiques côtiers en Afrique de l'Ouest des îles Bissagos au nord de la Mauritanie: description des types d'exploitation. *Cah. Orstom Sér. Océanogr.*, **18**, 209–28.
Fretwell, S.D. & Lucas, J.R. (1970) On territorial behaviour and other factors influencing habitat distribution in birds. I. Theoretical development. *Acta Biotheor.*, **19**, 16–36.
Freytag, G. & Karger, W. (1969) Investigations on noises caused by fishing vessels and their sources. *ICES CM 1969/B, 12*, pp. 1–3.
Fridriksson, A. & Aasen, O. (1950) The Norwegian–Icelandic herring tagging experiments. *FiskDir. Skr. Ser. Havunders.*, **9** (11), 1–44.
Frisch, K. von (1938) Zur Psychologie des Fish-Schwarmes. *Naturwissenschaften*, **26**, 601–6.
Frisch, K. von (1941) Ueber einen schreckstoff der fischhaut und seine biologische bedeutung. *Z. Vergl. Physiol.*, **29**, 46–145.
Fry, F.E.J. (1949) Statistics of a lake trout fishery. *Biometrics*, **5**, 26–67.
Fryer, G. & Iles, T.D. (1972) The cichlid fishes of the great lakes of Africa: their biology and evolution. Oliver and Boyd, Edinburgh.
Fuiman, L.A. & Magurran, A.E. (1994) Development of predator defences in fishes. *Rev. Fish Biol. Fisheries*, **4**, 145–83.
Fuiman, L.A. & Webb, P.W. (1988) Ontogeny of routine swimming activity and performance in zebra danois (Teleostei: Cyprinidae). *Anim. Behav.*, **26**, 250–61.
Fulton, T.W. (1896) Review of the trawling experiments of the 'Garland' in the Firth of Forth and St. Andrews Bay in the years 1886–1895. *Rpt. Fish. Bd. Scot.*, **14**, Part III, Scientific Investigations, 128–49.

Furusawa, M. & Takao, Y. (1992) Acoustic specifications of fisheries research vessel *Kaiyo-maru*. *ICES FAST & FTFB Working Groups Meeting*, Bergen, 16 June 1992.

Gaertner, D. & Medina-Gaertner, M. (1991) Factores ambientales y pesca atunera de superficie en el Mar Caribe. In: *Report of the Yellowfin Year Program*, ICCAT, Coll. Vol. Sci. Pap. 36, 523–50.

Gaertner, D. & Medina-Gaertner, M. (1992) Aperçu sur les relations entre les thons et les objets flottants dans le Sud de la mer des Caraïbes. In: *International Workshop on Fishing for Tunas Associated with Floating Objects* (ed. Hall, M.). Inter-American Tropical Tuna Commission Int. Rep., San Diego (mimeo).

Gaertner, D., Pagavino, M. & Marcano, J. (1996) Utilisation de modèles linéaires généralisés pour évaluer les stratégies de pêche thonière à la senne en présence d'espèces associées dans l'Atlantique ouest. *Aquat. Living Resour.*, **9**, 305–23.

Gallego, A. & Heath, M.R. (1994) The development of schooling behaviour in Atlantic herring *Clupea harengus*. *J. Fish. Biol.*, **45**, 569–88.

Gallego, A., Heath, M.R. & Fryer, R.J. (1995) Premature schooling of larval herring in the presence of more advanced conspecifics. *Anim. Behav.*, **50** (2), 333–41.

Galván-Magaña, F. & Olson, R.J. (1997) Trophic relations of apex predators in the bycatch of eastern Pacific purse-seiner fishery. In: *48th Tuna Conference at Lake Arrowhead* (eds Scott, M. & Oson, R.). Inter-American Tropical Tuna Commission Int. Rep., San Diego.

Garcia, S.M. (1994) Precautionary principle: its implications in capture fisheries management. *Ocean and Coastal Management*, **22**, 99–125.

Garnier, B., Beltri, E., Marchand, P. & Diner, N. (1992) Noise signature management of fisheries research vessels: a European survey. *ICES FTFB & FAST Working Groups Joint Session*, Bergen, Norway, 16 June 1992, pp. 1–5.

Garrod, D.J. (1969) Empirical assessments of catch effort relationship in the North Atlantic cod stocks. *Res. Bull. ICNAF*, **6**, 26–34.

Garstang, W. (1900) The impoverishment of the sea – a critical summary of the experimental and statistical evidence bearing upon the alleged depletion of the trawling grounds. *J. Mar. Biol. Ass.*, **6**, 1–69.

Gascuel, D. (1994) Modélisation de la dynamique des stocks exploités par la pêche artisanale sénégalaise: intérêt, limites et contraintes de l'approche structurale. In: *L'évaluation des Ressources Exploitables par la Pêche Artisanale Sénégalaise* (eds Barry-Gérard, M., Diouf, T. & Fonteneau, A.), pp. 385–403. Colloques et Séminaires, Orstom Editions, Paris.

Gauldie, R.W., Sharma, S.K. & Helsley, C.E. (1996) LIDAR applications to fisheries monitoring problems. *Can. J. Fish. Aquat. Sci.*, **53**, 1459–68.

Gauthiez, F. (1997) *Structuration spatiale des populations de poissons marins démersaux: caractérisation, conséquences biométriques et halieutiques*. Thèse de Doctorat, Université Claude Bernard Lyon I (France).

Gayanilo, F.C., Sparre, P. & Pauly, D. (1996) *FAO-ICLARM Fish Stock Assessment (FiSAT) User's Guide*. FAO Computerised Info. Series, (Fish) 7.

Genin, A. & Boehlert, G.W. (1985) Dynamics of temperature and chlorophyll structures above a seamount: an oceanic experiment. *J. Mar. Res.*, **43** (4), 907–24.

Gerlai, R. (1993) Can paradise fish (*Macropodus opercularis, Anabantidae*) recognise a natural predator? An ethological analysis. *Ethology*, **94** (2), 127–36.

Gerlai, R. & Csányi, V. (1990) Genotype-environment interaction and the correlation structure of behaviour elements in the paradise fish (*Macropodus opercularis*). *Physiol. Behav.*, **47**, 343–56.

Gerlotto, F. (1993) *Méthodologie d'observation et d'évaluation par hydroacoustique des stocks tropicaux de poissons pélagiques côtiers*. Thése Université de Bretagne Occidentale.

Gerlotto, F. & Fréon, P. (1990) *Review of avoidance reactions of tropical fish to a survey vessel*. ICES FAST Working Group, Rostock.

Gerlotto, F. & Fréon, P. (1992) Some elements on vertical avoidance of fish schools to a vesel during acoustic surveys. *Fish. Res.*, **14**, 251–9.

Gerlotto, F. & Stéquert, C. (1983) Une methode de simulation pour étudier la distribution des densitiés en poissons: application a deux cas reels. *FAO Fish. Rep.*, **300**, 278–92.

Gerlotto, F., Petit, D. & Fréon, P. (1990) Influence of the light of a survey vessel on TS distribution. ICES FAST Working Group, Rostock.

Gerlotto, F., Fréon, P., Soria, M., Cottais, P.H. & Ronzier, L. (1994) Exhaustive observation of 3D school structure using multibeam side scan sonar: potential use for school classification, biomass estimation and behaviour studies. *ICES CM/1994, B:26*.

Ghazai, M.E., Benech, V. & Paugy, D. (1991) L'alimentation de *Brycinus leuciscus* (Teleostei: Characidae) au Mali: aspects qualitatifs, quantitatifs et comportementaux. *Ichthyol. Explor. Fresh.*, **2** (1), 47–54.

Gibson, R.N. & Ezzi, I.A. (1992) The relative profitability of particulate-and filter-feeding in the herring, *Clupea harengus* L. *J. Fish Biol.*, **40**, 577–90.

Giles, N. & Huntingford, F.A. (1984) Predation risk and interpopulation variation in antipredator behaviour in the three-spined stickleback. *Anim. Behav.*, **32**, 264–75.

Gjestland, T. (1968) *Undervanns støygenerering ved båter, støyens karakter og dens betydning for fangst og bruk av sonaranlegg (Underwater noise generation by boats, noise characteristics and its importance for capture and use of sonar equipment)*. MSc. Thesis, Norw. Techn. Highschool. (Unpublished.)

Gjøsæter, H. & Korsbrekke, K. (1990) Schooling-by-size in the Barents Sea capelin stock. *ICES CM 1990/ D:28 Ref. H.*

Glantz, M.H. & Thompson, J.D. (1981) (eds) *Resource Management and Environmental Uncertainty: Lessons from Coastal Upwelling Fisheries.* John Wiley & Sons, New York.

Glass, C.W. & Wardle, C.S. (1989) Comparison of the reactions of fish to a trawl gear, at high and low light intensities. *Fish. Res.*, **7**, 249–66.

Glass, C.W., Wardle, C.S. & Mojsiewicz, W.R. (1986) A light intensity threshold for schooling in the Atlantic mackerel, *Scomber scombrus*. *J. Fish. Biol.*, **29**, 71–81.

Glass, C.W., Johnstone, A.D.F., Smith, G.W. & Mojsiewicz, W.R. (1992) The movements of saithe (*Pollachius virens* L.) in the vicinity of an underwater reef. In: *Wildlife telemetry: Remote monitoring and tracking of animals. Conference on Wildlife Telemetry* (eds Priede, I.G. & Swift, S.M.), pp. 328–340. Ellis Horwood, Chichester.

Glass, C.W., Wardle, C.S. & Gosden, S.J. (1993) Behavioural studies of the principles underlying mesh penetration by fish. *ICES Mar. Sci. Symp.*, **196**, 92–97.

Godin, J.-G.J. (1986) Antipredator function of schooling in teleost fishes: a selective review. *Nat. Can. (Quebec)*, **71**, 149–55.

Godin, J.-G.J. & Morgan, J. (1985) Predator avoidance and school size in a cyprindontid fish, the banded killifish (*Fundulus diaphanus* Lesueur). *Behav. Ecol. Sociobiol.*, **16**, 105–10.

Godø, O.R. (1994) Factors affecting the reliability of groundfish abundance estimates from bottom trawl surveys. In: *Marine Fish Behaviour in Capture and Abundance Estimation* (eds Fernö, A. & Olsen, S.) pp. 166–99. Fishing News Books, Oxford.

Godø, O.R. & Valdemarsen, J.W. (1993) A three level pelagic trawl for near surface sampling of juvenile fish. *ICES CM/1993, B:19, Ref. H.*

Godø, O.R. & Wespestad, V. (1993) Monitoring changes in abundance of gadoids with varying availability to surveys. *ICES J. Mar. Sci.*, **50**, 39–51.

Gomes, C., Mahon, R., Singh-Renton, S. & Hunte, W. (1994) The role of drifting objects in pelagic fisheries in the south-eastern Caribbean. *Caricom Fish. Res. Doc.*, Chap. 15.

Goncharov, S.E., Borisenko E.S. & Pyanov, A. (1989) Jack mackerel schools' defence reaction to a surveying vessel. *Proc. I.O.A.*, **11**, 74–8.

González-Ramos, A.J. (1989) Influencia de las condicionantes medio-ambientales en la pesquería superficial del atún listado (*Katsuwonus pelamis*) en aguas del Archipiélago Canario. *Col. Doc. Client. ICCAT.*, **30** (1), 138–49.

González-Ramos, A.J. (1992) *Bioecología del listado* (Katsuwonus pelamis, *linneatus 1758) en aguas del archipiélago Canario: modelo de gestíon y explotación mediante el uso de la teledetección.* Thesis Universidad de las Palmas de Gran Canaria (Spain).

Gooding, R.M. & Magnuson, J.J. (1967) Ecological significance of a drifting object to pelagic fishes. *Pacific Science*, **21**, 486–97.

Graham, M. (1935) Modern theory of exploiting a fishery, and application to North Sea trawling. *J. Cons. Int. Explor. Mer*, **10**, 264–74.

Graham, W.M. & Largier, J.L. (1997) Upwelling shadows as nearshore retention sites: the example of northern Monterey Bay. *Cont. Shelf Res.*, **17** (5), 509–32.

Grandperrin, R. (1975) *Structures trophiques aboutissant aux thons de longue ligne dans le Pacifique sud-ouest tropical.* Thèse de Doctorat, Université d'Aix-Marseille (France).

Graves, J. (1977) Photographic method for measuring spacing and density within pelagic fish schools at sea. *Fish. Bull.* (US) **74**, 230–34.

Gray, J.A.B. & Denton, E.J. (1991) Fast pressure pulses and communication between fish. *J. Mar. Biol. Assn.* (UK) **71**, 83–106.

Green, R.E. (1967) Relationship of the thermocline to success of purse seining for tuna. *Transactions of the Amer. Fish. Soc.*, **96** (2), 126–30.

Greenblatt, P.R. (1976) Factors affecting tuna purse seine fishing effort. *ICCAT SCRS 76/75*, 18–30.

Greenblatt, P.R. (1979) Associations of tuna with flotsam in the eastern tropical Pacific. *Fish. Bull.* (US) **77** (1), 147–55.

Gregory, R.S. (1993) Effect of turbidity on the predator avoidance behaviour of juvenile chinook salmon (*Oncorhynchus tshawytscha*). *Can. J. Fisheries Aquat. Sci.*, **50** (2), 241–6.

Gross, M.R. & Charnov, E.L. (1980) Alternative male life histories in bluegill sunfish. *Proc. Natl. Acad. Sci.* (USA) **77**, 6937–40.

Guerrero, A. & Yañez, E. (1986) Análisis de las principales pesquerías pelágicas desarrolladas en la zona de Talcahuano (37°S–73.°W) entre 1965 y 1984. Valparaíso. In: *La pesca en Chile* (ed. Arana, P.), pp. 223–241. Escuela de Ciencias del Mar.

Gulland, J.A. (1955) Estimation of growth and mortality in commercial fish populations. *Fish. Invest., Lond.*, Ser. 2, **18** (9).

Gulland, J.A. (1964) Catch per unit effort as a measure of abundance. *Rapp. P.-v. Réun. Cons. Int. Explor. Mer*, **155**, 8–14.

Gulland, J.A. (1965) Estimation of mortality rates. Annex to Arctic Fisheries. Report of the Working Group. *ICES CM/1965*, doc. 3 (mimeo).

Gulland, J.A. (1969) Manual of methods for fish stock assessment Part 1. Fish population analysis. *FAO Man. Fish. Sci.*, **4**.

Gunderson, D.R. (1993) *Surveys of Fisheries Resources*. John Wiley & Sons, New York.

Guthrie, D.M. (1986) Role of vision in fish behaviour. In: *The Behaviour of Teleost Fishes* (ed. Pitcher, T.J.), pp. 75–113. Croom Helm, London.

Guthrie, D.M. & Muntz,W.R.A. (1993) Role of vision in fish behaviour. In: *Behaviour of Teleost Fishes*, 2nd edn, (ed. Pitcher, T.J.), pp. 89–128. Chapman and Hall, London.

Guyomarc'h, J.-C. (1980) Abrégé d'éthologie. In: *Déterminisme, fonction, ontogénèse, évolution des comportements*. Masson, Paris.

Haegele, C.W. & Schweigert, J.F. (1985) Distribution and characteristics of herring spawning grounds and description of spawning behavior. *Can. J. Fish. Aquat. Sci.*, **42** (Suppl. 1), 39–55.

Hagstrøm, O. & Røttingen, I. (1982) Measurements of the density coefficient and average target strength of herring using purse seine. *ICES CM 1982/B:33*, pp. 1–13.

Hall, M. (1992a) The association of tunas with floating objects and dolphins in the Eastern Pacific ocean: VII. Some hypotheses on the mechanisms governing the association of tunas with floating objects and dolphins. In: *International workshop on the ecology and fisheries for tunas associated with floating objects and on assessment issues arising from the association of tunas with floating objects* (ed. Hall, M.). Inter-American Tropical Tuna Commission Int. Rep., San Diego.

Hall, M. (ed.) (1992b) *International Workshop on the ecology and fisheries for tunas associated with floating objects and on assessment issues arising from the assocition of tunas with floating objects*. Inter-American Tropical Tuna Commission Int. Rep., San Diego (mimeo).

Hall, M. & Garcia, M. (1992) The association of tunas with floating objects and dolphins in the Eastern Pacific ocean: IV. Study of repeated sets on the same object. In: *International workshop on the ecology and fisheries for tunas associated with floating objects and on assessment issues arising from the association of tunas with floating objects* (ed. Hall, M.). Inter-American Tropical Tuna Commission Int. Rep., San Diego.

Hall, M., Lennert, C. & Arenas, P. (1992a) The association of tunas with floating objects and dolphins in the Eastern Pacific ocean: II. The purse-seine fishery for tunas in the eastern Pacific Ocean. In: *International workshop on the ecology and fisheries for tunas associated with floating objects and on assessment issues arising from the association of tunas with floating objects* (ed. Hall, M.). Inter-American Tropical Tuna Commission Int. Rep., San Diego.

Hall, M., Garcia, M. Lennert, C. & Arenas, P. (1992b). The association of tunas with floating objects and dolphins in the Eastern Pacific ocean: III. Characteristics of floating objects and their attractiveness for tunas. In: *International workshop on the ecology and fisheries for tunas associated with floating objects and on assessment issues arising from the association of tunas with floating objects* (ed. Hall, M.). Inter-American Tropical Tuna Commission Int. Rep., San Diego.

Hall, S.J., Wardle, C.S. & MacLennan, D.N. (1986) Predator evasion in a fish school: test of a model for the fountain effect. *Mar. Biol.*, **91** (1), 143–8.

Halldorsson, O. & Reynisson, P. (1983) Target strength measurement of herring and capelin in-situ. *ICES CM 1983/H:36.*

Hallier, J.-P. (1986) Purse seining on debris associated schools in the Western Pacific Ocean. Indo Pacific Tuna Development and Management Programme. Expert consultation on stock assessment of tunas in the Indian Ocean in Colombo, Sri Lanka, 28 November–12 December 1985. *IPTP work. doc.*, pp. 150–56.

Hallier, J.P. (1995) Tropical tuna fishing with purse seine and log. *FAO Infofish International*, **4**, 53–8.

Hallier, J.-P. & Kulbicki, M. (1985) Analyse des résultats de la pêcherie à la canne de la Nouvelle-Calédonie. *Rapp. Sci. Tech. Cent. Nouméa Orstom*, 36 (mimeo).

Hallier, J.-P. & Parajua, J.I. (1992a) Tropical tuna at sea: what are they associated with? In *International workshop on the ecology and fisheries for tunas associated with floating objects and on assessment issues arising from the association of tunas with floating objects* (ed. Hall, M.). Inter-American Tropical Tuna Commission Int. Rep., San Diego (mimeo).

Hallier, J.-P. & Parajua, J.I. (1992b) Review of tuna fisheries on floating objects in the Indian Ocean. In: *International workshop on the ecology and fisheries for tunas associated with floating objects and on assessment issues arising from the association of tunas with floating objects* (ed. Hall, M.) Inter-American Tropical Tuna Commission Int. Rep., San Diego (mimeo).

Hallier, J.-P. & Parajua, J.I. (1992c) Fishing for tunas on the same floating object. In *International workshop on the ecology and fisheries for tunas associated with floating objects and on assessment issues arising from the association of tunas with floating objects* (ed. Hall, M.) Inter-American Tropical Tuna Comission Int. Rep., San Diego (mimeo).

Hamilton, W.D. (1971) Geometry for the selfish herd. *J. Theor. Biol.*, **31**, 295–311.

Hamilton, M.G., Nevarez-Martinez, M.O. & Rosales-Casian, J.A. (1991) Pacific sardine and northern anchovy in the Gulf of California, Mexico: current results of SARP MEXICO. *ICES CM/1991, H:20.*

Hampton, I. (1996) Acoustic and egg-production estimates of South African anchovy biomass over a decade: comparisons, accuracy, and utility. *ICES J. Mar. Sci.*, **53** (2), 493–500.

Hampton, I., Armstrong, M.J., Jolly, G.M. & Shelton, P.A. (1990) Assessment of anchovy spawner biomass off South Africa through combined acoustic and egg-production surveys. *Rapp. P.-v. Réun. Cons. Int. Explor. Mer*, **189**, 18–32.

Hampton, J. (1986) Effect of tagging on the condition of southern bluefin tuna. *Thunnus maccoyii* (Castlenau). *Aust. J. Mar. Freshwater Res.*, **37** (6), 699–705.

Hampton, J. & Bailey, K. (1992) Fishing for tunas associated with floating objects: Review of the western Pacific fishery. In: *International workshop on the ecology and fisheries for tunas associated with floating objects and on assessment issues arising from the association of tunas with floating objects* (ed. Hall, M.). Inter-American Tropical Tuna Commission Int. Rep., San Diego (mimeo).

Hamre, J. (1991) Interrelation between environmental changes and fluctuating fish population in the Barents Sea. In: *Long-term Variability of Pelagic Fish Populations and their Environment* (eds Kawasaki, T., Tanaka, S., Toba, Y. & Taniguchi, A.), pp. 259–70. Pergamon Press.

Hamre, J. & Dommasnes, A. (1994) Test experiments of target strength of herring by comparing density indices obtained by acoustic method and purse seine catches. *ICES CM 1984/B:17*, pp. 1–5.

Hanamoto, E. (1975) Fishery oceanography of bigeye tuna. II. Thermocline and dissolved oxygen content in relation to tuna longline fishing grounds in the eastern tropical Pacific Ocean. *Bull. Soc. Franco-Japonaise Oocéanogr.*, **13**, 58–71.

Hancock, J., Hart, P.J.B. & Antezana, T. (1995) Searching behaviour and catch of horse mackerel (*Trachurus murphyi*) by industrial purse-seiners off south-central Chile. *ICES J. Mar. Sci.*, **52**, 991–1004.

Hand, C.M. & Richards, L.J. (1991) Purse-seine surveys of post-larval lingcod (*Ophiodon elongatus*) in the Strait of Georgia, 1989–90. *Can. Manuscr. Rep. Fish. Aquat. Sci.*, **2106**.

Hansen, L.P. (1988) Effects of carlin tagging and fin clipping on survival of Atlantic salmon (*Salmo salar* L.) released as smolts. *Aquaculture*, **70** (4), 391–4.

Hansen, L.P. & Jonsson, B. (1994) Homing of Atlantic salmon: effects of juvenile learning on transplanted post-spawners. *Anim. Behav.*, **47** (1), 220–22.

Hansson, S. (1993) Variation in hydroacoustic abundance of pelagic fish. *Fish. Res.*, **16**, 203–22.

Hara, I. (1985) Shape and size of Japanese sardine school in the waters off the southeastern Hokkaido on the basis of acoustic and aerial surveys. *Bull. Japan. Soc. Sci. Fish.*, **51** (1), 41–6.

Hara, I. (1986) Stock assessment of Japanese sardine in the waters off the Southeast coast of Hokkaido using line transect method. *Bull. Jap. Soc. Sci. Fish.*, **52** (1), 69–73.

Hara, I. (1987) Swimming speed of sardine school on the basis of aerial survey. *Nippon Suisan Gakkaishi*, **53** (2), 223–7.
Hara, T.J. (1993) Role of olfaction in fish behaviour. In: *Behaviour of Teleost Fishes*, 2nd edn (ed. Pitcher, T.J.), pp. 171–99. Chapman and Hall, London.
Haralabous, J. & Georgakarakos, S. (1996) Artificial neural networks as a tool for species identification of fish schools. *ICES J. Mar. Sci.*, **53** (2), 173–80.
Harden Jones, F.R. (1963) The reaction of fish to moving backgrounds. *J. Exp. Biol.*, **40**, 437–46.
Harden Jones, F.R. (1968) *Fish Migration*. St Martin's Press, New York.
Harden Jones, F.R. (1978) Objectives and problems related to research into fish behaviour. In: *Sea Fisheries Research* (ed. Harden Jones, F.R.), pp. 261–75. John Wiley and Sons, New York.
Harden Jones, F.R. & McCartney, B.S. (1962) The use of electronic sector-scanning sonar for following the movements of fish shoals: sea trials on RRS Discovery II. *J. Cons. Int. Explor. Mer*, **27**, 141–9.
Harden Jones, F.R. & Scholes, P. (1981) The swimbladder and vertical movements and the target strength of fish. In: *Meeting on Hydroacoustical Methods for the Estimation of Marine Fish Populations*, 25–29 June 1979, vol. 2 (ed. Soumala, J.B.), pp. 157–82. The Charles Stark Draper Laboratory, Cambridge, Mass., USA.
Hardin, G. (1960) The competitive exclusion principle. *Science*, **131**, 1291–7.
Harris, J.H. (1989) Planning of mark-recapture experiments to estimate population abundance in freshwater fisheries. Tagging: solution or problem. Workshop, Australian Society for Fish Biology, Sydney, NSW (Australia), 21 July 1988. *Proc. Bur. Rural Resour., Australian Government Publishing Service, Canberra*, **5**, 43–4.
Hart, P.J.B. (1986) Foraging in Teleost Fishes. In: *Behaviour of Teleost Fishes*, 2nd edn. (ed. Pitcher, T.J.), pp. 211–35. Chapman and Hall, London.
Hattala, K., Stang, D., Kahnle, A. & Mason, W. (1988) *Beach seine survey of juvenile fishes in the Hudson River estuary*. Summary report for 1980–1986. New York State Dept. Environ. Conserv.
Hauser, J.W. & Sissenwine, M.P. (1991) The uncertainty in estimates of the production of larval fish derived from samples of larval abundance. *ICES J. Mar. Sci.*, **48**, 23–32.
Hawkins, A.D. (1973) The sensitivity of fish to sounds. *Oceanogr. Mar. Biol. Ann. Rev.*, **11**, 291–340.
Hawkins, A.D. (1981) The hearing ability of fish. In: *Hearing and Sound Communication in Fishes* (ed. Tavolga, W.N., Popper, A.N. & Fay, R.R.), pp. 109–133. Springer-Verlag, New York.
Hawkins, A.D. (1986) Underwater sound and fish behaviour. In: *The Behaviour of Teleost Fishes* (ed. Pitcher, T.J.), pp. 114–51. Croom Helm, London.
Hawkins, A.D. (1993) Underwater sound and fish behaviour. In: *Behaviour of Teleost Fishes*, 2nd edn (ed. Pitcher, T.J.), pp. 129–69. Chapman & Hall, London.
Hawkins, A.D. & Chapman, C.J. (1969) *A field determination of the hearing thresholds for the cod,* Gadus morhua L. International working group for fishing technology. Report of the 8th IF Meeting, Lowestoft, 29–30th April, 1969, pp. 89–97.
Hawkins, A.D. & Chapman, C.J. (1975) Masked auditory thresholds in the cod, *Gadus morhua* L. *J. Comp. Physiol.*, **103**, 209–26.
Hawkins, A.D. & MacLennan, D.N. (1976) An acoustic tank for hearing studies on fish. In: *Sound Reception in Fish* (eds Schuijf, A. & Hawkins, A.D.), pp. 146–169. Elsevier, Amsterdam.
Hawkins, A.D. & Sand, O. (1977) Directional hearing in the median vertical plane by the cod. *J. Comp. Physiol.*, **122**, 1–8.
Hay, D.E. (1985) Reproductive biology of Pacific herring (*Clupea harengus pallasi*). *Can. J. Fish. Aquat. Sci.*, **42** (Suppl. 1), 11–126.
Hay, T.F. (1978) Filial imprinting in the convict cichlid fish, *Cichlasoma nigrofasciatum*. *Behavior*, **65**, 138–60.
Hayasi, S. (1974) Effort and CPUE as measure of abundance. *ICATT Coll. Vol. Sci. Pap.*, **3**, 12–31.
He, P. (1991) Swimming endurance of Atlantic cod, *Gadus morhua*, at low temperatures. *Fish. Res.*, **12**, 65–73.
He, P. & Wardle, C.S. (1986) Tilting behaviour of the Atlantic mackerel, *Scomber scombrus*, at low swimming speeds. *J. Fish. Biol.*, **29**, 223–32.
Heape, W. (1931) *Emigration, Migration and Nomadism*. Heffer, Cambridge, UK.
Heath, M.R. (1992) Field investigations of the early life stages of marine fish. *Adv. Mar. Biol.*, **28**, 1–133.
Heath, M.R. (1993) An evaluation and review of the ICES herring larval surveys in the North Sea and

adjacent waters. In: *Advances in the Early Life History of Fishes. Part 2. Ichthyoplankton Methods for Estimating Fish Biomass* (eds Hunter, J.R., Lo, N.C.H. & Fuiman, L.A.), pp. 795–817. *Bull. Mar. Sci.* (US), **53** (2).

Hedgecock, D., Hutchinson, E.S., Li, G., Sly, F.L. & Nelson, K. (1989) Genetic and morphometric variation in the Pacific Sardine, *Sardinops sagax caerulea:* comparisons and contrasts with historical data and with variability in the northern anchovy, *Engraulis mordax*. *Fish Bull.* (US), **87** (3), 653–71.

Helfman, G.S. (1981) The advantage to fishes of hovering in shade. *Copeia*, **2**, 392–400.

Helfman, G.S. (1984) School fidelity in fishes: the yellow perch pattern. *Anim. Behav.*, **32** (3), 663–72.

Helfman, G.S. (1992) Fish behaviour by day, night and twilight. In: *Behaviour of Teleost Fishes*, 2nd edn, (ed. Pitcher, T.J.), pp. 479–512. Chapman and Hall, London.

Helfman, G.S. & Schultz, E.T. (1984) Social transmission of behaviour in a coral reef fish. *Anim. Behav.*, **32** (2), 379–84.

Helfman, G.S., Meyer, J.L. & McFarland, W.N. (1982) The ontogeny of twilight migration patterns in grunts (pisces: Haemulidae). *Anim. Behav.*, **30**, 317–26.

Hergenrader, G.L. & Hasler, A.D. (1968) Influence of changing seasons on schooling behaviour of yellow perch. *J. Fish. Res. Board Can.*, **25**, 711–16.

Hering, G. (1968) Avoidance of acoustic stimuli by the herring. *ICES, CM 1968/H:18*, pp. 1–5.

Hervé, A., Bard, F.X. & Gonzales Costas, F. (1991) Facteurs d'accroissement potentiels de la puissance de pêche des senneurs tropicaux français et espagnols entre 1985 et 1989. *ICCAT Collect. Vol. Sci. Pap.*, **35** (1), 8–13.

Hewitt, R.P., Smith, P.E. & Brown, J.C. (1976) Development and use of sonar mapping for pelagic stock assessment in the California current area. *Fish. Bull.*, **74**, 218–300.

Hilborn, R. (1979) Comparison of fisheries control systems that utilise catch and effort data. *J. Fish. Res. Board Can.*, **36**, 1477–89.

Hilborn, R. (1985) Apparent stock recruitment relationships in mixed stock fisheries. *Can. J. Fish. Aquat. Sci.*, **42**, 718–23.

Hilborn, R. (1991) Modeling the stability of fish schools: exchange of individual fish between schools of skipjack tuna (*Katsuwonus pelamis*). *Can. J. Fish. Aquat. Sci.*, **48**, 1081–91.

Hilborn, R. & Medley, P. (1989) Tuna purse-seine with fish-aggregating devices (FAD): models of tuna FAD interactions. *Can. J. Fish. Aquat. Sci.*, **46**, 28–32.

Hilborn, R. & Sibert, J. (1986) Is international management of tuna necessary? *South Pacific Comm. Fish. Newsl.*, **39**, 31–9.

Hilborn, R. & Walters, C.J. (1987) A general model for simulation of stock and fleet dynamics in spatially heterogeneous fisheries. *Can. J. Fish. Aquat. Sci.*, **44**, 1366–9.

Hilborn, R. & Walters, C.J. (1992) *Quantitative Fisheries Stock Assessment: Choice, Dynamics and Uncertainty*. Chapman and Hall, New York.

Hiramoto, K. (1991) The sardine fishery and ecology in the Joban and Boso waters of Central Japan. In: *Long-term Variability of Pelagic Fish Populations and their Environment* (eds Kawasaki, T., Tanaka, S., Toba, Y. & Taniguchi, A.), pp. 117–128. Pergamon Press, Oxford.

Hjort, J. (1908) *Some Results of the International Ocean Researches*. Scottish Ocean. Lab., Edinburgh.

Hjort, J., Jahn, G. & Ottestadt, P. (1933) The optimum catch. *Hvalrad. Skr.*, **7**, 92–127.

Hoar, W.S. & Randall, D.J. (1979) *Fish Physiology*, vol. V. Academic Press, New York.

Hobson, E.S. (1963) Selective feeding by the gafftopsail pompano *Trachinotus rhodopus* (Gill), in mixed schools of herring and anchovies in the Gulf of California. *Copeia*, **3**, 595–6.

Hobson, E.S. (1968) Predatory behaviour of some shore fishes in the Gulf of California. *Bur. Sport Fish. Wild. Res. Rep.*, **73**, 1–92.

Holeton, G.F. (1980) Oxygen as en environmental factor of fishes. In: *Environmental Physiology of Fishes*, series A (ed. Ali, M.A.), pp. 7–32. Plenum Press, New York.

Holeton, G.F., Pawson, M.G. & Shelton, G. (1982) Gill ventilation, gas exchange, and survival in the Atlantic mackerel (*Scomber scombrus* L.). *Can. J. Zool.*, **60** (5), 1141–7.

Holland, K.N. & Kajiura, S. (1997) Analysis of Hawaii sea mount tuna tagging program recaptures. In: *48th Tuna Conference at Lake Arrowhead* (eds Scott, M. & Oson, R.). Inter-American Tropical Tuna Commission Int. Rep., San Diego.

Holland, K.N., Brill, R.W. & Chang, R.K.C. (1990) Horizontal and vertical movements of Pacific blue marlin captured and released using sportfishing gear. *Fish. Bull.* (US), **88**, 397–402.

Holliday, D.V. (1977) Two applications of the Doppler effect in the study of fish schools. *Rapp. P.-v. Réun. Cons. Perm. Int. Explor. Mer*, **170**, 21–30.

Horne, J.K. & Schneider, D.C. (1995) Spatial variance in ecology. *Oikos*, **74**, 18–26.

Hourston, A.S. (1982) Homing by Canada's west coast herring to management units and divisions as indicated by tag recoveries. *Can. J. Fish. Aquat. Sci.*, **39**, 1414–22.

Hunter, J.R. (1966) Procedure for analysis of schooling behaviour. *J. Fish. Res. Board Can.*, **23**, 547–62.

Hunter, J.R. (1968) Effects of light on schooling and feeding jack mackerel, *Trachurus symmetricus. J. Fish. Res. Board Can.*, **25**, 393–407.

Hunter, J.R. (1969) Communication of velocity changes in jack mackerel (*Trachurus symmetricus*) schools. *Anim. Behav.*, **17**, 507–14.

Hunter, J.R. & Coyne, K.M. (1982) The onset of schooling in the northern anchovy larvae, *Engraulis mordax. Cal. Coop. Oceanic Fish. Invest. R.*, **23**, 246–51.

Hunter, J.R. & Lo, N.C.H. (1993) Ichthyoplankton methods for estimating fish biomass introduction and terminology. In: *Advances in the Early Life History of Fishes. Part 2. Ichthyoplankton Methods for Estimating Fish Biomass* (eds Hunter, J.R., Lo, N.C.H. & Fuiman, L.A.), pp. 723–7. *Bull. Mar. Sci.* (US) **53** (2).

Hunter, J.R. & Mitchell, C.T. (1967) Association of fishes with flotsam in the offshore waters of Central America. *Fish Bull.* (US), **66** (1), 13–29.

Hunter, J.R. & Mitchell, C.T. (1968) Field experiments on the attraction of pelagic fish to floating objects. *J. Cons. Perm. Int. Explor. Mer*, **41** (3), 427–34.

Hunter, J.R. & Nicholl, R. (1985) Visual threshold for schooling in northern anchovy, *Engraulis mordax. Fish. Bull.*, **83** (3), 235–42.

Hunter, J.R. & Wisby, W.J. (1964) Net avoidance behaviour of carp and other species of fish. *J. Fish. Res. Board Can.*, **21**, 613–33.

Hunter, J.R., Argue, A.W., Bayliff, W.H., Dizon, A.E., Fonteneau, A., Goodman, D. & Seckel, G.R. (1986) The dynamics of tuna movements: an evaluation of past and future research. *FAO Fish. Tech. Pap.*, **277**.

Hunter, J.R., Lo, N.C.H. & Fuiman, L.A. (eds) (1993) Advances in the early life history of fishes. *Bull. Mar. Sci.* (US) **53** (2).

Huntingford, F.A. (1986) Development of behaviour in fish. In: *Behaviour of Teleost Fishes*, 2nd edn, (ed. Pitcher, T.J.), pp. 57–83. Chapman and Hall, London.

Huntingford, F.A. (1993) Behaviour, ecology and teleost fishes. In: *Behavioural Ecology of Fishes* (eds Huntingford, F.A. & Torricelli, P.), 11, 1–5. Harwood Academic Publishers, Chur.

Huntingford, F.A. & Coulter, R.M. (1989) Habituation of a predator inspection in the three-spined stickleback, *Gasterosteus aculeatus* L. *J. Fish Biol.*, **35**, 153–4.

Huntingford, F.A. & Torricelli, P. (eds) (1993) *Behavioural Ecology of Fishes*. Ettore Majorana International Life Sciences Series, 11. Harwood Academic Publishers, Chur.

Huntingford, F.A. & Wright, P.J. (1992) Inherited population differences in avoidance conditioning in three-spined sticklebacks, *Gasterosteus aculeatus. Behaviour*, **122** (3–4), 264–73.

Huntingford, F.A., Metcalfe, N.B. & Thorpe, J.E. (1988) Feeding motivation and response to predation risk in Atlantic salmon parr adopting different life history strategies. *J. Fish Biol.*, **32**, 777–82.

Hurley, A.C. (1978) School structure of the squid *Loligo opalescens. Fish. Bull.* (US) **76**, 433–41.

Huse, I. & Ona, E. (1996) Tilt angle distribution and swimming speed of overwintering Norwegian spring spawning herring. *ICES J. Mar. Sci.*, **53**, 863–73.

Hutchings, J.A. (1996) Spatial and temporal variation in the density of northern cod and a review of hypotheses for the stock's collapse. *Can. J. Fish. Aquat. Sci.*, **53**, 943–62.

Hutchings, J.A. & Myers, R.A. (1994) Timing of cod reproduction: interannual variability and the influence of temperature. *Mar. Ecol-Progr. Ser.*, **108**, 21–31.

Huth, A. & Wissel, C. (1990) The movement of fish schools: a simulation model. In: *Lecture Notes in Biomathematics. Biological motion*, **89**, 577–590.

Huth, A. & Wissel, C. (1993) Analysis of the behaviour and the structure of fish schools by means of computer simulations. *Comm. Theor. Biol.*, **3** (3), 169–201.

Huth, A. & Wissel, C. (1994) The simulation of fish schools in comparison with experimental data. *Ecol. Model.*, **75/76**, 135–46.

Ianelli, J. (1987) *A method for estimating recruitment patterns of tunas to floating objects using removal data*. South Pacific Commission Int. Rep., 11 (mimeo).

Icochea, L., Chipollini, A. & Niquen, M. (1989) Analisis de pesquería de arrastre pelágica en la costa peruana durante 1983–1987 y su relación col el medio ambiente. Simposium Internacional de los Recursos Vivos y las Pesquerias en el Pacífico Sudeste, Vina Del Mar (Chile), 9–13 May 1988. *Rev. South Pacific Comm.* (Special Issue), 455–65.

Inoue, Y. & Arimoto, T. (1989) Scanning sonar survey on the capturing process of trapnet. In: *Proceedings of World Symposium on Fishing Gear and Fishing Vessel Design* (ed Marine Institute), pp. 417–21. St John's, Newfoundland, Canada.

Inoue, Y. Matsushita, Y. & Arimoto, T. (1993). The reaction behaviour of walleye pollock (*Theragra chalcogramma*) in a deep/low-temperature trawl fishing ground. *ICES Mar. Sci. Symp.*, **196**, 77–9.

Isshiki, T., Hanamoto, E., Mouri, M., Miyabe, N. & Takeuchi, S. (1997) The effect of dissolved oxygen content on the distribution of bigeye tuna, *Thunnus obesus*, in the Indian Ocean. *48th Tuna Conference at Lake Arrowhead.* Inter-American Tropical Tuna Commission, San Diego.

Ivanova, V.F. & Khmel'-nitskaya, G.N. (1991) On the distribution pattern of Peruvian horse mackerel in the south-east and south-west Pacific and its influence on the catchability of midwater trawls. Ecological Fishery Investigations in the South Pacific (ed. Nesterov, A.A.), pp. 28–35. *Rev. South Pacific Comm.* (special issue).

Iversen, R.T.B. (1969) Auditory thresholds of the scombrid fish, *Euthynnus affinis*, with comments on the use of sound in tuna fishing. *FAO Fish. Rep.*, **62** (3), 849–59.

Jacques, G. & Tréguer, P. (eds) (1986) *Ecosystèmes Pélagiques Marins.* Masson, Paris.

Jakobsson, J. (1983) Echo surveying of the Icelandic summer spawning herring 1973–1982. *FAO Fish. Rep.*, **300**, 240–48.

Jakobsson, J. (1962) On the migrations of the north coast herring during the summer season in recent years with special reference to the increased yield in 1961 and 1962. *ICES meeting, Doc. 98* (mimeo).

Jakobsson, J. (1963) Some remarks on the distribution and availability of the north coast herring of Iceland. *Rapp. P.-v. Réun. Cons. Int. Explor. Mer*, **154**, 73–82.

Jakobsson, J. (1968) Herring migrations east of Iceland during the summer and autumn 1966 and 1967. *Fisk Dir. Skr. Ser. Havunders.*, **15**, 17–22.

Jakobsson, J. (1985) Monitoring and management of the north-east Atlantic herring stocks. *Can. J. Fish. Aquat. Sci.*, **42** (Suppl. 1), 207–21.

Jakobsson, J. & Østvedt, O.J. (1996) A preliminary review of the joint investigations on the distribution of herring in the Norwegian and Iceland Seas 1950–1970. *ICES CM/1996, H:14.*

James, A.G. (1988) Are clupeid microphagists herbivorous or omnivorous? A review of the diets of some commercially important clupeids. *S. Afr. J. Mar. Sci.*, **7**, 161–77.

Johannessen, T., Fernö, A. & Løkkeborg, S. (1993) Behaviour of cod (*Gadus morhua*) and haddock (*Melanogrammus aeglefinus*) in relation to various sizes of long-line bait. *ICES Mar. Sci. Symp.*, **196**, 47–50.

Johannesson, K.A. & Robles, A.N. (1977) Echo surveys of Peruvian anchoveta. *Rapp. P.-v. Réun. Cons. Int. Explor. Mer*, **170**, 237–44.

John, K.R. (1964) Illumination, vision and schooling of *Astyanax mexicanus* Fillipi. *J. Fish. Res. Board Can.*, **21**, 1453–73.

Jones, J.W. (1959) *The Salmon.* Collins New Naturalist, London.

Jones, R. (1974) Assessing the long term effect of changes in fishing effort and mesh size from length composition data. *ICES CM/1974, F:33.*

Jones, R. (1981) The use of length composition data in fish stock assessments (with notes on VPA and cohort analysis). *FAO Fish. Cir.*, **734**.

Joseph, A. & Somvanshi, V.S. (1989) Application of swept volume method in pelagic stock assessment of *Megalaspsis cordyla* (Linnaeus) along north-west coast of India. *Spec. Publ. Fish. Surv. India*, **2**, 136–48.

Josse, E. (1992) Different ways of exploiting tuna associated with fish aggregating devices anchored in French Polynesia. In: *International workshop on the ecology and fisheries for tunas associated with floating objects and on assessment issues arising from the association of tunas with floating objects* (ed. Hall. M.). Inter-American Tropical Tuna Commission Int. Rep., San Diego (mimeo).

Josse, E., Bach, P. & Dagorn, L. (in press) Tuna/prey relationships studied by simultaneous sonic trackings and acoustic surveys. *Hydrobiologia.*

Karlsen, H.E. (1992a) Infrasound sensitivity in the plaice (*Pleuronectes platessa*). *J. Exp. Biol.*, **171**, 173–87.

Karlsen, H.E. (1992b) The inner ear is responsible for detection of the infrasound in the perch (*Perca fluviatilis*). *J. Exp. Biol.*, **171**, 163–72.

Karp, W.A. (ed) (1990) *Developments in fisheries acoustics*. Symposium held in Seattle, 22–26 June 1987. *Rapp. P.-v. Réun. Cons. Int. Explor. Mer*, **189**.

Kawasaki, T. (1992) Mechanisms governing fluctuations in pelagic fish populations. *S. Afr. J. Mar. Sci.*, **12**, 873–9.

Kawasaki, T., Tanaka, S., Toba, Y. & Taniguchi, A. (eds) (1990) *Long-term Variability of Pelagic Fish Populations and their Environment*. Pergamon Press, Oxford.

Kay, B.J., Jones, D.K. & Mitson, R.B. (1991) *FRV* Corystes: *a purpose-built fisheries research vessel*. The Royal Institution of Naval Architects, London. Spring Meeting 1991.

Keenleyside, M.H.A. (1955) Some aspects of the schooling behaviour of fish. *Behaviour*, **8**, 183–248.

Kemmerer, A.J. (1980) Environmental preferences and behaviour patterns of gulf menhaden (*Brevoortia patronus*) inferred from fishing and remotely sensed data. In: *Fish Behaviour and its use in the capture and culture of fishes: proceedings of the conference on the physiological and behavioural manipulation of food fish as production and management tools* (eds Bardach, J.E., Magnuson, J.J., May, R.C. & Reinhart, J.M.), pp. 345–70. Bellagio, Italy, 3–8 November 1977. ICLARM Conference Proceedings, Manila.

Kendall, A.W., Incze, L.S., Ortner, P.B., Cummings, S.R. & Brown, P.K. (1994) The vertical distribution of eggs and larvae of walleye pollock, *Theragra chalcogramma*, in Shelikof Strait, Gulf of Alaska. *Fish. Bull.* (US) **92** (3), 540–54.

Kennedy, M. & Gray, R.D. (1993) Can ecological theory predict the distribution of foraging animals? A critical analysis of experiments on the Ideal Free Distribution. *Oikos*, **68**, 158–66.

Kerstan, M. (1995) Ages and growth rates of Agulhas Bank horse mackerel *Trachurus trachurus capensis:* comparison of otolith ageing and length frequency analyses. *S. Afr. J. Mar. Sci.*, **15**, 137–56.

Kesteven, G.L. (ed) (1960) *Symposium on Fish Behaviour. Section III*. Proceedings of Indo-Pacific Fisheries Council, 8th session, Columbo, Ceylon, 6–22 December 1958. IPFC Secretariat, FAO Regional Office for Asia and the Far East, Bangkok.

Kieffer, J.D. & Colgan, P.W. (1992) The role of learning in fish behaviour. *Rev. Fish Biol. Fisheries*, **2**, 125–43.

Kikawa, S. & Nishikawa, Y. (1986) Predators of skipjack, *Katsuwonus pelamis*, in the Atlantic Ocean and a preliminary indication of skipjack distribution. In: *Proceedings of the ICCAT Conference on the International Skipjack Year Program* (eds Symons, P.E.K., Miyake, P.M. & Sakagawa, G.T.), pp. 296–8. ICCAT, Madrid.

Kim, M.-K., Arimoto, T., Matsushita, Y. & Inoue, Y. (1993) Migration behaviour of fish schools in set-net fishing grounds. *Bull. Japan Soc. Sci. Fish.*, **59** (3), 473–9.

Kimura, D.K. (1977) Statistical assessment of the age-length key. *J. Fish. Res. Board Can.*, **34** (3), 317–24.

Kimura, K. (1929) On the detection of fish-groups by an acoustic method. *J. Imp. Fish. Inst. Tokyo*, **24**, 41–5.

Kislalioglou, M. & Gibson, R. (1976) Prey 'handling time' and its importance in food selection by the 15-spined stickleback, *Spinachia spinachia*. *J. Exp. Mar. Biol. Ecol.*, **25**, 151–8.

Kleiber, P.M. & Hampton, J. (1994) Modelling effects of FADs and islands on movement of skipjack tuna (*Katsuwonus pelamis*): estimating parameters from tagging data. *Can. J. Fish. Aquat. Sci.*, **51**, 2642–53.

Klima, E.F. & Wickham, D.A. (1971) Attraction of coastal pelagic fishes with artificial structures. *Trans. Amer. Fish. Soc.*, **1**, 86–99.

Klimley, A.P. & Holloway, C. (1996a) Automated monitoring of yellowfin tuna at Hawaiian FADs. *NMFS Tuna Newsletter*, Nov. 1996.

Klimley, P.A. & Holloway, C. (1996b) Automated monitoring of yellowfin tuna at Hawaiian FADs. In: *Proceedings of the 47th tuna conference: Sustaining tuna fisheries: Issues and answers* (eds Jackson, A., Rasmussen, R. & Bartoo, N.). Southwest Fisheries Science Center/National Marine Fisheries Service, NOAA, La Jolla (mimeo).

Knudsen, F.R., Enger, P.S. & Sand. O. (1992) Awareness reactions and avoidance responses to sound in juvenile Atlantic salmon, *Salmo salar* L. *J. Fish. Biol.*, **40**, 523–34.

Knudsen, H.P. (1990) The Bergen Echo Integrator: an introduction. *J. Cons. Int. Explor. Mer*, **47**, 167–74.

Koester, F.W. (1989) Feeding Activity of Baltic Herring in the Bornholm Basin. ICES Baltic Fish Comm.

Kojima, S. (1956) Fishing for dolphins in the western part of the Japan Sea. II.Why do the fish take shelter under floating materials? *Bull. Japan. Soc. Sci. Fish.*, **21** (10), 1049–52.

Koltes, K.H. (1984) Temporal patterns in three-dimensional structure and activity of schools of the Atlantic silverside, *Menidia menidia*. *Mar. Biol.*, **78**, 113–22.

Koslow, J.A. (1981) Feeding selectivity in schools of northern anchovy in the southern Californian Bight. *Fish Bull.* (US) **79**, 131–42.

Kramer, D.L. (1987) Dissolved oxygen and fish behaviour. *Environ. Biol. Fish.*, **18** (2), 81–92.

Krause, J. & Tegeder, R.W. (1994) The mechanism of aggregation behaviour in fish shoals: individuals minimise approach time to neighbours. *Anim. Behav.*, **48**, 353–9.

Krebs, J.R. & Davies, N.B. (1978) *Behavioural Ecology: An Evolutionary Approach*. Blackwell Science, Oxford.

Krebs, J.R. & Davies, N.B. (1984) *Behavioural Ecology: An Evolutionary Approach*, 2nd edn. Blackwell Science, Oxford.

Krebs, J.R. & Davies, N.B. (1991) *Behavioural Ecology: An Evolutionary Approach*, 3rd edn. Blackwell Science, Oxford.

Krebs, J.R., MacRoberts, M.H. & Cullen, J.M. (1972) Flocking and feeding in the Great Tit *Parus major* – an experimental study. *Ibis*, **114**, 507–30.

Krechevsky, I. (1932) Hypothesis in rats. *Psychol. Rev.*, **39**, 516–32.

Kristjonsson, H. (ed) (1959) *Modern Fishing Gear of the World*. Fishing News Books, Oxford.

Kronman, M. (1992) LIDAR: Light detecting and ranging. Laser technology could revolutionise aerial fish finding. *Natl. Fisherman*, **72**, 40–42.

Kuhlmann, D.H.H. & Karst, H. (1967) Freiwasserbeobachtungen zum verhalten von tobiasfishschwarmen (*Ammodytidae*) in ther westlichen Ostsee. Z. *Tierpsychol*, **24**, 282–97.

Kulka, D.W., Wroblewski, J.S. & Narayanan, S. (1995) Recent changes in the winter distribution and movements of northern Atlantic cod (*Gadus morhua* Linnaeus, 1758) on the Newfoundland–Labrador Shelf. *ICES J. Mar. Sci.*, **52**, 889–902.

Kvalsvik, K. & Skagen, D.W. (1995) Trawl avoidance as a source of error in estimates of the prevalence of *Icthyophonushoferi* disease in Norwegian spring spawning herreing (*Clupea harengus* L.) in the feeding area. *ICES CM 1995/H:8*, pp. 1–8.

Laevastu, T. (1990a) UUSDYNE numerical dynamical ecosystem simulation (personal computer version of Dynumes). National Marine Fisheries Service. *Alaska Fisheries Science Center, Prog. Doc. 31.*

Laevastu, T. (1990b) FEBAP Fish Ecosystem Biomass Assessment Program (a bulk biomass simulation for resource assessment). National Marine Fisheries Service. *Alaska Fisheries Science Center, Progr. Doc.*

Laevastu, T. & Rosa, H.Jr. (1963) Distribution and relative abundance of tunas in relation to their environment. *FAO Fish. Rep.*, **6** (3), 1835–51.

Laloë, F. (1988) Un modèle global avec quantité de biomasse inaccessible liée aux conditions environnementales: application aux données de la pêche ivoiro-ghanéenne de *Sardinella aurita. Aquat. Living Resour.*, **1**, 289–98.

Laloë, F. (1989) Un modèle global avec quantité de biomasse inaccessible dépendant de la surface de pêche. Application aux données de la pêche d'albacore (*Thunnus albacares*) de l'Atlantique Est. *Aquat. Living Resour.*, **2**, 231–9.

Laloë, F. (1995) Should surplus production models be fishery description tools rather than biological models? *Aquat. Living Resour.*, **8**, 1–16.

Laloë, F. & Samba, A. (1991) A simulation model of artisanal fisheries of Senegal. *ICES Mar. Sci. Symp.*, **193**, 281–6.

Lamberth, S.J., Bennett, B.A. & Clark, B.M. (1995) The vulnerability of fish to capture by commercial beach-seine nets in False Bay, South Africa. *S. Afr. J. Mar. Sci.*, **15**, 25–31.

Langton, C.G. (ed) (1989) *Artificial Life*. Santa Fe Institute Studies in the Sciences of Complexity, vol. VI. Addison-Wesley Publishing Company, Reading, Masschusetts.

Larkin, P.A. & Walton, A. (1969) Fish school size and migration. *J. Fish. Res. Board Can.*, **26**, 1372–4.

Laurec, A. (1977) Analyse et estimation des puissances de pêche. *J. Cons. Int. Explor. Mer*, **37** (2), 173–85.

Laurec, A. (1993) Etalonnage de l'analyse des cohortes en halieutique. In: *Biométrie et Environnement* (eds Lebreton, J.D. & Asselain, J.), pp. 206–39. Masson, Paris.

Laurec, A. & Fonteneau, A. (1978) Estimation de l'abondance d'une classe d'âge: utilisation des CPUE de plusieurs engins, en différentes zones et saisons. *ICCAT, SCRS/78/86*, 79–100.

Laurec, A. & Le Gall, J.-Y. (1975) De-seasonalising of the abundance index of a species. Application to the albacore (*Thunnus alalunga*) monthly catch per unit effort (CPUE) by the Atlantic Japanese longline fishery. *Bull. Far Seas Fish. Res. Lab.*, **12**, 181–204.

Laurec, A. & Le Guen, J.-C. (1978) CPUE des senneurs et abondance: impact des structures fines. *ICCAT Coll. Vol. Sci. Pap.*, **7**, 30–54.

Laurec, A., Fonteneau, A. & Champagnat, C. (1980) A study of the stability of some stocks described by self-regenerating stochastic models. *Rapp. P.-v. Réun. Cons. Int. Explor. Mer*, **177**, 423–38.

Le Danois, E. (1928) Le sondeur ultra-sonore Langevin-Florisson. *Rev. Trav. Off. Pêch. Mar.*, **2** (1), 5–10.

Le Gall, J.-Y. (1975) Description, efficacité et sélectivité des engins et techniques de pêche de germon (*Thunnus alalunga*) du Nord-Est Atlantique. *La Pêche Maritime*, February, 94–8.

Le Guen, J.-C. (1971) Dynamique des populations de *Pseudotolithus (Fonticulus) elongatus* (Bowd, 1825) – Poisson Scianidae. *Cah. Orstom, Sér. Océanogr.*, **9** (1), 3–84.

Le Page, C. & Cury, P. (1996) How spatial heterogeneity influences population dynamics: simulations in SeaLab. *Adaptive Behaviour*, **4** (3/4), 255–81.

Lear, W.H. & Parsons, L.S. (1993) History and management of the fishery for northern cod in NAFO Divisions 2J, 3K and 3L. *Can. Bull. Fish. Aquat. Sci.*, **226**, 55–90.

Leatherland, J.F. Farbridge, K.J. & Boujard, T. (1992) Lunar and semi-lunar rhythms in fishes. In: *Rhythms in Fishes* (ed. Ali, M.A.), pp. 83–107. Life Sci., Séries A, 236.

Lebedev, N.V. (ed) (1969) *Elementary Populations in Fish*. Translated from Russian. Israel Program for Scientific Translations, Jerusalem.

Lebreton, J.D., Burnham, K.P., Clobert, J. & Anderson, D.R. (1992) Modelling survival and testing biological hypothesis using marked animals: a unified approach with case studies. *Ecol. Monogr.*, **62**, 67–118.

Lehodey, P., Bertignac, M., Hampton, J., Lewis, T. & Picaut, J. (1997) El Niño Southern Oscillation and Tuna in the Western Pacific. *Nature*, **389**, 16 October, 715–8.

Leroy, C. & Binet, D. (1986) Anomalies thermiques dans l'Atlantique tropical – conséquences possibles sur le recrutement du germon (*Thunnus alalunga*). *ICCAT SCRS/86/46*.

Lévénez, J.-J. (1982) Note préliminaire sur l'opération sénégalaise de tracking de listao. *Coll. Vol. Sci. Pap., ICCAT*, **17**, 189–94.

Lévénez, J., Samb, B. & Camarena, T. (1985) *Résultats de la Campagne Echosar VI*. Arch. Cent. de Rech. Océanogr, Inst. Sénég. Rech. Agr. Dakar-Thiaroye.

Lévénez, J.J., Gerlotto, F. & Petit, D. (1990) Reaction of tropical coastal pelagic species to artificial lighting and implications for the assessment of abundance by echo integration. *Rapp. P.-v. Reun. Cons. Int. Explor. Mer*, **189**, 128–34.

Levine, J.S., Lobel, P.S. & MacNichol, E.F. (1980) Visual communication in fishes. In: *Environmental Physiology of Fishes* (ed Ali, M.A.), pp. 447–75. Plenum Press, New York.

Levy, J. (1925) *Calcul des Probabilités*. Gauthier-Villars, Paris.

Licht, T. (1989) Discriminating between hungry and satiated predators: the response of guppies (*Poecilia reticulata*) from high and low predation sites. *Ethology*, **82**, 238–43.

Lima, S.L. & Dill, L.M. (1990) Behavioural decisions made under risk of predation: a review and prospectus. *Can. J. Zool.*, **68**, 619–40.

Littaye-Mariatte, A. (1990) Rendements de la pêche sardinière (*Sardina pilchardus*) *et conditions de vent, dans le nord du golfe de Gascogne. Aquat. Living Resour.*, **3**, 163–80.

Lleonart, J., Salat, J. & Roel, B. (1985) A dynamic production model. *Coll. Sci. Pap. Int. Comm. SE. Atl. Fish.*, **12** (1), 119–46.

Lluch-Belda, D., Crawford, R.J.M. Kawasaki, T., MacCall, A.D., Parrish, R.H., Schwartzlose, R.A. & Smith, P.E. (1989) Worldwide fluctuations of sardine and anchovy stocks: the regime problem. *S. Afr. J. Mar. Sci.*, **8**, 195–205.

Lluch-Belda, D., Lluch-Cota, D.B., Hernández-Vázquez, S. & Salinas-Zavala, C.A. (1992) Sardine population expansion in eastern boundary systems of the Pacific Ocean as related to sea surface temperature. In: *Benguela Trophic Functioning* (eds Payne, A.I.L., Brink, K.H., Mann, K.H. & Hilborn, R.). *S. Afr. J. Mar. Sci.*, **12**, 147–57.

Loew, E.R. & McFarland, W.N. (1990) The underwater visual environment. In: *The Visual System of Fish* (eds Douglas, R.H. & Djamgoz, M.B.A.), pp. 1–44. Chapman and Hall, London.

Longhurst, A.R. & Pauly, D. (1987) *Ecology of Tropical Oceans*. Academic Press, San Diego.

Lonzarich, D.G. & Quinn, T.P. (1995) Experimental evidence for the effect of depth and structure on the distribution, growth, and survival of stream fishes. *Can. J. Zool.*, **73** (12), 2223–30.

Lorentz, K. (1974) *Evolution et Modification du Comportement*. Payot, Paris. Translation of: Evolution and modification of behaviour (1966), the University of Chicago Press, Chicago.

Lorentz, K. (1984) *Les Fondements de l'Éthologie*. Flammarion. Traduit de l'ouvrage original: Vergleichende verhaltensforschung: grundlagen der Ethologie (1978), Springer-Verlag, Vienne.

Lucas, C.E. (1936) On the diurnal variation of size groups of trawl-caught herring. *J. Cons. Int. Explor. Mer*, **11**, 53–9.

Ludwig, D. & Walters, C.J. (1985) Are age-structured models appropriate for catch-effort data? *Can. J. Fish. Aquat. Sci.*, **42**, 1066–72.

Ludwig, D. & Walters, C.J. (1989) A robust method for parameter estimation from catch and effort data. *Can. J. Fish. Aquat. Sci.*, **46**, 137–44.

Luecke, C. & Wurtsbaugh, W.A. (1993) Effects of moonlight and daylight on hydroacoustic estimates of pelagic fish abundance. *Amer. Fish. Soc.*, **122**, 112–20.

Lutcavage, M. & Kraus, S. (1995) The feasibility of direct photographic assessment of giant bluefin tuna, *Thunnus thynnus*, in New England waters. *Fish. Bull.* (US), **93**, 495–503.

Lytle, D.W. & Maxwell, D.R. (1983) Hydroacoustic assessment in high density fish schools. *FAO Fish. Rep.*, **300**, 157–71.

Løkkeborg, S. (1994) Fish behaviour and longlining. In: *Marine Fish Behaviour in Capture and Abundance Estimation* (eds Fernö, A. & Olsen, S.), pp. 9–28. Fishing New Books, Oxford.

Løkkeborg, S. & Bjordal, Å. (1992) Species and size selectivity in longline fishing: a review. *Fish Res.*, **13**, 311–22.

Løkkeborg, S. & Johannessen, T. (1992) The importance of chemical stimuli in bait fishing: fishing trials with presoaked bait. *Fish. Res.*, **14**, 21–9.

Løkkeborg, S., Bjordal, Å. & Fernö, A. (1989) Responses of cod (*Gadus morhua*) and haddock (*Melanogrammus aeglefinus*) to baited hooks in the natural environment. *Can. J. Fisheries Aquat. Sci.*, **46**, 1478–83.

MacArthur, R.H. & Pianka, E.R. (1966) On the optimal use of a patchy environment. *Amer. Naturalist*, **100**, 603–9.

MacCall, A.D. (1976) Density dependence of catchability coefficient in the California Pacific sardine, *Sardinops sagax caerulea*, purse seine fishery. *Cal. Coop. Ocean Fish. Invest. R.* **18**, 136–48.

MacCall A.D. (1990) *Dynamic Geography of Marine Fish Populations. Books in Recruitment Fishery Oceanography*. Washington Sea Grant, University of Washington Press, Seattle.

MacCartney, B.S. & Stubbs, A.R. (1971) Measurements of the acoustic target strength of fish in dorsal aspect, including swimbladder resonance. *J. Sound Vib.*, **15**, 397–420.

MacFarland, W.N. & Moss, S.A. (1967) Internal behaviour in fish schools. *Science*, **156**, 260–62.

MacKenzie, B.R. & Leggett, W.C. (1991) Quantifying the contribution of small-scale turbulence to the encounter rates between larval fish and their zooplankton prey: effects of wind and tide. *Mar. Ecol. Prog. Ser.*, **73**, 149–60.

MacKinlay, D.D., Johnson, M.V.D. & Celli, D.C. (1994) Evaluation of stress of carbon dioxide anaesthesia. In: *Proceedings of an International Fish Physiology Symposium*, Vancouver, BC (Canada), 16–21 July 1994, pp. 421–4. University of British Columbia, Vancouver.

Mackinson, S., Sumaila, U.R. & Pitcher, T.J. (1997) Bioeconomics and catchability: Fish and fishers behaviour during stock collapse. *Fish. Res.*, **31** (1–2), 11–17.

MacLennan, D.N. (1990) Acoustical measurement of fish abundance. *J. Acoust. Soc. Am.*, **87**, 1–15.

MacLennan, D.N. & MacKenzie, I.G. (1988) Precision of acoustic fish stock estimates. *Can. J. Fish. Aquat. Sci.*, **45**, 605–16.

MacLennan, D.N. & Simmonds, E.J. (1992) *Fisheries Acoustics*. Chapman & Hall, London.

MacLennan, D.N., Hollingworth, C.E. & Armstrong, F. (1989) Target strength and the tilt angle distribution of caged fish. Proc. I.O.A., **11**, 11–21.

MacLennan, D.N., Magurran, A.E., Pitcher, T.J. & Hollingworth, C.E. (1990) Behavioural determinants of target strength. *Rapp. P.-v. Reun. Cons. Int. Explor. Mer*, **189**, 245–53.

Magnhagen, C. & Forsgren, E. (1991) Behavioural responses to different types of predators by sand goby, *Pomatoschistus minutus:* an experimental study. *Mar. Ecol.-Progr. Ser.*, **70** (1), 11–16.

Magnusson, J.J. (1969) Swimming activity in the scombrid fish *Euthynnus affinis* as related to search of food. FAO Conf. on Fish Behaviour as Related to Fishing Techniques and Tactics. *FAO Fish. Rep.*, **62** (2), 439–45.

Magnusson, J.J. & Prescott, J.H. (1966) Courtship, locomotion, feeding and miscellaneous behaviour of Pacific bonito (*Sarda chiliensis*). *Anim. Behav.*, **14**, 54–67.

Magnusson, J.J., Brandt, S.B. & Stewart, D.J. (1980) Habitat preferences and fishery oceanography. In: *Fish behaviour and its use in the capture and culture of fishes: proceedings of the conference on the physiological and behavioural manipulation of food fish as production and management tools*, Bellagio,

Italy, 3–8 November 1977 (eds Bardach, J.E., Magnusson, J.J., May, R.C. & Reinhart, J.M.), pp. 371–82. ICLARM Conference Proceedings (Manila).

Magurran, A.E. (1986) The development of shoaling behaviour in the European minnow, *Phoxinus phoxinus*. *J. Fish. Biol.*, **29A**, 159–70.

Magurran, A.E. (1989) Acquired recognition of predator odour in the European minnow, *Phoxinus phoxinus*. *Ethology*, **82** (3), 216–23.

Magurran, A.E. (1990) The inheritance and development of minnow anti-predator behaviour. *Anim. Behav.*, **39**, 834–42.

Magurran, A.E. & Higham, A. (1988) Information transfer across fish shoals under predator threat. *Ethology*, **78**, 153–8.

Magurran, A.E. & Pitcher, T.J. (1983) Foraging, timidity and school size in minnows and goldfish. *Behav. Ecol. Sociobiol.*, **12**, 147–52.

Magurran, A.E. & Pitcher, T.J. (1987) Provenance, shoal size and sociobiology of predator evasion behaviour in minnow shoals. *Proc. R. Soc. Lond. B*, **229**, 439–65.

Magurran, A.E. & Seghers, B.H. (1990) Population differences in the schooling behaviour of newborn guppies, *Poecilia reticulata*. *Ethology*, **84**, 334–42.

Magurran, A.E. & Seghers, B.H. (1991) Variation in schooling and aggression amongst guppy, *Poecilia reticulata*, populations in Trinidad. *Behaviour*, **118**, 214–34.

Magurran, A.E., Oulton, W.J. & Pitcher, T.J. (1985) Vigilant behaviour and shoal size in minnows. *Z. Tierpsychol.*, **67**, 167–78.

Magurran, A.E., Seghers, B.H., Carvalho, G.R. & Shaw, P.W. (1993) Evolution of adaptive variation in antipredator behaviour. In: *Behavioural Ecology of Fishes* (eds Huntingford, F.A. & Torricelli, P.), 11, pp. 29–44. Harwood Academic Publishers, Chur.

Magurran, A.E., Seghers, B.H., Shaw, P.W. & Carvalho, G.R. (1994) Schooling preferences for familiar fish in the guppy, *Poecilia reticulata*. *J. Fish Biol.*, **45**, 401–6.

Main, J. & Sangster, G.I. (1981) A study of the fish capture process in a bottom trawl by direct observations from an underwater vehicle. *Scott. Fish. Rep.*, **23**.

Mais, K.F. (1977) Acoustic surveys of northern anchovies in the California current system, 1966–1972. *Rapp. P.-v. Réun. Cons. Int. Explor. Mer*, **170**, 287–95.

Major, P.F. (1977) Predator–prey interactions in schooling fishes during periods of twilight: a study of the silversides, *Pranesus insularum*, in Hawaii. *Fish. Bull.* (US) **75**, 415–25.

Major, P.F. (1978) Predator–prey interactions in two schooling fishes, *Caranx ignobilis* and *Stolephorus purpureus*. *Anim. Behav.*, **26**, 760–77.

Major, P.F. & Dill, L.M. (1978) The three-dimensional structure of airborne bird flocks. *Behav. Ecol. Sociobiol.*, **4**, 111–22.

Mangel, M. & Beder, J.H. (1985) Search and stock depletion: theory and applications. *Can. J. Fish. Aquat. Sci.*, **42**, 150–63.

Mangel, M. & Smith, P.E. (1990) Presence–absence sampling for fisheries management. *Can. J. Fish. Aquat. Sci.*, **47** (10), 1875–87.

Maravelias, C.D. & Reid, D.G. (1995) Relationship between herring (*Clupea harengus*, L.) distribution and sea surface salinity and temperature in the northern North Sea. *Scientia Marina*, **59** (3–4), 427–8.

Maravelias, C.D. & Reid, D.G. (1997) Identifying the effects of oceanographic features on zooplankton on prespawning herring abundance using generalised additive models. *Mar. Ecol. Progr. Ser.*, **147**, 1–9.

Marchal, E. (1993) Biologie et écologie des poissons pélagiques côtiers du littoral ivoirien. In: *Environnement et ressources aquatiques de Côte d'Ivoire*. Vol. I. Le milieu marin (eds Le Loeuff, P., Marchal, E. & Amon-Kothias, J.-B.), pp. 237–69. Orstom Editions, Paris.

Marchal, E. & Lebourges, A. (1996) Acoustic evidence for unusual diel behaviour of a mesopelagic fish (*Vinciguerria nimbaria*) exploited by tuna. *Int. Coun. Explor. Sea J. Mar. Sci.*, **53** (2), 443–7.

Margetts, A.R. (ed) (1977) *Hydro-acoustics in Fisheries Research*. Symposium held in Bergen, 19–22 June 1973. *Rapp. P.-v. Réun. Cons. Int. Explor. Mer*, **170**.

Marsac, F. (1994) Yellowfin tuna fisheries in the past decade: Indian Ocean versus eastern Atlantic and eastern Pacific Oceans. 5th Expert Consultation on Indian Ocean Tunas, Mahe (Seychelles), 4–8 Oct 1993. FAO/UNDP Indo-Pacific Tuna Development and Management Programme, Colombo (Sri Lanka). *IPTP Coll.*, **8**, 168–82.

Marsac, F. & Cayré, P. (1998) Telemetry applied to behaviour analysis of yellowfin tuna (*Thunnus albacares*, Bonnaterre, 1788) movements in a network of fish aggregating devices. *Hydrobiologia* (in press).

Marsac, F. & Stéquert, B. (1986) La pêche des thons autour d'épaves ancrées dans l'Océan Indien. *La Pêche Maritime*, **66** (1311), 439–46.

Marsac, F., Cayré, P. & Conand, F. (1996) Analysis of small scale movements of yellowfin tuna around fish aggregating devices (FADs) using sonic tagging. In: *Proceedings of the Expert Consultation on Indian Ocean Tunas*. 6th Session, Colombo, Sri Lanka, 25–29 Sept. 1995 (eds Anganuzzi, A.A., Stobberup, K.A. & Webb, N.J.) *IPTP Coll.*, **9**, 151–9.

Marsac, F., Petit, M. & Stretta, J.M. (1988) *Radiometrie Aérienne et Prospection Thonière à l'Orstom. Orstom Editions*, **68**.

Marty, P. Méthodes Acoustiques. *Rapp. P.-v. Réun. Cons. Int. Explor. Mer*, **3**, 27–32.

Massé, J., Koutsikopoulos, C. & Patty, W. (1996) The structure and spatial distribution of pelagic fish schools in multispecies clusters: an acoustic study. *ICES J. Mar. Sci.*, **53** (2), 155–60.

Massé, J. Leroy, C., Halgand, D. & Beillois, P. (1995) Anchovy (*Engraulis encrasicolus*, L.) adult and egg distribution in the Bay of Biscay, in relation to environmental conditions, as observed during acoustic surveys (1990, 1991, 1992). *Actas del IV Coloquio Internacional sobre Oceanografia del Golfo de Viscaya*, pp. 281–3.

Mathis, A., Chivers, D.P. & Smith, R.J.F. (1996) Cultural transmission of predator recognition in fishes: intraspecific and interspecific learning. *Anim. Behav.*, **51**, 185–201.

Mathisen, O.A. (1989) Adaption of the anchoveta, *Engraulis ringens*, to the Peruvian upwelling system. In: *The Peruvian Upwelling Ecosystem: Dynamics and Interactions* (eds Pauly, D., Muck, P., Mendo, J. & Tsukayama, I.). ICCLARM Conference Proceedings 18, 220–34.

Matsumiya, Y. & Hayase, S. (1982) Estimation of stock biomass of anchovy on continental shelf off Argentina by echosounder. *Bull. Fac. Fish. Nagasaki. Univ.*, **52**, 1–17.

Matsuura, Y. & Hewitt, R. (1995) Changes in the spatial patchiness of Pacific mackerel, *Scomber japonicus*, larvae with increasing age and size. *Fish. Bull.* (US), **93**, 72–178.

Matzel, L.D., Collin, C. & Alkon, D.L. (1992) Biophysical and behavioural correlates of memory storage, degradation, and reactivation. *Behav. Neurosci.*, **106** (6), 954–63.

Mayr, E. (1970) *Populations, Species and Evolution*. Harvard University Press, Cambridge, Mass.

Mayr, E. (1976) *Evolution and the Diversity of Life*. Harvard University Press, Cambridge, Mass.

McNicol, R.G. & Noakes, D.L.G. (1984) Environmental influences in territoriality of juvenile brook char *Salvelinus fontinalis* in a stream environment. *Environ. Biol. Fish.*, **10**, 29–42.

McCleave, J.D., Arnold, G.P., Dodson, J.J & Neill, W.H. (1982) *Mechanisms of Migration in Fishes*. Plenum Press, New York.

McCullagh, P. & Nelder, J.A. (1989) *Generalised Linear Models*. Chapman and Hall, London.

McElligott, J.G., Weiser, M. & Baker, R. (1995) Effect of temperature on the normal and adapted vestibulo-ocular reflex in the goldfish. *J. Neurophysiol.*, **74** (4), 1463–72.

McFarland, W.N. (1967) Internal behaviour in fish schools. *Science*, **156**, 260–62.

McFarland, W.N. & Hillis, Z.-M. (1982) Observations on agonistic behaviour between members of juvenile French and white grunts – family Haemulidae. *Bull. Mar. Sci.*, **32**, 255–68.

McFarland, W.N., Ogden, J.C. & Lythgoe, J.N. (1979) The influence of light on the twilight migration of grunts. *Environ. Biol. Fish.*, **4**, 9–22.

Meaden, G.J. & Do Chi, T. (1996) Geographical information systems: applications to marine fisheries. *FAO Tech., Pap.*, **356**, 335.

Medina-Gaertner, M. (1985) *Etude du zooplancton côtier de la baie de Dakar et de son utilisation par les poissons comme source de nourriture*. Thèse de Doctorat, Université de Bretagne Occidentale, Brest (France).

Megrey, B.A. (1989) Review and comparison of age-structured stock assessment models from theoretical and applied points of view. In: *Mathematical Analysis of Fish Stock Dynamics* (eds Edwards, E.F. & Megrey, B.A.), pp. 8–48. American Fisheries Society Symposium, 6.

Mejuto, J. (1994) Standardised indices of abundance at age for swordfish (*Xiphias gladius*) from the Spanish longline fleet in the Atlantic, 1983–92. *Coll. Vol. Sci. Pap.*, **42** (1), 328–34.

Mejuto, J. & De La Serna, J.M. (1995) Standardised catch rates in number and weight for the swordfish (*Xiphias gladius*) from the Spanish longline fleet in the Mediterranean Sea, 1988–1993. *ICCAT Collect. Vol. Sci. Pap.*, **44** (1), 124–9.

Mendelssohn, R. & Roy, C. (1986) Environmental influences on the French, Ivory Coast, Senegalese and Moroccan tuna catches in the Gulf of Guinea. In: *Proceedings of the ICCAT Conference on the International Skipjack Year Program* (eds Symonds, P.E.K., Miyake, P.M. and Sakagawa, G.T.), pp. 170–88. ICCAT, Madrid.

Mertz, G. & Myers, R.A. (1996) An extended cohort analysis: incorporating the effect of seasonal catches. *Can. J. Fish. Aquat. Sci.*, **53**, 159–63.

Mesnil, B. (1980) Théorie et pratique de l'analyse de cohortes. *Rev. Trav. Instit. Pêch. Mar.*, **44** (2), 119–55.

Meyer, T.L., Cooper, R.A. & Langton, R.W. (1979) Relative abundance, behaviour and food habits of the American sand lance, *Ammodytes americanus*, from the gulf of Maine. *Fish. Bull.* (US) **1979**, 243–53.

Midtun, L. & Hoff, I. (1962) Measurements of the reflection of sound by fish. *FiskDir. Skr. Ser. Havunders*, **13**, 1–8.

Miklósi, Á. & Csányi, V. (1989) The influence of olfaction on exploratory behaviour in the paradise fish (*Macropodus opercularis* L.). *Acta. Biol. Hung.*, **40**, 195–202.

Miklósi, A., Berzsenyi, G., Pongrácz, P. & Csányi, V. (1995). The ontogeny of antipredator behaviour in paradise fish larvae (*Macropodus opercularis*): the recognition of eyespots. *Ethology*, **100**, 284–94.

Milinski, M. (1979) Can an experienced predator overcome the confusion of swarming prey more easily? *Anim. Behav.*, **27**, 1122–6.

Milinski, M. (1985) Risk of predation taken by parasited sticklebacks foraging under competition for food. *Behaviour*, **93**, 203–16.

Milinski, M. (1988) Games fish play: making decisions as a social forager. *Trends Ecol. Evol.*, **3**, 325–30.

Milinski, M. & Heller, R. (1978) Influence of a predator on the optimal foraging behaviour of stickleback *Gasterosteus aculeatus*. *Nature*, **275**, 642–4.

Misund, O.A. (1986) *Sonarobservasjonar av stimåtferd under ringnotfisket etter sild (Sonar observations of schooling behaviour during herring purse seining)*. Cand. Scient. thesis, University of Bergen, Bergen, Norway. (In Norwegian.)

Misund, O.A. (1990) Sonar observations of schooling herring: school dimensions, swimming behaviour and avoidance of vessel and purse seine. *Rapp. P.-v. Réun. Cons. Int. Explor. Mer*, **189**, 135–46.

Misund, O.A. (1993a) Abundance estimation of fish schools based on a relationship between school area and school biomass. *Aquat. Living Resour.* (FRA), **6**, 235–41.

Misund, O.A. (1993b) Dynamics of moving masses: variability in packing density, shape, and size among herring, sprat and saithe schools. *ICES J. Mar. Sci.*, **50**, 145–60.

Misund, O.A. (1993c) Avoidance behaviour of herring (*Clupea harengus*) and mackerel (*Scomber scombrus*) in purse seine capture situations. *Fish Res.*, **16**, 179–94.

Misund, O.A. (1994) Swimming behaviour of fish schools in connection with capture by purse seine and pelagic trawl. In: *Marine Fish Behaviour in capture and abundance estimation* (eds Fernö, A. & Olsen, S.), pp. 84–106.

Misund, O.A. (1997) Underwater acoustics in marine fisheries and fisheries research. *Rev. Fish. Biol. Fisheries*, **7**, 1–34.

Misund, O.A. & Aglen, A. (1992) Swimming behaviour of fish schools in the North Sea during acoustic surveying and pelagic trawl sampling. *ICES J. Mar. Sci.*, **49**, 325–34.

Misund, O.A. & Beltestad, A.K. (1995) Survival of herring after simulated net bursts and conventional storage in net pens. *Fish. Res.*, **22** (3–4), 293–7.

Misund, O.A. & Beltestad, A.K. (1996) Target strength estimates of schooling herring and mackerel using the comparison method. *ICES J. Mar. Sci.*, **53**, 281–4.

Misund, O.A. & Floen, S. (1993) Packing density structure of herring schools. *ICES Mar. Sci. Symp.*, **196**, 26–9.

Misund, O.A., Aglen, A., Johanessen, S.Ø., Skagen, D. & Totland, B. (1993) Assessing the reliability of fish density estimates by monitoring the swimming behaviour of schools during acoustic surveys. *ICES. Mar. Sci. Symp.*, **196**, 202–6.

Misund, O.A., Aglen, A. & Frønæs, E. (1995) Mapping the shape, size and density of fish schools by echo integration and a high-resolution sonar. *ICES J. Mar. Sci.*, **52**, 11–20.

Misund, O.A., Melle, W. & Fernö, A. (1997) Migration behaviour of Norwegian spring spawning herring when entering the cold front in the Norwegian Sea. *Sarsia*, **82**, 107–12.

Misund, O.A., Vilhjálmsson, H., Jákupsstovu, S.H., Røttingen, I., Belikov, S., Asstthorrsson, O.S., Blindheim, J., Jónssen, J., Krysov, A., Malmberg, S.A. & Sveinbjørnsson, S. (1998) Distribution, migrations and abundance of Norwegian spring spawning herring in relation to temperature and zooplankton biomass in the Norwegian Sea in spring and summer 1996. *Sarsia*, **83**, 117–127.

Misund, O.A., Øvredal, J.T. & Hafsteinsson, M.T. (1996) Reactions of herring schools to the sound field of a survey vessel. *Aquat. Living Resour.*, **9**, 5–11.

Mitson, R.B. (1983) *Fisheries Sonar*, Fishing News Books, Oxford.

Mitson, R.B. (1989) Ship noise related to fisheries research. *Proc. I.O.A.*, **11** (3), 61–67.

Mitson, R.B. (1992) Research vessel noise signatures. *ICES, FTFB & FAST Working Groups Joint Session*, Bergen, Norway, 16 June 1992, pp. 1–6.

Mitson, R.B. & Holliday, D.V. (1990) Future developments in fisheries acoustics. *Rapp. P.-v. Reun. Cons. Int. Explor. Mer*, **189**, 82–91.

Mohan, M. & Kunhikoya, K.K. (1985) Comparative efficiency of live-baits for skipjack tuna *Katsuwonus pelamis* fishery at Minicoy. *J. Mar. Biol. Assoc.* (India), **27** (1–2), 21–8.

Mohr, H. (1969) Observations of the Atlanto-Scandian herring with respect to schooling and reactions to fishing gear. *FAO Fish. Rep.*, **62** (3), 567–77.

Mohr, H. (1971) Behaviour patterns of different herring stocks in relation to ship and midwater trawl. In: *Modern Fishing Gear of the World*, Vol. 3, pp. 368–71 (ed. Kristjonsson, H.) Fishing News Books Ltd., Oxford.

Moksness, E. & Øiestad, V. (1987) Interaction of Norwegian spring-spawning herring larvae (*Clupea harengus*) and Barents Sea capelin larvae (*Mallotus villosus*) in a mesocosm study. *J. Cons. Int. Explor. Mer*, **33**, 32–42.

M'onstad, T. (1990) Distribution and growth of blue whiting in the North-East Atlantic. *ICES CM/1990, H:14.*

Morgan, M.J. (1988) The effect of hunger, shoal size and presence of a predator on shoal cohesiveness in bluntnose minnows, *Pimephales notatus* Rafinesque. *J. Fish. Biol.*, **32**, 963–71.

Morgan, M.J. & Godin, J.-G.J. (1985) Antipredator benefits of schooling behaviour in a cyprinodontid fish, the banded killifish (*Fundulus diaphanus*). *Z. Tierpsychol.*, **70**, 236–46.

Morgan, M.J. & Colgan, P.W. (1987) The effects of predator presence and shoal size on foraging in bluntnose minnows, *Pimephales notatus. Env. Biol. Fishes*, **20**, 105–11.

Moser, H.G. & Smith, P.E. (1993) Larval fish assemblages of the California current region and their horizontal and vertical distributions across a front. In: *Advances in the Early Life History of Fishes. Part 1. Larval Fish Assemblages and Ocean Boundaries* (eds Moser, H.G., Smith, P.E. & Fuiman, L.A.), pp. 645–691. *Bull. Mar. Sci.* (US), **53** (2).

Motos, L., Uriarte, A. & Valencia, V. (1996) The spawning environment of the Bay of Biscay anchovy (*Engraulis encrasicolus* L.). *Scientia Marina*, **60** (Suppl. 2), 117–40.

Muck, P. (1989) Relationships between anchoveta spawning strategies and the spatial variability of sea surface temperature off Peru. In: *The Peruvian Upwelling Ecosystem: Dynamics and Interactions* (eds Pauly, D., Muck, P., Mendo, J. & Tsukayama, I.), pp. 168–73. ICLARM, Manila.

Mullen, A.J. (1989) Aggregation of fish through variable diffusivity. *Fish Bull.*, **87**, 353–62.

Mullen, A.J. (1994) Effects of movement on stock assessment in a restricted-range fishery. *Can. J. Fish. Aquat. Sci.*, **51**, 2027–33.

Munck, J.C. de & Schellart, N.A.M. (1987) A model for the nearfield acoustics of the fish swim bladder and its relevance for directional hearing. *J. Acoust. Soc. Am.*, **81** (2), 556–60.

Munz, F.W. & McFarland, W.N. (1973) The significance of spectral position in the rhodopsins of tropical marine fishes. *Vision Res.*, **13**, 1829–74.

Murawski, S.A. & Finn, J.T. (1986) Optimal effort allocation among competing mixed-species fisheries, subject to fishing mortality constraints. *Can. J. Fish. Aquat. Sci.*, **43** (1), 90–100.

Murphy, G.I. (1980) Schooling and the ecology and management of marine fish. In: *Fish Behaviour and Its Use in the Capture and Culture of Fishes* (eds Bardach, J.E., Magnusson, J.J., May, R.C. & Reinhart, J.M.), pp. 400–14. ICLARM Conference Proceedings 5, Manila, Philippines.

Murray, R.W. (1971) Temperature receptors. In: *Fish Physiology*, vol. V (eds Hoar, W.S. & Randall, D.J.). Academic Press, New York.

Myers, R.A. & Cadigan, N.G. (1995) Was an increase in natural mortality responsible for the collapse of northern cod? *Can. J. Fish. Aquat. Sci.*, **52**, 1274–85.

Myers, R.A. & Stokes, K. (1989) Density-dependent habitat utilisation of ground-fish and the improvement of research surveys. *ICES CM/1989, D:15.*

Myers, R.A. Barrowman, N.J., Hoenig, J.M. & Qu, Z. (1996) The collapse of cod in eastern Canada: the evidence from tagging data. *ICES J. Mar. Sci.*, **53** (3), 629–40.

Myrberg, A.A. Jr. (1980) Sensory mediation of social recognition processes in fishes. In: *Fish behaviour and its use in the capture and culture of fishes* (eds Bardach, J.E., Magnusson, J.J., May, R.C. & Reinhart, J.M.), pp. 146–78. ICLARM Conference Proceedings, Manila, Philippines.

Myrberg, A.R. Jr., Gordon, C.R. & Klimley, P. (1976) Attraction of free ranging sharks by low frequency sound, with comments on its biological significance. In: *Sound Reception in Fish* (eds Schuijf, A. & Hawkins, A.D.), pp. 205–28. Elsevier, Amsterdam.

Nakamura, E.L. & Wilson, R.C. (1970) The Biology of the Marquesan sardine, *Sardinella marquesensis. Pacific Sci.*, **24** (3), 359–76.

Nakamura, H. (1969) *Tuna Distribution and Migration*. Fishing News Books, Oxford.

Nakano, H. & Bayliff, W.H. (1992) A review of the Japanese longline fishery for tunas and billfishes in the eastern Pacific Ocean, 1981–1987. *Bulletin of the Inter-American Tropical Tuna Commission*, **20** (5).

Nakashima, B.S. (1983) Aerial photography of capelin (*Mallotus villosus*) schools in the coastal waters of Newfoundland. *Proc. RNRF Symp. Application of Remote Sensing to Resource Management*, Seattle, Washington, 22–27 May 1983, pp. 655–60.

Nakashima, B.S. & Borstad, G.A. (1993) Detecting and measuring pelagic fish schools using remote sensing techniques. *ICES CM 1993/B:7.*

Nakashima, B.S. & Borstad, G.A. (1996) Relative abundance and behaviour of capelin (*Mallotus villosus*) schools from aerial surveys. In: *Proceedings of the Symposium on the Role of Forage Fishes*, Alaska, Nov. 1996.

Nakashima, B.S., Borstad, G.A., Hill, D.A. & Kerr, R.C. (1989) Remote sensing of fish schools: early results from a digital imaging spectrometer. *Proceedings of IGARRS' 89, 12th Canadian Symposium on Remote Sensing*, Vancouver, BC, 10–14 July, pp. 2044–6.

Nakken, O. & Michalsen, K. (1996) Year to year variations in horizontal and vertical distribution of north-east Arctic cod: influence on survey estimates of abundance. *ICES FAST Working Group Meeting*, Woods Hole, 17–19 April 1996.

Nakken, O. & Olsen, K. (1977) Target strength measurements of fish., *Rapp. P.-v. Reun. Cons. Int. Explor. Mer*, **170**, 52–69.

Neill, S.R. St. J. & Cullen, J.M. (1974) Experiments on whether schooling by their prey affects the hunting behaviour of cephalopod and fish predators. *J. Zool.*, **172**, 549–69.

Neill, W.H. & Gallaway, B.J. (1989) 'Noise' in the distributional responses of fish to environment: an exercise in deterministic modelling motivated by the Beaufort sea experience. *Biol. Pap. Univ. Alaska*, **24**, 123–30.

Neilson, J.D. & Perry, R.I. (1990) Diel vertical migrations of marine fishes: an obligate or facultative process? *Adv. Mar. Biol.*, **26**, 115–68.

Nelson, G. & Hutchings, L. (1983) The Benguela upwelling area. *Prog. Oceanogr.*, **12** (3), 333–56.

Nelson, G. & Hutchings, L. (1987) Passive transportation of pelagic system components in the southern Benguela area. In: *The Bengula and Comparable Frontal Systems* (eds Payne, A.I.L., Gulland, J.A. & Bring, K.H.S.), pp. 223–34. *Afr. J. Mar. Sci.*, **5**.

Nelson, W.R. Ingham, M.C. & Schaaf, W.E. (1977) Larval transport and year-class strength of Atlantic menhaden, *Brevoortia tyrannus. Fish Bull.*, **75** (1), 23–41.

Neproshin, A.Y. (1979) Behaviour of the Pacific mackerel, *Pneumatophorus japonicus*, when affected by vessel noise. *J. Ichthyol.*, **18**, 695–9.

Nestler, J.H., Ploskey, G.H. & Pickens, J. (1992) Responses of blueback herring to high-frequency sound and implications for reducing entrainment at hydropower dams. *North American Journal of Fisheries Management*, **12**, 667–83.

Nicholson, M.D., Rackham, B.D. & Mitson, R.B. (1992) Measuring the effect of underwater radiated noise on trawl catches. *ICES, FTFB & FAST Working Groups Joint Session*, Bergen, Norway, 16 June 1992, pp. 1–6.

Nicholson, W.R. (1978) Movements and population structure of Atlantic menhaden indicated by tag returns. *Estuaries*, **1** (3), 141–50.

Nieland, H. (1982) The food of *Sardinella aurita* (Val.) and *Sardinella eba* (Val.) off the coast of Senegal. *Rapp. P.-v. Réun. Cons. Int. Explor. Mer*, **180**, 369–73.

Nitta, E.T. & Henderson, J.R. (1993) A review of interactions between Hawaii's fisheries and protected species. *Mar. Fish. Rev.* (Washington DC) **55** (2), 83–92.

Noakes, D.L.G. (1992) Behaviour and rhythms in fishes. In: *Rhythms in Fishes* (ed Ali, M.A.), pp. 39–50. Plenum Press, New York.

Noakes, D.L.G. & Barlow, G.W. (1973) Ontogeny of parent-contacting in young *Cichlasoma citrinellum*. *Behaviour*, **46**, 221–57.

Noda, M., Gushima, K. & Kakuda, S. (1994) Local prey search based on spatial memory and expectation in the planktivorous reef fish, *Chromis chrysurus* (Pomacentridae). *Anim. Behav.*, **47** (6), 1413–22.

Nunnallee, E.P. (1991) An investigation of the avoidance reactions of Pacific whiting (*Merlucius productus*) to demersal and midwater trawl gear. *ICES CM 1991/B:5*.

Nøttestad, L., Aksland, M., Beltestad, A.K., Fernø, A., Johannessen, A. & Misund, O.A. (1996) Schooling dynamics of the Norwegian spring spawning herring (*Clupea harengus* L.) in a coastal spawning area. *Sarsia*, **80**, 277–84.

O'Brien, W.J., Slade, N.A. & Vinyard, G.L. (1976) Apparent size as the determinant of prey selection by Bluegill sunfish (*Lepomis macrochirus*). *Ecology*, **57**, 1304–11.

Ohshimo, S. (1996) Acoustic estimation of biomass and school character of anchovy *Engraulis japonicus* in the East China Sea and the Yellow Sea. *Fish. Sci.*, **62**, 344–9.

Ojak, W. (1988) Vibrations and waterborne noise on fishery vessels. *J. Ship. Res.*, **32** (2), 112–33.

Olla, B.L., Studholme, A.L. & Bejda, A.J. (1985) Behaviour of juvenile bluefish *Pomatomus saltatrix* in vertical thermal gradients: influence of season, temperature acclimatation and food. *Mar. Ecol-Progr. Ser.*, **23**, 165–77.

Olsen, K. (1967) A comparison of acoustic threshold in cod with recordings of ship noise. *FAO Fish. Rep.*, **62**, 431–8.

Olsen, K. (1969) Directional responses in herring to sound and noise stimuli. *ICES CM 1969/B:20*, pp. 1–5.

Olsen, K. (1971) Influence of vessel noise on the behaviour of herring. In: *Modern Fishing Gear of the World: 3* (ed. Kristjonsson, H.), pp. 291–4. Fishing News Books, Oxford.

Olsen, K. (1976) Evidence for localisation of sound by fish in schools. In: *Sound in fish. Proceedings of a symposium held in honour of Professor Dr. Sven Dijkgraaf*, Utrecht, The Netherlands, 16–18 April 1975, pp. 257–270. Elsevier, Amsterdam.

Olsen, K. (1979) Observed avoidance behaviour in herring in relation to passage of an echo survey vessel. *ICES CM 1979/B:53*.

Olsen, K. (1980) Echo surveying and fish behaviour. *ICES FTFB Working Group meeting*, Reykjavik, Iceland, May 1980.

Olsen, K. (1986) Sound attenuation within schools of herring. *ICES CM 1986/B:44* (mimeo).

Olsen, K. (1990) Fish behaviour and acoustic sampling. *Rapp. P.-v. Reun. Cons. Int. Explor. Mer*, **189**, 147–58.

Olsen, K. & Ahlquist, I. (1989) Target strength of fish at various depths, observed experimentally. *ICES CM 1989/B:35*.

Olsen, K., Angell, J. & Løvik, A. (1983a) Quantitative estimations of the influence of fish behaviour on acoustically determined fish abundance. *FAO Fish. Rep.*, **300**, 131–8.

Olsen, K., Angell, J., Pettersen, F. & Løvik, A. (1983b) Observed fish reactions to a surveying vessel with special reference to herring, cod, capelin and polar cod. *FAO Fish. Rep.*, **300**, 131–8.

Olsen, Y.A., Einarsdottir, I.E. & Nilssen, K.J. (1995) *Metomidate anaesthesia* in Atlantic salmon, *Salmo salar*, prevents plasma cortisol increase during stress. *Aquaculture*, **134** (1–2), 155–68.

Olson, F.C.W. (1964) The survival value of fish schooling. *J. Conseil*, **29**, 115–16.

Olson, R.J. & Boggs, C.H. (1986) Apex predation by yellowfin tuna (*Thunnus albacares*): Independent estimates from gastric evacuation and stomach contents, bioenergetics, and cesium concentrations. *Can. J. Fish. Aquat. Sci.*, **43** (9), 1760–75.

Olson, R.J. & Galván-Magaña, F. (1995) Trophic relations of yellowfin tuna and dolphins in the eastern Pacific Ocean. *46th Tuna Conference*, San Diego and Lake Arrowhead (mimeo).

Omori, M. & Jo, S.G. (1989) Plankton sampling system with a new submersible vortex pump and its use to estimate small-scale vertical distribution of eggs and larvae of *Sergia lucens*. *Bull. Plankton Soc. Jap.*, **36** (1), 19–26.

Ona, E. (1990) Physiological factors causing natural variations in acoustic target strength of fish. *J. Mar. Biol. Ass.* (UK), **70**, 107–21.

Ona, E. (1994) Recent development of acoustic instrumentation in connection with fish capture and abundance estimation. In: *Marine Fish Behaviour in Capture and Abundance Estimation* (eds Fernö, A. & Olsen, S.), pp. 200–16. Fishing News Books, Oxford.

Ona, E. & Beltestad, A.K. (1986) Use of acoustics in studies of fish reaction to imposed stimuli. *Model. Identifi. Control*, **7** (4), 219–26.

Ona, E. & Chruickshank, O.C. (1985) Haddock avoidance reactions during trawling. *ICES CM 11986/B:36*.

Ona, E. & Godø, O.R. (1990) Fish reactions to trawling noise: the significance for trawl sampling. *Rapp. P.-v. Reun. Cons. Int. Explor. Mer*, **189**, 159–66.

Ona, E. & Mitson, R.B. (1996) Acoustic sampling and signal processing near the seabed: the deadzone revisited. *ICES J. Mar. Sci.*, **53**, 677–90.

Ona, E. & Toresen, R. (1988a) Avoidance reactions of herring to a survey vessel, studied by scanning sonar. *ICES CM 1988/H:46*.

Ona, E. & Toresen, R. (1988b) Reactions of herring to trawling noise. *ICES CM 1988/B:36*.

Oxenford, H., Mahon, R. & Hunte, W. (eds) (1993) The eastern Caribbean flyingfish project. OECS: Cane Garden. *OECS Fish Rep. Kingstown*, **9**.

Paloheimo, J.E. (1958) A method of estimating natural and fishing mortalities. *J. Fish. Res. Board Can.*, **18**, 645–61.

Paloheimo, J.E. & Dickie, L.M. (1964) Abundance and fishing success surveys: eggs, larvae and young fish. *ICES Rep.*, **155** (28), 152–63.

Pankhurst, N.W. (1988) Spawning dynamics of orange roughy, *Hoplostethus atlanticus*, in mid-slope waters of New Zealand. *Environ. Biol. Fish.*, **21** (2), 101–16.

Parin, N.V. & Fedoryako, B.I. (1992) Pelagic fish communities around floating objects in the open ocean. In: *Fishing for Tunas Associated with Floating Objects. International Workshop* (ed. Hall, M.). Inter-American Tropical Tuna Commission Int. Rep., San Diego (mimeo).

Park, S.W. (1978) The fishing condition factors on the catch per set variation of the purse seine sets. *Bull. Nat. Fish. Univ. Busan*, **18** (1, 2), 1–11.

Parker, G.A. & Sutherland, W.J. (1986) Ideal free distributions when individuals differ in competitive abilities: phenotype-limited ideal free models. *Anim Behav.*, **34**, 1222–42.

Parker, R.O. Jr. (1972) An electric detector system for recovering internally tagged menhaden, genus *Brevoortia*. NOAA Technical Report, **65**.

Parmanne, R. & Sjöblom, V. (1988) Trends in the abundance, recruitment and mortality of Baltic herring and sprat off the coast of Finland according to exploratory fishing with a pelagic trawl in 1956–84. *Finn. Fish. Res.*, **7**, 18–23.

Parr, A.E. (1927) *A contribution to the theoretical analysis of the schooling behaviour of fishes*. Occasional Papers of the Bingham Oceanographic Collection, pp. 1–32.

Parrish, J.K. (1989a) Re-examining the selfish herd: are central fish safer? *Anim. Behav.*, **38**, 1048–53.

Parrish, J.K. (1989b) Layering with depth in a heterospecific fish aggregation. *Environ. Biol. Fish.*, **26**, 79–85.

Parrish, J.K. (1992a) Levels of diurnal predation on a school of flat-iron herring. *Harengula thrissina. Environ. Biol. Fish.*, **34**, 257–63.

Parrish, J.K. (1992b) Do predators 'shape' fish schools: interactions between predators and their schooling prey. *Netherlands Journal of Zoology*, **42** (2–3), 358–70.

Parrish, J.K. & Kroen, W.K. (1988) Sloughed mucus and drag-reduction in a school of Atlantic silversides, *Menidia menidia. Mar. Biol.*, **97**, 165–9.

Parrish, R.A., Bakun, A., Husby, D.M. & Nelson, C.S. (1983) Comparative climatology of selected environmental processes in relation to eastern boundary current pelagic fish reproduction. In: *Proceedings of the expert consultation to examine changes in abundance and species composition of neritic fish resources* (eds Sharp, G. & Csirke, J.), pp. 731–77. *FAO Fish. Rep.*, **291** (3).

Parrish, R.H., Nelson, C.S. & Bakun, A. (1981) Transport mechanisms and reproductive success of fishes in the California current. *Biolog. Oceanog.*, **1** (2), 175–203.

Partridge, B.L. (1980) The effect of school size on the structure and dynamics of minnow schools. *Anim. Behav.*, **28**, 67–77.

Partridge, B.L. (1981) Internal dynamics and interrelations of fish in schools. *J. Comp. Physiol.*, **144**, 313–25.

Partridge, B.L. (1982a) Rigid definitions of schooling behaviour are inadequate. *Anim. Behav.*, **30**, 298–9.

Partridge, B.L. (1982b) Structure and functions of fish schools. *Scient. Amer.*, **245**, 114–23.

Partridge, B.L. & Pitcher, T.J. (1980) The sensory basis of fish schools: relative roles of lateral lines and vision. *J. Comp. Physiol.*, **135**, 315–25.

Partridge, B.L., Johansson, J. & Kalish J. (1983) The structure of schools of giant bluefin tuna in Cape Cod Bay. *Environ. Biol. Fish.*, **9**, 253–62.

Partridge, B.L., Pitcher, T.J., Cullen, J.M. & Wilson, J. (1980) The three-dimensional structure of fish schools. *Behav. Ecol. Sociobiol.*, **6**, 277–88.

Partridge, L. (1978) Habitat selection. In: *Behavioural Ecology* (eds Krebs, J.R. & Davis, N.B.), pp. 351–76. Blackwell, Oxford.

Parzefall, J. (1986) Behavioural ecology of cave-dwelling fishes. In: *The Behaviour of Teleost Fishes* (ed. Pitcher, T.J.), pp. 433–58. Croom Helm, London.

Pati, S. (1981) Observations on the lunar and tidal influence on gill netting in the Bay of Bengal. *Fish. Tech.*, **18**, 25–7.

Patten, G.G. (1977) Body size and learned avoidance as factors affecting predation on coho salmon, *Oncorhynchus kisutch*, fry by torrent sculpin, *Cottus rhotheus*. *Fish. Bull.* (US), **75**, 457–59.

Patty, W. (1996) *Déterminisme de la répartition spatio-temporelle des populations de poissons pélagiques à partir d'observations acoustiques et environnementales.* Thèse de Doctorat, Université de Bretagne Occidentale, Brest (France).

Pauly, D. (1981) The relationships between gill surface area and growth performance in fish: a generalisation of von Bertalanffy's theory of growth. *Meeresforschung/Rep. Mar. Res.*, **28** (4), 251–82.

Pauly, D. (1995) Anecdotes and the shifting baseline syndrome of fisheries. *Trends Ecol. Evol.*, **10**, 1.

Pauly, D. (1997a) *Méthodes pour l'Évaluation des Ressources Halieutiques.* Coll. Polytech, Cépadués-éditions, Toulouse.

Pauly, D. (1997b) Points of view. Putting fisheries management back in places. *Rev. Fish Biol. Fisheries*, **7**, 125–7.

Pauly, D. & Palomares, M.L. (1989) New estimates of monthly biomass, recruitmenjt and related statistics of anchoveta (*Engraulis rigens*) off Peru (4–14°S), 1953–1985. In: *The Peruvian Upwelling Ecosystem: Dynamics and Interactions* (eds Pauly, D., Muck, P., Mendo, J. & Tsukayama, I.), pp. 189–206. ICLARM Conference Proceedings 18, Manila, Philippines.

Pauly, D. & Tsukayama, I. (eds) (1987) The Peruvian anchoveta and its upwelling ecosystem; three decades of change. *ICLARM Studies and Reviews*, **15**.

Pauly, D., Christensen, V., Dalsgaard, J., Froese, R. & Torres Jr, F. (1997) Fishing down marine food webs. *Science*, **279**, 860–3.

Pauly, D., Ingles, J. & Neal, R. (1984) Application to shrimp stocks of objective methods for the estimation of growth, mortality and recruitment-related parameters from length-frequency data (ELEFAN I and II). In: *Penaeid Shrimps – their Biology and Management* (eds Gulland, J.A. & Rotschild, B.J.), pp. 220–34. Fishing News Book, Oxford.

Pauly, D., Muck, P., Mendo, J. & Tsukayama, I. (eds) (1989) The Peruvian Upwelling Ecosystem: Dynamics and Interactions. ICLARM, Manila, Philippines.

Payne, A.I.L., Brink, K.H., Mann, K.H. & Hilborn, R. (eds) (1992) Benguela Trophic Functioning. *S. Afr. J. Mar. Sci.*, **12**.

Pella, J.J. & Tomlinson, P.K. (1969) A generalised stock production model. *Bulletin of the Inter-American Tropical Tuna Commission*,**13**, 419–96.

Pelletier, D. (1990) Sensitivity and variance estimators for virtual population analysis and the equilibrium yield per recruit model. *Aquat. Living Resour.*, **3**, 1–12.

Pelletier, D. & Parma, A.M. (1994) Spatial distribution of Pacific halibut (*Hippoglossus stenolepis*): an application of geostatistics to longline survey data. *Can. J. Fish. Aquat. Sci.*, **51** (7), 1506–18.

Perrin, W.F. (1968) The porpoise and the tuna. *Sea Frontiers*, **14** (3), 15–19.

Perrin, W.F. (1969) Using porpoise and tuna. *World Fish*, **18** (6), 42–5.

Petersen, C.H. (1976) Cruising speed during migration of the striped mullet (*Mugil cephalus* L.): an evolutionary response to predation? *Evolution*, **30**, 393–6.

Petit, M. (1991) *Contribution de la télédétection aérospatiale à l'élaboration des bases de l'haleutique opérationnelle: l'exemple des pêcheries thonières tropicales de surface (aspect évaluatif).* Thèse de Doctorat, Université Paris VI (France).

Petit, M., Stretta, J.-M. Simier, M. & Wadsworth, A. (1989) Anomalies de surface et pêche thonière: SPOT et la télédétection de zones de pêche par l'inventaire des hauts-fonds. *Mappe Monde*, **89** (3), 13–19.

Petit, M., Stretta, J.-M., Farrugio, H. & Wadsworth, A. (1992) Synthetic aperture radar imaging of sea surface life and fishing activities. *IEEE Trans. Geosci. Remote Sens.*, **30** (5), 1085–9.

Petit, M. Dagorn, L., Lena, P., Slépoukha, M., Ramos, A.G. & Stretta, J.-M. (1995) Oceanic landscape concept and operational fisheries oceanography. In: *Les Nouvelles Frontières de la Télédétection Océanique* (eds Doumenge & Toulemont A.). *Mém. Inst. Océanogr. Monaco*, **18**, 85–97.

Petitgas, P. (1993) Geostatistics for fish stock assessments: a review and acoustic application. *ICES J. Mar. Sci.*, **50**, 285–98.

Petitgas, P. (1994) Spatial strategy of fish populations. *ICES CM 1994/D:14.*

Petitgas, P. (1996) Geostatistics and their applications to fisheries survey data. In: *Computers in Fisheries Research* (eds Megrey, B.A. & Moksness, E.), pp. 113–42. Chapman and Hall, London.

Petitgas, P. (1997) Sole egg distributions in space and time characterised by a geostatistical model and its estimation variance. *ICES J. Mar. Sci.*, **54**, 213–25.

Petitgas, P. & Lévénez, J.J. (1996) Spatial organisations of pelagic fish: echogram structure, spatio-temporal condition, and biomass in Senegalese waters. *ICES J. Mar. Sci.*, **53**, 147–53.

Pillar, S.C., Moloney, C.L., Payne, A.I.L. & Shillington, F.A. (eds) (1997) Benguela dynamics. Impacts of variability on shelf-sea environment and their living resources. *S. Afr. J. Mar. Sci.*, **19**.

Pitcher, T.J. (1973) The three-dimensional structure of fish schools in the minnow, *Phoxinus phoxinus. Anim. Behav.*, **21**, 673–86.

Pitcher, T.J. (1979a) The role of schooling in fish capture. *ICES CM 1979/B:5*, pp. 1–12.

Pitcher, T.J. (1979b) Sensory information and the organisation of behaviour in a schooling cyprinid fish. *Anim. Behav.*, **27**, 126–49.

Pitcher, T.J. (1980) Some ecological consequences of fish school volumes. *Freshwater Biol.*, **10** 539–44.

Pitcher, T.J. (1983) Heuristic definitions of shoaling behaviour. *Anim. Behav.*, **31**, 611–13.

Pitcher, T.J. (1986) Functions of shoaling behaviour in teleosts. In: *The Behaviour of Teleost Fishes* (ed. Pitcher, T.J.), pp. 294–337. Croom Helm, London and Sydney.

Pitcher, T.J. (ed.) (1993) *Behaviour of Teleost Fishes*, 2nd edn. Chapman and Hall, London.

Pitcher, T.J. (1997) Fish shoaling behaviour as a key factor in the resilience of fisheries: shoaling behaviour alone can generate range collapse in fisheries. In: *Developing and sustaining world fisheries resources, proceedings of the 2nd World Fisheries Congress*, Brisbane, July 1996 (eds Hancock, D.A. & Beumer, J.P.), pp. 143–8. CSIRO, Collingwood, Australia.

Pitcher, T.J. & Magurran, A.E. (1983) Shoal size, patch profitability and information exchange in foraging goldfish. *Anim. Behav.*, **31**, 546–55.

Pitcher, T.J. & Parrish, J.K. (1993) Functions of shoaling behaviour in teleosts. In: *Behaviour of Teleost Fishes*, 2nd edn, (ed. Pitcher, T.J.), pp. 363–439. Chapman and Hall, London.

Pitcher, T.J. & Partridge, B.L. (1979) Fish school density and volume. *Mar. Biol.*, **54**, 383–94.

Pitcher, T.J. & Turner, J.R. (1986) Danger at dawn: experimental support for the twilight hypothesis in shoaling minnows. *J. Fish. Biol.*, **29** (Supp. A), 59–70.

Pitcher, T.J. & Wyche, C.J. (1983) Predator-avoidance behaviour of sand-eels schools: Why schools seldom split. In: *Predators and Prey in Fishes* (eds Noakes, D.L.G., Lindquist, B.G. Helfman, G.S. & Ward, J.A., pp. 193–204. Junk, the Hague.

Pitcher, T.J., Green, D.A. & Magurran, A.E. (1986) Dicing with death: predator inspection behaviour in minnow shoals. *J. Fish Biol.*, **28**, 439–48.

Pitcher, T.J., Magurran, A.E. & Edwards, J.L. (1985) Schooling mackerel and herring choose neighbours of similar size. *Mar. Biol.*, **86**, 319–22.

Pitcher, T.J., Magurran, A.E. & Winfield, I. (1982) Fish in larger shoals find food faster. *Behav. Ecol. Sociobiol.*, **10**, 149–51.

Pitcher, T.J., Misund, O.A., Fernø, A. Totland, B. & Melle, V. (1996) Adaptive behaviour of herring schools in the Norwegian Sea as revealed by high resolution sonar. *ICES J. Mar. Sci.*, **53**, 449–52.

Pitcher, T.J., Partridge, B.L. & Wardle, C.S. (1976) A blind fish can school. *Science*, **194**, 963–5.

Platt, C. & Popper, A.N. (1981) Fine structure and function of the ear. In: *Hearing and Sound Communication in Fishes* (eds Tavolga, W.N., Popper, A.N. & Fay, R.R.), pp. 4–38. Springer-Verlag, New York.

Polovina, J.J. (1984) Model of a coral reef ecosystem: I.: the ECOPATH model and its application to French frigate shoals. *Coral Reefs*, **3**, 3–11.

Polovina, J.J. (1989) A system of simultaneous dynamic production and forecast models for multispecies or multiarea applications. *Can. J. Fish.Aquat. Sci.*, **46**, 961–3.

Polovina, J.J. (1996) Decadal variation in the trans-Pacific migration of northern bluefin tuna (*Thunnus thynnus*) coherent with climate-induced change in prey abundance. *Fish. Oceanogr.*, **5** (2), 114–19.

Pope, J.G. (1977) Estimation of fishing mortality, its precision and implications for the management of fisheries. In: *Fisheries Mathematics* (ed. Steele, J.H.), pp. 63–76. Academic Press, New York.

Pope, J.G. (1982) User requirements for acoustic surveys. *International Symposium on Fisheries Acoustics*, Bergen, Norway, 21–24 June, 1982, Doc. no. 84.

Pope, J.G. & Shepherd, J.G. (1982) A simple method for the consistent interpretation of catch-at-age data. *J. Cons. Int. Explor. Mer*, **40**, 176–84.

Pope, J.G. & Shepherd, J.G. (1985) A comparison of the performance of various methods for tuning VPAs using effort data. *J. Cons. Int. Explor. Mer*, **42**, 129–51.

Potier, M. & Sadhotomo, B. (1995) Seiners fisheries in Indonesia. In: *BIODYNEX: Biology, Dynamics, Exploitation of the Small Pelagic Fishes in the Java Sea* (eds Potier, M. & Nurhakim, S.), pp. 49–66. Pelfish, Jakarta.

Potier, M., Petitgas, P. & Petit, D. (1997) Interaction between fish and fishing vessels in the Javanese purse seine fishery. *Aquat. Living Resour.*, **10** (4), 149–56.

Potter, D.C., Lough, R.C., Perry, R.I. & Neilson, J.D. (1990) Comparison of the MOCNESS and IYGPT pelagic samplers for the capture of O-group cod (*Gadus morhua*) on Georges Bank. *J. Cons. Int. Expl. Mer*, **46** (2), 121–8.

Powell, G.V.N. (1974) Experimental analysis of the social value of flocking by starlings (*Sturnus vulgaris*) in relation to predation and foraging. *Anim. Behav.*, **23**, 504–8.

Priede, I.G. & Watson, J.J. (1993) An evaluation of the daily egg production method for estimating biomass of Atlantic mackerel (*Scomber scombrus*). In: *Advances in the Early Life History of Fishes. Part 2. Ichthyoplankton Methods for Estimating Fish Biomass* (eds Hunter, J.R., Lo, N.C.H. & Fuiman, L.A.), pp. 891–911. *Bull. Mar. Sci.* (US), **53** (2).

Priede, I.G., Raid, T. & Watson, J.J. (1995) Deep-water spawning of Atlantic mackerel *Scomber scombrus*, west of Ireland. *J. Mar, Biol. Assoc.* (UK) **75** (4), 849–55.

Pristas, P.J. & Willis, T.D. (1973) Menhaden tagging and recovery: part I – Field methods for tagging menhaden, *genus Brevoortia*. *Marine Fisheries Review* (USA), **35** (5–6), 31–5.

Pullen, G., Williams, H. & Jordan, A. (1992) Development and applicability of fishery independent methods for assessment of the Tasmanian jack mackerel fishery. *Newsl. Aust. Soc. Fish Biol.*, **22** (2), 48–9.

Pyanov, A.I. (1993) Fish learning in response to trawl fishing. *ICES Mar. Sci. Symp.*, **196**, 12–16.

Pyanov, A.I. & Zhujkov, A. Yu. (1993) Conditioned reflex in fishes for avoidance of active fishing gear. *J. Ichthyol.*, **33** (8), 40–50.

Quinn, T.J. II, Deriso, R.B. & Neal, P.R. (1990) Migratory catch-age analysis. *Can. J. Fish. Aquat. Sci.*, **47** (12), 2315–27.

Quinn, T.P., Brannon, E.L. & Dittman, A.H. (1989) Spatial aspects of imprinting and homing in coho salmon, *Oncorhynchus kisutch*. *Fish. Bull.* (US) **87** (4), 769–74.

Radakov, D.V. (1973) *Schooling in the Ecology of Fish*. Israel Program for Scientific Translations, Jerusalem. John Wiley and Sons, New York.

Radovich, J. (1979) Managing pelagic schooling prey species. In: *Predator–prey Systems in Fisheries Management* (ed. Clepper, E.), pp. 365–75. Sport Fish. Inst., Washington DC.

Raid, T. (1994) Structure of Estonian herring catches in the Gulf of Finland. *ICES CM 1994/J:21*.

Ranta, E. & Kaitala, V. (1991) School size affects individual feeding success in three-spined sticklebacks (*Gasterosteus aculeatus* L.). *J. Fish. Biol.*, **39**, 733–7.

Ranta, E., Juvonen, S.-K. & Peuhkuri, N. (1992) Further evidence for size-assortative schooling in sticklebacks. *J. Fish. Biol.*, **41**, 627–30.

Reid, D.G. (1995) Relationship between herring school distribution and sea bed substrate derived from RoxAnn. *Communication to ICES Int. Symp. on Fisheries and Plankton Acoustics*, Aberdeen (UK), 12–16 June 1995.

Reid, D.G. & Simmonds, E.J. (1993) Image analysis techniques for the study of fish school structure from acoustic survey data. *Can. J. Fish. Aquat. Sci.*, **50**, 886–93.

Reid, G.R., Turrell, W.R., Walsh, M. & Corten, A. (1997) Cross-shelf processes north of Scotland in relation to the southerly migration of western mackerel. *Int. Coun. Explor. Sea J. Mar. Sci.*, **54**, 168–78.

Reiner, F. & Lacerda, M. (1989) Note on the presence of *Pseudorca crassidens* in Azoran waters. Bocagiana. *Funchal*, 130.

Reuter, H. & Breckling, B. (1994) Selforganisation of fish schools: an object-oriented model. *Ecol. Model.*, **75/76**, 147–59.

Revie, J., Weston, D.E., Harden Jones, F.R. & Fox, G.P. (1990) Identification of fish echoes located at 65 km range by shorebased sonar. *J. Cons. Int. Explor. Mer*, **46** (3), 313–24.

Reynisson, P. (1993) *In situ* target strength measurements of Icelandic summer spawning herring in the period 1985–1992. *Int. Coun. Explor. Sea*, C.M. 1993/B:40 Ref. H, pp. 1–15.
Reynolds, W.W. & Casterlin, M.E. (1980) The role of temperature in the environmental physiology of fishes. In: *Environmental Physiology of Fishes* (ed. Ali, M.A.), pp. 497–518. NATO Advanced Study Institutes Series. Series A: Life Sci. Plenum Press, New York.
Richards, L.J. & Schnute, J.T. (1986) An experimental and statistical approach to the question: is CPUE an index of abundance? *Can. J. Fish. Aquat. Sci.*, **43**, 1214–27.
Richardson, I.D. (1960) Observations on the size and numbers of herring taken by the herring trawl. *J. Cons. Int. Explor. Mer*, **25**, 204–17.
Ricker, W.E. (1954) Stock and recruitment. *J. Fish. Res. Board Can.*, **11**, 559–623.
Ricker, W.E. (1971) Derzavin's biostatistical method of population analysis. *J. Fish. Res. Board Can.*, **28**, 1666–72.
Ricker, W.E. (1975) Computation and interpretation of biological statistics of fish populations. *Fish. Res. Board Can. Bull.*, **191**.
Rivard, D. & Bledsoe, L.J. (1978) Parameter estimation for the Pella-Tomlinson stock production model under nonequilibrium conditions. *Fish. Bull.* (US) **76** (3), 523–34.
Rivault, C. & Cloarec, A. (1992) Foraging in *Blattella germanica* (L.) and aggregation pheromone. *Etologia*, **2**, 33–9.
Rivest, L.-P., Potvin, F., Crepeau, H. & Daigle, G. (1995) Statistical methods for aerial surveys using the double-count technique to correct visibility bias. *Biometrics*, **51** (2), 461–70.
Roberts, C.M. (1995) Rapid build-up of fish biomass in a Caribbean marine reserve. *Conserv. biol.*, **9** (4), 816–26.
Roberts, C.M. (1997) Ecological advice for the global fisheries crisis. *TREE*, **12**, 35–8.
Robertson, D.R., Sweatman, H.P.A., Fletcher, E.A. & Cleland, M.G. (1976) Schooling as a mechanism for circumventing the territoriality of competitors. *Ecology*, **57**, 1208–20.
Robinson, C.J. & Pitcher, T.J. (1989a) The influence of hunger and ration level on shoal density, polarisation and swimming speed of herring, *Clupea harengus* L. *J. Fish. Biol.*, **34**, 631–3.
Robinson, C.J. & Pitcher, T.J. (1989b) Hunger motivation as a promoter of different behaviours within a shoal of herring: selection for homogenity in fish shoal? *J. Fish. Biol.*, **35**, 459–60.
Robson, D.S. (1966) Estimation of the relative fishing power of individual ships. *ICNAF Res. Bull.*, (3), 5–14.
Rocha, C.A.S., Stokes, T.K., Pope, J.G. & Kell, L.T. (1991) Defining species specific effort. *ICES CM/1991*, Statistics Comm. (mimeo).
Roepke, A. (1989) Day/night differences in determination of growth rate of mackerel *Scomber scombrus* larvae during a patch-study in the Celtic Sea. *Mar. Biol.*, **102** (4), 439–43.
Roff, D.A. & Fairbairn, D.J. (1980) An evaluation of Gulland's method for fitting the Schaefer model. *Can. J. Fish. Aquat. Sci.*, **37**, 1229–35.
Roger, C. (1993) On feeding conditions for surface tunas (yellowfin, *Thunnus albacares* and skipjack, *Katsuwonus pelamis*) in the Western Indian Ocean. Proceedings of the 5th Expert Consultation on Indian Ocean Tunas. *FAO*, **8**, 131–5.
Rogers, P.H., Popper, A.N., Hastings, M.C. & Saidel, W.N. (1988) Processing of acoustic signals in the auditory system of bony fish. *J. Acoust. Soc. Am.*, **83**, 338–49.
Romey, W.L. (1996) Individual differences make a difference in the trajectories of simulated schools of fish. *Ecol. Model.*, **92** (1), 65–77.
Rose, G.A. & Leggett, W.C. (1988) Atmosphere-ocean coupling and Atlantic cod migrations: effects of wind-forced variations in sea temperature and currents on nearshore distributions and catch rates of *Gadus morhua*. *Can. J. Fish. Aquat. Sci.*, **45**, 1234–43.
Rose, G.A. & Leggett, W.C. (1991) Effects of biomass-range interactions on catchability of migratory demersal fish by mobile fisheries: an example of Atlantic cod (*Gadus morhua*). *Can. J. Fish. Aquat. Sci.*, **48**, 843–8.
Rosenzweig, M.L. (1981) A theory of habitat selection. *Ecology*, **62**, 327–35.
Rosenzweig, M.L. (1985) Some theoretical aspects of habitat selection. In: *Habitat selection in birds* (ed. Cody, M.L.), pp. 517–40. Academic Press, Orlando.
Ross, R.M. & Backman, T.W.H. (1992) Mechanisms and function of school formation in subyearling American shad (*Alosa sapidissima*). *J. Appl. Ichthyol.*, **8**, 143–53.
Rothschild, B.J. (1978) Fishing effort. In: *Fish Population Dynamics* (ed. Gulland, J.A.), pp. 96–115. John Wiley and Sons, New York.

Rothschild, B.J. & Osborn, T.R. (1988) The effect of turbulence on planktonic contact rates. *J. Plankton Res.*, **10**, 465–74.
Rountree, R.A. (1989) Association of fishes with fish aggregation devices: effects of structure size on fish abundance. *Bull. Mar. Sci.*, **44** (2), 960–72.
Roy, C. (1998) Upwelling-induced retention area: a mechanism to link upwelling and retention processes. In Benguela Dynamics (eds Pillar, S.C., Moloney, C., Payne, A.I.L. & Shillington, F.A.). *S. Afr. J. Mar. Sci.* (in press).
Roy, C., Cury, P. & Kifani, S. (1992) Pelagic fish recruitment success and reproductive strategy in upwelling areas: environmental compromise. *S. Afr. J. Mar. Sci.*, **12**, 135–46.
Russel, E.S. & Bull, H.O. (1932) A selected bibliography of fish behaviour. *J. Cons. Int. Explor. Mer*, **7**, 255–83.
Ruzzante, D.E., Taggart, C.T. & Cook, D. (1996a) Spatial and temporal variation in the genetic composition of a larval cod (*Gadus morhua*) aggregation: cohort contribution and genetic stability. *Can. J. Fish. Aquat. Sci.*, **53**, 2695–705.
Ruzzante, D.E., Taggart, C.T., Cook, D. & Goddard, S. (1996b). Genetic differentiation between inshore and offshore Atlantic cod (*Gadus morhua* L.) off Newfoundland: microsatellite DNA variation and antifreeze level. *Can. J. Fish. Aquat. Sci.*, **53**, 634–45.
Ryer, C.H. & Olla, B.L. (1991) Information transfer and the facilitation and inhibition of feeding in a schooling fish. *Environ. Biol. Fish.*, **30**, 317–23.
Ryer, C.H. & Olla, B.L. (1992) Social mechanisms facilitating exploitation of spatially variable ephemeral food patches in a pelagic marine fish. *Anim. Behav.*, **44**, 69–74.
Røttingen, I. (1976) On the relation between echo intensity and fish density. *FiskDir. Skr. Ser. Havunders*, **16**, 301–14.
Røttingen, I. (1990) A review of variability in the distribution and abundance of Norwegian spring spawning herring and Barents Sea capelin. In: *What Determines the Distribution of Seabirds at Sea* (eds Erikstad, K.E., Barrett, R.T. & Mehlum, F.), pp. 33–42. *Polar Res.*, **8**.
Røttingen, I. (1992) Recent migration routes of Norwegian spring spawning herring. *ICES CM/1992, H:18.*
Røttingen, I., Foote, K.G., Huse, I. & Ona, E. (1994) Acoustic abundance estimation of wintering Norwegian spring-spawning herring, with emphasis on methodological aspects. *ICES CM 1994/(B+D+G+H):1*, pp. 1–17.
Saila, S.B. (1996) Guide to some computerised artificial intelligence methods. In: *Computers in Fisheries Research* (eds Megrey, B.A. & Moksness, E.), pp. 8–40. Chapman and Hall, London.
Samples, K.C. & Sproul, J.T. (1985) Fish aggregating devices and open-access commercial fisheries. *Bulletin of the Inter-American Tropical Tuna Commission*, **1**, 27–56.
Sampson, D.B. (1991) Fishing tactics and fishing abundance, and their influence on catch rates. *ICES J. Mar. Sci.*, **48**, 291–301.
Sand, O. & Enger, P.S. (1973) Evidence for an auditory function of the swim bladder in the cod. *J. Exp. Biol.*, **59**, 405–14.
Sand, O. & Hawkins, A.D. (1973) Acoustic properties of the cod swim bladder. *J. Exp. Biol.*, **58**, 797–820.
Sand, O. & Karlsen, H.E. (1986) Detection of infrasound by the Atlantic cod. *J. Exp. Biol.*, **25**, 197–204.
Sarno, B., Glass, C.W., Smith, G.W., Johnstone, A.D.F. & Mojsiewicz, W.R. (1994) A comparison of the movements of two species of gadoid in the vicinity of an underwater reef. *J. Fish Biol.*, **45**, 811–17.
SAS Institute (1989) *SAS/STAT®* User's Guide, version 6, 4th Edition, 2. Cary, NC.
Saville, A. & Bailey, R.S. (1980) The assessment and management of the herring stocks in the North Sea and to the west of Scotland. *Rapp. P.-v. Réun. Cons. Int. Explor. Mer*, **177**, 112–42.
Scalabrin, C., Diner, N., Weill, A., Hillion, A. & Mouchot, M.C. (1996). Narrowband acoustic identification of monospecific fish shoals. *ICES J. Mar. Sci.*, **53** (2), 181–8.
Schaefer, M.B. (1954) Some aspects of the dynamics of populations important to the management of the commercial marine fisheries. *Bulletin of the Inter-American Tropical Tuna Commission* (USA), **1** (2), 26–56.
Schaefer, M.B. (1957) A study of the dynamics of the fishery for yellowfin tuna in the eastern tropical Pacific Ocean. *Bulletin of the Inter-American Tropical Tuna Commission* (USA), **2** (6), 246–68.
Schaefer, M.B. (1967) Fishery dynamics and present status of the yellowfin tuna population of the eastern Pacific Ocean. *Bulletin of the Inter-American Tropical Tuna Commission*, **12** (3), 89–112.
Schaefer, M.B. & Beverton, R.J. (1963) Fishery dynamics: their analysis and interpretation. In: *The Sea* (ed. Hill, M.N.), pp. 464–83. John Wiley and Sons, New York.

Schellart, N.A.M. & Buwalda, R.J.A. (1990) Directional variant and invariant hearing thresholds in the rainbow trout (*Salmo gairdneri*). *J. Exp. Biol.*, **149**, 113–31.

Schellart, N.A.M. & Munck, J.C. de (1987) A model for directional and distance hearing in swim bladder-bearing fish based on the displacement orbits of the hair cells. *J. Acoust. Soc. Am.*, **82** (3), 822–9.

Schneider, D.C. (1989) Identifying the spatial scale of density-dependent interaction of predators with schooling fish in the southern Labrador current. *J. Fish Biol.*, **35** (Suppl. A), 109–15.

Schnute, J.T. (1977) Improved estimates from the Schaefer production model: theoretical considerations. *J. Fish. Res. Board Can.*, **34**, 583–603.

Schnute, J.T. (1985) A general theory for analysis of catch and effort data. *Can. J. Fish. Aquat. Sci.*, **42**, 414–29.

Schnute, J.T. (1987) A general fishery model for a size-structured fish population. *Can. J. Fish. Aquat. Sci.*, **44**, 924–40.

Schnute, J.T. (1989) The influence of statistical error on stock assessment: illustration from Schaefer's model. In: *Effects of Ocean Variability on Recruitment and an Evaluation of Parameters used in Stock Assessment Models* (eds Beamish, R.J. & McFarlane, G.A.), pp. 101–9. *Can. Spec. Publ. Fish. Aquat. Sci.*, **108**.

Schuijf, A. (1974) *Field studies of directional hearing in marine teleosts*. PhD. Thesis, Univ. of Utrecht, The Netherlands (unpublished).

Schuijf, A. (1975) Directional hearing of cod (*Gadus morhua*) under approximate free field conditions. *J. Comp. Physiol.*, **98**, 307–32.

Schuijf, A. (1976) The phase model of directional hearing in fish. In: *Sound Reception in Fish* (eds Schuijf, A. & Hawkins, A.D.), pp. 63–86. Elsevier, Amsterdam.

Schuijf, A. & Buwalda, R.J.A. (1975) On the mechanism of directional hearing in cod (*Gadus morhua* L.). *J. Comp. Physiol.*, **98**, 333–43.

Schuijf, A. & Hawkins, A.D. (1983) Acoustic distance discrimination by the cod. *Nature*, **302**, 143–4.

Schuijf, A. & Siemelink, M.E. (1974) The ability of cod (*Gadus morhua*) to orient towards a sound source. *Experientia*, **30**, 773–4.

Schwarz, A.L. (1985) The behaviour of fishes in their acoustic environment. *Environ. Biol. Fish.*, **13** (1), 3–15.

Schwarz, A.L. & Greer, G.L. (1984) Responses of Pacific herring, *Clupea harengus pallasi*, to some underwater sounds. *Can. J. Fish. Aquat. Sci.*, **41**, 1183–92.

Schweigert, J.F. (1993) A review and evaluation of methodology for estimating Pacific herring egg deposition. In: *Advances in the Early Life History of Fishes. Part 2. Ichthyoplankton Methods for Estimating Fish Biomass* (eds. Hunter, J.R., Lo, N.C.H. & Fuiman, L.A.), pp. 818–41. *Bull. Mar. Sci.* (US), **53** (2).

Scott, J.M. & Flittner, G.A. (1972) Behaviour of bluefin tuna schools in the eastern north Pacific Ocean as inferred from fishermen's logbooks, 1960–67. *Fish. Bull.* (US), **70**, 915–27.

Scott, M., Loson, R., Chivers, S. & Holland, K. (1995) Tracking of spotted dolphins and yellowfin tuna in the eastern tropical Pacific. *46th Tuna Conference*. San Diego and Lake Arrowhead (mimeo).

Seghers, B.H. (1974a) Schooling behaviour in the guppy (*Poecillia reticulata*): an evolutionary response to predation. *Evolution*, **28**, 486–9.

Seghers, B.H. (1974b) Geographic variation in the response of guppies (*Poecilia reticulata*) to aerial predators. *Oeceologia*, **14**, 93–8.

Serebrov, L.I. (1974) Relationship between school density and size of fish. *J. Ichthyol.*, **16**, 135–40.

Serebrov, L.I. (1984) On the density of distribution and orientation of capelin schools. In: *Proceedings of the Soviet–Norwegian Symposium on The Barents Sea Capelin*, Bergen, Norway, 14–17 August 1984, pp. 157–69.

Serra, R. (1991) Important life history aspects of the Chilean jack mackerel, *Trachurus symmetricus murphyi. Investigación Pesquera*, **36**, 67–83.

Sette, O.E. (1943) Studies on the Pacific pilchard or sardine (*Sardinops caerulea*). I – Structure of a research program to determine how fishing affects the resource. United States Department of the Interior, Fish and Wildlife Service. *Special Scientific Reports*, **19**, 1–27.

Shackleton, L.Y. (1987) A comparative study of fossil fish scales from three upwelling regions. *S. Afr. J. Mar. Sci.*, **5**, 79–84.

Shackleton, L.Y. (1988) Scale shedding: an important factor in fossil fish scale studies. *J. Cons. Int. Explor. Mer*, **44**, 259–63.

Shapiro, D.Y. (1983) On the possibility of kin groups in coral reef fishes. *NOAA Symp. Ser. Undersea Research*, **1**, 39–45.

Sharp, G.D. (1978a) Colonization: modes of opportunism in the ocean. In: *Workshop on the effect of environmental variation on the survival of larval pelagic fishes*, 20 April–5 May 1980, Lima, Peru. *IOC Workshop Report Series 28*, pp. 125–48. IOC/UNESCO, Paris.

Sharp, G.D. (1978b) Behaviour and physiological properties of tunas and their effects on vulnerability to fishing gear. In: *The Physiological Ecology of Tunas* (eds Sharp. G.D. & Dizon, A.E.), pp. 397–449. Academic Press, New York.

Sharp, G.D. & Dizon, A.E. (eds) (1978) *The Physiological Ecology of Tunas*. Academic Press, New York.

Shashar, N., Addessi, L. & Cronin, T.W. (1995) Polarization vision as a mechanism for detection of transparent objects. In: *Ultraviolet Radiation and Coral Reefs* (eds Gulko, D. & Jokiel, P.L.), pp. 207–11. Proceedings of the Workshop on Measurement of Ultraviolet Radiation in Tropical Coastal Ecosystems, Honolulu, HI (USA), 3–5 August 1994.

Shaw, E. (1960) The development of schooling behaviour in fishes I. *Physiol. Zool.*, **32**, 79–86.

Shaw, E. (1961) The development of schooling behaviour in fishes II. *Physiol. Zool.*, **34**, 263–72.

Shaw, E. (1969) Some new thoughts on the schooling of fishes. *FAO Fisheries Report* (ITA), **62** (2), 217–31.

Shaw, E. (1970) Schooling in fishes: critique and review. In: *Development and Evolution of Behaviour* (eds Aronsen, L.R., Tobach, E., Lehrman, D.S. & Rosenblat, J.S.), pp. 452–80. W.H. Freeman and Co., San Francisco.

Shaw, E. (1978) Schooling fishes. *Am. Scient.*, **66**, 166–75.

Shelton, P.A. & Armstrong, M.J. (1983) Variations in parent stock size and recruitment of pilchard and anchovy populations in the southern Benguela system. In: *Proceedings of the Expert Consultation to Examine Changes in Abundance and Species Composition of Neritic Fish Resources* (eds Sharp, G.D. & Csirke, J.), pp. 1113–32. *FAO Fish. Rep.*, **291** (3).

Shelton, P.A. & Hutchings, L. (1982) Transport of anchovy, *Engraulis capensis* Gilchrist, eggs and early larvae by a frontal jet current. *J. Cons. Int. Explor. Mer*, **40** (2), 185–98.

Shelton, P.A., Armstrong, M.J. & Roel, B.A. (1993). An overview of the application of the daily egg production method in the assessment and management of anchovy in the south-east Atlantic. In: *Advances in the Early Life History of Fishes. Part 2. Ichthyoplankton Methods for Estimating Fish Biomass* (eds Hunter, J.R., Lo, N.C.H. & Fuiman, L.A.), pp. 778–94. *Bull. Mar. Sci.* (US) **53** (2).

Shepherd, J.G. & Nicholson, M.D. (1986) Use and abuse of multiplicative models in the analysis of fish catch-at-age data. *Statistician*, **35**, 221–7.

Sherwood, N.M., Kyle, A.L., Kreiberg, H., Warby, C.M., Magnus, T.H., Carolsfeld, J. & Price, W.S. (1991) Partial characterisation of a spawning pheromone in the herring *Clupea harengus* pallasi. *Can. J. Zool.*, **69** (1), 91–103.

Shiohama, T. (1978) Overall fishing intensity and yield by the Atlantic longline fishery of albacore (*Thunnus alalunga*), 1956–1975. *ICCAT Res. Doc. Sci.*, **7** (2), 217–24.

Shiraishi, M., Azuma, N. & Aoki, I. (1996) Large schools of Japanese sardines, *Sardinops melanostictus*, mate in single pair units at night. *Environ. Biol. Fish.*, **45** (4), 405–9.

Sibly, R.M. (1983) Optimal group size is unstable. *Anim. Behav.*, **31**, 947–8.

Sigler, M.F. & Fujioka, J.T. (1988) Evaluation of variability in sablefish, *Anoplopoma fimbria*, abundance indices in the Gulf of Alaska using the bootstrap method. *Fish. Bull.* (US) **86** (3), 445–52.

Simmonds, J.E. & MacLennan, D.N. (eds) (1996) Fisheries and plankton acoustics. *ICES J. Mar. Sci.*, **53** (2), 129–535.

Simmonds, E.J., Armstrong, F. & Copland, P. (1996) Species identification using wideband backscatter with neural network and discriminant analysis. *ICES J. Mar. Sci.*, **53** (2), 189–95.

Simmonds, E.J., Williamson, N.J., Gerlotto, F. & Aglen, A. (1992) Survey design and analysis procedures: a comprehensive review of good practice. *ICES Coop. Res. Rep.*, **187**.

Sivasubramaniam, K. (1961) Relation between soaking time and catch of tuna, in long line fisheries. *Bull. Japan. Soc. Sci. Fish.*, **27**, 835–45.

Slotte, A. & Johannessen, A. (1997) Exploitation of Norwegian spring-spawning herring (*Clupea harengus* L.) before and after the stock decline; towards a size selective fishery. In: *Developing and sustaining world fisheries. Proceedings of the 2nd World Fisheries Congress*, Brisbane, Australia, July 1996 (eds Hancock, D.A., Smith, D.C., Grant, A. & Beumer, J.P.), pp. 103–8. CSIRO, Collingwood, Australia.

Smith, G.W., Glass, C.W., Johnstone, A.D.F. & Mojsiewicz, W.R. (1993) Diurnal patterns in the spatial relationships between saithe, *Pollachius virens*, schooling in the wild. *J. Fish. Biol.*, **43** (Suppl. A), 315–25.

Smith, M.F.L. & Warburton, K. (1992) Predator shoaling moderates the confusion effect in blue-green chromis, *Chromis viridis. Behav. Ecol. Sociobiol.*, **30**, 103–7.

Smith, P.E. (1977) The effects of internal waves on fish school mapping with sonar in the California current area. *Rapp. P.-v. Reun. Cons. Int. Explor. Mer*, **170**, 223–31.

Smith, P.E., Flerx, W. & Hewitt, R.P. (1985) The CalCOFI vertical egg tow (CalVET) net. In: *An egg production method for estimating spawning biomass of pelagic fish: application to the Northern anchovy*, Engraulix mordax. NOAA Tech. Rep. NMFS 36.

Smith, P.J., Francis, R.I.C.C. & McVeagh, M. (1991) Loss of genetic diversity due to fishing pressure. *Fish. Res.* (US) **10**, 309–16.

Smith, R.J.F. (1992) Alarm signals in fishes. *Rev. Fish Biol. Fisheries*, **2**, 33–63.

Smith, S.J. & Gavaris, S. (1993) Improving the precision of abundance estimates of Eastern Scotian Shelf Atlantic cod from bottom trawl surveys. *N. Amer. J. Fish. Manag.*, **13**, 35–47.

Smith, S.J. & Page, F.H. (1996) Association between Atlantic cod (*Gadus morhua*) and hydrographic variables: implications for the management of the 4VsW cod stock. *ICES J. Mar. Sci.*, **53**, 597–614.

Smith, T.D. (1988) Stock assessment methods: the first fifty years. In: *Fish Population Dynamics* (ed. Gulland, J.A.), pp. 1–33. John Wiley and Sons, New York.

Smith, T.D. (1994) *Scaling Fisheries: the Science of Measuring the Effects of Fishing*, pp. 1855–955. Cambridge University Press, Cambridge.

Snyder, D.E. (1983) Fish eggs and larvae. In: *Fisheries Techniques* (eds Nielsen, L.A. & Johnson, D.L.), pp. 165–98. Am. Dish. Soc., Bethesda, MD, US.

Soemarto (1960) Fish behaviour with special reference to pelagic schooling species: Lajang (*Decapterus* spp.). *8th Proc. Indo-Pacific Fish. Coun.*, **3**, 89–93.

Soldal, A.V., Engås, A. & Isaksen, B. (1993) Survival of gadoids that escape from a demersal trawl. *ICES Mar. Sci. Sym.*, **196**, 122–7.

Somerton, D.A. & Kikkawa, B.S. (1995) A stock survey technique using the time to capture individual fish on longlines. *Can. J. Fish. Aquat. Sci.*, **52** (2), 260–7.

Soria, M. (1994) *Structure et stabilité des bancs et agrégations de poissons pélagiques côtiers tropicaux: application halieutique*. Thèse de Doctorat, Université de Rennes (France).

Soria, M. & Dagorn, L. (1992) Rappels sur le comportement grégaire. In: *Action Incitative Comportement Agrégatif (AICA)*, Montpellier, June 1992, (ed. Stretta, J.-M.), pp. 5–9. Doc. Int. Centre, Orstom, Montpellier.

Soria, M., Fréon, P. & Gerlotto, F. (1996) Analysis of vessel influence on spatial behaviour of fish schools using a multi-beam sonar and consequences for biomass estimates by echo-sounder. *ICES J. Mar. Sci.*, **53** (2), 453–8.

Soria, M., Gerlotto, F. & Fréon, P. (1993) Study of learning capability of a tropical clupeoids using an artificial stimulus. *ICES Mar. Sci. Symp.*, **196**, 17–20.

Sorokin, M.A. (1987) Perception of sound by the eastern mackerel, *Scomber japonicus. Voprosy Ikhtiologii*, **2**, 329–35.

Sorokin, M.A. (1989) Directional hearing in clupeidae. *Soviet J. Mar. Biol.*, **14** (4), 224–8.

Souid, P. (1988) *Automatisation de la description et de la classification des détections acoustiques de bancs de poissons pélagiques pour leur identification.* Thèse de Doctorat de l'Université de bancs de poissons pélagiques pour leur identification. Thèse de Doctorat de l'Université d'Aix-Marseille II (France) Spécialité Ecologie.

Soutar, A. & Isaacs, J.D. (1974) Abundance of pelagic fish during the 19th and 20th centuries as recorded in anaerobic sediment off the Californias. *Fish. Bull.* (US) **72** (2), 257–73.

Squire, J.L. (1972) Apparent abundance of some pelagic marine fishes off the southern and central California coast as surveyed by an airborne monitoring program. *U.S. Fish. Wild. Serv. Fish. Bull.*, **70**, 1005–19.

Squire, J.L. Jr. (1978) Northern anchovy school shapes as related to problems in school size estimation. *Fish. Bull.* (US) **76**, 443–8.

Stacey, N.E. (1987) Roles of hormones and pheromones in fish reproductive behaviour. In: *Psychobiology of Reproductive Behaviour. An Evolutionary Perspective* (ed. Crews, D.), pp. 28–60. Prentice Hall, Englewood Cliffs.

Stacey, N.E. & Hourston, A.S. (1982) Spawning and feeding behavior of captive Pacific herring (*Clupea harengus pallasi*). *Can. J. Fish. Aquat. Sci.*, **39** (3), 489–98.

Stephens, D.W. & Krebs, J.R. (1987) *Foraging Theory*. Princeton University Press, Princeton, NJ.

Stephenson, R.L. (1991) Comparisons of tuning methods used in herring stock assessments in the north-east and north-west Atlantic. *ICES CM/1991/H:39.*

Stéquert, B. & Marsac, F. (1989) Tropical tuna surface fisheries in the Indian Ocean. *FAO Fish. Tech. Pap.*, **282**.

Stevens, J.D., Hansfield, H.F. & Davenport, S.R. (1984) Observations on the biology, distribution and abundance of *Trachurus declivis*, *Sardinops neopilchardus* and *Scomber australasicus* in The Great Australian Bight. *Rep. CSIRO Mar. Lab.* (Australia), **164**.

Stokes, T.K. & Pope, G. (1987) The detectability trends from catch-at-age and commercial effort data. *ICES CM/1987/D:14.*

Strand, S.W. & Hamner, W.N. (1990) Schooling behaviour of Antarctic krill (*Euphausia superba*) in laboratory aquaria: reactions to chemical and visual stimuli. *Mar. Biol.*, **106**, 355–9.

Strasburg, D.W. (1961) Diving behaviour of Hawaiian skipjack tuna. *J. Cons. Int. Explor. Mer*, **26**, 223–9.

Strasburg, D.W. & Yuen, H.S. (1958) Preliminary results of underwater observations of tuna schools and practical applications of these results. In: *Proceedings of the Indo-Pacific Fisheries Council*, 8th session, Colombo, Ceylon, 6–22 Dec. 1958 (III), pp. 84–9.

Stretta, J.-M. (1988) Environnement et pêche thonière en Atlantique tropical oriental. In: *Ressources, pêche et biologie des thonidés tropicaux de l'Atlantique centre-est* (eds Fonteneau, A. & Marcille, J.), pp. 269–316. *FAO Fish. Tech. Pap.*, **292**.

Stretta, J.-M. (1991) *Contribution de la télédétection aérospatiale à l'élaboration des bases de l'halieutique opérationnelle: l'exemple des pêcheries thonières tropicales de surface (aspect prédictif)*. Doctoral dissertation, Paris VI, Grenoble (France).

Stretta, J.-M. & Petit, M. (1989) Relation capturabilité température de surface. *Tech. Pap. FAO*, **302**, 49–50.

Stretta, J.-M. & Petit, M. (1992) Thonidés tropicaux: la synthèse écologique. *ICCAT Collect. Vol. Sci. Pap.*, **39** (1), 307–21.

Stretta, J.-M. & Slépoukha, M. (1986) Analyse des facteurs biotiques et abiotiques associés aux bancs de thons. In: *Proceedings of the ICCAT Conference on the International Skipjack Year Program* (eds Symons, P.E.K., Miyake, P.M. & Sakagawa, G.T.), pp. 161–9. ICCAT, Madrid.

Stretta, J.-M., Delgado de Molina, A., Ariz, J., Domalain, G. & Santana, J.C. (1996) *Les espèces associées aux pêches thonières tropicales.* Orstom/IEO rapport scientifique (2éme partie) + annexe. European community project (mimeo).

Strømme, T. & Sætersdal, G. (1987) Consistency of acoustic biomass assessments tested by repeated survey coverage. *International Symposium on Fisheries Acoustics*, Seattle, Washington, USA, 22–26 June.

Stuntz, W.E. (1981) The tuna–dolphin bond: a discussion of current hypotheses. Southwest Fisheries Center, National Marine Fisheries Service, NOAA La Jolla, CA. *Administrative Rep. LJ*, 81-19.

Suboski, M.D. & Templeton, J.J. (1989) Life skills training for hatchery fish: social learning and survival. *Fish. Res.*, **7**, 343–52.

Sullivan, P.J. (1991) Stock abundance estimation using depth-dependent trends and spatially correlated variation. *Can. J. Fish. Aquat. Sci.*, **48** (9), 1691–1703.

Sund, O. (1935) Echo sounding in fishery research. *Nature*, **135**, 953.

Sund, P.N., Blackburn, M. & Williams, F. (1981) Tunas and their environment in the Pacific Ocean: a review. *Oceanogr. Mar. Biol. Ann. Rev.*, **19**, 443–512.

Suomala, J.B. (1981) *Meeting on hydroacoustical methods for the estimation of marine fish populations*, 25–29 June 1979. II: contributed papers, discussion, and comments. The Charles Stark Draper Laboratory, Cambridge, Mass., US.

Sutherland, W.J. (1983) Aggregation and the ideal free distribution. *J. Anim. Ecol.*, **52**, 821–8.

Suuronen, P. (1995) Conservation of young fish by management of trawl selectivity. *Fin. Fish. Res.*, **15**, 97–116.

Suuronen, P. & Miller, R.B. (1992) Size selectivity of diamond and square mesh codends in pelagic herring trawls: only small herring will notice the difference. *Can. J. Fish. Aquat. Sci.*, **49** (10), 2104–17.

Suuronen, P., Erickson, D.L. & Orrensalo, A. (1996a) Mortality of herring escaping from pelagic trawl codends. *Fish. Res.* (US), **25**, 305–21.

Suuronen, P., Lehtonen, E. & Wallace, J. (1997) Avoidance and escape behaviour by herring encountering midwater trawls. *Fish. Res.* (US) **29**, 13–24.

Suuronen, P., Lehtonen, E., Tschernij, V. & Orrensalo, A. (1993) Survival of Baltic herring (*Clupea harengus* L.) escaping from a trawl codend and through a rigid sorting grid. *ICES CM 1993/B:14.*

Suuronen, P., Perez-Comas, J., Lehtonen, E. & Tschernij, V. (1996b) Size-related mortality of herring

(*Clupea harengus* L.) escaping through a rigid sorting grid and trawl codend meshes. *Int. Coun. Explor. Sea J. Mar. Sci.*, **53**, 691–700.

Suyehiro, Y. (1952) *Textbook of Ichthyology* (in Japanese). Iwanami Shoten, Tokyo.

Suzuki, Z. (1992) General description on tuna biology related to fishing activities on floating objects by Japanese purse seine boats in the western and the central Pacific. In: *International Workshop on the Ecology and Fisheries for Tunas Associated with Floating Objects and on Assessment Issues Arising from the Association of Tunas with Floating Objects* (ed. Hall, M). Inter-American Tropical Tuna Commission Int. Rep., San Diego (mimeo).

Suzuki, Z. (1994) A review of the biology and fisheries for yellowfin tuna (*Thunnus albacares*) in the western and central Pacific Ocean. In: *Interactions of Pacific Tuna Fisheries* (eds Shomura, R.S., Majkowski, J. & Langi, S.), pp. 108–37. *FAO Pap. Biol. Fish.*, **2**.

Swain, D.P. & Kramer, D.L. (1995) Annual variation in temperature selection by Atlantic cod *Gadus morhua* in the southern Gulf of St. Lawrence, Canada, and its relation to population size. *Mar. Ecol-Progr. Ser.*, **116**, 11–23.

Swain, D.P. & Morin, R. (1996) Relationships between geographic distribution and abundance of American plaice (*Hippoglossoides platessoides*) in the southern Gulf of St. Lawrence. *Can. J. Fish. Aquat. Sci.*, **53**, 106–19.

Swain, D.P. & Sinclair, A.F. (1994) Fish distribution and catchability: what is the appropriate measure of distribution? *Can. J. Fish. Aquat. Sci.*, **51**, 1046–54.

Swain, D.P. & Wade, E.J. (1993) Density-dependent geographic distribution of Atlantic cod (*Gadus morhua*) in the southern Gulf of St. Lawrence. *Can. J. Fish. Aquat. Sci.*, **50**, 725–33.

Swain, D.P., Nielsen, G.A., Sinclair, A.F. & Chouinard, G.A. (1994) Changes in catchability of Atlantic cod (*Gadus morhua*) to an otter-trawl fishery and research survey in the southern Gulf of St. Lawrence. *ICES J. Mar. Sci.*, **51**, 493–504.

Swartzman, G. (1997) Analysis of the summer distribution of fish schools in the Pacific Eastern Boundary Current. *ICES J. Mar. Sci.*, **54**, 105–16.

Swartzman, G., Silverman, E. & Williamson, N. (1995) Relating trends in walleye pollock (*Theragra chalcogramma*) abundance in the Bering Sea to environmental factors. *Can. J. Fish. Aquat. Sci.*, **52** (2), 369–80.

Swartzman, G., Stuetzle, W., Kulman, K. & Powojowski, M. (1994) Relating the distribution of pollock schools in the Bering Sea to environmental factors. *ICES J. Mar. Sci.*, **51** (4), 481–92.

Sætre, R. & Gjøsæther, J. (1975) Ecological investigations on the spawning grounds of the Barents Sea capelin. *FiskDir. Skr. Ser. HavUnders.*, **16**, 203–27.

Tamura, T. Hanyu, I. & Niwa, H. (1972) Spectral sensitivity and color vision in skipjack tuna and related species. *Bull. Japan. Soc. Sci. Fish.*, **38**, 799–802.

Thomas, R.M. & Schülein, F.H. (1988) The shoaling behaviour of pelagic fish and the distribution of seals and gannets off Namibia as deduced from routine fishing reports, 1982–1985. *S. Afr. J. Mar. Sci.*, **7**, 179–91.

Thomson, W.F. & Bell, F.H. (1934) Biological statistics of the Pacific halibut fishery (2). Effect of changes in intensity upon total yield and yield per unit of gear. *Rep. Int. Fish. Comm.*, **8**, 1–49.

Thorne, R.E. & Thomas, G.L. (1990) Acoustic observations of gas bubble release by Pacific herring (*Clupea harengus pallasi*). *Can. J. Fish. Aquat. Sci.*, **47**, 1920–28.

Tinbergen, N. (1950) *The Study of Instinct*. Clarendon Press Publishers, Oxford.

Tinbergen, N. (1963) On aims and methods of ethology. *Zeit. Tierpsychol.*, **20**, 410–33.

Tomlinson, P.K. (1970) A generalisation of the Murphy catch equation. *J. Fish. Res. Board Can.*, **27**, 821–5.

Toresen, R. (1991) Absorption of acoustic energy in dense herring schools studied by attenuation in the bottom echo signal. *Fish. Res.*, **10**, 317–27.

Traynor, J.J. & Williamson, N.J. (1983) Target strength measurements of walleye pollock (*Theragra chalcogramma*) and a simulation study of the dual beam method. *FAO Fish. Rep.*, **300**, 112–24.

Tregenza, T. (1994) Common misconceptions in applying the ideal free distribution. *Anim. Behav.*, **47**, 485–7.

Treherne, J.E. & Foster, W.A. (1981) Group transmission of predator avoidance in a marine insect: the Trafalgar effect. *Anim. Behav.*, **29**, 911–17.

Treschev, A.I., Efanov, S.F., Shevtsov, S.E. & Klavsons, U.A. (1975) Traumatism and survivability of Baltic herring which passed through the mesh of trawl cod end. *ICES CM 1975/B:9*.

Troadec, J.-P., Clark, W.G. & Gulland, J.A. (1980) A review of some pelagic fish stocks in other areas. *Rapp. P.-v. Réun Cons. Int. Explor. Mer*, **177**, 252–77.

Trout, G.C., Lee, A.J., Richardson, I.D. & Harden Jones, F.R. (1952) Recent echosounder studies. *Nature*, **170** (4315), 71–2.

Tshernij, V. (1988) *An investigation on the orientation of Baltic herring towards artificial light*. Thesis, Institute of Fisheries, Pargas, Finland (in Swedish, mimeo).

Tsukagoshi, T. (1983) Some peculiar phenomena on killer whale, *Orcinus orca*. *Bull. Japan Soc. Sci. Fish.*, **44**, 127–31.

Turner, G. (1986) Teleost mating: systems and strategies. In: *The Behaviour of Teleost Fishes* (ed. Pitcher, T.J.), pp. 253–74. Croom Helm, London.

Turner, G.F. & Huntingford, F.A. (1986) A problem for games theory analysis: assessment and intention in male mouthbrooder contests. *Anim. Behav.*, **34**, 1961–70.

Turner, G.F. & Pitcher, T.J. (1986) Attack abatement: a model for group protection by combined avoidance and dilution. *Am. Nat.*, **128**, 228–240.

Tyler, J.A. & Rose, K.A. (1994) Individual variability and spatial heterogeneity in fish population models. *Rev. Fish Biol. Fisheries*, **4** (1), 91–123.

Tyler, J.E. (1969) The natural light field underwater. *FAO Fish.Rep.*, **62** (3), 479–97.

Uda, M. (1933) Types of skipjack schools and their fishing qualities. *Bull. Japan Soc. Sci. Fish.*, **2** (3), 107–11.

Uhler, R.S. (1980) Least squares regression estimates of the Schaefer production model: some Monte Carlo simulation results. *Can J. Fish. Aquat. Sci.*, **37**, 249–60.

Ulltang, Ø. (1976) Catch per unit of effort in the Norwegian purse seine fishery for Atlanto-scandian (Norwegian spring spawning) herring. FAO Series, *FIRS/T155, 91–101.*

Ulltang, Ø. (1980) Factors affecting the reaction of pelagic fish stocks to exploitation and requiring a new approach to assessment and management. *Rapp P.-v. Réun. Cons. Int. Explor. Mer*, **177**, 489–504.

Uotani, I., Iwakawa, T. & Kawaguchi, K. (1994) Experimental study on the formation mechanisms of shirasu (postlarval Japanese anchovy) fishing grounds with special reference to turbidity. *Bull. Japan. Soc. Sci. Fish. Tokyo*, **60** (1), 73–8.

Uriarte, A., Prouzet, P. & Villamor, B. (1996) Bay of Biscay and Ibero Atlantic anchovy populations and their fisheries. *Scientia Marina*, **60** (Suppl. 2), 237–55.

Urick, R.J. (1983) *Principles of Underwater Sound*, 3rd edn. McGraw-Hill, New York.

Vabø, R., Olsen, K. & Huse, I. (in prep.) The effect of vessel avoidance of wintering Norwegian spring spawning herring.

Valdemarsen, J.W. & Misund, O.A. (1995) Trawl designs and techniques used by Norwegian research vessels to sample fish in the pelagic zone. In: *Proceedings of the Sixth IMR-PINRO Symposium*, Bergen, 14–17 June, 1994 (ed. Hylen, A.), pp. 135–44.

Valdes, S.E., Shelton, P.A., Armstrong, M.J. & Field, J.G. (1987) Cannibalism in South African anchovy: egg mortality and egg consumption rates. *S. Afr. J. Mar. Sci.*, **5**, 613–22.

Valdes, S.E. (1993) The energetics and evolution of intraspecific predation (egg cannibalism) in the anchovy *Engraulis capensis*. *Mar. Biol.*, **115** (2), 301–8.

Van Olst, J.C. & Hunter, J.R. (1970) Some aspects of the organisation of fish schools. *J. Fish. Res. Board Can.*, **27**, 1225–38.

Venema, S.C. & Nakken,O. (eds) (1983) Symposium on fisheries acoustics. ICES/FAO, *FAO Fish. Rep.*, **300**.

Viedeler, J.J. & Wardle, C.S. (1991) Fish swimming stride by stride: speed limits and endurance. *Rev. Fish. Biol. Fisheries*, **1**, 23–40.

Vignaux, M. (1996) Analysis of spatial structure in fish distribution using commercial catch and effort data from the New Zealand hoki fishery. *Can. J. Fish. Aquat. Sci.*, **53**, 963–73.

Vilhjálmsson, H. (1994) The Icelandic capelin stock. *Rit Fisk.*, **13**, 1–281.

Vilhjálmsson, H., Reynisson, P., Hamre, J. & Røttingen, I. (1983) Acoustic abundance estimation of the Icelandic stock of capelin 1978–1982. *FAO Fish. Rep.*, **300**, 208–16.

Villanueva, R. (1970) The Peruvian Eureka programme of rapid acoustic surveys using fishing vessels. Technical Conference on Fish Finding, Purse Seining and Aimed Trawling. *FAO, FII:FF/70/5.*

Voituriez, B. & Herbland, A. (1982) Comparaison des systèmes productifs de l'atlantique tropical est: dômes thermiques, upwellings côtiers et upwelling équatorial. *Rapp. P.-v. Cons. Int. Explor. Mer*, **180**, 114–30.

Wada, T. & Mutsumiya, Y. (1990) Abundance index in purse seine fishery with searching time. *Nippon Suisan Gakkaishi*, **56** (5), 725–8.

Walker, M.M. (1984) Learned magnetic field discrimination in yellowfin tuna, *Thunnus albacares*. *J. Comp. Physiol. A*, **155**, 673–9.

Walker, M.M., Dizon, A.E. & Kirschvink, J.L. (1982) Geomagnetic field detection by yellowfin tuna. *Anonyme Oceans '82 Conference record*, pp. 755–8. IEEE Press, New York.

Walsh, M. & Martin, J.H.A. (1986) Recent changes in the distribution and migrations of the western mackerel stock in relation to hydrographic changes. *ICES, CM 1986/H:17*.

Walsh, M., Reid, D.G. & Turrell, W.R. (1995) Understanding mackerel migration off Scotland: tracking with echosounders and commercial data and including environmental correlates and behaviour. *ICES J. Mar. Sci.*, **52**, 925–39.

Walsh, S.J. (1991) Diel variation in availability and vulnerability of fish to a survey trawl. *J. Appl. Ichthyol.*, **7**, 147–59.

Walsh, S.J. & Hickey, W.M. (1993) Behavioural reactions of demersal fish to bottom trawls at various light conditions. *ICES Mar. Sci. Symp.*, **196**, 68–76.

Walter, G.G. (1986) A robust approach to equilibrium yield curves. *Can. J. Fish. Aquat. Sci.*, **43**, 1332–9.

Walters, C. (1986) *Adaptive Management of Renewable Resources*. Macmillan, New York.

Walters, C., Christensen, V. & Pauly, D. (1997) Structuring dynamic models of exploited ecosystems from trophic mass-balance assessments. *Rev. Fish. Biol. Fish.*, **7**, 139–72.

Warburton, K. (1990) The use of local landmarks by foraging goldfish. *Anim. Behav.*, **40** (3), 500–5.

Wardle, C.S. (1980) Effects of temperature on the maximum swimming speed of fishes. In: *Environmental Physiology of Fishes* (ed. Ali, M.A.), pp. 519–31. NATO Advanced Study Institutes Series. Series A. Plenum Press, New York.

Wardle, C.S. (1983) Fish reactions to towed fishing gears. In: *Experimental Biology at Sea* (eds MacDonald, A.G. & Priede, I.G.), pp. 167–95. Academic Press, London.

Wardle, C.S. (1993) Fish behaviour and fishing gear. In: *Behaviour of Teleost Fishes*, 2nd edn, (ed. Pitcher, T.J.), pp. 609–43. Chapman and Hall, London.

Wardle, C.S. & Hollingworth, C.E. (eds) (1993) Fish behaviour in relation to fishing operations. *ICES. Mar. Sci. Symp.*, **196.**

Ware, D.M. (1980) Bioenergetics of stock and recruitment. *Can. J. Fish. Aquat. Sci.*, **37**, 1012–24.

Warren, W.G., Bohm, M. & Link, D. (1992) A statistical methodology for exploring elevation differences in precipitation chemistry. *Atmos. Environ.*, **26A** (1), 159–69.

Wassersug, R.J., Lum, A.W. & Potel, M.J. (1981) An analysis of school structure for tadpoles (*Anura: Amphibia*). *Behav. Ecol. Sociobiol.*, **9**, 15–22.

Watanabe, Y., Yokouchi, K., Oozeki, Y. & Kikuchi, H. (1991) Preliminary report on larval growth of the Japanese sardine spawned in the dominant current area. *ICES CM 1991/L:33*.

Watanabe, Y., Zenitani, H. & Kimura, R. (1996) Offshore expansion of spawning of the Japanese sardine, *Sardinops melanostictus*, and its implication for egg and larval survival. *Can. J. Fish. Aquat. Sci.*, **53**, 55–61.

Webb, B.F. (1977) Moon's influence on schoaling of pelagic fish species. *Austr. Fish*, Sept. 8–11.

Weihs, D. (1973) Hydromechanics of fish schooling. *Nature*, **241**, 290–1.

Weihs, D. (1975) Some hydromechanical aspects of fish schooling. In: *Symposium on Swimming and Flying in Nature* (eds Wu, T.Y., Broklaw, C.J. & Brennan, C.), pp. 703–18. Plenum Press, New York.

Weill, A., Scalabrin, C. & Diner, N. (1993) MOVIES-B: an acoustic detection description software. Application to shoal species classification. *Aquat. Liv. Resour.*, **6**, 255–67.

Welch, C. & Colgan, P. (1990) The effect of contrast and position on habituation to models of predators in eastern banded killifish (*Fundulus diaphanus*). *Behav. Process.*, **22**, 61–71.

Wenz, G.M. (1962) Acoustic ambient noise in the ocean: spectra and sounds. *J. Acoust. Soc. Am.*, **34** (12), 1936–56.

Werner, E.E. (1986) Ontogenic habitat shifts in the bluegill sunfish (*Lepomis macrochirus*) the role of the foraging rate-predation risk tradeoff. In: *The Behaviour of Fishes*. The Fisheries Society of the British Isles, Bangor.

Werner, E.E. & Hall, D.J. (1974) Optimal foraging and the size selection of prey by the Bluegill sunfish (*Lepomis macrochirus*). *Ecology*, **55**, 1042–52.

Werner, E.E., Gilliam, J.F., Hall, D.J. & Mittelbach, G.G. (1983) An experimental test of the effects of predation risk on habitat use in fish. *Ecology*, **64**, 1540–8.

Werner, R. (1977) A reasonableness algorithm for hydroacoustic estimation of biomass density. *Rapp. P.-v. Réun. Cons. Int. Explor. Mer*, **170**, 219–22.

Wespestad, V.G. & Megrey, B.A. (1990) Assessment of walleye pollock stocks in the eastern North Pacific Ocean: an integrated analysis using research survey and commercial fisheries data. *Rapp. P.-v. Réun. Cons. Int. Explor. Mer*, **189**, 33–49.

Wickham, D.A., Watson, J.W. Jr. & Ogren, L.H. (1973) The efficacy of midwater artificial structures for attracting pelagic sport fish. *Trans. Am. Fish. Soc.*, **3**, 563–72.

Wiebe, P.H., Burt, K.H., Boyd, S.H. & Morton, A.W. (1976) A multiple opening/closing net and environmental sensing system for sampling zooplankton. *J. Mar. Res.*, **34**, 313–26.

Williams, H. (1981) Aerial survey of pelagic fish resources off South East Australia 1973–1977. *Austr. CSIRO Div. Fish Oceanogr. Rep.* **130**.

Williams, H. & Pullen, G. (1993) Schooling behaviour of jack mackerel, *Trachurus declivis* (Jenyns), observed in the Tasmanian purse seine fishery. *Aust. J. Mar. Freshwater Res.*, **44**, 577–87.

Williams, K. (1981) Aerial survey of pelagic fish resources off south-east Australia 1973–1977. *Rep. Div. Fish. Oceanogr. CSIRO*, Cronulla, **130**.

Wilson, E.O. (1975) *Sociobiology: the New Synthesis*. Harvard University Press, Cambridge, Mass.

Winger, P.D., He, P. & Walsh, S.J. (1997) Preliminary analysis of the swimming endurance of Atlantic cod (*Gadus morhua*) and American plaice (*Hippoglossoides platessoides*). *ICES CM 1997/W:13.*

Winn, H.E. & Olla, B. (1972) *Behaviour of Marine Animals: Current Perspectives in Research*. Plenum Press, New York.

Winters, G.H. & Wheeler, J.P. (1985) Interaction between stock area, stock abundance, and catchability coefficient. *Can. J. Fish. Aquat. Sci.*, **42**, 989–98.

Wolf, N.G. (1985) Odd fish abandon mixed-species groups when threatened. *Behav. Ecol. Sociobiol.*, **17**, 47–52.

Wolf, N.G. (1987) Schooling tendency and foraging benefit in the ocean surgeonfish. *Behav. Ecol. Sociobiol.*, **21** (1), 59–63.

Woodhead, P.M.J. (1964) A change in the normal diurnal pattern of capture of soles during the severe winter of 1963. *Rapp. Cons. Int. Explor. Mer*, **155** (8).

Wootton, R.J. (1990) *Ecology of Teleost Fishes*. Chapman and Hall, London.

Woznitza-Mendo, C. & Espino, M. (1986) The impact of El Nino on recruitment in the Peruvian hake (*Merluccius gayi peruanus*). *Meeresforschung*, **31**, 47–51.

Wyatt, T. & Larrañeta, M.G. (eds) (1988) Long term changes in marine fish populations. *Symposium International*. Consejo Superior de Investigaciones Cientéficas, Vigo.

Wyatt, T., Cushing, D.H. & Junquera, S. (1986) Stock distinctions and evolution of European sardine. In: *Long-term Variability of Pelagic Fish Populations and their Environment* (eds Kawasaki, T., Toba, Y., Tanaka, S. & Taniguchi, A.), pp. 135–45. Pergamon Press, Oxford.

Wysokinski, A. (1987) Catch per unit of effort in the pelagic trawl fishery as illustrated by the horse mackerel fishery off Namibia. *Coll. Sci. Pap. ICSEAF/Recl. Doc. Sci.*, **14** (2), 297–307.

Yáñez, E.R. (1998) *Fluctuations des principaux stocks de poissons pélagiques exploités au Chili et leurs relations avec les variations de l'environnement*. Thèse Doctorat, Université de Bretagne Occidentale, Brest.

Yáñez, E.R., Canales, C., Barbieri, M.A., González, A. & Catasti, V. (1993) Estandarización del esfuerzo de pesca y distribución espacial e interanual de la CPUE de anchoveta y de sardina en la zona norte de Chile entre 1987–1992. *Investigaciones Marinas*, **21**, 111–32.

Yáñez, E.R., Gonzalez, A. & Barbieri, M.A. (1995) Estructura termica superficial del mar asociada a la distribucion espacio-temporal de sardina y anchoveta en la zona norte de Chile entre 1987 y 1992. *Investigaciones Marinas*, **23**, 123–47.

Yoshihara, T. (1954) Distribution of cath of tuna longline. IV. On the relation between k and ϕ with a table and diagram. *Bull. Japan Soc. Fish.*, **19**, 1012–14.

Yu, L.S. (1992) Characteristics of formation and behaviour of associated aggregations of tunas in the Western Indian Ocean. In: *International Workshop on the Ecology and Fisheries for Tunas Associated with Floating Objects and on Assessment Issues Arising from the Association of Tunas with Floating Objects* (ed. Hall, M.). Inter-American Tropical Tuna Commission Int. Rep., San Diego (mimeo).

Yuen, H.S. (1959) Variability of skipjack tuna (*Katsuwonus pelamis*) to live bait. *Fish. Bull.* (US) **60**, 147–60.

Yuen, H.S. (1969) Response of skipjack tuna (*Katsuwonus pelamis*) to experimental changes in pole and line fishing operations. In: *Proceedings of the FAO conference on fish behaviour in relation to fishing techniques and tactics* (eds Ben Tuvia, A. & Dickson, W.), pp. 607–18. *FAO Fish. Rep.* **62** (3).

Yuen, H.S. (1970) Behaviour of skipjack tuna, *Katsuwonus pelamis*, as determined by tracking with ultrasonic devices. *J. Fish. Res. Board Can.*, **27** 2071–9.

Zaferman, M.L. & Serebrov, L.I. (1989) On fish injuries when escaping through the trawl mesh. *ICES, CM 1989/B:18.*

Zavala-Camin, L.A. (1986) Predadores y areas de occurrencia de listado (*Katsuwonus pelamis* L.): revision de estudios sobre contenido estomacal. In: *Proceedings of the ICCAT Conference on the International Skipjack Year Program* (eds Symons, P.E.K., Miyake, P.M. & Sakagawa, G.T.), pp. 291–5. ICCAT, Madrid.

Zhang, X.M. & Arimoto, T. (1993) Visual physiology of walley pollock (*Theragra chalcogramma*) in relation to capture by trawl net. *ICES Mar. Sci. Symp.*, **196**, 113–16.

Zhujkov, A. Yu. (1995) Fast- and slow-learning individuals: an overflow in terms of behavioural ecology and the theory of learning. *Biol. Adv. Current Biol.*, **115** (4), 396–405.

Zhujkov, A. Yu. & Pyanov, A.I. (1993a) Differences in behaviour of fish with different learning ability as demonstrated with a model of a trap net. *J. Ichthyol.*, **33** (9), 141–6.

Zhujkov, A. Yu. & Pyanov, A.I. (1993b) Interaction of individuals with different learning abilities in the exploratory behaviour of *Hemigrammus caudovittatus* (Characinidae). *Russ. J. Aquat. Ecol.*, **2** (2), 131–4.

Zhujkov, A. Yu. & Trunov, V.L. (1994) Interspecific and intraspecific comparisons of motor activity and learning in fish. *J. Ichthyol.*, **34** (6), 152–3.

Zijlstra, J.J. & Boerema, L.K. (1964) Some remarks on effects of variations in fish density in time and space upon abundance estimates, with special reference to the Netherlands herring investigations. *Int. Coun. Explor. Sea Rep.*, **155**, 71–3.

Østvedt, O.J. (1965) The migration of Norwegian herring to Icelandic waters and the environmental conditions in May–June, 1961–1964. *FiskDir. Skr. Ser. Havunders.*, **13** (8), 29–47.

Index

abiotic factors 27–44, 86
abundance estimates 257
abundance indices 219
  see also catch per unit of effort
  avoidance influence 235
  correction 193–5
  mixed-species schools 230–32
acoustic shadowing 256–7
acoustic surveys 251–73, 278, 284
acoustics, school organisation 71
active space 49
adaptive sampling 279–80
aerial surveys 6, 276–9, 284
age
  catchability 214
  composition 5
  depth-dependent distribution 42
  floating object association 135, 143
  school structure 85
age-length key, aggregation influence 232–5
age-structured models 179–82, 286
age-structured stock assessment (ASA) models
    180–82, 200, 204, 214–15, 286
aggregation 249
  aerial surveys 278
  annual changes 227–30
  lunar phase influence 226–7
  pheromones 47
  population dynamic models 215–35
  target strength 255
aggregation curve 185
Agulhas Bank horse mackerel 42, 214
alarm substances 164
Alaska pollock 13
Alaskan walleye pollock 94
albacore 14, 132, 168, 202, 242
All-Union Conference 4
ambient noise 102–3, 126
American plaice 34, 45
American shad 85
amplitude discrimination 116
anaerobic metabolism 34
anchoveta 101
  catch 10
  depth 42
  dissolved oxygen 35
  mixed-species schools 89, 93
  school structure 86
  seasonal migration 23
  stock collapse 186, 188
anchovy
  see also Northern anchovy
  aggregation 228–9
  Cape anchovy 281, 282
  circadian cycles 224
  European anchovy 12
  packing density 80
  predator effects 53
  salinity 33
  school shape 81
  schooling 59
  seasonal migration 24
  spawning grounds 32
  vessel avoidance 261
  world catch 13
annual changes, aggregation 227–30
annual egg production method (AEPM) 280
annual larval production method (ALPM) 280
Arauchanian herring 12
Arctic cod 27
ASA *see* age-structured stock assessment
association 128–58
associative behaviour 133–58, 236–9
Atlantic cod 34, 275
Atlantic herring
  catch 11–12
  schooling 59
  seabed 42
Atlantic mackerel
  catch 12
  tidal cycle 39
Atlantic menhaden

schooling 59
seasonal migration 23–4
Atlantic silverside 59, 63, 86
attack abatement 61
attraction 128–58, 236–9
auditory detection, floating objects association 153
autocorrelation distance 94
aversive learning 159, 169
avoidance 102–27
abundance indices 235
gear 3, 124–5, 169, 270
instruments 269–70
learning 159
model 263
stock assessment 258–70
vessels 117–19, 168, 258–67, 272–3

bait
attraction to 131–3, 236
movement 133
Bali sardinella 13
ball packing 84
Baltic herring 240
dissolved oxygen 35
light attraction 129
spawning 64–5
basin approach 192
basin model 46
beach-seine surveys 276
behaviour studies, history 1–5
behavioural ecology 1–2, 4–5
Benguela pilchard 27
big gulp hypothesis 41
bigeye tuna 14, 50
bait attraction 132
circadian variation 206–7
dissolved oxygen 34
floating object association 141
night-time swimming depth 37
seamounts 42
billfishes, catch 14
biomass dynamic models 174
biomass quantification 7
biotic factors, habitat selection 44–53
birds
predation 53
tuna association 156
bivariate detection function 278
blackchin shiners 95
blue whiting 13, 18, 27, 235
blueback herring 116
bluefin tuna 234, 244
boat-log effect 222
body length, floating objects association 141, 151
body length-dependence 35
bongo nets 280
bonitos 14
Bonneville ciscoes 37
bottom depth 41–4
bottom trawl 6
bream 168
brook sticklebacks 164
bullet tuna 14

California Vertical Egg Tow (CalVet) 280
Californian sardines 89
CalVet *see* California Vertical Egg Tow
Cape anchovy 281, 282
Cape horse mackerel 268–9
capelin
catch 12
Doppler shift 259, 263
habitat selection 22
packing density 79–80
range excursion 26
schools 47, 65–6
seabed 42
semelparity 57
spawning grounds 30–31
surveys 278
vessel avoidance 263
capture-recapture 282–4
cardiac conditioning 115, 117
carp 66, 168
catch per unit of effort (CPUE) 175–7, 193, 195, 212, 215–17, 248, 286, 287
gLM 209
light intensity 37
mixed-species schools 230–31
sardine 23
surplus models 202, 203–4
tuna 223, 236–7
weekly changes 225, 226
catch-effort models *see* surplus-production models
catch-effort relationship 248
catchability 6, 249
aggregation influence 215–30
circadian variation 204–8
habitat selection influence 183–215
seasonal variability 200–204

spatial variability 208–14
yearly variability 183–200
catching efficiency 121
cavitation noise 104
chemical stimuli 132, 171, 236
Chilean jack mackerel 10–11, 17, 221
chub mackerel 11
circadian habitat selection 23
circadian migration 37
circadian variation 235
catchability 204–8
patchiness 224–6
vessel avoidance 265
cleaning station hypothesis 146
closed programmes 2
clustering 94–5
cod
acoustic surveys 253
Arctic cod 27
Atlantic cod 34, 275
direction detection 110–11
distance detection 112
distribution 26
frequency range 114, 115
gear avoidance 270
nearest neighbour distance 72
Northern cod 188
playback experiments 119–20
reaction to noise 117
school structure 75
vessel avoidance 168, 258, 260, 263, 264, 267, 268
vessel noise 121
coho salmon 164, 171–2
cohort analysis 180
comfortability stipulation hypothesis 148–9, 154
communication 95–9
communities 48
compact airborne spectrographic imager (CASI) 7, 278–9
comparative psychology 1
competition
among fishing units 221–4
interspecific 48
intraspecific 67
competitive exclusion principle 48
concentration of food supply hypothesis 146, 148, 158
conditioning 159, 164
confusion effect 61, 63, 93
conspecifics 44–7
contagious distribution hypothesis 155
continuous plankton recorders (RPCs) 280
continuous underway fish egg sampler (CUFES) 280
cooperative fishing 221–4
CPUE *see* catch per unit of effort
critical school biomass 151
current, habitat selection 38–40

dace 48
daily egg production method (DEPM) 280–81
daily fecundity reduction method (DFRM) 280
daily production method 5
density
distribution function 254
yearly changes 183–93
density-dependence 184, 203, 234
density-dependent habitat selection (DDHS) 44–6
depth
depth-dependent distribution 41–2
habitat selection 40–1
diffusion models 45
dilution effect 61
direction detection 110–12
dissolved oxygen
habitat selection 33–6
school structure effects 86
distance detection 110–12
distance discrimination 168
distribution function 254
diurnal predation 53
diurnal variation
aerial surveys 278
target strength 255
DNA fingerprinting 2
dolphins 18, 91, 93, 154–6, 158
Doppler shift 259
double-oblique tow 280
drag 64

echo integration 252, 253, 256
echo sounders 8, 107
echointegration 7
ECOPATH model 52
eels, lunar rhythms 23
egg deposition method (EDM) 280
egg surveys 5, 279–81
emigration 24
ethology 1
European anchovy 12

European minnow 58, 59, 62, 163
European pilchard 12
extinction problem 256

facultative schoolers 57
FADs *see* fish aggregating devices
FAO congress 3–4
fathead minnow 160, 163–4
feeding 3
  see also foraging; prey
  concentration of food supply hypothesis 146, 148, 158
  school size 66–7
  school structure 85
  schooling benefits 63
filtering mechanism 64
Fish Acoustic Sciences Technology (FAST) Group 4
fish aggregating devices (FADs) 133–5, 137–9, 141–54, 157
  attraction 237–8
  homing behaviour of tuna 285
  tuna 91
Fish Technology and Fish Behaviour (FTFB) Group 4
Fishing Gear Technology Working Group 4
flash expansion 61, 84
flat-bottom basin hypothesis 46–7
flatiron herring, mixed-species schools 89
flight transmission speed 61
floating objects 44, 128
  see also fish aggregating devices
  association 18, 133–54
  attraction 237
  light attraction 128
  tuna 91, 93
flying fish 3, 15, 134
following response 59
food finding hypothesis 155, 158
foraging 131
  see also feeding
  interactive learning 172
  UV photoreception 37
fountain effect 61, 84, 123
fractal distribution 94–5
French grunts 171
frequency discrimination 116
frequency range 113–16
frigate tuna 14
front priority rule 77
frontal gradients 29

Garland experiment 5
gear
  avoidance 3, 124–5, 169, 270
  learning about 167–70
  saturation 236
  surveys 274–6, 284
general linear models (GLM) 177, 207, 209, 236, 239
generalised additive models 52
generalised linear models (GLM) 176–7
generic-log hypothesis 149, 154, 157
genetic basis for schooling 57–8
geometric patterns 73–5
geostatistical model 280
gill net fisheries, circadian variation 207
gilt sardine 43
  mixed-species schools 89
  salinity 32
  seasonal migration 24
GLM *see* general linear models
goldfish 160, 171
growth estimates
  aggregation influence 232–5
  habitat selection influence 214
growth overfishing 6
grunts 53, 171
gudgeon 48
Gulf Menhaden 12
guppy
  predator discrimination 162
  schooling behaviour 57–8, 59
  schooling preferences 78
  Trinidadian guppy 160

habitat, yearly changes 183–93
habitat selection 3, 21–55, 183–215
  acoustic surveys 252–4
habituation 159
haddock
  amplitude discrimination 116
  bait attraction 132
  thermocline 30
  vessel avoidance 258, 260, 264, 268
hearing 99, 108–16, 126–7, 153
herring
  abundance estimates 257
  acoustic surveys 253, 271–2
  aerial survey 276–7
  age 5
  Arauchanian herring 12
  Atlantic herring 11–12, 42, 59

avoidance 124, 125, 235
Baltic herring 35, 64–5, 129, 240
blueback herring 116
circadian variation 204–5
direction detection 110
feeding behaviour 49
flatiron herring 89
frequency range 114, 115
fungal infection 78
gear avoidance 270
hearing 99
learning 163
light attraction 129–30
nearest neighbour distance 72, 75
net burst experiments 241
North Sea herring 38, 269
Pacific herring 33, 42, 47, 119
packing density 79–80
playback experiments 119–20
prey 52
purse seiners 100
school shape 82–3
school size 67–8
seasonal migration 24–6
shoals 60
social facilitation 171
sonar avoidance 269–70
spatial distribution 75
spawning grounds 31, 32, 64–5
stock collapse 186
swimbladders 41, 254
thread herring 66, 168, 169
vessel avoidance 258, 260, 261, 263–4, 265, 267–8
heterogeneous stimuli summation 166
homing behaviour 32, 147, 171, 172, 173, 244, 285
homogeneity 87–9
horizontal migration
dissolved oxygen 35
prey selection 52
horse mackerel 29, 205, 206
see also Cape horse mackerel
Agulhas Bank 42, 214
Cape horse mackerel 268–9
circadian cycles 224
cooperative fishing of 222
depth-dependent distribution 42
growth estimate 214
mixed-species schools 231
swept volume 275
hydroacoustic assessment method 251–2
hydroacoustic instruments 15–17
hydroacoustic surveys 102
hydroacoustics 7
hydrodynamic advantage 63–4, 88

IBM *see* individual-based models
ichthyoplankton surveys 279–82
ideal free distribution (IFD) 3, 44–6, 48
imaging LIDAR 277
imprinting 32, 159, 171, 173
individual preferences, within schools 77–8
individual recognition 172
individual-based models (IBM) 45, 71, 77
innate schooling behaviour hypothesis 155
inshore-offshore migration 202
instantaneous habitat selection 22
instrument avoidance 269–70
Inter-American Tropical Tuna Commission (IATTC) 136, 139, 238
interactive learning 167, 172
International Commission for the Conservation of Atlantic Tunas (ICCAT) 141
International Council for the Exploration of the Sea (ICES) 3, 4, 5
International Fishing Technology Working Group 4
interspecific competition 48
intraspecific competition 67
intraspecific diversity, learning influence 244–7
intrinsic aggregation rate 237
iteroparity 57

jack mackerel
Chilean jack mackerel 10–11, 17, 221
lunar phase 278
schooling 59
seasonal variability 202
vessel avoidance 261, 265
Japanese amberjack, catch 13
Japanese anchovy 11, 36
Japanese pilchard 11
Japanese sardine 24, 52, 84, 192, 241
Japanese Spanish mackerel 14

Kawakawa 14
kernel smoothing method 278
key stimuli 164–7
killifish 61
kin recognition 241

larval surveys 279–81
lateral eye-like spots 166
lateral line 97, 98, 101
learning 159–73
  in schools 66
  stock assessment influence 239–49
leks 66
length-dependence relationship 29
length-frequency distribution 232
LIDAR *see* light detecting and ranging
light 122–3, 127
  attraction to 17, 128–31, 236
  avoidance 266–7
  intensity 36–8, 40, 96
light detecting and ranging (LIDAR) 6, 277, 284
line fishing 19–20
linear production model 230
linearity principle 256
lingcod 275
log association *see* floating objects association
log boat tactic 128–9, 148
log fishing 139, 145
longlining 19–20, 29
  attraction influence 236
  bait attraction 132
  circadian variation 206–7
  saturation 216
  surveys 276
longtail tuna 14
lunar influence 267
lunar phase 37
  aerial surveys 278
  aggregation influence 226–7
lunar rhythms 23
lures 133

mackerel
  see also horse mackerel; jack mackerel
  Atlantic mackerel 12, 39
  chub mackerel 11
  Pacific mackerel 265
  spatial distribution 73
  spawning 66
  visual stimuli 124
  Western mackerel 26, 33, 39, 94
  world catch 14–15
Marquesan sardine 36
maximum effort 174
maximum surplus 174
maximum sustainable yield (MSY) 174, 196–8, 230, 248
meeting point hypothesis 150–54, 157
memory traces 167
menhaden
  Atlantic menhaden 23–4, 59
  circadian habitat selection 23
  Gulf Menhaden 12
  salinity 32
  schooling 64
  tagging 5
midwater trawling 6, 18–19, 275
  avoidance 124–5, 235
  circadian variation 204–5
  fish escapes 240
migration 21–55, 64
  learning 171–2, 173
  schooling 88
minimum approach distance 71
minnow 48, 73
  European minnow 58, 59, 62, 163
  fathead minnow 160, 163–4
mixed-species schools 88, 89–93, 230–2
MOCNESS *see* multiple opening-closing net and environmental sensing system
mortality 6, 28, 182, 214–15, 240
motivational status 67, 88
moving mass dynamic hypothesis 80
MSY *see* maximum sustainable yield
mullet 270
multiple opening-closing net and environmental sensing system (MOCNESS) 280

naked goby 56
narrow barred Spanish mackerel 14
natal homing 32, 171, 173, 244
natural selection 100–101
nearest neighbour distance 59, 71–2, 75, 78, 79, 88
night
  foraging 131
  light attraction 128
  trawling 124
  vessel avoidance 266–7
noise
  ambient 102–3, 126
  fish reactions to 116–22
  fishing gear 168
  vessels 103–8
North Sea herring
  passive transport 38
  vessel avoidance 269
Northern anchovy 43, 278

aerial surveys 278
mixed-species schools 89
prey 49
school shape 82
schooling 59
vertical migration 36
Northern bluefin 28, 52
Northern cod 188
Northern sardine 186

object-oriented models 71, 77, 78
obligate schoolers 57
obstinate reproductive strategy hypothesis 32
offshore sea ranching 128
olfaction 160
coho salmon 172
floating objects association 153
odour of fishing gear 168
prey detection 49–50
Olsen model 117–19
omnidirectional multibeam sonar 8
one-trial learning 163, 168, 239, 283
ontogeny
antipredator behaviour 160
schooling 58–9
open programmes 2, 54
optical plankton counter (OPC) 280
optimal foraging theory 1
optimality theory 2
optimum effort 174
optomotor response 124
orange roughy 47, 247
orientation 39–40, 255, 272
otoliths 108–9, 126
overexploitation 289
overfishing 6, 24, 99–100, 186, 188, 244, 247

Pacific bonito 66
Pacific herring
pheromones 47
playback experiments 119
salinity 33
seabed 42
Pacific mackerel 265
Pacific sardine 282
Pacific whiting 269
packing density 78–81, 85, 86, 101, 257
packing patterns 73–5
paradise fish 160, 161, 162–3, 166, 167, 168–9
parasitism 2, 3
passive avoidance 168–9
passive gear, avoidance 169
passive transport 38–9
patchiness 216–21
circadian changes 224–6
ichthyoplankton surveys 281–2
patchiness in distribution 94
pelagic trawling *see* midwater trawling
Peruvian anchovy *see* anchoveta
Peruvian hake 35
pheromones 47, 164
phototaxis 129
physoclists 41, 253
pilchard 27, 73, 224
plasticity 2
playback experiments 119–20, 268
pole-and-line 19, 20, 29
attraction influence 236
bait attraction 133
cooperative fishing 222
tuna fisheries 128
pollack 44, 253
population modelling 174–250
associative behaviour 236–9
attraction 236–9
population parameters, habitat selection influence 183–215
predation/predators
avoidance 3, 57, 97–8, 155
circadian 226
floating objects association 135, 151
habitat selection 52–3
inspection visit 160, 163
learning 159–70
pressure 101
school size 66–7
school structure effects 84–5
schooling effects 88–9
schooling function 60–3
shelter from predator hypothesis 146–7, 158
pressure 40
prestratification 279
prey, habitat selection 48–52
production models *see* surplus-production models
proximate factors 22
pumpkinseed sunfish 37
purse seining 15–18, 20, 29, 275–6
avoidance 124
circadian variation 205–6
cooperative fishing 222
fish escapes 240

herring 100
log association 145

radiometric LIDAR 277
rainbow trout 37, 111
random sampling 279
range collapse 46, 185, 186
recruitment overfishing 6
reefs 43–4
replicability, acoustic survey estimates 271–2
reproduction, schooling 64–6
reproductive isolation 57
Reynolds number 58
rise and glide swimming strategy 41, 254
rosy barb 168
RPCs *see* continuous plankton recorders
rumpons 134

saithe 13, 43–4
light attraction 129
nearest neighbour distance 72, 78
reefs 43–4
spatial distribution 75
subgroups 86
vessel noise 121
salinity, habitat selection 32–3
salmon
circadian variation 207
coho salmon 164, 171–2
lunar rhythms 23
sampling gear avoidance 270
sand goby 162
sand lance 13, 60
SAR 279
sardine
gear avoidance 3
gilt sardine 24, 32, 43, 89
Japanese sardine 24, 52, 84, 192, 241
learning 164
Marquesan sardine 36
migration 26
Northern sardine 186
Pacific sardine 282
predator effects 53
salinity 32
seasonal migration 23
southern Canary current 190
Spanish sardine 42
spawning grounds 22–3
vessel avoidance 261
sardinella
Bali sardinella 13
catch 12
circadian variation 205
depth-related distribution 41–2
floating objects 134
light intensity 37
salinity 32
schooling 85
temperature effects 28
turbidity 36
saturation of fishing unit 215–16
scads 12
schooling 1, 3, 56–101
age composition 5
capelin 47
companion hypothesis 146
dissolved oxygen 35
fidelity 241
school definition 56–7
school organisation 68–89
school shape 81–4, 101
school size 66–8
school structure 84–9
target strength 255
SCUBA egg deposition surveys 279
sea bed 41–4
seamounts 42, 149
search model 218
seasonal habitat selection 23–6
seasonal migration 52, 54
seasonal variation
avoidance 235
catchability 200–204
patchiness 224–6
school structure 86
selectivity curve 184–5
selfish herd principle 62–3
semelparity 57
semi-lunar rhythms 23
sex, schools 47, 65–6
shadow method 69–70
sharks 112, 134, 135
shelf platforms 43
shelf-break 42
shelter from predator hypothesis 146–7, 158
shiner 63
shoaling 59–60, 63, 163
silversides 59, 63, 64, 86
simulations 7
size of fish
mixed-species schools 91, 93

schools 47
segregation 77, 88
skipjack tuna 14–15, 68
dissolved oxygen 34
distribution 33
floating objects association 137, 139, 141, 143, 144–5, 148, 151–2
migration 40
night-time swimming depth 37
seamounts 42
tagging 283
thermocline 29
upwelling 209
snowballing effect 150–51
social aggregation 272
social behaviour, acoustic surveys 254–7
social facilitation 66, 159, 164, 169, 171, 172, 173, 244
sociobiology 2
sonar 8, 16–17, 100, 107, 259, 265, 269–70
sound intensity 263
South American pilchard 10–11
southern bluefin 28, 278, 283
Spanish sardine 42
spatial analysis 220–21
spatial distribution 73–5, 94–5
spatial habitat selection 22–7
spatial reference hypothesis 154
spatial references 147
spatial variability, catchability 208–14
spawning
habitat selection 30–32
pheromones 47
salinity 32–3
spawning grounds
colonisation 247
herring 64–5
learning 171–2
sardine 22–3
split-beam sounders 259–60
sprat 39, 261
sticklebacks 66, 76–7
stochastic dynamic modelling 2
stock abundance, fishing influence 183–93
stock assessment 274–84
acoustic surveys 251–73
history 5–8
population dynamic models 174–250
stock identification, learning influence 244–7
stock-recruitment relationship 182–3
stout-body chromis 171
structural models 6, 247–8
subgroups 77, 86
substitute environment hypothesis 146
substocks 244–5, 247
sunfish 52
surface temperature, aerial surveys 278
surplus-production models 6, 174–9, 195–200, 230–31, 237, 239, 240–44, 247–8
swept volume 275
swept-area method 6
swimbladder
acoustic surveys 253–4
depth effects 40–41
echoes 16
sound sensing 109, 110, 113, 127
swordfish 37, 50, 132, 226, 236
synchrony 76–7, 87–9

tagging 5, 260, 282–4
target strength 254–5
temperature
aerial surveys 278
burst speed 34
dissolved oxygen relationship 33, 35
habitat selection 22, 27, 28–32, 54
migration 39
spawning 31, 33
thermoregulation 29
vertical distribution 40
temporal habitat selection 22–7
temporal structure of sound 116
territoriality 21
tetra 168, 169
thermocline 29–30, 266
thermoregulation 29
thread herring 66, 168, 169
tilt angle 255
topsmelt, spatial distribution 73
Trafalgar effect 61, 97
trawls, vessel avoidance 267–9
trigger fish 191
Trinidadian guppy 160
tuna 249–50
see also bigeye tuna; skipjack tuna; yellowfin tuna
attraction 237
bait attraction 131, 132–3
bird association 156
bluefin tuna 234, 244
chemical cues 49–50
clustering 94

cooperative fishing of 222, 223
diffusion models 45
dolphin association 18, 91, 93, 154–6
floating objects association 44, 133, 135, 137–54, 157–8, 239
frigate tuna 14
habitat selection 27, 54–5
homing behaviour 147, 285
light attraction 128, 130
longlining 19
longtail tuna 14
migration 39–40
mixed-species schools 91, 93, 231
pole-and-line fishing 20
prey 50
purse seining 17–18
salinity 33
saturation 216
seamounts 42
temperature gradients 39
temperature selection 28
thermocline 29
upwellings 40
vertical migration 41
vessel noise 121
whale association 156–7
world catch 14–15
turbidity 36
turbot 270
turbulence 38–40

ultimate factors 22
upwellings 35, 36, 38–40, 43, 209
UV photoreception 37

vertical distribution/migration 40–41, 54
aerial surveys 278
body length-dependence 35
circadian 23
light attraction 129–30
light intensity 36–8
physoclists 41
predator effects 53
prey 50
thermocline 29–30
walleye pollock 30
vessels
avoidance 117–19, 168, 258–66, 272–3
light avoidance 266–7
noise 103–8, 116–22
vigilance 63
virtual population analysis (VPA) 180, 181, 200, 212, 247
viscosity 58
vision 96–8, 101, 122–6
visually elicited avoidance 122–6
volumetric LIDAR 277

walleye pollock 30, 50, 94, 123, 125, 172, 255
water transparency 36
Weberian ossicles 113
Western mackerel
clustering 94
currents 39
migration 26
salinity 33
whale sharks 156–7
whales 156–7
white perch 61
whitefish 23
wind stress 27
world catch
large mackerel 14–15
small species 10–13
tuna 14–15
wrasse 270

yellow perch 68, 86
yellowfin tuna 14–15, 209
bait attraction 131–2
cooperative fishing of 223
dissolved oxygen 34
dolphin association 154–6
floating objects association 135, 137, 139, 141, 143–5, 149
homing behaviour to FADs 147
migration 40
night-time swimming depth 37
seabed 43
seamounts 42
thermocline 29
yield per recruit 179–82

zebra danio 164
zeitgeber 23

Compiled by Indexing Specialists, Hove, UK